A Concrete Approach to Mathematical Modelling

A Concrete Approach to Mathematical Modelling

Michael Mesterton-Gibbons

Department of Mathematics
Florida State University, Tallahassee

A Wiley-Interscience Publication
JOHN WILEY & SONS, INC.
New York • Chichester • Brisbane • Toronto • Singapore

This text is printed on acid-free paper.

This is a revised edition of the work originally published in 1989.

Copyright © 1995, 1989 by John Wiley & Sons, Inc.

Library of Congress Cataloging in Publication Data:

Mesterton-Gibbons, Michael.
    A concrete approach to mathematical modelling / Michael Mesterton
-Gibbons.
        p.   cm.
    "A Wiley-Interscience publication."
    Includes bibliographical references and index.
    ISBN 0-471-10960-6
    1. Mathematical models.   I. Title.
QA401.M5138   1995
511'.8—dc20   94-35535
                                                              CIP

Printed in the United States of America

10 9 8 7 6 5 4 3 2

To three cherished nonmathematicians:

to my Mother

who once said, "If they pay you to do all those squiggles,
small wonder the world's in a mess."

to my Father

who said it was better than working in a factory.

and to Karen

who says if you want to know the meaning of infinity, just ask someone
whose spouse is writing a math book.

PREFACE

Controversy has persisted in recent years over how to teach mathematical modelling. There are two extreme schools of thought. One believes that a modelling course should have little structure and be dominated by "case studies," open-ended exercises requiring data collection and research. These might lead students down a great many unforeseen avenues of creativity but might also lead them nowhere. The other school believes that a modelling course should be highly structured and dominated by what, for want of a better phrase, I will refer to as "modelling exercises." These are exercises that involve formulation and interpretation but bear a more certain relationship to the principal content of the course. Between these two extremes, of course, lie many compromises, one of which forms the basis for this textbook.

Before describing it, however, let me argue that there is an important sense in which the second school of thought should take precedence over the first. Most college majors in mathematics do not become professional mathematicians. Rather, they join the graduate workforce as businesspeople, as financial planners, as politicians, or in a number of other careers, and it is far more important that they can understand, interpret, criticize, and appreciate the mathematical models of others than that they are able to develop their own models. Just as drivers need not build and repair their cars but should know how they work, so consumers of mathematical models need not produce them but should understand both their usefulness and their limitations. Naturally, it remains *desirable* that math majors be able to develop their own models, but this is *essential* only to the few who subsequently make model development their career. On the other hand, the capacity to understand, interpret, criticize, and appreciate models is essential to all.

With this in mind, I have written this text for the deserving many, not simply a chosen few. Yet I have not neglected the specific needs of those who aspire to building their own models, although I have not *explicitly* addressed them until relatively late in the book. This is deliberate. Between the two extreme schools of thought this book has adopted a dynamic compromise, in which it begins by espousing the second extreme and gradually evolves toward the first. This approach enables the book to emphasize both the validation of mathematical models and the rationale behind improving them. Each of these features is novel, as will be clear to anyone familiar with Murthy and Rodin's (1987) review of books on mathematical modelling.

This is a book about the process of modelling. Its approach is heuristic—but systematic—and embodies my belief that the three most fundamental ideas in mathematical modelling are transience, permanence, and optimality. Because models incorporating these ideas may adopt either a deterministic or a probabilistic viewpoint, there are six combinations of idea and viewpoint. These define the structure of Chapters 1–3 and Chapters 5–7 of the text, which proceed in parallel as depicted in the following diagram.

View \ Idea	Transience	Permanence	Optimality
Deterministic	**Chapter 1** Growth and Decay. Dynamical Systems	**Chapter 2** Equilibrium	**Chapter 3** Optimal Control and Utility
Probabilistic	**Chapter 5** Birth and Death. Probabilistic Systems	**Chapter 6** Stationary Distributions	**Chapter 7** Optimal Decision and Reward

In applying these ideas, mathematical modellers are called upon to play conflicting roles. On the one hand, they must be creative artists, assuming boldly perhaps what no one has assumed before. On the other hand, they must be critics, scrupulously doubting whether their models provide an adequate description of reality. In practice, of course, the distinctions I have drawn between ideas, roles, and even viewpoints are artificial; but it greatly benefits a beginning student of mathematical modelling to focus on a single aspect at a time, even though the other aspects cannot be entirely ignored. Accordingly, the critical aspect of modelling is the subject of Chapter 4, its creative aspect the subject of Chapter 9, and the interface between the two the subject of Chapter 8. The remainder of the book, Chapters 10–12, is a reinforcement of the ideas developed in the first nine chapters.

The goal of this book is to let readers acquire both critical and creative modelling skills and the confidence to use them. All other matters are regarded as secondary. Thus, as far as is possible, technical mathematical details are explored in exercises, or occasionally in footnotes, rather than in the main body of text. To distinguish such purely technical exercises from those that are more directly related to the goals of the book, modelling exercises are denoted by a single asterisk and case studies by a double asterisk. But not every detail has found its way into a footnote or exercise, and the book cannot be used profitably unless pencil and paper at all times accompany the reader. The book is to be studied, not simply read.

The book's goal is pursued through a layered approach, with frequent revisitations to earlier sections, so that even the most sophisticated models are perceived as merely natural outgrowths of less sophisticated ones. As proclaimed in the title, this layered approach is unashamedly concrete. Philosophical points are not discussed until several illustratory examples have already been introduced. Even then the discussion is brief, as at the beginning of Chapter 4, or in Chapter 9, where the art of adapting, extending, and combining is first discussed formally, even though it characterizes every model developed in the text. At all times, I have tried to avoid unnecessary abstractions. I have taken particular care never to introduce a utility function as $F(x)$, with constraints on the signs of the derivatives of F; instead, I give the dependence of F on x explicitly, so that properties of the derivatives are self-evident. The loss of generality is more than compensated by the gain in comprehension. Indeed my experience suggests that the educational value of a modelling course to the average student is decided by this factor more than by any other.

This book is primarily intended for a senior level course that gives equal weight to deterministic and probabilistic modelling. Accordingly, minimal mathematical prerequisites for mastery of its *entire* contents are the standard calculus sequence (including the Newton–Raphson method) and first courses in linear algebra, ordinary differential equations, and probability and statistics. Probability and statistics are reviewed in Appendix 1. A few sections require access to computer packages for solving linear programs or ordinary differential equations (now almost universally available), while some knowledge of numerical analysis is desirable (though by no means necessary). A modelling course need not include the book's entire contents, however, and alternative uses are described in an accompanying instructor's manual. In particular, there is ample material in Chapters 1–4 and Chapters 8–11 for a course that requires no probability and statistics.

But this book is designed to be also suitable for study alone, perhaps by beginning graduate students or by professionals whose background in mathematical modelling is weak. For this reason, if an exercise is crucial to later developments and of even moderate difficulty, then its solution (or a possible solution) appears at the back of the book. I therefore assume that my readers are mature enough not to consult a solution until a serious attempt has been made at the problem. Solutions to most other exercises appear in the instructor's manual.

By using plurals and genderless singulars, I have thus far avoided the vexed question of whether my reader is male or female. The lack of epicene pronouns and possessive adjectives in the English language is regrettable, and I have long advocated the use of "their" to mean "his or her." But grammarians may be offended by this, and the continual use of "his or her" is deplorably inelegant. I therefore felt that odd-numbered chapters should have one gender, even-numbered chapters another, and the toss of a coin decided that males would be odd, females even. Thus Chapters 1, 3, 5, 7, 9, and 11 are male, and Chapters 2, 4, 6, 8, 10, and 12 are female. This convention is not without minor inconveniences—for example, the forester we meet in Chapter 3 has to change his sex before she enters Chapter 4—but this is surely preferable to the alternatives.

The reviews of the first edition of this book have all been very positive and pleasing. I haven't fixed what isn't broken. Nevertheless, for this revised edition I have corrected all known errors, I have added a few exercises, and I have updated both the list of references and other time-sensitive material.

Finally, this is a substantially original book. The models are based on the scientific literature, specifically, on the sources identified in Appendix 2. But I have made it a point of honor to personalize each model by deriving it from scratch, starting with a blank piece of paper and working through all the details myself. This has led not only to differences of emphasis and approach but also to new extensions and variations. Moreover, not only are models freely adapted, they are presented here in a fresh perspective, as the interlocking fragments of a mosaic. This is my mosaic of the modelling process, carefully assembled from my teaching experience, in which every fragment contributes to the synthesis of a coherent picture. What the picture shows me is that even a modest amount of mathematics—no more than should be expected of every college major in business, science, or engineering—is enough to describe a wide variety of phenomena, offer penetrating insights, and contribute effectively to rational decision making. If that is also what the picture shows my readers, then I shall be satisfied with what I have written.

Acknowledgement

I am deeply indebted to the many scientists whose mathematical models have formed a basis for this textbook. I wish to express my thanks to all of them, and to each in a measure proportional to my indebtedness.

I wish to thank Lou Gross for constructive and valuable criticism of the original manuscript, Steve Blumsack and Christopher Fye for discovering errors, and Charlie Nam for suggestions concerning the literature. I wish also to thank John Wiley & Sons for their support, cooperation, and professionalism in producing this revised edition.

Most of all, I would like to express my deepest gratitude to Karen, my wife, in whose space and time this book was written. Without her patience, encouragement and humor, it would never have been possible to complete the manuscript.

Tallahassee, Florida   MICHAEL MESTERTON-GIBBONS

CONTENTS

An ABC of modelling   xix

I The Deterministic View

1 Growth and decay. Dynamical systems   3

   1.1 Decay of pollution. Lake purification   5
   1.2 Radioactive decay   7
   1.3 Plant growth   7
   1.4 A simple ecosystem   8
   1.5 A second simple ecosystem   11
   1.6 Economic growth   13
   1.7 Metered growth (or decay) models   21
   1.8 Salmon dynamics   23
   1.9 A model of U.S. population growth   26
   1.10 Chemical dynamics   29
   1.11 More chemical dynamics   30
   1.12 Rowing dynamics   32
   1.13 Traffic dynamics   34
   1.14 Dimensionality, scaling, and units   35
   Exercises   40

2 Equilibrium   46

   2.1 The equilibrium concentration of contaminant in a lake   52
   2.2 Rowing in equilibrium   53
   2.3 How fast do cars drive through a tunnel?   57
   2.4 Salmon equilibrium and limit cycles   58

2.5   How much heat loss can double-glazing prevent?   63
2.6   Why are pipes circular?   66
2.7   Equilibrium shifts   71
2.8   How quickly must drivers react to preserve an
      equilibrium?   76
      Exercises   83

3   Optimal control and utility   91

3.1   How fast should a bird fly when migrating?   93
3.2   How big a pay increase should a professor
      receive?   95
3.3   How many workers should industry employ?   103
3.4   When should a forest be cut?   104
3.5   How dense should traffic be in a tunnel?   109
3.6   How much pesticide should a crop grower
      use—and when?   111
3.7   How many boats in a fishing fleet should be
      operational?   115
      Exercises   119

II   Validating a Model

4   Validation: accept, improve, or reject   127

4.1   A model of U.S. population growth   127
4.2   Cleaning Lake Ontario   128
4.3   Plant growth   129
4.4   The speed of a boat   130
4.5   The extent of bird migration   132
4.6   The speed of cars in a tunnel   136
4.7   The stability of cars in a tunnel   138
4.8   The forest rotation time   142
4.9   Crop spraying   146
4.10 How right was Poiseuille?   148
4.11 Competing species   151
4.12 Predator–prey oscillations   154
4.13 Sockeye swings, paradigms, and complexity   157
4.14 Optimal fleet size and higher paradigms   159
4.15 On the advantages of flexibility in prescriptive
      models   161
      Exercises   163

III   The Probabilistic View

5   Birth and death. Probabilistic dynamics   175

  5.1   When will an old man die? The exponential
      distribution   180
  5.2   When will N men die? A pure death process   183
  5.3   Forming a queue. A pure birth process   185
  5.4   How busy must a road be to require a pedestrian
      crossing control?   187
  5.5   The rise and fall of the company executive   189
  5.6   Discrete models of a day in the life of an elevator   193
  5.7   Birds in a cage. A birth and death chain   198
  5.8   Trees in a forest. An absorbing birth and death
      chain   200
      Exercises   202

6   Stationary distributions   208

  6.1   The certainty of death   210
  6.2   Elevator stationarity. The stationary birth and
      death process   213
  6.3   How long is the queue at the checkout? A first
      look   215
  6.4   How long is the queue at the checkout? A second
      look   217
  6.5   How long must someone wait at the checkout?
      Another view   219
  6.6   The structure of the work force   225
  6.7   When does a T-junction require a left-turn lane?   227
      Exercises   234

7   Optimal decision and reward   237

  7.1   How much should a buyer buy? A first look   237
  7.2   How many roses for Valentine's Day?   243
  7.3   How much should a buyer buy? A second look   245
  7.4   How much should a retailer spend on
      advertising?   247
  7.5   How much should a buyer buy? A third look   253
  7.6   Why don't fast-food restaurants guarantee service
      times any more?   258
  7.7   When should one barber employ another?
      Comparing alternatives   263
  7.8   On the subjectiveness of decision making   267
      Exercises   268

IV The Art of Application

8 Using a model: choice and estimation — **275**

8.1 Protecting the cargo boat. A message in a bottle — 276
8.2 Oil extraction. Choosing an optimal harvesting model — 279
8.3 Models within models. Choosing a behavioral response function — 281
8.4 Estimating parameters for fitted curves: an error control problem — 285
8.5 Assigning probabilities: a brief overview — 291
8.6 Empirical probability assignment — 293
8.7 Choosing theoretical distributions and estimating their parameters — 304
8.8 Choosing a utility function. Cautious attitudes to risk — 316
Exercises — 322

9 Building a model: adapting, extending, and combining — **327**

9.1 How many papers should a news vendor buy? An adaptation — 328
9.2 Which trees in a forest should be felled? A combination — 329
9.3 Cleaning Lake Ontario. An adaptation — 334
9.4 Cleaning Lake Ontario. An extension — 337
9.5 Pure diffusion of pollutants. A combination — 345
9.6 Modelling a population's age structure. A first attempt — 350
9.7 Modelling a population's age structure. A second attempt — 360
Exercises — 373

V Toward More Advanced Models

10 Further dynamical systems — **383**

10.1 How does a fetus get glucose from its mother? — 383
10.2 A limit-cycle ecosystem model — 389
10.3 Does increasing the money supply raise or lower interest rates? — 393
10.4 Linearizing time: The semi-Markov process. An extension — 398
10.5 A more general semi-Markov process. A further extension — 406

10.6 Who will govern Britain in the twenty-first century? A combination   409
Exercises   412

11 Further flow and diffusion   **416**

11.1 Unsteady heat conduction. An adaptation   417
11.2 How does traffic move after the train has gone by? A first look   421
11.3 How does traffic move after the train has gone by? A second look   423
11.4 Avoiding a crash at the other end. A combination   429
11.5 Spreading canal pollution. An adaptation   433
11.6 Flow and diffusion in a tube: a generic model   436
11.7 River cleaning. The Streeter–Phelps model   440
11.8 Why does a stopped organ pipe sound an octave lower than an open one?   446
Exercises   454

12 Further optimization   **458**

12.1 Finding an optimal policy by dynamic programming   458
12.2 The interviewer's dilemma. An optimal stopping problem   465
12.3 A faculty hiring model   470
12.4 The motorist's dilemma. Choosing the optimal parking space   475
12.5 How should a bird select worms? An adaptation   479
12.6 Where should an insect lay eggs? A combination   496
Exercises   507

Epilogue   **514**

Appendix 1: A review of probability and statistics   **516**

Appendix 2: Models, sources, and further reading arranged by discipline   **531**

Solutions to selected exercises   **539**

References   **583**

Index   **591**

AN ABC OF MODELLING

is for Assume.

If your mathematical experience has taught you never to assume what you are not given, then this book will startle you. What you are given is rarely enough, so you will have to make assumptions—about what is important and what is not, about what is assured beyond reasonable doubt and what is still open to question. Indeed, in a very real sense, a model is simply the assumptions you make. Mathematics enables you to deduce, from those assumptions, conclusions which (a) might otherwise not be so readily apparent and (b) can be compared with observations of the real phenomenon that your abstract model attempts to explain. The degree of correspondence determines the value of the model. Poor agreement does not (or should not!) suggest that the mathematics is wrong, however, but rather that one (or more) of the assumptions you made is of doubtful validity. Then you modify your model (i.e., modify your assumptions), and the merry-go-round begins again.

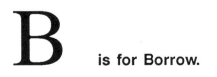

is for Borrow.

Why borrow? A mathematical model is an attempt to capture, in abstract form, the essential characteristics of an observed phenomenon. The success of the attempt depends as much (if not more) on the modeller's empirical knowledge of that phenomenon as on her or his mathematical ability. What do you do if you have neither the knowledge nor the time to acquire it? The answer is that you borrow your assumptions, from the scientific literature or from more experienced colleagues (being sure, of course, to acknowledge your sources). With due respect to geniuses, it is much more practical to build upon existing models than it is to start from scratch. So you must be prepared to borrow freely. Yet therein lies a danger, the danger that you will accept too readily the authority of the printed word. Hence ...

C is for Criticize.

You must be prepared to criticize, too; prepared to criticize not only your own assumptions but also those you have borrowed from other people. Who is to say if they are right or wrong? The answer is you! Modelling is an iterative process. You begin by assuming or borrowing; with the help of mathematics, you reach conclusions; you criticize them; if you are not satisfied then you assume or borrow again, conclude and criticize again, and so on, until eventually you are satisfied (that the model explains the observations). Don't forget the ABC. Assume. Borrow. Criticize.

D is for, among other things, Decay.

Which brings us nicely to Chapter 1.

A Concrete Approach to Mathematical Modelling

I THE DETERMINISTIC VIEW

1 GROWTH AND DECAY. DYNAMICAL SYSTEMS

One of the most useful concepts in mathematical modelling is that of net specific growth rate. We'll begin this chapter by defining and deriving an expression for it. At the same time, we'll introduce two of the most useful notations in applied mathematics, namely "big oh" and "little oh." Then, in several examples, we'll see how useful net specific growth rate can be.

Let $x(t)$ denote an appropriate numerical measure, at time t, of the size or magnitude of a certain population. Examples of population magnitudes would be the number of millions of human beings in the world, the number of millions of dollars in a firm's capital stock, the dry weight in kilograms of a plant (which is just a population of cells), the biomass in tonnes of a fish population, and so on. Let b denote the rate at which units are born or added to the population *for each existing unit of population*; b is called the instantaneous specific birth rate. Then the number of units actually added to the population, in an infinitesimal interval of length δt, is given approximately by $b \cdot \delta t \cdot x$.

Why is it not exactly $b \cdot \delta t \cdot x$? We supplied the adjective "instantaneous" above because b is generally changing with time; i.e., $b = b(t)$. Thus, during the infinitesimal interval $[t, t + \delta t)$, b increases from $b(t)$ to $b(t + \delta t)$. Similarly, x increases from $x(t)$ to $x(t + \delta t)$. On the other hand, the interval $[t, t + \delta t)$ is so short that, during it, b will not differ very much from $b(t)$ and x will not differ very much from $x(t)$. We are assuming, of course, that b and x are at least continuous functions of t; and we are about to assume that x is also differentiable. In a pure mathematics text, we might feel obliged to state such assumptions explicitly whenever they were made. In the process of modelling, however, the repetition of such purely mathematical assumptions is apt to be distracting. We shall therefore assume, once and for all, that all functions used in this text have derivatives of as high an order as is necessary, the existence of the first derivative implying

3

continuity. Hence

$$b = b(t) + \text{terms that tend to zero as } \delta t \longrightarrow 0;$$
$$x = x(t) + \text{terms that tend to zero as } \delta t \longrightarrow 0.$$

But this is rather clumsy, so we introduce some new notation. We define "big oh" by

$$O(\delta t) = \text{terms that tend to zero as } \delta t \longrightarrow 0. \tag{1.1}$$

Thus, in "big oh" notation, $b = b(t) + O(\delta t)$ and $x = x(t) + O(\delta t)$. We can now be more precise about the number of units actually added to the population in the infinitesimal interval $[t, t + \delta t)$. It is

$$\{b(t) + O(\delta t)\} \cdot \delta t \cdot \{x(t) + O(\delta t)\} . \tag{1.2}$$

Let d denote the rate at which units die or are removed from the population *for each existing unit of population*; d is called the instantaneous specific death rate. Then the number of units actually removed from the population in the infinitesimal interval $[t, t + \delta t)$ is

$$\{d(t) + O(\delta t)\} \cdot \delta t \cdot \{x(t) + O(\delta t)\} , \tag{1.3}$$

by an argument similar to that used above.

The difference between (1.2) and (1.3) is the net number of units added to the population in the infinitesimal interval $[t, t + \delta t)$. But this is just the amount by which $x(t)$ increases. Hence, writing b in place of $b(t)$,

$$x(t + \delta t) - x(t) = \{b + O(\delta t)\} \cdot \delta t \cdot \{x + O(\delta t)\} - \{d + O(\delta t)\} \cdot \delta t \cdot \{x + O(\delta t)\} . \tag{1.4}$$

We now define the net specific growth rate μ to be the *net* rate at which units are added to the population for each existing unit. Thus μ is the difference between the instantaneous specific birth rate b and the instantaneous specific death rate d; i.e., $\mu = b - d$. But from (1.4) we have

$$\frac{x(t + \delta t) - x(t)}{\delta t} = \{b + O(\delta t)\} \{x + O(\delta t)\} - \{d + O(\delta t)\} \{x + O(\delta t)\} . \tag{1.5}$$

Taking the limit as $\delta t \to 0$ (assumed to exist), we have

$$\frac{dx}{dt} = bx - dx = (b - d) x. \tag{1.6}$$

Hence

$$\mu = \frac{1}{x} \frac{dx}{dt}. \tag{1.7}$$

Note that $x(t)$ is assumed to be differentiable and hence continuous. An appropriate unit of magnitude is therefore very much larger than a typical increase (or decrease) in population. For example, we might count the U. S. population in millions, because a million people is very much more than the (discrete) number of people that can be added (are born) or removed (die) at any given instant.

Note also that (1.4), though considerably tidier than if we had written $O(\delta t)$ in words, is still rather clumsy. We can tidy it up a little bit by expanding the right-hand side to obtain $(b - d)x\delta t + (b - d + x)\delta t \cdot O(\delta t) + \delta t \cdot O(\delta t^2)$. (In obtaining this expression, you should bear in mind that $O(\delta t) - O(\delta t)$ is not zero but still $O(\delta t)$; whereas $O(\delta t) \cdot O(\delta t) = O(\delta t^2)$.) We can tidy it up even more by writing it as $(b - d)x\delta t + O(\delta t)$. Unfortunately, this last expression, although correct, would not enable us to deduce (1.5), because we cannot be sure that the result of dividing $O(\delta t)$ by δt would tend to zero as $\delta t \to 0$. This leads us naturally to define "little oh" by

$$o(\delta t) = \frac{\text{Terms that are so small that we can divide them by}}{\delta t \text{ and the result will still tend to zero as } \delta t \to 0.} \tag{1.8}$$

It follows immediately that $\delta t \cdot O(\delta t) = o(\delta t)$. This allows us to write the right-hand side of (1.5) very succinctly as $(b - d)x + o(\delta t)$—and still retain sufficient information to deduce equation (1.6). Little oh notation will be very useful later, particularly in Chapter 5.

By itself, (1.7) is not a mathematical model. It is simply a definition of the net specific growth rate μ, which might be a function of x, t, and a whole host of other variables and parameters. It may become a mathematical model, however, as soon as assumptions are made about the form of that function (e.g., linear or quadratic in x, independent of t, etc.). In this book, the source of these assumptions will usually be either borrowed knowledge or native wit (conceptual approach); but it could just as easily be supporting data (empirical approach), or some ad hoc combination of the two. No source is invariably best in the initial stages of modelling a phenomenon, though the ultimate test of a model's worth is often its ability to predict observable data. We'll discuss that last point in Chapter 4.

Now let's see the examples.

1.1 Decay of Pollution. Lake Purification

Let's consider a population of particles that contaminate a lake. Assume that the lake has constant volume V cubic meters. If the lake is well mixed, then contamination will be almost uniform throughout, and an appropriate measure of the extent of pollution at time t will be the number of grams of contaminant per cubic meter of lake. We'll denote this magnitude by $x(t)$ and follow convention by calling it the *concentration* (of contaminant). Let r denote the rate at which water flows out of the lake in cubic meters per day. Because the lake volume is constant, this must also be the rate at which water flows in; there might, for example, be a balance between rainfall and evaporation. If all pollution input suddenly ceases, how much time will lapse before the level of pollution is reduced to 5% of its initial value? We have

$$\begin{array}{c}
\text{Rate of Change} \\
\text{of Pollution}
\end{array} = \begin{array}{c}
\text{Pollution} \\
\text{Inflow}
\end{array} - \begin{array}{c}
\text{Pollution} \\
\text{Outflow}
\end{array}$$

$$\frac{d}{dt}\{x(t) \cdot V\} = 0 - rx(t),$$

(1.9a)

whence

$$\frac{1}{x}\frac{dx}{dt} = -\frac{r}{V}.$$

(1.9b)

This is a special case of (1.7) with $\mu = -d = -r/V < 0$. The net specific growth rate is strictly negative, corresponding to the decay of the pollution.

Let's assume that r is constant. Then (1.9) becomes a model, and (1.9b) is readily solved to yield

$$x(t) = x(0)\exp(-rt/V) = x(0)e^{-rt/V},$$

(1.10)

where exp denotes the exponential function and $e = \exp(1)$. According to the model, the pollution decays to $0.05x(0)$ in time

$$t_{0.05} = -\ln(0.05)\frac{V}{r} = \ln(20)\frac{V}{r} \approx \frac{3V}{r},$$

(1.11)

where ln denotes the natural logarithm (see Exercise 1.1).

We could apply this result to the Great Lakes of North America. According to Rainey (1967, p. 1242), Lake Erie, Lake Michigan, and Lake Superior have volumes 458×10^9, 4871×10^9 and $12,221 \times 10^9$ cubic meters, respectively; and their respective mean outflows are 479,582,208; 433,092,096; and 178,619,904 cubic meters per day. (Rainey quotes the outflows in terms of liters per second, so you must multiply his figures by $24 \times 60 \times 60 \div 10^3 = 86.4$ for conversion to cubic meters per day.) Thus, for Lake Erie, we have $t_{0.05} \approx 2861$ days or 7.8 years. Similarly, $t_{0.05} \approx 33,693$ days or 92 years for Lake Michigan, and $t_{0.05} \approx 204,965$ days or 562 years for Lake Superior. These values must, of course, be interpreted in the light of the initial pollution levels $x(0)$.

Our model is riddled with assumptions. We have assumed that r is constant; maybe it varies seasonally. We have assumed that x does not vary across the lake; maybe mixing is insufficient to justify this assumption. We would not expect seasonal variation to matter much, however (see Exercise 1.1), and we would expect poor mixing to prolong the cleaning. It's therefore reasonable to suppose that (1.11) represents a lower bound—the most optimistic cleaning time—and we have derived this useful piece of information from a very simple model.

Observe that, despite the cessation of pollution input, (1.10) predicts that our lake can never be thoroughly purified in a finite time. Does this make sense to you? Perhaps you would like to think about that. We'll return to the subject of lake purification in Chapters 4 and 9.

1.2 Radioactive Decay

Let's consider the population of radioactive atoms in some isotope of a chemical element, i.e., in a form of the element with a given number of neutrons in the nucleus (all forms of the nucleus have the same number of protons). The conventional measure of that population's size at time t is simply the number of such atoms present. Let this be denoted by $x(t)$. We borrow from physical chemistry the result that radioactive isotopes decay at a constant specific rate, say λ. Then (1.7) becomes a mathematical model with $\mu = -d = -\lambda$. The number of atoms is reduced by a factor of 2 in time

$$t_{0.5} = \frac{\ln(2)}{\lambda}. \qquad (1.12)$$

This time is known as the half-life of the isotope. For example, the half-life of carbon-14, the isotope of carbon with 8 neutrons (and the obligatory 6 protons) in its nucleus, is about 5,700 years.

Note that the models introduced in Sections 1.1 and 1.2 have a common mathematical structure. They are both examples of what is often referred to as *exponential decay*.

1.3 Plant Growth

Let $x(t)$ denote the dry weight at time t of a plant that feeds off a fixed amount of a single substrate. Let the weight of substrate that remains at time t be denoted by $S(t)$. The more substrate there is, the greater the specific growth rate of the plant; the less substrate, the slower the growth. The simplest hypothesis that incorporates this observation is that

$$\mu = kS, \qquad (1.13)$$

where k is a constant.

Note that (1.13) is not the only hypothesis incorporating the observation. Two of the infinitely many other possibilities are $\mu = kS^{3/2}$ and $\mu = kS^{4/7}$. In the absence of further information about the plant, however, we have no way of distinguishing between these possibilities. We therefore adopt the simplest. It might be a good hypothesis; if not, then we can alter it later. The rationale behind this strategy will be studied in Chapter 4.

Adopting (1.13) and substituting into (1.7):

$$\frac{dx}{dt} = kSx. \qquad (1.14)$$

Let's assume no material is lost when S is converted into x, i.e., that $dx = -dS$; then $S = x_f - x$, where x_f is simply the value of x corresponding to $S = 0$. Substitution in (1.14) now yields

$$\frac{dx}{dt} = k(x_f - x)x; \qquad (1.15)$$

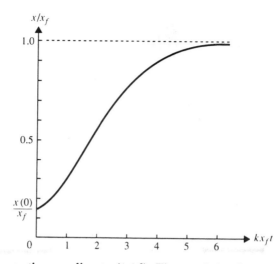

Fig. 1.1 Plant growth according to (1.16). The graph has been plotted in dimensionless variables, which will be discussed in Section 1.14; see (1.91). Meanwhile note that for given values of k, x_f, and $x(0)$, the value of x can be deduced from this diagram for any value of t.

and (1.7) becomes a plant growth model, with $\mu = k(x_f - x)$. The solution of this differential equation (Exercise 1.2) is

$$x(t) = \frac{x(0) \, x_f}{x(0) + (x_f - x(0)) \exp(-kx_f t)} \tag{1.16}$$

and is depicted in Fig. 1.1. Note that the graph is (mildly) S-shaped. It bends upward toward an inflection point, where $x(t) = 0.5x_f$, and then bends downward, approaching the horizontal asymptote $x = x_f$ as $t \to \infty$. Growth according to (1.15) or (1.16) is generally described as *logistic*. Observe that a real plant actually attains its maximum dry weight within a finite time; whereas (1.16) predicts that $x(t)$ merely approaches x_f as $t \to \infty$. As we remarked in our ABC of modelling, however, a model attempts only to capture the essential features of a phenomenon, not mimic it in every respect.

Empirical data will determine the ultimate worth of the model contained in (1.15) or (1.16), but we were able to derive it from a purely conceptual approach. In fact (1.16) gives a fairly good description of annual plant growth for appropriate values of k and x_f; see Thornley (1976).

1.4   A Simple Ecosystem

Let $x(t)$ denote the number of thousands (say) of individuals, at time t, in a population of herbivores. Let $y(t)$ be the corresponding size of a carnivore population. The carnivores eat the herbivores, and the herbivores eat grass. This is the simple ecosystem that we shall attempt to model.

By definition, the specific growth rates for the two populations are

$$\mu_1 = \frac{1}{x}\frac{dx}{dt}, \qquad \mu_2 = \frac{1}{y}\frac{dy}{dt}. \tag{1.17}$$

Let's borrow from biology: populations tend to grow *exponentially*, i.e., at a constant specific rate, until some other factor (e.g., crowding, disease, predators) interferes with that growth. Hence, in the absence of carnivores, we might propose the following model for herbivore growth: $\mu_1 = a_1$, where $a_1 > 0$ is a constant, this positive *net* growth rate for the herbivore population being the difference between a positive birth rate and a smaller positive death rate. But there *are* some carnivores, whose effect will be to diminish the growth rate of the herbivore population. The simplest assumption we can make is that this decrease will be proportional to the number of carnivores. Hence

$$\mu_1 = a_1 - b_1 y, \tag{1.18}$$

where $b_1 > 0$ is a constant.

Similarly, in the absence of herbivores, the carnivore population would die off exponentially (assuming that carnivores don't eat grass). The appropriate model for this would be $\mu_2 = -a_2$, where $a_2 > 0$ is constant. But there *are* some herbivores, whose effect will be to enlarge the growth rate of the carnivore population. Let us again make the simplest assumption, i.e., that this increase will be proportional to the number of herbivores. Then

$$\mu_2 = -a_2 + b_2 x, \tag{1.19}$$

where $b_2 > 0$ is a constant. From (1.17)–(1.19), we obtain the following pair of ordinary differential equations for the growth and decay of the two populations in our simple ecosystem:

$$\frac{1}{x}\frac{dx}{dt} = a_1 - b_1 y, \qquad \frac{1}{y}\frac{dy}{dt} = -a_2 + b_2 x. \tag{1.20}$$

These equations, known as the Lotka–Volterra equations, constitute a mathematical model in which a_1, a_2 are pure growth and decay rates and b_1, b_2 are interaction parameters. As simple as they look, equations (1.20) cannot be solved analytically for x and y as explicit functions of t in the same way that we solved (1.15) to obtain (1.16). They can, however, be solved "numerically," i.e., by utilizing a computer package, for any given values of a_1, b_1, a_2, b_2, $x(0)$ and $y(0)$. For $a_1 = 3$, $a_2 = 5/2$, $b_1 = 2$, $b_2 = 1$, $x(0) = 1$, and $y(0) = 1$, the solution is plotted in Fig. 1.2. The IMSL computer package DVERK produced these graphs, and it might be a useful exercise for you to duplicate them (your computer center will have the package, or a similar one, and will be able to show you how to use it).

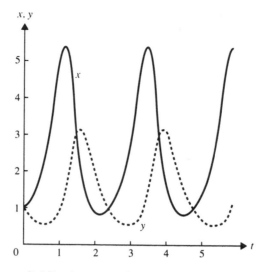

Fig. 1.2 Solutions to (1.20) when $a_1 = 3$, $a_2 = 5/2$, $b_1 = 2$, and $b_2 = 1$. The solid curve is $x(t)$; the dashed curve, $y(t)$.

Note that the graphs of x and y repeat themselves every 2.4 units of time. We say that x and y are periodic or cyclic, with period 2.4.

Pairs of values $(x(t), y(t))$, representing Fig. 1.2's solution in the x-y plane, form the outermost of the closed curves depicted in Fig. 1.3. The arrows denote the direction in which time increases. The remaining curves were obtained by repeating the entire procedure for different values of $(x(0), y(0))$, namely, $(3/2, 1)$, $(2, 1)$, $(5/2, 2)$, and $(3, 3/2)$. The dot denotes the degenerate closed curve $(x(t), y(t)) = (5/2, 3/2)$.

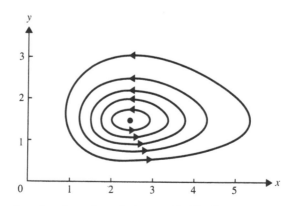

Fig. 1.3 Phase-plane topology for solutions to (1.20). Curves were drawn for $a_1 = 3$, $a_2 = 5/2$, $b_1 = 2$, $b_2 = 1$, and for various values of $x(0)$, $y(0)$. Time increases in the direction of the arrows. See Exercise 1.3 for more details.

Although this diagram was drawn for $a_1 = 3$, $a_2 = 5/2$, $b_1 = 2$, and $b_2 = 1$, it would look the same for any other set of parameter values in the sense that all diagrams would be topologically equivalent. That's rather fortunate, because the particular values chosen for illustration have no ecological significance! Fig. 1.3 is an example of a *phase-plane diagram*. Another one appears in the following section, and they are further discussed in Chapter 2. For any values of a_1, b_1, a_2, and b_2, the phase-plane diagram contains a degenerate closed curve $(x(t), y(t)) = (a_2/b_2, a_1/b_1)$, which represents a possible state of the ecosystem. More generally, however, the model suggests a continual cycle of growth and decay.

How good is this model—how realistic are its assumptions? That's an interesting and important question, but we'll put it aside until Chapter 4. Meanwhile, if you are unfamiliar with diagrams like Fig. 1.3 (or Fig. 1.4.), then try Exercise 1.5.

1.5 A Second Simple Ecosystem

This time, let $x(t)$ and $y(t)$ denote the numbers of thousands (say) of individuals, at time t, in populations of two species that compete for the same food supply (herbivores for grass, say). Definitions (1.17) still apply, of course, but the model will have to be altered. This time, each species would grow exponentially in the absence of the other; whereas, put together, each species would have a restraining effect on the food supply of the other. We would expect this effect to manifest itself as a decrease in growth rate. Let's assume that the decrease for one population is proportional to the size of the other population. The model we propose is therefore

$$\mu_1 = a_1 - b_1 y, \qquad \mu_2 = a_2 - b_2 x, \qquad (1.21)$$

where a_1, a_2 are pure growth rates and b_1, b_2 are interaction (crowding) parameters. The growth or decay of the two populations is thus described by the pair of ordinary differential equations

$$\frac{1}{x}\frac{dx}{dt} = a_1 - b_1 y, \qquad \frac{1}{y}\frac{dy}{dt} = a_2 - b_2 x, \qquad (1.22)$$

often called Gause's equations; see, for example, Pielou (1977, p. 75). The first of the equations agrees with (1.20), but the second is different. Again, these seemingly innocuous equations can be solved completely only by numerical means.

Nevertheless, by a similar procedure to the one used in Section 1.4, it can be shown that pairs of values $(x(t), y(t))$, representing the solutions to (1.22) in the x–y plane, form curves like those depicted in Fig. 1.4 (see Exercise 1.4). This phase-plane diagram was drawn for $a_1 = 3$, $a_2 = 5/2$, $b_1 = 2$, and $b_2 = 1$; but the topology would be unaltered if we changed the parameter values. Notice that all trajectories below the dashed curve

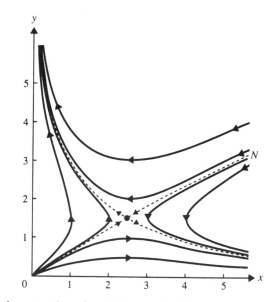

Fig. 1.4 Phase-plane topology for solutions to (1.22). Curves were drawn for $a_1 = 3$, $a_2 = 5/2$, $b_1 = 2$, $b_2 = 1$, and for various values of $x(0)$, $y(0)$. The dashed curve is known as a separatrix. As in Fig. 1.3, time increases in the direction of the arrows. See Exercise 1.4 for more details.

ON ultimately approach the x-axis, whereas all trajectories above ON ultimately approach the y-axis. Because it's really too remarkable a coincidence that the point $(x(0), y(0))$ would just happen to lie on the special curve ON (isn't it?), our model suggests the ultimate extinction of one or other of the two species, depending on the initial population magnitudes.

How good is this model—how realistic are its assumptions? Perhaps you would like to consider that question; we'll return to it in Chapter 4.

Meanwhile, before proceeding, we pause to observe that our simple ecosystem model is easily generalized to describe interactions among an arbitrary number, n, of species, which we shall label as $i = 1, \ldots, n$. Let $x_i(t)$ denote the population in thousands (say) of species i at time t, $i = 1, \ldots, n$. Then the specific growth rate of species i is

$$\mu_i = \frac{1}{x_i} \frac{dx_i}{dt}, \qquad i = 1, \ldots, n. \tag{1.23}$$

These n equations are still just definitions, but they become a model as soon as we make assumptions about the form of μ_i. Let's suppose that $\mu_i = \mu_i(\mathbf{x}, t, \alpha)$, where $\mathbf{x} \equiv (x_1, x_2, \ldots, x_n)^T$ is an n-dimensional column vector, $\alpha = (\alpha_1, \alpha_2, \ldots, \alpha_m)^T$ is an m-dimensional vector of parameters, and a superscript T denotes transpose. Then our n-species model has the mathematical structure of a system of n ordinary differential equations:

$$\frac{dx_i}{dt} = f_i(x, t, \boldsymbol{\alpha}) \equiv x_i \mu_i(\mathbf{x}, t, \boldsymbol{\alpha}), \qquad i = 1, \ldots, n. \qquad (1.24)$$

The generic term for growth and decay is dynamics, and therefore the model (1.24) is said to form a *dynamical system*. In exceptional circumstances, an equivalent mathematical model of the form

$$x_i(t) = \xi_i(t; \mathbf{x}^0, \boldsymbol{\alpha}), \qquad i = 1, \ldots, n, \qquad (1.25)$$

where $\mathbf{x}^0 = \mathbf{x}(0)$ is an arbitrary vector of initial conditions and ξ_i a suitable function, may be deduced from (1.24). We saw this happen in Section 1.1, with $x_1 = x$, $n = 1$, $m = 1$, and $\alpha_1 = r/V$. We also saw it happen in Section 1.3, with $x_1 = x$, $n = 1$, $m = 2$, $\alpha_1 = k$, and $\alpha_2 = x_f$. In this and the previous section, however, with $n = 2$, $m = 4$, and $\boldsymbol{\alpha}^T = (a_1, a_2, b_1, b_2)$, we saw only the form (1.24); (1.25) was implicit in Fig. 1.3 and in Fig. 1.4, but we were unable to deduce its form explicitly.

Although the differential equations (1.24) always form a dynamical system, dynamical systems need not have the form (1.24). A *nondifferential* dynamical system will first appear in Section 1.6. Dynamical systems expressed in terms of recurrence relations will first appear in Sections 1.7 and 1.8. A dynamical system expressed in terms of *differential-delay* equations will appear in Section 1.13.

1.6 Economic Growth

As we said at the beginning of the chapter, a model's assumptions may be suggested either conceptually (by native wit) or empirically (by supporting data). Hitherto we have been following the conceptual approach. Now let's take the empirical one.

For the most part, the quantity of goods produced by a society increases with time; or, which is exactly the same thing, the society experiences *economic growth*. Why? Because manufacturers increase their output by hiring more workers and acquiring more buildings and machinery for them. But if this increased output must be shared by an increased work force, will anyone really be better off? Let's attempt to answer this question by constructing a simple mathematical model.

Let $L(t)$ denote a suitable measure of the amount of labor employed at time t, where t is measured in years, in the manufacturing sector of a certain economy; for example, the economy of Massachusetts. Let $Q(t)$ denote the *instantaneous annual output* at time t, i.e., the quantity of goods *per year* produced at time t by this labor force. What do we mean by goods per year at time t? We mean that if the production rate could somehow be held constant for a whole year beginning at time t, then the amount of goods produced in that year would be $Q(t)$. Provided, as we have assumed, that time is measured in years, this is exactly the same thing as saying that the amount of goods produced in the short interval $[t - \delta t/2, t + \delta t/2]$ is $Q(t)\delta t + o(\delta t)$, or that the amount of goods produced in s years beginning at time t is

$$\int_t^{t+s} Q(\tau)d\tau, \tag{1.26}$$

the error term vanishing in the limit of integration. Because "instantaneous annual output" is rather a mouthful, we'll refer to Q simply as *output*.

In producing this output, workers need buildings, raw materials, machinery, tools, and a variety of other aids. We'll refer to these aids collectively as *capital stock*, or simply *capital*. Let $K(t)$ denote capital at time t.

How should we measure the amount of labor—in numbers of workers, in numbers of hours worked, or in terms of some other unit? And how should we measure the economy's output? Because we are primarily interested in how things are growing rather than how big they are, we can temporarily avoid these questions by defining suitable indices. We define a *labor index* by

$$i_L(t) \equiv \frac{L(t)}{L(0)}. \tag{1.27a}$$

It compares the amount of labor being used at time t to the amount being used at time zero. The index i_L is independent of the unit of measurement for labor. Massachusetts, for example, estimated that its amount of labor increased by 30% between 1899 and 1905 (Douglas, 1934, p. 160; see also Table 1.1). If we assume, arbitrarily, that $t = 0$ corresponds to 1899, then $i_L(6) = 1.3$, regardless of the amount of labor actually employed at either time. In a similar vein, we can define capital and output indices by

$$i_K(t) \equiv \frac{K(t)}{K(0)}, \qquad i_Q(t) \equiv \frac{Q(t)}{Q(0)}. \tag{1.27b}$$

For example, Massachusetts estimated that capital increased by 37% and output by 42% between 1899 and 1905. Thus $i_K(6) = 1.37$ and $i_Q(6) = 1.42$. Like i_L, the indices i_K and i_Q are independent of their corresponding units of measurement. All three indices are said to be "dimensionless" (we'll discuss dimensionality in Section 1.14).

Indices for the years 1890 to 1926 are presented in Table 1.1. Notice from the final column of this table that, *on the whole*, output increased with time, even though there were years like 1908 when it fell temporarily. Labor and capital also increased on the whole. But how was this increased output related to labor and capital? Were these data the manifestation of some economic law?

In the early part of this century, this question was addressed by economists C. W. Cobb and P. H. Douglas, who came to the conclusion that there was indeed such a law; for details, see Douglas (1934), on which this

Table 1.1 Variation with time of capital, labor, and
output indices.

t	$i_K(t)$	$i_L(t)$	$i_Q(t)$	t	$i_K(t)$	$i_L(t)$	$i_Q(t)$
-9	0.95	0.78	0.72	10	2.05	1.43	1.60
-8	0.96	0.81	0.78	11	2.51	1.58	1.69
-7	0.99	0.85	0.84	12	2.63	1.59	1.81
-6	0.96	0.77	0.73	13	2.74	1.66	1.93
-5	0.93	0.72	0.72	14	2.82	1.68	1.95
-4	0.86	0.84	0.83	15	3.24	1.65	2.01
-3	0.82	0.81	0.81	16	3.24	1.62	2.00
-2	0.92	0.89	0.93	17	3.61	1.86	2.09
-1	0.92	0.91	0.96	18	4.10	1.93	1.96
0	1.00	1.00	1.00	19	4.36	1.96	2.20
1	1.04	1.05	1.05	20	4.77	1.95	2.12
2	1.06	1.08	1.18	21	4.75	1.90	2.16
3	1.16	1.18	1.29	22	4.54	1.58	2.08
4	1.22	1.22	1.30	23	4.54	1.67	2.24
5	1.27	1.17	1.30	24	4.58	1.82	2.56
6	1.37	1.30	1.42	25	4.58	1.60	2.34
7	1.44	1.39	1.50	26	4.58	1.61	2.45
8	1.53	1.47	1.52	27	4.54	1.64	2.58
9	1.57	1.31	1.46				

Data refer to the Massachusetts economy in the years from
1890 ($t = -9$) until 1926 ($t = 27$); see Douglas (1934,
p. 160).

section is based. We can appreciate their work most readily by defining
two new variables:

$$\xi(t) \equiv \ln\left(\frac{i_L}{i_K}\right), \qquad \psi(t) \equiv \ln\left(\frac{i_Q}{i_K}\right). \tag{1.28}$$

The usefulness of these definitions will shortly emerge. For integer values
of t between -9 and 27, $\xi(t)$ and $\psi(t)$ may be derived from Table 1.1, and
these values are plotted in Fig. 1.5. Notice that, with the exception of a few
outlying points such as that representing 1917, the points in this diagram
lie fairly close to the line drawn through the origin.

Now, suppose that we wish to capture the *essential features* of the way
in which labor and capital determined economic growth in Massachusetts
in the years between 1890 and 1926. In other words, suppose that we wish
to have a mathematical model of that growth. Wouldn't it be reasonable to
approximate the information in the data points in Fig. 1.5 by the straight
line through the origin? Because the line has slope 3/4, we would thus
obtain $\psi(t) = 3\xi(t)/4$ or, in view of (1.28),

$$i_Q(t) = \{i_L(t)\}^{3/4}\{i_K(t)\}^{1/4}. \tag{1.29}$$

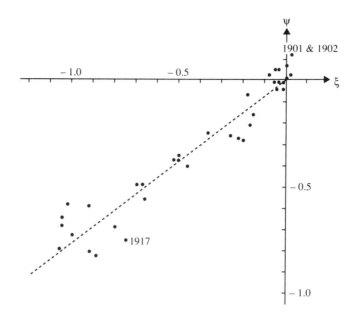

Fig. 1.5 Plot of ψ against ξ for the data in Table 1.1; ξ and ψ are defined by (1.28). There are 36, not 37, data points because the uppermost point represents both 1901 and 1902.

Indeed, (1.29) might even be a better description of *overall* economic growth than the diagram itself, because even the briefest of glances at Fig. 1.5 suggests that output is abnormal in certain years (e.g., 1917); and such abnormalities may be associated with the outlying points. We might therefore propose (1.29) as a model of economic growth in Massachusetts.

A possible criticism of this proposal immediately springs to mind. Suppose we accept that there is some line that captures the essential features of the distribution of points in Fig. 1.5. Then the line must certainly pass through the origin, as a direct consequence of definitions (1.27) and (1.28). But, having fitted the line in Fig. 1.5 to the data by eye, how do we know that slope 3/4 is the one that fits best? The answer is that we don't; perhaps the best slope is as small as 0.7 or as big as 0.8. Therefore, let's simply call it γ, where $0 < \gamma < 1$. Then, instead of (1.29), we have

$$i_Q(t) = \{i_L(t)\}^\gamma \{i_K(t)\}^{1-\gamma} . \tag{1.30}$$

The true value of the best slope, in an appropriate sense to be defined in Section 8.4, and assuming that the data in Table 1.1 are exact, is actually 0.734; see (8.20). But the data in Table 1.1 are themselves only estimates, and greater accuracy than we determined by eye may not be relevant.

A second criticism might be that, however good this model is for Massachusetts, it will fail to describe economic growth in Vermont or New

Hampshire. But if these economies were qualitatively similar to that of Massachusetts, then wouldn't we still expect them to yield a diagram like Fig. 1.5, though the points would lie in different places and the line through the origin would have a different slope? Thus, by varying the value of γ, we could use (1.30) to describe economic growth in a variety of economies. Let us define

$$a \equiv Q(0)L(0)^{-\gamma}K(0)^{\gamma-1}. \qquad (1.31)$$

Then, from (1.27) and (1.30), we obtain

$$Q(t) = aL(t)^{\gamma}K(t)^{1-\gamma}. \qquad (1.32)$$

In the jargon of economics, capital and labor are *factors* of production, and relationships between output and its factors are known as production functions. Thus $Q(K, L) = aL^{\gamma}K^{1-\gamma}$ is a production function. It is known as the Cobb–Douglas production function, in honor of the men who introduced it.

Taking logarithms of (1.32) and differentiating with respect to time, we find the following expression for the specific growth rate of the economy's manufacturing sector, in terms of the specific growth rates of labor and capital:

$$\frac{1}{Q}\frac{dQ}{dt} = \frac{\gamma}{L}\frac{dL}{dt} + \frac{1-\gamma}{K}\frac{dK}{dt}. \qquad (1.33)$$

Notice that when γ is close to zero, the growth rate of output is much less responsive to the growth rate of labor than to the growth rate of capital; whereas if γ is close to 1, then the output growth rate is much more responsive to the labor growth rate than to the capital growth rate. We can therefore interpret the parameter γ as measuring the desirability of labor. In the jargon of economics, if F denotes a factor, then the quantity $(F/Q)\partial Q/\partial F$ is called the elasticity of production with respect to that factor. Setting $F = L$ in (1.32), we see that γ is the elasticity of production with respect to labor. For further elaboration, see a text on microeconomics, for example Nicholson (1979).

We have thus obtained an expression, namely, (1.33), for the specific growth rate of an economy. But what good is it? Unless we know the specific growth rates of labor and capital, how can we use (1.33) to obtain an explicit expression for the growth rate of output?

Before answering this criticism, let us first observe that (1.32) is a single relation among three functions, namely, K, L, and Q. It is because there are fewer relations than functions that we are unable to obtain an explicit expression for any of them. More generally, this situation would arise if we had n relations among $n + m$ functions. If the word "closed" were used to mean explicit, then such a model would not be closed, because explicit expressions could not be deduced for all $n+m$ functions. We might therefore call the model *open*. If the n relations did not involve derivatives of the $n + m$ functions with respect to t, then the model would also be *nondifferential*. Thus (1.32) is a special case of a nondifferential, open, dynamical system.

Two answers to the above criticism are now possible. The first is simply that, in certain circumstances, we might wish to know how fast output *would* grow, if capital and output were to grow at certain rates. Or, given the labor growth rate, we might like to know how fast capital must grow to keep output rising at a certain rate. More generally, we might like to know how n of the $n + m$ functions, called the *endogenous* functions, would respond to hypothetical movements of the remaining m functions, called the *exogenous* ones. For these purposes, nondifferential, open, dynamical systems can be very useful, as we will demonstrate more convincingly in Chapter 10.

The second answer is that an open dynamical system can always be closed by combining it with other relations (which might be either differential or nondifferential). Let's attempt to close the model (1.32). Where does a firm obtain its capital stock? It buys it from other firms, so that capital stock, in a certain sense, is part of the economy's output. There is one crucial difference, however. Suppose that a company buys a tool that was made this year. Then, next year, the tool will still be part of the company's capital stock—but it won't be part of the economy's output, because it was counted as part of this year's. More generally, suppose you are standing on the bank of a river, watching it flow around a bend. Silt gathers on a bend; imagine that the silt belongs to you. Now, most of the silt that the river carries will flow right past you. It's like the output of an economy: once it's gone you can't include it, but there will always be more of it. On the other hand, some of the silt that the river carries will be deposited on your bend, and there it will remain, to be augmented by the later flow of the river. It's like your capital stock, and the silt on all the bends in the river is like the capital of the whole economy.

The part of its output that an economy retains for its capital stock is said to be *invested*. Let $I(t)$ denote the instantaneous annual rate of investment, i.e., the amount of goods per year retained at time t for capital stock. Thus, in the short interval $[t - \delta t/2, t + \delta t/2]$, the amount of goods retained is $I(t)\delta t + o(\delta t)$. From now on, we'll refer to I as the *investment rate*. Furthermore, we'll refer to $I(t)/K(t)$ as the *specific investment rate*, i.e., the amount retained per year per unit of existing stock. Let us make the bold assumption that capital lasts forever. Then by analogy with (1.9):

$$\begin{array}{ccccc} \text{Rate of Change of} & = & \text{Capital} & - & \text{Capital} \\ \text{Capital Stock} & & \text{Inflow} & & \text{Outflow} \end{array}$$

$$\frac{dK}{dt} \quad = \quad I(t) \quad - \quad 0. \tag{1.34}$$

Here is a further relation, a differential one. But by also introducing an extra function, namely, $I(t)$, we have moved no nearer to closing our model!

We must therefore devise a new relation without introducing new functions. A new relation requires a fresh assumption. Perhaps the simplest we can make is that the economy retains a constant proportion, σ, of output for its capital stock. Thus

$$I(t) = \sigma Q(t). \tag{1.35}$$

With (1.32), (1.33), and (1.35), we now have three relations (one differential and two nondifferential) among the four functions $I(t)$, $K(t)$, $L(t)$, and $Q(t)$. To close our model we must devise a fourth. But that fourth relation will close our model only if no new functions are introduced—and ensuring this is not always easy, as we have just discovered!

To make matters simple, therefore, we will simply assume (see Exercise 1.6) that the labor force is growing at constant specific rate ρ, i.e.,

$$L(t) = L(0)e^{\rho t}. \tag{1.36}$$

Then on substituting (1.32), (1.35), and (1.36) into (1.34), we obtain the first order ordinary differential equation

$$\frac{dK}{dt} = \sigma a \{L(0)\}^{\gamma} e^{\rho \gamma t} K^{1-\gamma}. \tag{1.37}$$

The solution of this equation (Exercise 1.7) is readily shown to be given by

$$\{K(t)\}^{\gamma} = \{K(0)\}^{\gamma} + \frac{\sigma a}{\rho} \{L(0)\}^{\gamma} \left(e^{\rho \gamma t} - 1\right). \tag{1.38}$$

We can now return to the question of whether economic growth makes anyone better off. Let us assume that our measure, $L(t)$, of labor is simply the number of workers employed, as was actually the case for the Massachusetts study summarized in Table 1.1, and define the output per worker by

$$Z(t) \equiv \frac{Q(t)}{L(t)}. \tag{1.39}$$

If economic growth is going to improve the lot of the labor force, shouldn't Z increase with time? Taking logarithms of (1.39), differentiating, and using (1.33), we find that:

$$\frac{1}{Z}\frac{dZ}{dt} = \frac{1}{Q}\frac{dQ}{dt} - \frac{1}{L}\frac{dL}{dt} = (1-\gamma)\left\{\frac{1}{K}\frac{dK}{dt} - \frac{1}{L}\frac{dL}{dt}\right\}. \tag{1.40}$$

Using (1.36)–(1.38), (1.31), and (1.35) in (1.40), we now find (Exercise 1.7):

$$\frac{1}{Z}\frac{dZ}{dt} = (1-\gamma)\{I(0) - \rho K(0)\}\{K(0)\}^{\gamma-1}\{K(t)\}^{-\gamma}. \tag{1.41}$$

Thus, because $0 < \gamma < 1$, our model predicts that output per worker rises if

$$\frac{I(0)}{K(0)} = \left\{ \frac{1}{K} \frac{dK}{dt} \right\}_{t=0} > \rho, \tag{1.42}$$

i.e., if the specific rate of investment is higher, initially, than that of the labor force. Early investment is seen to be crucial. Because (1.41) tends to zero as $t \to \infty$ (Exercise 1.7), however, our model also predicts that output per worker ultimately settles down to a constant value.

How good is this model—how realistic are its assumptions? Perhaps you would like to consider that question, though you'll be in a better position to answer it after studying Chapter 4.

Before changing the subject, however, and in preparation for Sections 3.3, 3.4, and 10.3, we pause to remark upon the way in which output is measured, a matter we have managed to avoid until now by defining the output index $i_Q(t)$. The output of an economy may consist of products as diverse as knives, bookcases, and automobiles; therefore it would be meaningless to count the number of items produced. That would create the illusion that a factory producing 1000 knives had equalled the output of one producing 1000 cars. It would be equally meaningless to measure the weight of the output in tonnes, because a factory producing 1000 glass paperweights would have equalled the output of one producing several thousand scientific calculators. In avoiding such absurdities, the traditional method of incorporating diverse products into a single measure of output has been to count each according to its monetary value. Even here there's a difficulty, however, because it's common knowledge that a dollar won't buy as much in the year 2000 as it would have bought in 1950; or, more generally, that it won't buy as much today as it would have bought t years ago. Therefore, if we wish to make a fair comparison between an economy's output in one year and its output in a different year, then we must assign to the output of the second year the monetary value that it *would have had* in the first; or vice versa; or else assign to each the monetary value that it would have had in some other year, which we shall call the *base year* for comparison. It is usually most convenient to take $t = 0$ as the base year.

Now suppose that a dollar at time zero has the buying power of $P(t)$ dollars at time t, so that $P(0) = 1$; and suppose that the monetary value of an economy's output at time t is $Y(t)$. Then because each dollar at time t is equivalent to $1/P(t)$ dollars at time zero, the economy's output would have been assigned monetary value $Y(t)/P(t)$ if it had been produced at $t = 0$. This is a true measure of the economy's output, the one we have been denoting by $Q(t)$. Hence $Q = Y/P$, or

$$Y(t) = P(t)Q(t). \tag{1.43}$$

You might like to think of $Q(t)$ as being measured in real dollars and $Y(t)$ as being measured in paper dollars, with $P(t) : 1$ being the exchange rate between a paper dollar and a real dollar. Economists call Y the nominal output, Q the real output, and P the *price index* (compare with (1.27), remembering that $P(0) = 1$). Taking logarithms of (1.43) and differentiating with respect to t, we obtain

$$\frac{1}{Y}\frac{dY}{dt} = \frac{1}{P}\frac{dP}{dt} + \frac{1}{Q}\frac{dQ}{dt}. \tag{1.44}$$

The term on the left-hand side of this relation is called the nominal (specific) growth rate of output. The second term on the right-hand side is called the real (specific) growth rate of output. The first term on the right-hand side, the specific growth rate of the price index, is called the *rate of inflation*.

Thus true output, $Q(t)$, is measured by assigning to output its nominal monetary value $Y(t)$ and dividing by the price index, a procedure known as *deflation*. The true measure of capital stock, namely $K(t)$, is likewise obtained by assigning to capital stock its monetary value and dividing by the price index (and similarly for true investment, $I(t)$). In the present section, however, we were able to work exclusively in terms of real measures, because the data in Table 1.1 had already been deflated.

1.7 Metered Growth (or Decay) Models

Let's return to Section 1.1's polluted lake, in which $x(t)$ denotes the concentration of contaminant at time t, V lake volume, and r the rate at which water flows through. Suppose that, for economic reasons, the environmental police measure the lake's pollution level only once every unit of time. They obtain a set of readings $x(0)$, $x(1)$, $x(2)$, ..., $x(n)$.... To test the model of Section 1.1, we could compare these readings with the predictions $x(1) = x(0)e^{-r/V}$, $x(2) = x(0)e^{-2r/V}$, $x(3) = x(0)e^{-3r/V}$, and so on of equation (1.10); which we can write more succinctly as $x(n + 1) = e^{-r/V}x(n)$, $n = 0$, 1, 2,.... If we found acceptable agreement then we'd be satisfied with our model. If not, we would either reject it or try to improve it. (We'll discuss this at length in Chapter 4.)

But notice, now, that what has been happening in the intervals $0 < t < 1$, $1 < t < 2$, $2 < t < 3$, ..., $n < t < n + 1$, ... is of no importance. It might happen, for example, that in every time unit some mischievous polluter dumps fresh pollution into the lake but always removes it before the next measurement is taken. Then the true pollution level would fluctuate according to the solid curve in Fig. 1.6; whereas the model's predictions would be given by the dashed curve. In these circumstances, the model would not be a faithful guide to what was actually happening—yet we wouldn't be aware of this! Indeed we might as well replace the model of Section 1.1 by the *discrete* dynamic model

$$x(n + 1) = e^{-r/V}x(n), \qquad n = 0, 1, 2, \ldots. \tag{1.45}$$

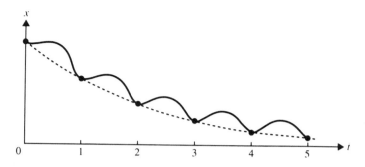

Fig. 1.6 A mischievous polluter deceives the environmental police.

Whenever integer values of t are the only relevant ones, it is customary to write $x(t)$ as x_t. Hence the more usual form of (1.45) is

$$x_{n+1} = e^{-r/V}x_n, \qquad n = 0, 1, 2, \ldots. \qquad (1.46)$$

Once x_0 is known, we can determine the sequence of values x_1, x_2, x_3, ..., by repeated, or recurrent, application of (1.46). Thus (1.46) is dubbed a *recurrence relation*. You can see that the measurable predictions of the continuous model (1.10) are exactly the same as those of the discrete model (1.46). But a cautious person—and scientists are supposed to be cautious people—would certainly prefer the model (1.46), because it says just as much with far less risk that it might actually be wrong. (Incidentally, I am unable to resist pointing out that a cautious person is also discreet.)

If we're interested only in the long-term pollution dynamics of this extraordinary lake, then (1.46) suffices as a model. If we're interested in the short-term dynamics, however, then (1.46) tells us nothing—and (1.10) is false! What can we do about this? Notice from Fig. 1.6 that the short-term dynamics appears cyclic. This raises the intriguing possibility of constructing a separate model to describe pollution dynamics *within* each unit of time, independently of the value of n, although the long-term dynamics would still be described by (1.46). We might regard the short-term model as imbedded in the long-term one. A possible form for the imbedded model would be

$$y(t_a) = A(x_n);$$
$$\frac{dy}{dt} = g(y), \qquad n < t_a \le t \le t_b < n+1; \qquad (1.47)$$
$$x_{n+1} = B(y(t_b)),$$

where y denotes pollution within a time unit and appropriate functions A, B, and g are to be defined. The time that elapses between $t = t_a$ and $t = t_b$ in (1.47) is independent of n. Perhaps it is easier to appreciate this by writing $t_a = n + \epsilon_1$, $t_b = n + 1 - \epsilon_2$, where $\epsilon_1, \epsilon_2 > 0$ and $\epsilon_1 + \epsilon_2 < 1$;

then $t_b - t_a = 1 - \epsilon_1 - \epsilon_2$. Note that the possible dependence of A, B, and g on any parameters of the problem, e.g., r/V, has been suppressed in the notation of (1.47).

Just suppose, for example, that our mischievous polluter achieves his goal by doubling the pollution level one tenth of a time unit after every reading, letting it decay at the natural rate and then halving it again one tenth of a time unit before the next reading. Then (1.47) would describe this process if we took $\epsilon_1 = 0.1 = \epsilon_2$ and defined $A(x) = 2xe^{-0.1r/V}$, $g(y) = -ry/V$, and $B(y) = 0.5ye^{-0.1r/V}$. You can easily check that (1.46) is implied. (Note that the solid curve in Fig. 1.6 would no longer be smooth but would have four corners in each time unit; two where x rises vertically after one-tenth of the time unit, and two where x falls vertically four-fifths of a time unit later.)

The example above is very fanciful and was introduced purely for ped-agogical purposes. But the idea of modelling growth and decay by imbed-ding a short-term dynamic model like (1.47) in a long-term dynamic model of the form

$$x_{n+1} = F(x_n) \tag{1.48}$$

can be a powerful one. We will follow Clark (1976) in calling such models *metered models*. The essential prerequisite for using such models is that the system being modelled is *qualitatively* the same within each unit of time. The system is said to be memoryless, because it is incapable of knowing from its own behavior which unit of time it is in. We will discuss the property of memorylessness more fully in Chapter 5.

Now let's see a more realistic example.

1.8   Salmon Dynamics

Suppose we wish to know the size of a salmon population at the end of each spawning cycle. We might expect a metered model to be useful here, because the spawning process is qualitatively the same within each cycle. There are, to be sure, some quantitative differences. But those that are due to initial conditions are part of the long-term dynamics described by (1.48). Other quantitative differences are due to changes in the environment. To make matters simple, let's assume that the environment is changing so slowly as to make such differences negligible. It is then legitimate to ex-press the short-term dynamics in an imbedded model of the form (1.47). The following model, due to Ricker (1954), is adapted from Clark (1976, p. 229).

Let x_n be the number of hundreds of millions of salmon at the end of cycle n and therefore also at the beginning of cycle $n+1$. These salmon are all adults. Referring to (1.47), let $y(t)$ denote the larval population (again in hundreds of millions) during the subsequent cycle (i.e., cycle $n + 1$). How is the initial larval population, $y(t_a)$, related to x_n? The more salmon there are, the more females there are, the more eggs they lay, and hence the more larvae there are. The simplest assumption incorporating this observation is that the number of larvae is proportional to the number of adult salmon at the end of the previous cycle, or

$$y(t_a) = \alpha x_n, \tag{1.49}$$

where α is a constant. Hence $A(x) = \alpha x$ in (1.47).

What happens to these larvae? The adults cannibalize them. The more adults there are, the more larvae that are eaten. The simplest assumption incorporating this observation is that the larval population decays at a rate proportional to the adult population—which, *in the short term,* is fixed. Thus

$$\frac{1}{y}\frac{dy}{dt} = -\beta x_n, \tag{1.50}$$

where β is a constant. In terms of (1.47), $g(y) = -\beta x_n y$, where x_n is constant in the embedded model. Integrating (1.50) from t_a to t_b gives the number of (hundreds of millions of) larvae that survive to mature into young salmon:

$$y(t_b) = y(t_a)\exp(-\beta x_n(t_b - t_a)) = \alpha x_n \exp(-\beta x_n(t_b - t_a)). \tag{1.51}$$

What happens to these young salmon? The ocean is fraught with risk for them and not all will survive to breed. Let's assume that a fraction γ survives. Then, at the end of the cycle, the salmon population is $\gamma y(t_b)$ *plus* the surviving adults. How many adults survive? In the case of Pacific salmon, we can safely assume that the answer is zero, because the adults die soon after spawning. On the other hand, Atlantic salmon may return to the sea after spawning, spend a year or two in the open water, then return upstream to spawn again. To keep things simple, let's assume that we're interested in Pacific salmon. Then

$$x_{n+1} = \gamma y(t_b). \tag{1.52}$$

Thus $b(y) \equiv \gamma y$ in (1.47). On defining

$$a = \gamma\alpha, \qquad b = \beta(t_b - t_a) \tag{1.53}$$

and combining (1.51) with (1.52), we obtain

$$x_{n+1} = ax_n e^{-bx_n}. \tag{1.54}$$

It isn't possible to express the solution to this recurrence relation as an explicit function of n, for all values of the parameters a and b, in the way that the solution to (1.46) was expressed as $x_n = x_0 e^{-nr/V}$, for all values of the parameter V/r. We must therefore solve (1.54) numerically, for particular values of a and b. Suppose, for example, that there are 10^8 salmon initially; then $x_0 = 1$, because we are counting in hundreds of millions. Suppose that, on average, each female lays 10^5 eggs for each individual, male or female, in the existing population (thus, when $n = 0$, all the females together lay 10^{13} eggs). Then, if we assume that eggs are just young larvae (salmon cannibalize both), $\alpha = 10^5$. Assume, moreover, that the effect of this cannibalism on the first generation is to reduce the larval population by 90%. Then, from (1.51), $0.1 = y(t_b)/y(t_a) = \exp(-\beta x_0(t_b - t_a))$, whence $b = \beta(t_b - t_a) = \ln(10)/x_0 \approx 2.303$. For these values of α, β and γ, (1.54) was solved numerically for various values of γ (the fraction of young surviving until the next spawning). These values were $\gamma = 0.5 \times 10^{-4}$, $\gamma = 1.0 \times 10^{-4}$, $\gamma = 1.1 \times 10^{-4}$, and $\gamma = 1.5 \times 10^{-4}$. Because $a = \gamma\alpha = 10^5\gamma$, the corresponding values of a were 5, 10, 11, and 15. The results are presented in Table 1.2, and you should verify them for yourself (Exercise 1.9).

Notice that the four values of γ (or a) correspond to four completely different kinds of behavior. For $a = 5$, the population settles down to the steady value $x^* = 0.699$, 70% of the initial value. For $a = 10$, the population is absolutely static. For $a = 11$, the population bounces back and forth between the values $x_L = 0.3568$ and $x_U = 1.726$. For $a = 15$, it

Table 1.2 Variation with time of salmon population.

n	$a = 5$	$a = 10$	$a = 11$	$a = 15$	n	$a = 11$	$a = 15$
0	1.000	1.000	1.000	1.000	21	1.725	2.396
1	0.500	1.000	1.100	1.500	22	0.3572	0.1443
2	0.7906	1.000	0.9611	0.7115	23	1.726	1.552
3	0.6402	1.000	1.156	2.074	24	0.3567	0.6526
4	0.7329	1.000	0.8876	0.2625	25	1.726	2.178
5	0.6778	1.000	1.265	2.151	26	0.3569	0.2167
6	0.7117	1.000	0.7562	0.2278	27	1.726	1.973
7	0.6912	1.000	1.458	2.022	28	0.3568	0.3147
8	0.7037	1.000	0.5584	0.2882	29	1.726	2.287
9	0.6961	1.000	1.698	2.226	30	0.3569	0.1771
10	0.7007	1.000	0.3744	0.1983	31	1.726	1.767
11	0.6979	1.000	1.739	1.884	32	0.3568	0.4531
12	0.6996	1.000	0.3488	0.3691	33	1.726	2.394
13	0.6986	1.000	1.719	2.367	34	0.3568	0.1449
14	0.6992	1.000	0.3614	0.1526	35	1.726	1.557
15	0.6988	1.000	1.730	1.611	36	0.3568	0.6481
16	0.6991	1.000	0.3545	0.5919	37	1.726	2.186
17	0.6989	1.000	1.724	2.272	38	0.3568	0.2137
18	0.6990	1.000	0.3581	0.1821	39	1.726	1.960
19	0.6989	1.000	1.727	1.796	40	0.3568	0.3225
20	0.6990	1.000	0.3562	0.4308			

The four values of a correspond to four values of the survival rate of young, when $b = 2.3026$; $x_n = 0.6990$ for $n > 20$ when $a = 5$.

seems that the population tries to settle into a pattern that repeats itself every twelve units of time, but without much success.

Do populations of salmon really behave in any of these ways? Perhaps you would like to consider that question. We'll return to it in Chapter 4; (1.54) is further discussed in Chapter 2.

1.9   A Model of U.S. Population Growth

Having discussed economic growth in Section 1.6, we now discuss a further example of the empirical approach to modelling. The number of millions in the U.S. population from 1790 until 1850 appears in the third column of Table 1.3. The unit of time is a decade, and $x(t)$ denotes magnitude at time t. Notice that population increases with time. The simplest hypothesis that might explain this observation is that x is proportional to t; then dx/dt would be constant. But the fourth column of Table 1.3 records values of the quantity

$$D(t) \equiv \frac{1}{2}\{x(t+1) - x(t-1)\}. \tag{1.55}$$

This is a good approximation to dx/dt if d^3x/dt^2 is small, which we shall simply assume (although see Exercise 1.12). Because D also increases with t, we reject the hypothesis that x is proportional to time. On the other hand, the final column of Table 1.3 shows that D/x is almost constant. Indeed

$$\frac{1}{x}\frac{dx}{dt} = 0.3 \tag{1.56}$$

is correct to one significant figure. In other words, the data suggest that we model growth of the population by setting $\mu = 0.3 = $ constant in (1.7). This empirically derived model predicts

$$x(t) = x(0)e^{0.3t} \approx 3.9e^{0.3t}. \tag{1.57}$$

Table 1.3   U.S. population in millions, 1790–1850.

Year	t	$x(t)$	$D(t)$	$D(t)/x(t)$
1790	0	3.929		
1800	1	5.308	1.66	0.31
1810	2	7.24	2.17	0.3
1820	3	9.638	2.81	0.29
1830	4	12.87	3.72	0.29
1840	5	17.07	5.16	0.3
1850	6	23.19		

Data for $x(t)$ are taken from *Historical Statistics of the United States, Colonial Times to 1970*, Bicentennial Edition, Part I, p. 8 (U.S. Bureau of the Census, Washington, D.C., 1975).

You can easily check that (1.57) predicts the population correctly to two significant figures for the years 1790–1840, except for 1810, when it predicts 7.1 instead of 7.2 for $x(2)$.

Can the same model predict population magnitude in later years? In the third and fourth columns of Table 1.4, the predictions of (1.57) are compared with observed values for the years 1850–1970. You can see that (1.57) soon fails to describe population growth with acceptable accuracy. On the other hand, $D/x \approx 0.24$ for the years 1860–1880, suggesting the improved model

$$x(t) = \begin{cases} 3.9e^{0.3t}, & 0 \le t < 6.5; \\ 31e^{0.24(t-7)}, & t > 6.5. \end{cases} \qquad (1.58)$$

But although this is correct to two significant figures for the years 1860–1880, and even to one significant figure for 1850 and 1890, it seriously overestimates the magnitude of the population in the present century. We must try again!

To find a better description empirically, the values for D/x in Tables 1.3 and 1.4 were plotted against those for x. The result appears in Fig. 1.7. Notice that points corresponding to the years 1800–1940 are clustered around the dashed line, which intercepts the vertical axis at about 0.31 and the horizontal axis at about 198. Hence the data suggest the improved model

$$\frac{1}{x}\frac{dx}{dt} = 0.31 \left(1 - \frac{x}{198}\right). \qquad (1.59)$$

Table 1.4 U.S. population magnitude and related data, 1850–1970.

Year	t	(1.57)	$x(t)$	$D(t)$	$D(t)/x(t)$	(1.60)
1850	6	24	23.19	7.19	0.31	23
1860	7	32	31.44	7.68	0.24	30
1870	8	43	38.56	9.36	0.24	39
1880	9	58	50.16	12.2	0.24	49
1890	10	78	62.95	12.9	0.2	61
1900	11	110	75.96	14.5	0.19	75
1910	12	140	91.97	14.9	0.16	90
1920	13	190	105.7	15.4	0.15	110
1930	14	260	122.8	13.0	0.11	120
1940	15	350	131.7	14.0	0.11	130
1950	16	470	150.7	23.8	0.16	150
1960	17	640	179.3	26.2	0.15	160
1970	18	860	203.2			170

Source of fourth column as in Table 1.3.

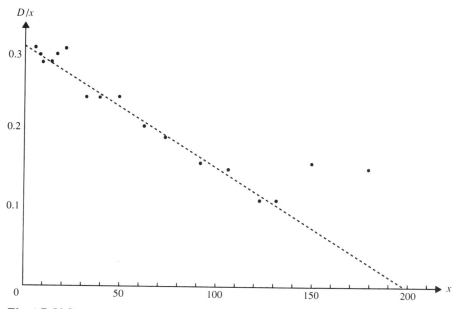

Fig. 1.7 U.S. population (horizontal axis) plotted against specific growth rate (vertical axis) for years 1800–1960. The unit of time is a decade; the unit of population, a million.

The solution of this differential equation, subject to $x(0) = 3.93$, is

$$x(t) = \frac{198}{1 + 49.4e^{-0.31t}}. \tag{1.60}$$

The predictions of (1.60), rounded to two significant figures, are recorded in the final column of Table 1.4. Notice that they are correct to at least one significant figure until 1950. In fact, the percentage error, that is,

$$100 \times \left\{ 1 - \frac{\text{predicted value}}{\text{observed value}} \right\} \%, \tag{1.61}$$

is less than 2.5% throughout this entire period, except for the year 1860, when the error is a little over 5%. But the model fails to account for the postwar rise in specific growth rate (appearing in the final column of Table 1.4). Indeed the current population of the United States exceeds the least upper bound of (1.60) by about 20%.

The model (1.60), which we derived empirically, could also have been conceptually derived by the following simple argument. Assume that there is a maximum population K, the capacity, that the land can sustain. When the population is x, the unused fraction of the capacity is $1 - x/K$. This is a measure of the potential for growth, and we might expect the specific growth rate to be proportional to it. Hence

$$\frac{1}{x}\frac{dx}{dt} = R\left(1 - \frac{x}{K}\right), \qquad (1.62)$$

where R is a constant of proportionality. We may interpret R strictly as an upper bound on the specific growth rate. On the other hand, if x is much smaller than K, say $x/K \leq \epsilon$, where ϵ is sufficiently small, then (1.62) yields

$$\frac{1}{x}\frac{dx}{dt} \approx R. \qquad (1.63)$$

Thus R may be regarded more intuitively as the early or maximum specific growth rate.

For the U.S. population we have found $\epsilon \approx 0.1$, $R = 0.31$, and $K = 198$, as you should check for yourself. But if the capacity of the United States is 198 million people, why are there over 235 million people here today? Does this mean that the model is useless—or can we improve it to resolve the discrepancy between it and the observations? We'll return to this question in Chapter 4. Meanwhile, try to answer the question yourself (Exercise 1.11); then try Exercises 1.10 and 1.13.

1.10   Chemical Dynamics

A chemical reaction is a process in which one or more substances, the reactants, combine or separate to form one or more other substances, the products. Mass is conserved in such a process (energy is not). The conservation of mass in one such reaction is represented by the symbolic equation

$$A + B \longrightarrow P. \qquad (1.64)$$

This says that one molecule of substance A combines with one molecule of substance B to form one molecule of substance P. For example, A could be carbon ($A = C$), B could be oxygen in its diatomic form ($B = O_2$), and P could be carbon dioxide ($P = CO_2$). Then (1.64) would describe mass conservation in the combustion of graphite (which releases heat energy at the rate of about 0.664 Joules per molecule of P).

Molecules, however, are such tiny things that we prefer to count them in units of approximately 6.024×10^{23}, just as we preferred to count the U.S. population in units of one million people. We call 6.024×10^{23} molecules of substance X a *mole* of X[1]. Then (1.64) may also be read as saying that a mole of A combines with a mole of B to produce a mole of P. Now the molecules are constantly whizzing about—they combine because they collide—and the more they collide, the more they combine. But the greater the concentration of each substance, i.e., the more of it that there is, then

---

[1] See Supplementary Note 1.1. The Supplementary Notes sections can be found at the ends of the chapters.

the greater the number of collisions between respective molecules. It therefore seems reasonable to assume that the rate at which A and B are converted into P will be proportional to the concentration of each. Let $[X]$ denote the concentration of substance X, in moles per unit volume. Then we have

Rate of Increase of Product = Rate of Conversion of Reactant

$$\frac{d}{dt}[P] \qquad = \qquad k[A][B], \qquad (1.65)$$

where k is a constant.

Suppose that $\xi(t)$ moles of A and B are converted to P in time t, i.e., $\xi(t)$ moles of A plus $\xi(t)$ moles of B become $\xi(t)$ moles of P—which is perfectly consistent, because the molecules of P are heavier than those of A or B. Then, if a_0 and b_0 denote the initial concentrations of A and B, (1.65) becomes

$$\frac{d\xi}{dt} = k(a_0 - \xi(t))(b_0 - \xi(t)). \qquad (1.66)$$

This may, at first, appear to be a new mathematical equation. But you can easily check that the substitutions

$$x = \max(a_0, b_0) - \xi, \qquad R = k(a_0 - b_0), \qquad K = a_0 - b_0 \qquad (1.67)$$

convert (1.66) into (1.62). The complete solution can thus be deduced from Exercise 1.10, and we leave that to you. You should check that, as $t \to \infty$ in your solution, the product concentration ξ tends to $\min(a_0, b_0)$; whereas the reactant concentrations tend to 0 and $\max(a_0, b_0) - \min(a_0, b_0)$. In other words, the less abundant of the two reactants is completely converted into a constituent of the product.

Recalling Exercise 1.10, notice how many phenomena are described by a single differential equation. Plant growth, U.S. population growth, and a chemical reaction—all are described mathematically by (1.62). That diverse phenomena may have a common mathematical structure is a recurring theme of applied mathematics, and further examples will appear throughout the text. The first such example appears in the following section.

1.11   More Chemical Dynamics

The chemical reaction (1.64) is *irreversible*: P does not separate into its constituents A and B. The simplest *reversible* reaction would be represented symbolically by

$$A \longleftrightarrow B. \qquad (1.68)$$

This says that A is converted into B, a process known as the forward reaction; at the same time, B is converted into A, a process known as the backward reaction. For the sake of definiteness, we will continue to call the term on the right-hand side the product and the term on the left-hand side the reactant, though, because of symmetry, this terminology is arbitrary and purely for convenience. To make matters simple, suppose that the substances A and B are diluted in water. Then A and B are *solutes*, and water is the *solvent*. Suppose, moreover, that the amount of water (in moles) per unit volume is extremely large compared with the amount of solute. We sometimes describe this by saying that the substances are in weak solution. Then the reaction takes place because molecules of A and B collide with water molecules. Perhaps, therefore, the reaction (1.68) should really be written as

$$H_2O + A \longleftrightarrow H_2O + B. \tag{1.69}$$

If so then, for reasons given in the previous section, the rate of conversion from A to B, i.e., the *forward rate*, would be $K_1[H_2O][A]$, where K_1 is a constant; whereas the rate of conversion from B to A, or the *backward rate*, would be $K_2[H_2O][B]$, where K_2 is another constant. But the concentration of H_2O in this weak solution is so extremely large that it does not alter appreciably during the lifetime of the reaction. Hence, as an approximation, $k_1 = K_1[H_2O]$ and $k_2 = K_2[H_2O]$ may be regarded as constants. The forward rate is then $k_1[A]$ and the backward rate $k_2[B]$.

Let a_0 and b_0 be the initial concentrations of substances A and B, respectively, in number of moles per unit volume of water. Suppose that, per unit volume of water, $\xi(t)$ moles of A are converted into B in time t. Then

$$\frac{d}{dt}[\text{Product}] = \text{Forward Rate} - \text{Backward Rate}$$
$$\frac{d}{dt}\{b_0 + \xi\} = k_1(a_0 - \xi) - k_2(b_0 + \xi); \tag{1.70}$$

$$\frac{d}{dt}[\text{Reactant}] = \text{Backward Rate} - \text{Forward Rate}$$
$$\frac{d}{dt}\{a_0 - \xi\} = k_2(b_0 + \xi) - k_1(a_0 - \xi). \tag{1.71}$$

Either equation is equivalent to

$$\frac{d\xi}{dt} = k_1 a_0 - k_2 b_0 - (k_1 + k_2)\xi. \tag{1.72}$$

Equation (1.72) is a mathematical model for the dynamics of the reversible chemical reaction that is described symbolically by (1.68).

At first, this may look like a new mathematical equation; but you've seen it before, haven't you? See Exercise 1.18.

1.12   Rowing Dynamics

What determines the speed of a rowing boat? By pulling on their oars the crew push against the water far from the boat. This exerts a force, which *tries* to make the boat gain speed; we'll call it the tractive force and denote it by T. The water adjacent to the sides of the boat, however, exerts a force that *tries* to make the boat lose speed; we'll call this the drag force and denote it by D. If T exceeds D, then the boat really will gain speed, or *accelerate*; and the greater the difference between T and D, the greater this acceleration. On the other hand, if D exceeds T, then the boat will actually lose speed; and the greater the difference, the greater the speed loss.

Let's try to turn these ideas into mathematics. Let $u(t)$ be the speed of the boat at time t. Then du/dt is the rate of change of speed, which we have agreed to call the acceleration. What we have found is that

$$\frac{du}{dt} > 0 \ \ \text{if } T > D, \quad \frac{du}{dt} < 0 \ \ \text{if } T < D, \quad \left|\frac{du}{dt}\right| \ \text{increases with } |T-D|.$$

The simplest hypothesis that explains all this is that du/dt is proportional to $T - D$. Let us therefore assume that

$$\frac{du}{dt} = \frac{1}{M}\{T - D\}, \tag{1.73}$$

where $1/M$ is a constant of proportionally. Notice that, for given tractive force and drag, large M is associated with small acceleration, and small M is associated with large acceleration. M is therefore a measure of the boat's reluctance to change speed, or *inertia*. We call M the *mass*.

Before proceeding, we pause to record a detail from physics. Heavy objects have greater reluctance to change speed than light ones (try pushing), hence greater mass. But heavier objects, by definition, have greater weight, denoted by W. Indeed the relationship between weight and mass is a simple one of proportionality, namely,

$$W = Mg. \tag{1.74}$$

The constant of proportionality, g, is known as the *acceleration due to gravity*. The justification for this terminology will emerge in Section 1.14.

Further to develop our model (1.73), we borrow again from the science of physics. In an infinitesimal time δt, the boat will move a distance $u\delta t$; and maintaining a tractive force T for this distance requires mechanical energy $Tu\delta t$. Where does this energy come from? From the muscular exertions of the crew. The rate at which energy is supplied by them is known as the *power*. But since humans are only about 25% efficient at converting the chemical energy of food into mechanical energy for useful work, the power actually supplied to the boat is only about a quarter of the rate at which oarsmen burn energy, the remaining three quarters being dissipated as heat. For this reason, some people prefer to call the rate of energy supply the *effective power.* Now, assume that there are eight in the boat's crew, and that they have entered a race. In training, they will have determined empirically the maximum effective power that an oarsman can sustain for the full length of the course; let's denote it by P. To make matters simple, let's assume that the oarsmen are identical, at least for the purpose of rowing a boat. Then, assuming that they would like to win the race, they will supply power $8P$ for the entire length of the course. Therefore, because power is rate of energy supply, the energy supplied in an infinitesimal time δt will be $8P\delta t$. This must equal $Tu\delta t$. Hence

$$\text{Effective Power} = \text{Tractive Force} \times \text{Velocity}$$
$$8P \quad = \quad T \times u. \tag{1.75}$$

From fluid dynamics, we borrow the result that drag is proportional both to the square of the boat's velocity and to its wetted surface area, i.e., the area in contact with the water. Let's denote wetted surface area by S. Then

$$D = bSu^2, \tag{1.76}$$

where b is a constant of proportionality. From (1.73), (1.75), and (1.76):

$$\text{Mass} \times \text{Acceleration} = \text{Tractive Force} - \text{Drag Force}$$
$$M \times \frac{du}{dt} \quad = \quad \frac{8P}{u} \quad - \quad bSu^2. \tag{1.77}$$

This is a first order ordinary differential equation for $u(t)$, which we propose as a model for the dynamics of rowing. The solution of this differential equation, subject to the initial condition $u(0) = 0$, is defined implicitly (Exercise 1.19) by

$$\ln\left\{\frac{u^2 + uV + V^2}{(u-V)^2}\right\} + \frac{\pi}{\sqrt{3}} - 2\sqrt{3}\tan^{-1}\left\{\frac{V + 2u}{V\sqrt{3}}\right\} = \frac{6bSVt}{M}, \tag{1.78}$$

where we have defined

$$V = 2\left(\frac{P}{bS}\right)^{1/3}. \tag{1.79}$$

The function $u(t)$ defined implicitly by (1.78) is plotted in Fig. 1.8. Notice that the initial acceleration is theoretically infinite according to (1.77), when in practice it must be finite; but this should not concern us any more than the fact that, in theory, the boat will actually reach speed V only as $t \to \infty$. A model is intended only to capture the essential features of reality, not to be an identical replica.

Yet how good is this model—does it fit the facts? Fig. 1.8 suggests that oarsmen supplying power at a constant rate would achieve maximum acceleration at the start of a race and subsequently reduce acceleration, until their speed levelled off at the constant value defined by (1.79). My intuition suggests that this is more or less what happens in a race; except perhaps that the oarsmen slow down slightly towards the end of it. But in

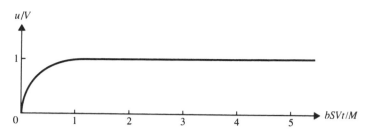

Fig. 1.8 Speed history of a rowing boat. Both dependent and independent variables are dimensionless; see Section 1.14 and Exercise 1.20.

science it's generally a bad idea to rely exclusively upon one's intuition. Therefore, in Chapter 4, we will test this model more objectively.

1.13   Traffic Dynamics

Suppose that there are N vehicles in a lane of traffic. We will label them $j = 1, \ldots, N$, with $j = 1$ corresponding to the leading vehicle. Let us assume that their masses and lengths are equal. This might be at least approximately true on a bridge or in a tunnel from which trucks have been barred. Let the front bumper of the jth vehicle have displacement $z_j(t)$ at time t, measured from the beginning of the road. Then the jth vehicle has velocity dz_j/dt and acceleration d^2z_j/dt^2; whereas its displacement and velocity *relative to the vehicle in front of it* are, respectively, $z_j(t) - z_{j-1}(t)$ and $dz_j/dt - dz_{j-1}/dt$.

Suppose that the road is quite congested, which bridges and tunnels usually are. Thus the average value of $|z_j(t) - z_{j-1}(t)|$ is not too large. The smaller it is, the more congested the road. Now, on a busy road, vehicles are prevented from crashing into one another, at least in theory, because drivers brake when they see that the gap in front of them is closing. The faster that gap closes, i.e., the greater the value of $dz_j/dt - dz_{j-1}/dt$, the harder they brake. Moreover, the greater the congestion, i.e., the smaller the value of $|z_j(t) - z_{j-1}(t)|$, the harder they brake. The simplest hypothesis incorporating these two observations is that

$$\text{Braking Force} = A \frac{z_j'(t) - z_{j-1}'(t)}{|z_j(t) - z_{j-1}(t)|}, \tag{1.80}$$

where $A(>0)$ is a constant and a prime denotes differentiation. Now, borrowing from physics, the braking force is $-M \times \text{ACCELERATION}$, where M is the mass of a vehicle. You can see this by identifying the braking force with D in (1.73) above and setting $T = 0$. But drivers take a certain time, say τ, to respond to changing traffic conditions. Hence

$$-M\frac{d^2z_j(t+\tau)}{dt^2} = A\frac{z'_j(t) - z'_{j-1}(t)}{|z_j(t) - z_{j-1}(t)|}; \tag{1.81}$$

or, because $z_{j-1}(t) > z_j(t)$,

$$\frac{d^2z_j(t+\tau)}{dt^2} = \lambda\frac{d}{dt}\{\ln|z_j(t) - z_{j-1}(t)|\}, \tag{1.82}$$

where $\lambda = A/M$. Integrating, we obtain

$$\frac{dz_j(t+\tau)}{dt} = \lambda\ln|z_j(t) - z_{j-1}(t)| + \alpha_j, \tag{1.83}$$

where α_j is independent of t. Because this equation is valid for $2 \le j \le N$, we have a system of $N - 1$ nonlinear ordinary differential equations. This system is even more complex than (1.20) and (1.22) because of the delay, τ, in driver response to the stimulus of changing traffic conditions.

If we weren't able to solve (1.20) or (1.22) analytically, then what hope do we have of solving (1.83)? To be sure, we can always find solutions by numerical approximation. As we shall discover repeatedly throughout this text, however, there is a satisfaction in the simplicity but generality of analytic solutions which makes them always desirable and always preferable, when available, to numerical approximations. But the question still remains: how do we hope to obtain analytic solutions from equations as complicated as (1.83)?

The answer lies in the observation that, in the real world, nothing can grow or decay forever: ultimately, growth or decay must cease. What happens then? An equilibrium or steady state is reached. In the present model, this steady state is a special solution of (1.83) which, as we shall see in Section 2.3, is readily found. But already we're talking about equilibrium, and this is the topic of Chapter 2.

1.14 Dimensionality, Scaling, and Units

The use of dynamic models is greatly facilitated by an understanding of three closely related concepts, namely, those of dimensionality, units, and scaling. We conclude our first chapter by discussing them.

Suppose that a friend of mine drove 20,000 miles last year and earned 20,000 dollars. Would I be right to say that he drove as much as he earned? No, of course not, because the equation

$$\$20,000 = 20,000 \text{ miles} \tag{1.84}$$

does not make sense. Distance is measured in miles, income is measured in dollars, and no amount of dollars can ever equal a mile. We formalize this idea by saying that distance has the dimension of length or, in shorthand notation,

$$[\text{distance}] = L;$$

while income has the dimension of value, or

$$[\text{income}] = V.$$

In these terms, (1.84) is dimensionally inconsistent because $[\$20,000] = V$, and $[20,000 \text{ miles}] = L$. If, however, my friend spent 50¢ for every mile he drove, then I would be absolutely right to say that

$$\$10,000 = 20,000 \text{ miles} \times 0.5 \text{ dollars per mile}. \tag{1.85}$$

This equation is dimensionally consistent, because [dollars per mile]$= V/L$. Thus the right-hand side has the dimensions of $L \times V/L = V$, agreeing with the left-hand side.

All parameters and variables used in this book have characteristic dimensions, which are derived from products or quotients (or both) of the six basic dimensions listed in Table 1.5. Dimensions are given for a selection of quantities used in this book in Table 1.6.

All equations between variables and parameters must be dimensionally consistent. Consider, for example, the lake purification equation (1.9a), repeated here for convenience:

$$V\frac{dx}{dt} = -rx, \tag{1.86}$$

where V denotes volume, x concentration, t time, and r rate of water flow. On using rows three through five of Table 1.6, the left-hand side of (1.86) has dimensions $[V] \cdot [x]/[t] = L^3 \cdot ML^{-3}/T = M/T$; whereas the right-hand side has dimensions $[r] \cdot [x] = L^3/T \cdot M/L^3 = M/T$. Hence the equation is dimensionally consistent. Similarly, $y(t_a) = \alpha x_n$ in (1.49) is dimensionally consistent, because the proportion α is dimensionless, and the pure numbers $y(t_a)$ and x_n are also dimensionless.

The fact that all equations must be dimensionally consistent can be used to determine the dimensions of new parameters as they arise. For example, in (1.50), i.e., in the equation

$$\frac{1}{y}\frac{dy}{dt} = -\beta x_n,$$

Table 1.5   The six basic dimensions and their symbols.

length	L	temperature	C
mass	M	value	V
time	T	no dimensions	1

Table 1.6 Dimensions of some quantities arising frequently in mathematical modelling.

Quantity	Derivation	Dimensions	S.I. units	CGS units
Velocity	[Velocity] = [Length]/[Time]	L/T	m/s	cm/s
Acceleration	[Acceleration] = [Velocity]/[Time]	L/T^2	m/s²	cm/s²
Concentration of pollutant	[Concentration] = [Mass]/[Volume]	M/L^3	kg/m³	gm/cm³
Water flow	[Water flow] = [Volume]/[Time]	L^3/T	m³/s	cm³/s
Pollution inflow	[Pollution inflow] = [Water inflow] · [Concentration]	M/T	kg/s	gm/s
Force	[Force] = [Mass] · [Acceleration]	ML/T^2	kg · m/s² (Newton)	gm · cm · /s² (Dyne)
Energy	[Energy] = [Force] · [Distance]	ML^2/T^2	Newton · m (Joule)	Dyne · cm (Erg)
Power	[Power] = [Energy]/[Time] or [Power] = [Force] · [Velocity]	ML^2/T^3	Joule/s	Erg/s
Pressure	[Pressure] = [Force]/[Area]	$ML^{-1}T^{-2}$	Newton/m² (Pascal)	Dyne/cm²
Shear viscosity	See Section 2.6	$ML^{-1}T^{-1}$	kg · m⁻¹ · s⁻¹	gm · cm⁻¹ · s⁻¹
Kinematic viscosity	$[\nu] = [\mu]/[\rho]$; see Section 2.6	L^2/T	m²/s	cm²/s
Thermal conductivity	See Section 2.5	$MLT^{-3}C^{-1}$	J · m⁻¹ · s⁻¹/°C	Erg · cm⁻¹ · s⁻¹/°C

x_n is dimensionless, y is dimensionless, and dy/dt has dimensions $1/T$. Hence the dimensions of β must also be $1/T$. Similarly, the quantities λ and α_j appearing in (1.83) must have the dimensions of velocity, or L/T.

After a quantity's dimensionality has been settled, the number that determines its actual value will still depend upon the *units* in which those basic dimensions are measured. For example, velocity has the dimensions of length per unit of time. But a car that travels at 60 m.p.h., if lengths are measured in miles and time in hours, will travel at 88 feet per second if lengths are measured in feet and time in seconds. The car's velocity has not changed; velocity's dimensions have not changed; only the units chosen for the basic dimensions have changed. Two frequent choices for the basic dimensions of mass, length, and time are kilogram, meter, and second (Systeme Internationale, or S.I.) and gram, centimeter, and second (CGS). For the quantities listed in Table 1.6, the appropriate S.I. units or CGS units are given next to the respective dimensions. Temperature is usually measured in degrees Celsius (°C) and value in dollars. But other units are sometimes chosen for the basic dimensions. Moreover, units are sometimes mixed within a problem so that numbers are neither too large nor too small. In modelling lakes, for example, we might choose to measure flow of contaminant in tonnes per day, because tonnes per second would give too small a number, and kg per day would be too large. In brief, units for the basic dimensions may be chosen arbitrarily. Once they have been chosen, however, the units for quantities with derived dimensions must always be consistent with them.

Dimensionless parameters are useful in mathematical modelling for a variety of reasons. First, as we shall see in Sections 4.10 and 9.5, they are important measures of the relative strengths of competing effects. Second, they are important in the design of experiments. Being primarily concerned with the conceptual approach to modelling, however, we shall not pursue this matter; if you are interested, see Chapter 10 of Hansen (1967). Third, by using dimensionless parameters, we can reduce the number of arbitrary constants associated with a mathematical model. The process by which we achieve this is known as *scaling*. A few examples should suffice to explain how scaling works.

Equation (1.86) contains a single arbitrary constant, namely, $\tau = V/r$. We will call this the drainage time, because volume V of water would take time τ to flow out at rate r. Let us define a new variable by

$$\hat{t} = \frac{t}{\tau}. \tag{1.87}$$

Then, on using the chain rule, we have

$$\frac{dx}{d\hat{t}} = \frac{dt}{d\hat{t}}\frac{dx}{dt} = \tau\frac{dx}{dt} = \tau\left(\frac{-rx}{V}\right) = -\left(\frac{\tau r}{V}\right)x = -x,$$

whence

$$x(\bar{t}) = x(0)e^{-\bar{t}}. \tag{1.88}$$

The arbitrary constant has disappeared, because $\bar{t}$ measures the passage of time in units of drainage time. The expression for x corresponds, of course, to (1.10).

A dependent variable may also be scaled. Equation (1.15), i.e.,

$$\frac{dx}{dt} = k(x_f - x)x,$$

contains two arbitrary constants, namely, k and x_f. Now, if x were so small that x^2 could be neglected, then this equation would be approximately $(1/x)dx/dt = kx_f$, an exponential growth equation for which the doubling time, i.e., the growth equivalent of the half-life, would be $t_{2.0} = (1/kx_f)\ln(2)$; see Section 1.2. Let's use this doubling time to scale t and x_f to scale x, i.e., let's define

$$\bar{t} = \frac{t}{t_{2.0}}, \qquad \bar{x} = \frac{x}{x_f}. \tag{1.89}$$

Then, on using the chain rule, we have

$$\frac{d\bar{x}}{d\bar{t}} = \frac{1}{x_f}\frac{dx}{d\bar{t}} = \frac{t_{2.0}}{x_f}\frac{dx}{dt} = \frac{t_{2.0}}{x_f}kx(x_f - x)$$
$$= t_{2.0}kx_f\bar{x}(1 - \bar{x}) = \ln(2)\bar{x}(1 - \bar{x}), \tag{1.90}$$

an equation from which the arbitrary constants have disappeared. The solution

$$\bar{x}(\bar{t}) = \frac{2^{\bar{t}}\bar{x}(0)}{(1 - \bar{x}(0) + \bar{x}(0)2^{\bar{t}})} \tag{1.91}$$

corresponds, of course, to (1.16).

Equation (1.54), namely, $x_{n+1} = ax_n \exp(-bx_n)$, contains two arbitrary constants, a and b. By using a recurrence relation, we have, in effect, already scaled the time variable. It remains only to scale the dependent variable, x. Let us define

$$\bar{x}_n = \frac{x_n}{x^*}, \tag{1.92}$$

where $x^* = \ln(a)/b$. The significance of x^* will emerge in Section 2.4. Then x_n is measured in units of x^*, just as x was measured in units of x_f above. If we now define $r = \ln(a)$, then we easily obtain the dimensionless recurrence relation

$$\overline{x_{n+1}} = \frac{x_{n+1}}{x^*} = \frac{ax_n e^{-bx^*(x_n/x^*)}}{x^*}$$
$$= a\bar{x}_n e^{-\ln(a)\bar{x}_n} = \bar{x}_n e^{r(1-\bar{x}_n)}, \tag{1.93}$$

which contains a single arbitrary constant, r.

As a final example, let's consider equations (1.20), which contain the four arbitrary constants a_1, a_2, b_1, and b_2. Let's scale x with respect to $x^* = a_2/b_2$ and y with respect to $y^* = a_1/b_1$. The significance of x^* and y^* will emerge in the following chapter. Suppose that you cannot think of an appropriate characteristic magnitude for t; then just call it T, and determine its value later. Hence the new dimensionless variables are

$$\bar{x} = \frac{x}{x^*} = \frac{b_2 x}{a_2}, \qquad \bar{y} = \frac{y}{y^*} = \frac{b_1 y}{a_1}, \qquad \bar{t} = \frac{t}{T}. \qquad (1.94)$$

By methods that should now be familiar, we obtain

$$\frac{d\bar{x}}{d\bar{t}} = \frac{T}{x^*}\frac{dx}{dt} = \frac{T}{x^*}(x^*\bar{x})(a_1 - b_1(y^*\bar{y})) = a_1 T\bar{x}(1 - \bar{y});$$

$$\frac{d\bar{y}}{d\bar{t}} = \frac{T}{y^*}\frac{dy}{dt} = \frac{T}{y^*}(y^*\bar{y})(-a_2 + b_2(x^*\bar{x})) = -a_2 T\bar{y}(1 - \bar{x}), \qquad (1.95)$$

on using (1.20). Because T is at our disposal, we choose $T = 1/a_1$. The resultant set of differential equations contains only a single arbitrary constant, a_2/a_1, instead of the original four. You should check that (1.95) would contain the same single arbitrary parameter if we had chosen $T = 1/a_2$ instead.

Scaling variables is by no means compulsory and not always useful. Though it always simplifies the mathematics, it sometimes obscures important details of the process being modelled. Therefore, scale if it helps you; otherwise don't. That's what I've done throughout the text.

You can test your understanding of dimensionality and scaling by attempting Exercise 1.20.

Exercises

1.1 (i) Obtain (1.10) and (1.11).

(ii) Show that, in the lake purification model of Section 1.1, an annual variation of r about its mean would not significantly alter the value predicted for $t_{0.05}$. Hint: Take $r(t) = \rho(1 + \epsilon \sin(2\pi t))$, where ρ is the mean outflow (and inflow), t is the number of years since pollution input ceased, and $|\epsilon| < 1$ (otherwise r could be negative).

1.2 Derive (1.16).

1.3 Suppose that the phase-plane diagram in Fig. 1.3 is drawn for arbitrary values of the parameters a_1, a_2, b_1, and b_2. Deduce from (1.20) that the solution curves belong to the family

$$a_2 \ln(x) - b_2 x + a_1 \ln(y) - b_1 y = \ln(c),$$

where c is a constant. To two significant figures, the values of c in Fig. 1.3 are, starting with the degenerate innermost curve denoted by a dot, $c = 0.14$, $c = 0.13$, $c = 0.12$, $c = 0.10$, $c = 0.08$, and $c = 0.05$.

1.4 Suppose that the phase-plane diagram in Fig. 1.4 is drawn for arbitrary values of the parameters a_1, a_2, b_1, and b_2. Deduce from (1.22) that the solution curves belong to the family

$$-a_2 \ln(x) + b_2 x + a_1 \ln(y) - b_1 y = \ln(c)$$

where c is a constant. In Fig. 1.4, the value $c = 0.207$ corresponds to the two branches of the dashed curve (separatrix). For values of c exceeding this value, the two branches of the corresponding curve lie above and below the separatrix; for $c < 0.207$, to the right and left of it.

1.5 Two plants are growing independently off independent fixed supplies of substrate. Let $x(t)$ be the dry weight at time t of the first plant and $y(t)$ that of the second. Then, according to Section 1.3, we may regard $x(t) \equiv (x(t), y(t))^T$ as the state at time t of a two-dimensional system governed by the equations

$$dx/dt = kx(x_f - x), \qquad dy/dt = ky(y_f - y).$$

(i) Write down the solution $x(t)$ to these equations.
(ii) For $k = 0.1$, $x_f = 7$, and $y_f = 10$, plot solution curves in the x–y plane for a variety of initial values $x(0)$ in the rectangle $0 < x < 13$, $0 < y < 16$. Not all initial values are physically meaningful, but this will help you to understand the phase-plane.

1.6 Use the data in Table 1.1 to plot $\ln(i_L)$ against t. Use your diagram to comment upon the validity of (1.36). Do you consider this to be an acceptable labor growth model for the entire period 1890–1926? For part of it? If so, can you provide a rough estimate for the best value of ρ?

1.7 Solve (1.37) to obtain (1.38). Verify (1.41), and show that this expression approaches zero as $t \to \infty$.

*1.8 (i) Suppose that a lake is being polluted at the rate of P_{in} tonnes per day. How must the model of Section 1.1 be modified to account for this pollution input? Obtain a new expression for $x(t)$.
(ii) Everyone knows that machines wear out and buildings deteriorate. Suppose that, if there were no investment, then an economy's capital stock would decay in value at constant specific rate α. This decay is known as *depreciation*. How must the model of Section 1.6 be modified to account for depreciation?

1.9 Reproduce Table 1.2 for yourself by solving (1.54) on a calculator or computer.

1.10 *Deduce* the solution of the differential equation (1.62) from (1.16).

*1.11 According to the 1981 *Yearbook* of *Encyclopedia Britannica*, the U.S. population in 1980 was 226.5 million. Using Fig. 1.7, can you suggest a model that describes U.S. population growth from 1940 onward? Can you interpret the apparent change in the capacity of the land?

1.12 *Mathematical* assumptions can often be justified subsequently *without* recourse to data by checking the consistency of some mathematical prediction (whereas modelling assumptions can ultimately be justified only by recourse to data). We call this an *a posteriori* justification of the mathematical assumption. In Section 1.9, for example, we had to assume that d^3x/dt^2 was small in order to arrive at equation (1.59). Show *a posteriori* that the error in approximating dx/dt by $D(t)$, to obtain (1.59), could not have exceeded 10^{-3}.

*1.13 In this exercise, we will derive a model empirically and then interpret it conceptually. The model concerns the decay of a population of houseflies, not with respect to an increase in time but rather with respect to an increase in dosage of insecticide.

    Let ϕ denote the concentration in milligrams per cubic centimeter of pyrethrin insecticide. Let d denote the fraction of the population of houseflies that would be killed by that dosage. The following data are given by Finney (1952, p. 149):

ϕ	0.50	0.75	1.00	1.50	2.00
d	0.20	0.35	0.53	0.80	0.88

(i) Let $x = 1 - d$ denote the surviving fraction of the population when the concentration of insecticide is ϕ. Plot the values of $-\ln(x)$ against those of ϕ. What possible model is suggested by the data?
(ii) Can you interpret this model conceptually? *Hint*: Consider the effect on the survivors of an infinitesimal increase in concentration of insecticide.

*1.14 In Section 1.8 we imagined that the fraction, γ, of young salmon surviving until spawning was 1 in 10^4, or thereabouts. What could cause this fraction to be so small? A possible answer is predation. Let $x(t)$ denote the number of units (hundreds of millions) of young salmon at time t, *during the survival period* (thus $t_b \le t \le n+1$, where $n+1-t_b = \epsilon_2$ is independent of n). Let $h(t)$ denote the harvest rate, i.e., the number of units of young salmon captured and eaten per unit of time. Then, clearly,

$$\frac{dx}{dt} = -h(t).$$

(i) The more young salmon there are, the more that are captured. The more predators there are, the more young salmon that are captured. If $u(t)$ denotes the predator population, what is the simplest form of $h(t)$ consistent with these two observations?

(ii) The model described only short-term dynamics, because everything happens within the salmon's breeding cycle. What is the simplest form of $u(t)$ consistent with this observation?

(iii) Solve the resulting differential equation for $x(t)$ in the interval $t_b \le t \le n + 1$. Hence express γ as a function of the other parameters. Does it now seem reasonable that γ might be as small as 10^{-4}?

*1.15 In Exercise 1.14, we introduced a *short-term* model for preying on fish (salmon, in that particular example). Perhaps you imagined the predator to be another species of fish. It might be so, but it might just as easily be the human species. Then $u(t)$ denotes number of fishermen. Since fishermen who hope to sustain their livelihoods will be interested mainly in the *long-term* dynamics of harvesting, we must modify our model to account for this.

Now, in Section 1.8, the long-term dynamics were described by a recurrence relation, but they might also be described by a differential equation— witness (1.59), which describes U.S. population growth from 1790 onward. Indeed the main reason that the model of Exercise 1.14 is a short-term one is that growth of the fish population is ignored. Let's therefore include it by writing

$$\frac{dx}{dt} = F(x(t)) - h(t), \qquad (a)$$

where $F(x)$, the *net productivity*, is the number of units of fish added to the population per unit of time (and $h(t)$ remains the harvest rate). Exercise 1.14 is the special case of (a) for which $F(x) \equiv 0$. Equation (1.62) is the special case of (a) for which $F(x) \equiv Rx(1 - x/K)$ and $h(t) \equiv 0$.

(i) Suppose that in the absence of harvesting the (long-term) growth of a fish population would be given by (1.62), and observation (i) of Exercise 1.14 would remain valid. Write down the ordinary differential equation that describes harvesting of fish under these conditions.

(ii) Suppose that the number of fishermen, $u(t)$, is constant (even in the long term). Solve your differential equation for $x(t)$ as a function of t in the interval $0 \le t < \infty$.

*1.16 (i) We have seen in Section 1.9 that a population may adequately be described by (1.63) in the early stages of growth. Indeed this observation is implicit in (1.20) and (1.22). Let $x(t)$ denote the size of an insect population, which acts as a pest on an agricultural crop and is controlled by applying pesticide. Why might (1.62) actually be inappropriate?

(ii) Assume that the population dynamics would be described by (1.63) if it weren't for the fact that, enticed by the crop, additional pests of the same species immigrate into the area of cropland at a constant rate q per unit of time. Write down a differential equation that describes the growth of the total pest population before the crop is sprayed.

(iii) Show that

$$x(t) = \left(x(0) + \frac{q}{R}\right) e^{Rt} - \frac{q}{R},$$

at least until the crop is sprayed.

*1.17 (i) In Exercise 1.15, we defined the net productivity of a species of fish to be the number of units added to the population per unit of time. Let us suppose that the fish are not being harvested (i.e., $h = 0$). Then what is the relationship between net productivity and net specific growth rate? In particular, what is the relationship of $\mu(x)$ to the graph of $F(x)$?

(ii) What is the value of $F(0)$? If K is the population's capacity, i.e., the value that x cannot exceed, then what must be the value of $F(K)$?

(iii) In (1.62), with $\mu(x) = R(1 - x/K)$, we assumed that specific growth rate was largest when the population was least. Why might this assumption be unrealistic?

(iv) Why might it be more realistic to assume that $\mu(x)$ is small when x is small, increases to a maximum and then decreases to zero as x approaches the capacity, K?

(v) In view of (i) and (ii), what is the shape of the graph of F if it embodies (iv)?

(vi) Verify that $F(x) = ax^2(1 - x/K)$, where a is a constant, embodies (ii) and (iv).

1.18 Show that model (1.72) for a reversible chemical reaction is mathematically equivalent to the one you developed in Exercise 1.8(i). Hence obtain an expression for $\xi(t)$.

1.19 (i) Derive (1.78).

(ii) Using the relation $u \, du/dx = du/dt$, where x is distance travelled, solve (1.77) for u as a function of x and show that u approaches V as distance travelled becomes infinitely large.

(iii) Suppose that the length of the course is d. If the crew were magically able to attain speed V instantly, how long would it take them to row the course?

(iv) How long does it actually take them? Call this time t_f and find an expression of the form $t_f = G(q)d/V$, where $q \equiv M/6bSd$.

(v) Show that G is an increasing function of q, which approaches 1 as q approaches 0 from above.

(vi) If $q = 0.2$, by what percentage is t_f longer than the time given in (iii)? What if $q = 0.02$?

1.20 (i) What are the dimensions of the constant b in (1.77)?

(ii) By suitably defining dimensionless variables, show that (1.77) may be rewritten in a form that does not contain any arbitrary constants.

(iii) Integrate this differential equation and compare your answer with the one you obtained in Exercise 1.19.

(iv) By scaling t with respect to $1/R$ and x with respect to K, show that the logistic equation (1.62) may be written in the form

$$\frac{d\bar{x}}{d\bar{t}} = \bar{x}(1 - \bar{x}).$$

There are, of course, no arbitrary constants; but note that the factor $\ln(2)$, which appeared in (1.90) has also disappeared.

*1.21 Two plants are feeding off the same substrate, whose weight at time t is $S(t)$. Can you adapt the model of Section 1.3 to devise a system of two differential equations that describes their growth? Can you solve these equations? What if the plants are identical in all respects but one, namely, that they were of different sizes when they began to feed off the substrate?

Supplementary Notes

1.1 The number 6.024×10^{23}, known as Avogadro's number, was originally chosen so that a mole of hydrogen would have a mass of 1 gram. Then the mass of a mole of any other substance could be counted in terms of hydrogen molecule equivalents (or hydrogen atom equivalents, since atoms and molecules are the same thing for elements). Masses quoted in these units are known as molecular weights. Carbon dioxide (CO_2), for example, has a molecular weight of 44 because the carbon atom in each molecule has six protons and six neutrons and each of the two oxygen atoms has eight protons and eight neutrons, making a grand total of 44 protons and neutrons (each of which is counted as a hydrogen atom equivalent). The problem with this approach is that it pretends electrons have zero mass, when in fact their mass is $1/1840$ of a proton's. For greater accuracy, therefore, it is customary nowadays to replace "hydrogen atom equivalent" by "equivalent of one twelfth of carbon atom" (in terms of mass); but if you round molecular weights to the nearest integer then it makes no difference.

2 EQUILIBRIUM

Two more of the most useful ideas in mathematical modelling are those of equilibrium, or steady state, and stability. The two are closely related. Without stability, equilibrium has virtually no practical significance. Without equilibrium (or steady state), stability has absolutely no meaning, either practical or theoretical. To understand these important ideas, it will be convenient to introduce a little fiction.

Imagine that, while travelling through a distant region of the universe, you have accidentally discovered an infinite two-dimensional world. This world, called Phaseplane, is inhabited by microscopic creatures who call themselves Phasies. They are so tiny that we may think of them as points. Because their world is absolutely flat, they are like points moving about on an infinite page.

What kind of a world is Phaseplane? The first thing I must tell you is that it's an extremely windy one! The wind in Phaseplane moves in curious ways; yet its pattern never changes. Thus maps of it are extremely useful, and every Phasie is given a wind-atlas on her first day in kindergarten. A page from this atlas appears in Fig. 2.1. It shows the wind pattern over the rectangle $0 \leq x \leq 90$, $0 \leq y \leq 60$; x and y are measured in phasemiles. (The origin was chosen by the Phaseplane government and is jocularly known as the Big Saddle.)

A curve in Fig. 2.1 shows the path along which a Phasie would get blown by the wind if she were to stand at any point along that curve; the direction of movement is indicated by arrows. This path is known as the Phasie's *phase trajectory*. Because Phaseplane is such a windy place, a curve like this could in principle be drawn through any point of the rectangle in Fig. 2.1. The resulting infinity of curves would fill the entire diagram. But nobody can draw an infinite number of curves; and the exercise would be pointless even if possible, because a solid black diagram would give no information about the wind. Thus a page of a Phasie's atlas shows only

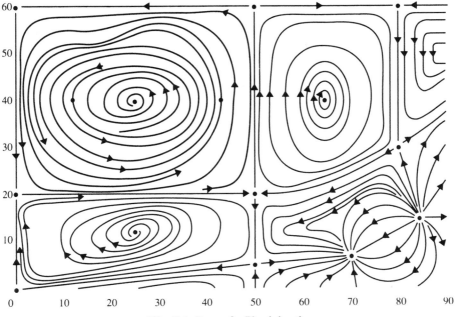

Fig. 2.1 Page of a Phasie's atlas.

a few phase trajectories. From these, a Phasie can infer the entire wind pattern by using her imagination.

Because Phasies are so tiny and the wind so strong, shouldn't they always go where the wind takes them? Isn't a Phasie's path in life completely determined by where she began it? There is much truth in this, but not the whole truth. The second thing I must tell you about is the wicked monster, Perturbation. Although the wind is almost always blowing, there are times when it disappears completely, just for a few seconds. When that happens, Perturbation comes out of her lair and takes great delight in blowing the Phasies all over the place. But the wind always comes back within a few seconds and, furthermore, resumes precisely the same pattern that it had before it mysteriously disappeared. Then Perturbation retires to her lair, and the Phasies return to following the destinies mapped out for ⸤hem in the atlas. Phasies have a name for these sudden experiences: they call them perturbations. Perturbations may be either large or small, depending on how far you get blown from where you were; and, as Phasies know well, even a small perturbation can be very disruptive. We'll discuss this further below.

Though Phaseplane is such a windy world, there are points where the wind velocity is always zero. These special points are known officially as static equilibria; though Phasies often omit the word static and refer to them merely as equilibria. Such points are marked on Fig. 2.1 by a small dot. There is one, for example, at the point $(70, 7)$ and another at $(25, 12)$. Notice how the wind pattern in the immediate vicinity of one equilibrium may differ from that in the immediate vicinity of another. Near $(70, 7)$, for example, the wind is always blowing toward the equilibrium. Thus a small perturbation would scarcely be noticed by a Phasie at $(70, 7)$; the wind, when it returned, would blow her back toward where she came from. For that reason, the equilibrium at $(70, 7)$ is called *locally stable*. Even a small perturbation, however, would be enough to nudge a Phasie at $(25, 12)$ onto a path that spirals outward, toward the curve consisting of four straight-line segments linking equilibria at $(0, 0)$, $(50, 5)$, $(50, 20)$, and $(0, 20)$. Because, following a perturbation, a Phasie at $(25, 12)$ would end up so far from where she began, this equilibrium is called *unstable*.

From this point of view, the equilibrium at $(25, 40)$ is identical to the one at $(70, 7)$. Both are locally stable. From another viewpoint, however, they differ: a Phasie being blown toward $(25, 40)$ will spiral toward the equilibrium; whereas a Phasie being blown toward $(70, 7)$ will approach it head on. We distinguish the two by calling the equilibrium that is approached circuitously a *focus* but the equilibrium that is approached directly a *node*. Hence $(25, 40)$ is a (locally) *stable focus*, and $(70, 7)$ is a (locally) *stable node*. Then what is $(25, 12)$? Notice that if the arrows were reversed then the wind currents in its vicinity would be the same topologically as $(25, 40)$. For that reason, $(25, 12)$ is called an *unstable focus*. Similarly, $(85, 15)$ is an *unstable node*.

We said that a small perturbation would scarcely be noticed by a Phasie who had been occupying $(70, 7)$. Nevertheless life, though not disrupted, would never be quite the same for her. Even if she lived forever, she would never again return to $(70, 7)$. The longer she lived, the closer she would approach; but she would never actually reach it. Why? The answer is that the wind speed keeps dropping as an equilibrium is approached—always just a little below the speed that would be necessary to blow the Phasie back home. (To be just a little more precise, the one way in which the Phasie could return all the way home before the end of eternity would be if the wicked monster blew her back during another perturbation; but that never happens, because monsters are so wicked.)

The wind in Phaseplane does vary in strength from place to place. Indeed the wind velocity can be described by a vector $\mathbf{u} = (u(x, y), v(x, y))^T$, where u is the component in the x direction, v the component in the y direction and both u and v are differentiable functions. Here, as throughout the book, bold typeface will denote a column vector and superscript T its transpose, which is a row vector. Thus $\mathbf{u}^T = (u(x, y), v(x, y))$ is the transpose of $\mathbf{u}$. If $\mathbf{x}(t) = (x(t), y(t))^T$ denotes the Phasie's phase trajectory, then her velocity is $d\mathbf{x}/dt \equiv (dx/dt, dy/dt)^T$; and, *because the Phasie is carried by the wind*, we have $d\mathbf{x}/dt = \mathbf{u}$ or, which is exactly the same thing,

$$\frac{dx}{dt} = u(x, y), \qquad \frac{dy}{dt} = v(x, y). \tag{2.1}$$

Let $\mathbf{x}^* = (x^*, y^*)^T$ denote a static equilibrium point. Then, because the wind velocity is zero at $\mathbf{x}^*$, we have $u(x^*, y^*) = 0 = v(x^*, y^*)$. Because u and v are both continuous, you can see that $(dx/dt, dy/dt)$ will approach the zero vector as the Phasie nears the equilibrium. The wind just isn't quite strong enough to take her all the way back home in a finite time. The best it can do for her is ensure that

$$\lim_{t \to \infty} x(t) = x^*, \qquad \lim_{t \to \infty} y(t) = y^* \tag{2.2}$$

(provided there are no further perturbations).

However disgruntled this Phasie might be, she has less cause for complaint than one who resided at $(65, 40)$ before a (small) perturbation. Notice that the wind currents in this vicinity form closed curves around the equilibrium. A Phasie perturbed from $(65, 40)$ will spend the rest of her life circling it. She will not approach it even in infinite time! On the other hand she at least remains in close proximity to her old equilibrium. Her plight is not as dire as that of the Phasie who once resided at $(25, 12)$. Thus $(65, 40)$ is far from being unstable; yet it isn't quite as stable as $(25, 40)$. How, then, should we distinguish between the two? Some mathematicians prefer to say that both are stable but that, in addition, $(25, 40)$ is asymptotically stable. Others, and we are among them, prefer to say that $(25, 40)$ is stable, but $(65, 40)$ is only *metastable*, or *neutrally stable*. All are agreed that $(65, 40)$ should be called a *center*, because the nearby trajectories form closed curves around it.

There is also some disagreement as to the best description for the equilibria at $(50, 5)$, $(50, 20)$, and $(80, 30)$, the neighborhoods of which have the same topology. The equilibria at $(0, 20)$, $(0, 60)$, $(50, 60)$, $(80, 60)$, and the Big Saddle itself are also of this form, though some of the trajectories nearby fall on different pages of the Phasie atlas. What distinguishes these equilibria is that each has two trajectories of approach; whereas all other trajectories emanating from the equilibrium are ones of departure. Thus a Phasie who once resided at $(50, 20)$, for example, and just happened to be perturbed in the x-direction, would have a similar experience to a Phasie perturbed from $(70, 7)$. For this reason, some mathematicians prefer to call $(50, 20)$ only partially unstable; for them, $(85, 15)$ would be totally unstable. Others, and we are among them, prefer to call $(50, 20)$ simply unstable, because why would the wicked monster be so considerate as to perturb the Phasie in just exactly the right direction? All mathematicians, however, agree that equilibria like $(50, 20)$ are called *saddle-points*.

We have thus far distinguished four categories of equilibria on the basis of neighborhood topology: center, focus, node, and saddle-point. A center must be metastable; a saddle-point must be unstable; and nodes and foci may be stable or unstable. All six forms of equilibrium have one thing in common, however: they are *static*. But a Phasie can live in a *steady state* without being absolutely static. This is what happens to the Phasies on the closed curves near (65, 40). It is also what happens to the Phasies who live on the closed curves through (12, 40) and (43, 40); these two points are marked by large dots, the only dots in Fig. 2.1 that do not represent static equilibria. But there is a difference. A Phasie who lived on a curve near a center, if slightly perturbed, would continue to live life on a closed curve. But a Phasie on the closed curve through (12, 40), if perturbed, would either spiral in toward the focus at (25, 40) or spiral out toward the closed curve through (43, 40). She would never again find herself on a closed curve. Likewise, a Phasie on the closed curve through (43, 40), if perturbed, would never again find herself on a closed curve; rather, depending on the nature of the perturbation, she would spend the rest of her life either spiralling in or spiralling out toward the steady state that she used to enjoy. For this reason, the latter trajectory is called a stable equilibrium cycle. The curve through (12, 40) is called an unstable equilibrium cycle. The more common term for a stable equilibrium cycle is *limit cycle*.

Until now, we have been talking exclusively about small perturbations and local stability. If the wind disappeared for quite a while, however, then the wicked monster might blow the Phasies a long way from home; or, as the Phasies prefer to say, cause them to suffer a large perturbation. It is clear from Fig. 2.1 that, even if a Phasie occupied a stable equilibrium, such a perturbation might land her on a trajectory that would never return to it, even in infinite time. We would say that her old equilibrium was locally stable, but not globally stable. For example, consider a Phasie who occupied (70, 7) when the wind disappeared. Suppose that the perturbation shifted her to (49, 21). Then, when the wind came back, she would find herself being blown toward the stable limit cycle in the upper left of Fig. 2.1. In the conventional jargon, she would find herself in the limit cycle's *domain of attraction* (the region between the unstable limit cycle and the upper left rectangle with a saddle-point at each corner). Unfortunately, the mathematics associated with large perturbations and global stability of an equilibrium or steady state is beyond the scope of this text. Therefore, we shall continue to assume that any perturbation is a small perturbation, and stability synonymous with local stability.

But the universe is an enormous place, and the world of Phaseplane is not unique. Rather, there are many such worlds, one for every dynamical system of the form (2.1). Because there are so many of them, let's drop the capital *P* and use *phase-plane* to denote the "wind" pattern associated with a pair of equations of the form (2.1). Different choices of $u(x, y)$ and $v(x, y)$ will lead to different patterns. Indeed we met two systems of the form (2.1) in Chapter 1, and it will be convenient now to give them a phase-plane interpretation.

From (1.20), the simple ecosystem of Section 1.4 corresponds to (2.1) with $u(x, y) \equiv x(a_1 - b_1 y)$ and $v(x, y) \equiv y(-a_2 + b_2 x)$. There is therefore an equilibrium at the point $\mathbf{x}^* = (a_2/b_2, a_1/b_1)^T$, because u and v both vanish there; this point is marked with a dot in Fig. 1.3. (There is also an equilibrium at $\mathbf{x}^* = (0, 0)^T$, but we paid no attention to it in Section 1.4, because neither species would then exist.) From the pattern in Fig. 1.3, we immediately deduce that the equilibrium is a center.

From (1.22), the simple ecosystem of Section 1.5 corresponds to (2.1) with $u(x, y) \equiv x(a_1 - b_1 y)$ and $v(x, y) \equiv y(a_2 - b_2 x)$. There is therefore an equilibrium at the point $\mathbf{x}^* = (a_2/b_2, a_1/b_1)^T$, because u and v both vanish there; this point is marked with a dot in Fig. 1.4. (Again, we ignore the equilibrium at $\mathbf{x}^* = (0, 0)^T$.) From the pattern in Fig. 1.4, we immediately deduce that the equilibrium is a saddle-point.

By drawing the phase-plane of a dynamical system of the form (2.1), we can always find its (static) equilibria and limit cycles; we can also characterize them as stable or unstable. But, as we saw in Chapter 1, not every dynamical system consists of a pair of differential equations. A system might, for example, be governed by differential-delay equations or a recurrence relation. It is therefore desirable to frame more general definitions of equilibrium and stability than are afforded by the Phasie atlas.

Accordingly, we define an equilibrium (or steady state) of a dynamical system to be a state (or cycle) that will persist forever if undisturbed. An equilibrium may be unstable, stable, or metastable. It is unstable if there exists a small perturbation that will cause the system to move further and further away from equilibrium. It is stable if any small perturbation will cause the system to spend the rest of its life returning to equilibrium (in the absence of further perturbation). And it is metastable if a small perturbation will neither send the system further and further away, nor return it to where it was, but cause it to persist forever in a state or cycle nearby. By perturbation we mean a sudden change of state due to circumstances *outside the control* of the system itself (i.e., a change that cannot be predicted by the differential equations or recurrence relations that govern the system).

Of these three kinds of equilibrium, a stable one is by far the most important, because equilibria in nature are subject to so many random perturbations that we would not even observe them as steady states unless they were quite stable. The word "observe" is crucial here, because nothing is truly constant in nature; but states that change appreciably only over very long time scales, such as evolutionary or geological ones, will *appear* to be steady—at least to mere mortals—and we model things as we observe them. Moreover, even from a purely mathematical viewpoint, things often appear steady when in fact they are not (consider equation (1.16) for very large values of t). Thus a system that, following a perturbation, will take infinitely long in theory to return to stable equilibrium can be regarded, in practice, as resuming its former state within a finite time.

Stable equilibria can be identified in two ways: either *directly* by solving an equation or equations whose solution can only be an equilibrium, or *indirectly* by following the system through a sequence of nonequilibrium states which converges to the equilibrium. (Unstable equilibria can be identified only directly. Why?) This is a chapter on equilibrium models; therefore we are primarily concerned with the first of the two methods. But we'll begin, in Sections 2.1 and 2.2, by understanding the interrelationship between the two methods. From Section 2.3 onward we'll concentrate on the direct method.

2.1 The Equilibrium Concentration of Contaminant in a Lake

Suppose that, at time $t = 0$, the concentration of contaminant in a certain lake of volume V is $x(0) = x_0$ tonnes per cubic meter. A new polluter now begins to pollute the lake at the rate P_{in} tonnes per day. There are no other polluters. What will be the equilibrium concentration of contaminant in the lake?

First, let's answer this question by using the direct method; i.e., let's construct an equation whose solution can only be an equilibrium. Let x^* tonnes per cubic meter be this equilibrium concentration. Then, if water flows out at the rate r cubic meters per day, the daily outflow of contaminant is rx^* tonnes per day. In equilibrium, this daily outflow must balance the daily inflow P_{in}. Hence $rx^* = P_{in}$ or

$$x^* = \frac{P_{in}}{r}. \tag{2.3}$$

Now let's use the indirect method. From Exercise 1.8:

$$\begin{array}{ccc} \text{Rate of change} & = & \text{Pollution} - \text{Pollution} \\ \text{of pollution} & & \text{inflow} \quad \text{outflow} \end{array}$$

$$\frac{d}{dt}\{Vx(t)\} \quad = \quad P_{in} \quad - \quad rx(t). \tag{2.4}$$

Note that (2.3) is the particular solution of this equation for which $dx/dt = 0$. You can easily show that the general solution is

$$x(t) = \frac{P_{in}}{r} + \left(x(0) - \frac{P_{in}}{r}\right) e^{-rt/V}. \tag{2.5}$$

If it just happens that $x(0) = x^*$, then you can see from (2.4) and (2.5) that the new polluter makes no difference; the lake remains forever in the state $x(0)$. If $x(0) \neq x^*$, however, then we can follow the sequence of states

$\{x(t): t \geq 0\}$, which converges to the equilibrium concentration as $t \to \infty$. If $x^* < x(0)$, then the pollution continues to fall; if $x^* > x(0)$, then the pollution rises. (If $P_{in} = 0$, of course, then the pollution just decays at the rate determined in Section 1.1).

2.2 Rowing in Equilibrium

In Section 1.12 we developed the following model for determining the speed of a rowing boat:

$$\text{Mass} \times \text{Acceleration} = \frac{\text{Tractive}}{\text{force}} - \frac{\text{Drag}}{\text{force}}$$

$$M \times \frac{d\xi}{dt} = \frac{8P}{\xi} - bS\xi^2, \tag{2.6}$$

where $\xi(t)$ denotes the speed of the boat at time t, M its mass, P the effective power supplied by an oarswoman, S the wetted surface area and b a constant.[1] We may regard $\xi(t)$ as the state of a dynamical system.

This system has a "static" equilibrium, in which the boat is moving but its speed does not change, if $\xi(t) = \xi^* = $ constant. Then $d\xi/dt = 0$, and the tractive force exactly balances the drag. Let us define

$$f(\xi) \equiv \frac{1}{M} \left\{ \frac{8P}{\xi} - bS\xi^2 \right\}. \tag{2.7}$$

Then, using the direct method, the equilibrium speed is given by $f(\xi^*) = 0$ or

$$\xi^* = V = 2 \left(\frac{P}{bs} \right)^{1/3}, \tag{2.8}$$

where V is the speed defined by (1.79).

This equilibrium can also be found indirectly by solving equation (2.6) and following the sequence of states $\{\xi(t): t \geq 0\}$ as $t \to \infty$. We did this in Chapter 1 and saw from Fig. 1.8 that $\xi \to V$ as $t \to \infty$. But this result also follows from (2.6). For if $\xi(t) < V$ then the right hand side of (2.6) is positive, implying that $d\xi/dt > 0$ and ξ increases with t; whereas if $\xi(t) > V$ then the right-hand side of (2.6) is negative, implying that $d\xi/dt < 0$ and ξ decreases with t. Thus ξ is always approaching V—even if the oarswomen began by rowing faster, the excess of drag over tractive force would soon reduce their speed to V.

These arguments are sufficient to show that the equilibrium $\xi = \xi^* = V$ is stable. But it will be instructive to verify this result by an alternative

---

[1] The speed is denoted by ξ rather than u as in (1.77) to emphasize its role as the state of a dynamical system.

method. Suppose that the oarswomen are rowing at the equilibrium speed V when their boat suffers an external shock, such as a sudden gust of wind. This perturbs their speed from V to $V + \eta_0$, where η_0 is small. Without loss of generality, let's suppose that this all took place at time $t = 0$. Then the subsequent motion of the boat is governed by (2.6) with $\xi(0) = V + \eta_0$. Let us define relative velocity $\eta(t) \equiv \xi(t) - V$, so that $\eta(0) = \eta_0$. Then, on using (2.6), (2.7), and the Taylor expansion of f about V, the motion is governed by

$$\frac{d}{dt}\{V + \eta\} = f(V + \eta) = f(V) + \eta f'(V) + O(\eta^2), \qquad (2.9)$$

where $O(\eta^2)$ is defined by (1.1). Now $dV/dt = 0 = f(V)$. Moreover, because η_0 is small and η continuous, $\eta(t)$ must remain small at least for t close to zero. Then $O(\eta^2)$ is negligible, and (2.9) yields the approximation

$$\frac{d\eta}{dt} \approx f'(V)\eta = -\frac{3bVS\eta}{M}, \qquad (2.10)$$

from which you can readily deduce that

$$\eta(t) \approx \eta_0 e^{f'(V)t} = \eta_0 e^{-3bVSt/M} = \eta_0 e^{-\lambda t}, \qquad (2.11)$$

where $\lambda = 3bSV/M$. Thus $\eta(t)$ remains small, not just for small t but for all t, and dies away completely as $t \to \infty$ (though note that η would increase without bound if λ were negative). The rapidity of return to equilibrium is clearly measured by the rate $|\lambda|$ at which (2.11) decays to zero. (You should verify that $|\lambda|$ is indeed a rate, i.e., has the dimensions of 1/TIME). The larger this rate, the more rapid the return. Moreover, the larger this rate, the sooner a crew reaches equilibrium speed from rest, and the better an approximation it is to estimate their time in a race by assuming they row at equilibrium speed for its entire duration (see Exercise 1.20). We shall make use of this observation in Section 4.4.

We have thus used analytical methods to establish the stability of the equilibrium $\xi^* = V$ (whereas Fig. 1.8 established this graphically). Likewise, we can determine analytically whether equilibria of the system

$$\frac{dx}{dt} = u(x, y), \qquad \frac{dy}{dt} = v(x, y) \qquad (2.12)$$

are stable or not (we do this graphically by sketching its phase-plane). Let $\mathbf{x}^* = (x^*, y^*)^T$ be a static equilibrium of this system. Let relative displacement from this equilibrium be defined by $\boldsymbol{\eta}(t) \equiv (\eta(t), \zeta(t))^T$, where $\eta(t) \equiv x(t) - x^*$ and $\zeta(t) \equiv y(t) - y^*$. Then, on using Taylor's theorem for functions of two variables, we deduce from (2.12) that

$$\frac{d}{dt}\{x^* + \eta\} = u(x^* + \eta, y^* + \zeta)$$

$$= u(x^*, y^*) + \eta \frac{\partial u}{\partial x}(x^*, y^*) + \zeta \frac{\partial u}{\partial y}(x^*, y^*) + O(\epsilon^2)$$

$$\frac{d}{dt}\{y^* + \zeta\} = v(x^* + \eta, y^* + \zeta)$$

$$= v(x^*, y^*) + \eta \frac{\partial v}{\partial x}(x^*, y^*) + \zeta \frac{\partial v}{\partial y}(x^*, y^*) + O(\epsilon^2),$$

(2.13)

where ϵ is defined to be the maximum of η and ζ. If η and ζ are small initially, then so is ϵ. Thus, at least for a short time after the system has suffered the perturbation $\boldsymbol{\eta}(0) = \boldsymbol{\eta}_0 = (\eta_0, \zeta_0)^T$, the evolution of the relative displacement $\boldsymbol{\eta}(t)$ is well approximated by ignoring the $O(\epsilon^2)$ terms in (2.13). Moreover, $\mathbf{x}^*$ is a static equilibrium; therefore we have $dx^*/dt = 0 = dy^*/dt$ and $u(x^*, y^*) = 0 = v(x^*, y^*)$. Hence, on defining

$$g_{11} = \frac{\partial u}{\partial x}(x^*, y^*), \qquad g_{12} = \frac{\partial u}{\partial y}(x^*, y^*),$$

$$g_{21} = \frac{\partial v}{\partial x}(x^*, y^*), \qquad g_{22} = \frac{\partial v}{\partial y}(x^*, y^*),$$

(2.14)

the immediately subsequent motion of the system is approximately governed by the linear equations

$$\frac{d\eta}{dt} = g_{11}\eta + g_{12}\zeta, \qquad \frac{d\zeta}{dt} = g_{21}\eta + g_{22}\zeta. \tag{2.15}$$

Multiplying the second of (2.15) by g_{12} and using the first of (2.15) to replace $g_{12}\zeta$ by $d\eta/dt - g_{11}\eta$, we easily obtain

$$\frac{d^2\eta}{dt^2} - (g_{11} + g_{22})\frac{d\eta}{dt} + (g_{11}g_{22} - g_{12}g_{21})\eta = 0. \tag{2.16}$$

From the first of (2.15), and because $\eta(0) = \eta_0$, we must solve this equation for $\eta(t)$ subject to

$$\eta(0) = \eta_0, \qquad \eta'(0) - g_{11}\eta(0) = g_{12}\zeta_0. \tag{2.17}$$

It is known from the theory of ordinary differential equations that *any* solution of (2.16) must have the form

$$\eta(t) = A\eta_1(t) + B\eta_2(t), \tag{2.18}$$

where A, B are constants and $\eta_1(t)$, $\eta_2(t)$ are any two nonzero solutions that are not constant multiples of one another; see, for example, Theorem 3.9 of Boyce and DiPrima (1965, p. 97). It follows, in particular, that the solution satisfying (2.17) must have the form (2.18). Now consider the quadratic equation

$$\omega^2 - (g_{11} + g_{22})\omega + g_{11}g_{22} - g_{12}g_{21} = 0. \qquad (2.19)$$

The roots, ω_1 and ω_2, of this equation must belong to one of the following three categories:
1. real and unequal;
2. real and equal, i.e., $\omega_1 = \omega_2 = \omega$;
3. complex conjugate, i.e., $\omega_1 = p + iq$, $\omega_2 = p - iq$, where $i = \sqrt{-1}$.

You can verify by simple substitution (Exercise 2.1) that appropriate forms for η_1 and η_2 corresponding to the three categories are, respectively:

$$
\begin{aligned}
&\textbf{1. } \eta_1(t) = e^{\omega_1 t}, \qquad \eta_2(t) = e^{\omega_2 t} \\
&\textbf{2. } \eta_1(t) = e^{\omega t}, \qquad \eta_2(t) = t e^{\omega t} \qquad (2.20) \\
&\textbf{3. } \eta_1(t) = e^{pt}\cos(qt), \qquad \eta_2(t) = e^{pt}\sin(qt).
\end{aligned}
$$

It follows immediately from (2.18) and (2.20) that $\eta(t)$ will decay to zero as $t \to \infty$ if *both* ω_1 and ω_2 have negative real parts, will increase without bound as $t \to \infty$ for some η_0 if *either* ω_1 or ω_2 has a positive real part, and will oscillate with constant amplitude if both ω_1 and ω_2 are purely imaginary, where ω_1 and ω_2 are defined as the roots of (2.19). Moreover (Exercise 2.2), you can show that ζ obeys the same differential equation as η, so that what we have just said about $\eta(t)$ is also true of $\zeta(t)$. Combining our results, we find that $\mathbf{x}^*$ is (locally) stable if ω_1 and ω_2 are both negative, metastable if ω_1 and ω_2 are purely imaginary, and unstable if either ω_1 or ω_2 has a positive real part. If case (3) applies, then trajectories in the phase-plane near $\mathbf{x}^*$ will spiral around it. In conclusion, the nature of the static equilibrium $\mathbf{x}^*$ may be classified as follows from knowledge of the roots of (2.19):

ROOTS OF (2.19)	EQUILIBRIUM	
Real, both positive	Unstable node	
Real, both negative	Stable node	
Real, opposite signs	Saddle-point	(2.21)
Complex, purely imaginary	Center	
Complex, $p > 0$	Unstable focus	
Complex, $p < 0$	Stable focus	

Limit cycles are somewhat more difficult to identify and classify analytically. We will proceed ad hoc whenever we meet them.

Now try Exercise 2.3.[2]

---

[2] See Supplementary Note 2.1.

2.3   How Fast Do Cars Drive Through a Tunnel?

The more congested a tunnel is, the slower the cars move; the less congested, the faster.[3] Thus the speed of cars through a tunnel is a decreasing function of congestion. How does this function decrease? In this section, we'll attempt to answer that question by using an equilibrium model.

First we need a suitable measure of traffic congestion. One such measure is the traffic density, x, i.e., the number of cars per unit length of road. In general, the density x varies with time, t, and with distance, z, along the road; hence $x = x(z, t)$. We measure the density by taking aerial photographs at fixed times. Let one such time be $t = t_0$. Then, to measure the traffic density at a particular station $z = z_0$ on the road (at time t_0), we use the photograph to count the number of cars in a suitable interval of the form $z_0 - \epsilon/2 < z \leq z + \epsilon/2$ and divide the answer by ϵ; this is our estimate of $x(z_0, t_0)$. The interval length ϵ must be chosen carefully. It cannot, of course, be less than a few car-lengths; neither can it be too long, for important spatial variations of density would be smoothed out. It is probably best to choose ϵ somewhere between 30 and 50 meters, and certainly much less than a mile (even though traffic density is traditionally quoted as number of cars per mile). Because z_0 is arbitrary, and because we may repeat this procedure for any value of z_0, a single photograph will enable us to plot $x(z, t_0)$ as a function of z. A series of such photographs taken at different times will enable us to estimate $x(z, t)$.

If speed v is a decreasing function of traffic congestion, and if density x is our chosen measure of congestion, then $v = v(x)$ where

$$v'(x) \leq 0. \tag{2.22}$$

At one extreme, if traffic is very light, then cars are free to move at the speed limit, which we will denote by v_{max}. This may be the official speed limit if police are watching, or whatever speed drivers deem safe, if not; all that matters is that the limit exists and may be assumed the same for all drivers. Hence there exists some critical traffic density x_c such that

$$v(x) = v_{max} \qquad \text{if } 0 \leq x \leq x_c. \tag{2.23}$$

The condition $v(0) = v_{max}$ should be interpreted as meaning that a car that enters an empty road is free to drive at the speed limit. At the other extreme, if traffic is so heavy that cars cannot move, then traffic density has reached its maximum, which we will denote by x_{max}. Thus

$$v(x_{max}) = 0. \tag{2.24}$$

Now assume that $x_c < x < x_{max}$, so that there is some degree of congestion; and suppose that the cars have reached an equilibrium, in which

---

[3]You may wish to review Section 1.13 before proceeding.

all are cruising at the constant speed v, at the same distance d from the car in front. To make matters simple, we'll also assume that all cars have the same length, L. Then the traffic density is

$$x = \frac{1}{d+L} \tag{2.25}$$

(why?). Note that, although equilibrium density, x, and equilibrium velocity, $v = v(x)$, are independent of both z and t, their values may vary from tunnel to tunnel. In this section, we'll assume that our equilibrium is stable. We'll discuss whether that assumption is reasonable or not in Section 2.8.

Let's assume that the dynamics of approach to this equilibrium are governed by the model introduced in Section 1.13. Then, from (1.83),

$$\frac{dz_j(t+\tau)}{dt} = \lambda \ln |z_j(t) - z_{j-1}(t)| + \alpha_j, \tag{2.26}$$

where $z_j(t)$ is the displacement of the jth car in a platoon of N cars, and where τ is the driver response time. Now, we have just assumed that all cars are moving at the same speed v, with front bumpers a distance $d+L$ apart. Therefore, for all j and at all times, $dz_j(t+\tau)/dt = v$, $|z_j(t) - z_{j-1}(t)| = d+L$ and α_j is independent of j. Hence (2.26) yields $v = \lambda \ln(d+L) + \alpha$. From (2.25), and on using (2.24) to determine α, we deduce

$$v(x) = \lambda \ln \left(\frac{x_{max}}{x} \right), \qquad x_c < x \leq x_{max}. \tag{2.27}$$

The value of λ is determined from the assumption that $v(x)$ is a continuous function throughout the interval $0 \leq x \leq x_{max}$ and hence, in particular, at $x = x_c$. Thus

$$v(x) = \begin{cases} v_{max}, & 0 \leq x \leq x_c \\ v_{max} \dfrac{\ln(x_{max}/x)}{\ln(x_{max}/x_c)}, & x_c < x \leq x_{max}. \end{cases} \tag{2.28}$$

You can easily see that (2.22) is satisfied.

How good is this model—does it fit the facts? A possible objection to this model is that traffic density is not allowed to vary *across* the road, as it surely would if a road had several lanes. In a tunnel, however, cars are invariably prohibited from switching lanes, so that each lane behaves as though it were a separate tunnel and density varies only *along* the road. But does it vary along the road as predicted by (2.28)? Perhaps you would like to think about that—we'll return to this point in Chapter 4. Meanwhile, try your hand at Exercises 2.4, 2.5, and 2.10.

2.4   Salmon Equilibrium and Limit Cycles

The salmon population described in Section 1.8 will be in static equilibrium if the population (in hundreds of millions) persists forever at the level x^*.[4] Recall from (1.54) that the population's dynamics are governed by the recurrence relation

$$x_{n+1} = ax_n e^{-bx_n}, \qquad (2.29)$$

where x_n denotes the population at the end of cycle n. Hence, in equilibrium, we have $x_{n+1} = x_n = x^*$ for all n. On using (2.29), we find that

$$x^* = ax^* e^{-bx^*}; \qquad (2.30)$$

whence, since $x^* \neq 0$, we have

$$x^* = \frac{1}{b} \ln(a). \qquad (2.31)$$

Equation (2.30) is one whose solution can only be an equilibrium. Thus (2.31) has been found by the direct method.

In principle, if x_n could be expressed explicitly as a function of n, then we could find x^* indirectly as the limit, as $n \to \infty$, of the sequence $\{x_n\}$. But no such expression is known. Thus the direct method is the only one that enables us to identify the equilibrium for arbitrary values of a and b. For particular values of a and b, however, we can identify the equilibrium by solving (2.29) numerically. You did this in Exercise 1.9 for $b = \ln(10) \approx 2.3026$ and for various values of a. The first value was $a = 5$, yielding $x^* = \ln(5) \div \ln(10) \approx 0.6990$; and you can see from Table 1.2 that this is the limiting value of the sequence $\{x_n\}$.

Now, the indirect method works only for stable equilibria, so that (2.31) must be a stable equilibrium for $a = 5$ and $b = \ln(10)$. It is possible to verify this by analogous methods to those of Section 2.2. Recall from (1.48) that (2.29) is just the special case of the recurrence relation

$$x_{n+1} = F(x_n) \qquad (2.32)$$

for which $F(x) = axe^{-bx}$. Any static equilibrium x^* of (2.32) must satisfy $x_{n+1} = x_n = x^*$ for all values of n; whence $x^* = F(x^*)$. Combining with (2.32) and using Taylor's theorem, we find that

$$x_{n+1} - x^* = F(x_n) - F(x^*) = F'(x^*) \cdot (x_n - x^*) + o(|x_n - x^*|), \qquad (2.33)$$

where little oh is defined by (1.8). If x_n is very close to x^*, then $o(|x_n - x^*|)$ is so small as to be negligible; let's therefore neglect it. Now, if $|F'(x^*)| < 1$, then $|x_{n+1} - x^*| < |x_n - x^*|$; i.e., x_{n+1} is closer to x^* than x_n. By an extension of this argument, we see that the sequence $\{x_n\}$ converges to x^* if

$$|F'(x^*)| < 1. \qquad (2.34)$$

Thus x^* is a stable equilibrium if (2.34) is satisfied. On the other hand, if $|F'(x^*)| > 1$, then the sequence $\{x_n\}$ diverges from x^*; i.e., the equilibrium

---

[4]You may wish to review Section 1.8 before proceeding.

is unstable. For $F(x) = axe^{-bx}$, we obtain $|F'(x^*)| = ax^* e^{-bx^*}|1 - bx^*| = |1 - \ln(a)|$, on using (2.31). Thus (2.31) is stable or unstable according to whether $a < e^2$ (≈ 7.4) or $a > e^2$. It is therefore stable for $a = 5$. (Moreover, it is actually unstable for $a = 10$, so that the slightest perturbation would cause x_n to move successively further and further away from $x^* = 1$. We identified this unstable equilibrium by the indirect method in Table 1.2 only because we were fortunate enough to pick $x_0 = 1 = x^*$ in the numerical iteration.)

According to (2.34), (2.31) is unstable when $a = 11$, because $11 > e^2$. But look at the column of Table 1.2 corresponding to $a = 11$. You can see that the sequence of values $\{x_n\}$ converges toward a stable equilibrium cycle, or limit cycle, in which the population magnitude alternates for the rest of time between the values $x_L = 0.3568$ and $x_U = 1.726$. We will denote this limit cycle by $\{x_L, x_U\}$.

This particular limit cycle was identified indirectly; but the direct method identifies all such limit cycles, i.e., all limit cycles with a period of two time units. First, recall from (1.92) that, by defining

$$\overline{x_n} = \frac{x_n}{x^*}, \tag{2.35}$$

i.e., by measuring population in units of $100 \ln(a)/b$ millions, we can scale the parameter b out of recurrence relation (2.29) and write it in the dimensionless form

$$\overline{x_{n+1}} = \overline{x_n} e^{r(1 - \overline{x_n})}, \tag{2.36}$$

where $r = \ln(a)$. Let us define $y = \hat{x}_L = bx_L/r$ and $z = \hat{x}_U = bx_U/r$. Then, because $\{y, z\}$ constitutes a limit cycle with period 2, (2.36) yields

$$z = ye^{r(1-y)}; \qquad y = ze^{r(1-z)}. \tag{2.37}$$

Eliminating z from these equations, we obtain $ye^{r(1-y)} = 2 - y$. Eliminating y from these equations, we obtain $ze^{r(1-z)} = 2 - z$ (Exercise 2.6). Thus y and z must both be roots of the equation $g(w) = 2 - w$, where

$$g(w) \equiv we^{r(1-w)}. \tag{2.38}$$

The function $g(w)$ has a maximum $(e^r - 1)/r$ at $w = 1/r$ and an inflection point at $w = 2/r$ (verify). The roots of $g(w) = w$ can be found graphically as the points where the curve $W = g(w)$ meets the straight line $W = 2 - w$. It is clear from (2.38) that $w = 1$ is always a root. Moreover, from Fig. 2.2, this is the only root if $g'(1) > -1$; whereas there are two other roots if $g'(1) < -1$. The condition for an equilibrium cycle with period 2 is therefore $1 - r < -1$ or $r > 2$. The corresponding values of y and z can be determined from (2.38) for any specific value of r. When $a = 11$, for example, so that $r = \ln(11) = 2.398$, the Newton–Raphson iterative technique finds the roots of (2.38) to be $y = 0.342658$, $z = 1.65734$ (and 1).

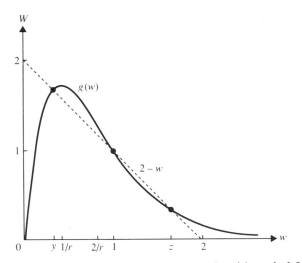

Fig. 2.2 Graphical determination of equilibrium cycle with period 2.

These values correspond to $x_L = 0.3568$ and $x_U = 1.726$ (hundreds of millions) in Table 1.2.

Let the period-2 equilibrium cycle for arbitrary $r\ (> 2)$ be denoted by $\{y(r), z(r)\}$. Then the functions $y = y(r)$ and $z = z(r)$ are defined implicitly by (2.38). For the sake of definiteness, $y(r)$ is the smallest of the three roots and $z(r)$ the largest (so that $0 < y(r) < 1$, $1 < z(r) < 2$). The two functions are sketched in Fig. 2.3. As r becomes indefinitely large, the smallest root approaches zero; whereas the largest approaches 2. As r approaches 2, on the other hand, the equilibrium cycle collapses to the static equilibrium $\bar{x}^* = 1$.

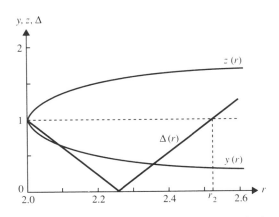

Fig. 2.3 Period-2 equilibrium cycles $\{y(r), z(r)\}$, which are limit cycles for $r < r_2$.

We have established the existence of a period-2 equilibrium cycle for all $r > 2$, but this cycle is a limit cycle only if it is stable. The condition for this is

$$\Delta(r) < 1, \tag{2.39}$$

where

$$\Delta(r) \equiv |(1 - ry(r))(1 - 2r + ry(r))| = |(1 - rz(r))(1 - 2r + rz(r))|. \tag{2.40}$$

To see this, observe that for any n, we have

$$\tilde{x}_{n+2} = \tilde{x}_{n+1} e^{r(1 - \tilde{x}_{n+1})} = \tilde{x}_n e^{r\{2 - \tilde{x}_n - \tilde{x}_n e^{r(1 - \tilde{x}_n)}\}}, \tag{2.41}$$

on using (2.36). Hence two recursions of (2.36) are equivalent to a single recursion of the recurrence relation $X_{n+1} = G(X_n)$, where

$$G(x) \equiv xe^{r(2 - x - xe^{r(1-x)})}; \tag{2.42}$$

this implies (verify) that

$$G'(x) = (1 - rx) \cdot (1 - rxe^{r(1-x)}) \cdot e^{r\{2 - x - xe^{r(1-x)}\}}. \tag{2.43}$$

Let v denote *either* $y(r)$ *or* $z(r)$. Then, from (2.38), $ve^{r(1-v)} = 2 - v$; so that $G'(v) = (1 - rv)(1 - r(2 - v))$. Because the stability condition is $|G'(v)| < 1$, from (2.34), we arrive immediately at (2.39).

The function Δ is also sketched in Fig. 2.3. Its graph appears to be two straight lines but in fact has a slight curvature, upward to the left of the cusp at $r = 2.2564$ and downward to the right of it. (The cusp occurs at the root of the equation $e^{(r-1)} = 2r - 1$ and corresponds to the value of r for which the maximum of g in Fig. 2.2 is at y.) We deduce from the graph that the period-2 equilibrium cycle is a limit cycle if $r < r_2$, where $\Delta(r_2) = 1$. It can be shown (Exercise 2.7) that $r_2 = 2.5265$.

The period-2 equilibrium cycle is unstable if $r > r_2$, but limit cycles with period higher than 2 may still exist (see Exercise 2.8). Indeed there is a bounded increasing sequence $r_0 < r_1 < r_2 < r_3 < \cdots$, with limit point $r^* = 2.6924$, such that the population $\tilde{x}_n$ will settle down to a limit cycle with period 2^k whenever $r_k < r < r_{k+1}$; see Clark (1976, pp. 215–216) or May (1976, pp. 12–14), by whom the model of Section 1.8 is denoted "Form B."[5] The first three terms in the sequence, already found, are $r_0 = 0$, $r_1 = 2$, and $r_2 = 2.5265$. The static equilibrium $\tilde{x}^* = 1$ is regarded as a period-1 limit cycle, so that convergence to it is guaranteed whenever $0 < r < r_1$. The steady state is still $\tilde{x}^* = 1$ when $r = r_1$, but convergence

---

[5]In theory, this may not be true if the initial condition just happens to place the population in an unstable equilibrium cycle. In practice, however, there are bound to be perturbations, the slightest of which will destroy the unstable equilibrium; see Exercise 2.9.

is then extremely slow. For $r > r^*$, the population may never settle into any finite cycle, as suggested by the columns in Table 1.2 corresponding to $a = 15$ or $r = 2.71$.

But what does all this have to do with salmon? Do salmon populations fluctuate in a limit cycle? Perhaps you would like to think about that—we'll return to the matter in Chapter 4. Meanwhile, you should not proceed to Section 2.5 before attempting Exercise 2.10.

2.5   How Much Heat Loss Can Double-Glazing Prevent?

The wall of a building is in (thermal) equilibrium if it transfers heat at a constant rate, in much the same way as a rowing boat is in (mechanical) equilibrium if it moves at a constant speed. In winter, the air outside a building is considerably colder than the air inside, so that heat energy is transferred outward; i.e., the building loses heat. Much of the heat is lost through windows, and double-glazing can reduce this loss. By how much? In this section, we'll use an equilibrium model to explore that question.

Let T_1 be the temperature of air inside the building and T_2 the temperature of the outside air; it's winter, so $T_1 > T_2$. Let d be the thickness of a pane of glass and L the distance between panes of a double-glazed window, as in Fig. 2.4 (a). Let x, increasing outward, measure distance perpendicular to the window. It is convenient to let the inside pane meet the inside air at $x = 0$. Then the outside pane meets the outside air at $x = 2d + L$, and air occupies the gap $d < x < d + L$. Let $T(x)$ denote the temperature at station x, so that $T(0) = T_1$ and $T(2d + L) = T_2$. We'll denote $T(d)$ by T_A and $T(d + L)$ by T_B.

Let $F(x)$ denote the heat flux *per unit area* at station x, i.e., the heat energy lost per unit time from left to right across unit area of the plane with coordinate x. Then, if F does not vary within any $x =$ constant plane, the actual heat flux across area A at station x is $AF(x)$. How does $F(x)$ depend upon temperature? If you attempted Exercise 2.10, then you have already tried to answer this question. Let ϵ be some very small number. Then the temperature gradient across the interval from $x - \epsilon/2$ to $x + \epsilon/2$ is

$$\left| \frac{T(x - \epsilon/2) - T(x + \epsilon/2)}{\epsilon} \right| \approx -T'(x). \tag{2.44}$$

If $T(x - \epsilon/2) = T(x + \epsilon/2)$ then no heat flows across the interval. If $T(x - \epsilon/2) > T(x + \epsilon/2)$, then heat flows from hot to cold, i.e., from left to right in Fig. 2.4 (a); and the steeper the gradient, the more heat that flows. It therefore seems reasonable to assume that $F(x)$ is proportional to (2.44); i.e., that

$$F(x) = -k\frac{dT}{dx}. \tag{2.45}$$

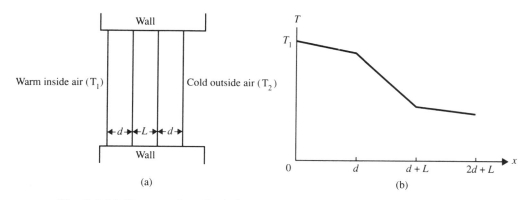

Fig. 2.4 (a) Cross section of window. (b) Steady temperature distribution.

This approximation has been found acceptable and is used by physicists. The constant of proportionality is called the *thermal conductivity* and depends upon the medium through which the heat is flowing. Because F has the dimensions of energy per unit area per unit of time (say, Joules per second per cm^2), it is clear from the dimensional consistency of (2.45) that k has the dimensions of energy per unit time per length per degree (say, Joules/cm · sec · °C). We will denote the thermal conductivities of air and glass by k_A and k_G, respectively.

Let us now assume that the glass and the air between the panes are in thermal equilibrium. Then the heat flux is constant. Hence $T'(x)$ will be constant in each of the intervals $0 < x < d$, $d < x < d + L$, and $d + L < x < 2d + L$; but will be discontinuous at $x = d$ and $x = d + L$, because of the discontinuity in k. This is depicted in Fig. 2.4 (b). Thus

$$k_G\frac{T_1 - T_A}{d} = k_A\frac{T_A - T_B}{L} = k_G\frac{T_B - T_2}{d} = F. \tag{2.46}$$

On solving these equations we find (verify)

$$T_A = \frac{(\rho + 1)T_1 + T_2}{\rho + 2}, \qquad T_B = \frac{T_1 + (\rho + 1)T_2}{\rho + 2}, \tag{2.47}$$

where

$$\rho \equiv \frac{k_G L}{k_A d}. \tag{2.48}$$

Hence

$$F = \frac{k_G}{d}\frac{T_1 - T_2}{\rho + 2} = \frac{k_A}{L}\frac{\rho}{\rho + 2}(T_1 - T_2). \tag{2.49}$$

Now, if the window were single-glazed with the same amount of glass, then the heat loss would clearly be

$$F_S = \frac{k_G}{2d}(T_1 - T_2). \tag{2.50}$$

Hence the relative reduction of heat loss is the proportion

$$\Delta \equiv 1 - \frac{F}{F_S} = \frac{\rho}{\rho + 2}. \qquad (2.51)$$

According to Landau et al. (1967, p. 321), the thermal conductivity of glass at room temperature is between about 4×10^{-3} and 8×10^{-3} J/cm · sec · °C. (Thermal conductivity varies slightly with temperature, but the effect is negligible in the circumstances being considered.) According to Batchelor (1967, p. 594), the thermal conductivity of dry air at the same temperature is about 2.5×10^{-4} J/cm · sec · °C. It therefore appears that k_G/k_A lies between about 16 and 32. A manufacturer of double-glazing would surely use glass with the best available thermal properties, and so let's assume that $k_G/k_A = 16$. Then, defining the gap aspect ratio to be $h = L/d$, (2.51) reduces to $\Delta(h) = h/(h + 1/8)$.

This heat loss reduction factor is plotted in Fig. 2.5 and clearly shows the insulating effect of a narrow gap of air. Indeed Δ increases so sharply with h that a gap of four pane–widths achieves a heat loss reduction of 97%. Thereafter, $\Delta(h)$ increases very slowly; so that, to two significant figures, the reduction factor is 0.98 for all values of h between 5 and 8. It therefore appears that very little is to be gained from increasing the separation of the window panes beyond a value of about four pane-widths. Manufacturers evidently agree. Double-glazing is advertised regularly in the pages of *Architectural Digest*; and you can see from the diagram between pages 64 and 65 of the April, 1986 issue, for example, that the value of h is just a little bit bigger than 4.

If cheap air is such a good insulator, as Fig. 2.5 suggests, then why is expensive synthetic fiber used to insulate cavities in walls? The answer is that air is an efficient insulator only if it is both dry and stagnant, as we have tacitly assumed (with good reason, because the cavity between the panes of a double-glazed window is perfectly sealed). Air containing water vapor has a higher thermal conductivity, and air that is moving can

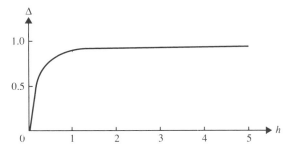

Fig. 2.5 Heat loss reduction factor as a function of the gap aspect ratio $h = L/d$ when $k_G = 16k_A$.

remove heat rapidly by a mechanism known as *convection* (transport by a moving medium), which is fundamentally different from the mechanism of *conduction* (transport through a medium at rest) described in this section. Filling the gap between the inner and outer brickwork of a building's external walls with synthetic fiber prevents convection within the gap from transporting heat to places that are especially vulnerable to heat loss (i.e., holes!).

Mathematical models of convection will be explored in Chapters 9 and 11. Moreover, by introducing a dynamic model for heat conduction in Chapter 11, we will show that the indirect method can also be used to identify the (stable) equilibrium temperature distribution that we found directly in the present section. In preparation for this, observe from (2.45) that, because heat flux is constant in equilibrium, the steady temperature distribution must satisfy

$$\frac{d}{dx}\left\{k\frac{dT}{dx}\right\} = 0. \qquad (2.52)$$

Meanwhile, try your hand at Exercises 2.11 and 2.12.

2.6   Why Are Pipes Circular?

The pipes that bring water into your house have a circular cross section. Why? Why not elliptical? The answer is that, for a given cross-sectional area, a circular pipe carries a greater quantity of water per unit of time than any elliptical one. To see this, we will have to develop a mathematical model of water flowing through an elliptical pipe. In keeping with the title of this chapter, we will assume that this flow is in equilibrium.

Let a typical cross section of our elliptical pipe have area A, major axis $2a$, and minor axis $2b$. Let Oxy be a Cartesian coordinate system, with O at the center of the pipe, Ox along the major axis, and Oy along the minor one (Fig. 2.6). Then the cross-sectional area is given by

$$A = \pi ab, \qquad (2.53)$$

and the inner wall of the pipe has equation

$$\frac{x^2}{a^2} + \frac{y^2}{b^2} = 1. \qquad (2.54)$$

Let w be the velocity of water along the pipe, perpendicular to the cross section in Fig. 2.6 and out of the page. The flow is in equilibrium, and so w is independent of time. The velocity of flow will vary across the pipe, however, with faster flow in the center and slower flow near the wall, so that w is a function of x and y. At the wall itself, the velocity will be zero, because the outermost particles of water adhere to the pipe. Hence

$$w(x, y) = 0 \text{ for all } x, y \text{ satisfying } (2.54). \qquad (2.55)$$

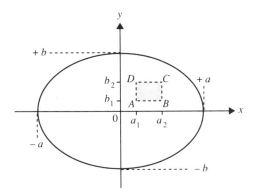

Fig. 2.6 Pipe cross section.

The constraint (2.55) is known in fluid dynamics as a *boundary condition.*

The force that reduces the velocity of the outermost particles to zero is known as friction. Friction is present throughout the flow, though it is stronger at the wall than anywhere else. The effect of friction within the flow is that water near the walls attempts to retard the adjacent water; this water, in turn, attempts to retard the water adjacent to it; and so on across the flow, with slower water exerting a drag on the faster water adjacent to it. But, according to Newton (or, if you prefer, borrowing from physics), every action has an equal and opposite reaction. Thus an equivalent picture is that the water near the center of the pipe attempts to accelerate the adjacent water; this water, in turn, attempts to accelerate the water adjacent to it; and so on across the flow, with the faster water exerting a tractive force on the slower water adjacent to it. But the attempts to accelerate are unsuccessful, because the water is flowing in equilibrium.

Another way of describing all this is to say that the faster water is losing its momentum to slower water; that slower water is losing its momentum, in turn, to slower water still; and so on across the pipe, with each layer of water losing its momentum to the slower water adjacent to it. Therefore, in equilibrium, there is a steady flux of momentum across the flow. (Each layer does not, of course, lose all its momentum, or there would be no flow; except for the outermost layer, which loses all its momentum to the wall. The wall experiences this momentum transfer as a tractive force. The equal and opposite reaction is the wall's drag on the water, often dubbed the skin friction, which, as explained toward the end of this section, is ultimately responsible for the pipe's resistance to the flow.)

This steady outward flux of momentum is reminiscent of the previous section, where a temperature gradient caused a steady outward flux of heat. Now a velocity gradient is causing a steady outward flux of momentum. By analogy with (2.45), it seems fair to assume that the flux of momentum per unit area across the plane with coordinate x is

$$-\nu\frac{\partial}{\partial x}\{\text{Momentum per unit volume}\}, \tag{2.56}$$

where ν is a constant of proportionality. We use the symbol for partial differentiation rather than that for ordinary differentiation as in (2.45), because the momentum flux across the plane with coordinate x will vary with y. Let ρ denote the density of water, i.e., its mass per unit volume. Then momentum per unit volume = {mass × velocity} per unit volume = {mass per unit volume} × velocity = $\rho w(x, y)$. Hence, denoting momentum flux per unit area across the plane with coordinate x by $M(x, y)$, we have

$$M(x, y) = -\frac{\partial}{\partial x}(\mu w), \tag{2.57}$$

where

$$\mu = \rho\nu. \tag{2.58}$$

Similarly, the flux of momentum per unit area across the plane with coordinate y is given by

$$N(x, y) = -\frac{\partial}{\partial y}(\mu w). \tag{2.59}$$

The constant μ is known in fluid dynamics as the *shear viscosity* of the fluid. The constant ν is known as the *kinematic viscosity*. It varies with temperature and (Exercise 2.13) has the dimensions of area per unit of time. At 20°C the kinematic viscosity of water is about $10^{-2}\text{cm}^2/\text{sec}$; see Batchelor (1967, p. 597).

Now suppose that L is the length of pipe over which the water pressure drops from p_2 to p_1. (In practice, you would create such a pressure drop by turning on a faucet, and without it there would be no flow.) Consider (Fig. 2.7) the cylinder of water, with rectangular cross section and length

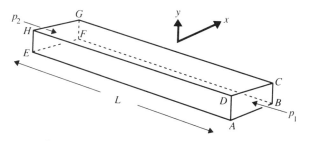

Fig. 2.7 Elementary slice of flow with rectangular cross section.

L, which occupies the region

$$a_1 \leq x \leq a_2, \qquad b_1 \leq y \leq b_2 \tag{2.60}$$

shaded in Fig. 2.6. Pressure is just a force per unit area, which acts in all directions; but the only directions affecting the flow are perpendicular to the faces $ABCD$ and $EFGH$ of the rectangular cylinder sketched in Fig. 2.7. Hence the net pressure force on the fluid in this diagram is the force on $EFGH$ in the direction of EA minus the force on $ABCD$ in the direction of AE; i.e.,

$$\int_{b_1}^{b_2} \int_{a_1}^{a_2} (p_2 - p_1)dxdy. \tag{2.61}$$

According to Newton, however (or again, if you prefer, borrowing from physics), force equals rate of change of momentum. The flow is in equilibrium, and so the water cannot speed up; so that the rate of change of momentum is simply the *net* rate at which momentum flows out of Fig. 2.7 across the four rectangular faces that are parallel to the direction of flow. Now, the flux of momentum across the face $BCGF$ (in the direction of increasing x) is

$$\int_{b_1}^{b_2} M(a_2, y)Ldy; \tag{2.62}$$

and the flux of momentum across the face $DCGH$ (in the direction of increasing y) is

$$\int_{a_1}^{a_2} N(x, b_2)Ldx. \tag{2.63}$$

Thus the total flux of momentum *out of* the rectangular cylinder is the sum of (2.62) and (2.63), i.e.,

$$L\left\{ \int_{b_1}^{b_2} M(a_2, y)dy + \int_{a_1}^{a_2} N(x, b_2)dx \right\}. \tag{2.64}$$

The total flux of momentum *into* the rectangular cylinder, on the other hand, is the sum of the fluxes across $EADH$ and $EABF$, i.e.,

$$L\left\{ \int_{b_1}^{b_2} M(a_1, y)dy + \int_{a_1}^{a_2} N(x, b_1)dx \right\}. \tag{2.65}$$

The *net* momentum flux out of the rectangular cylinder is the difference between (2.64) and (2.65), i.e.,

$$L \left\{ \int_{b_1}^{b_2} (M(a_2, y) - M(a_1, y)) \, dy + \int_{a_1}^{a_2} (N(x, b_2) - N(x, b_1)) \, dx \right\}$$

$$= L \left\{ \int_{b_1}^{b_2} \int_{a_1}^{a_2} \frac{\partial M(x, y)}{\partial x} \, dx \, dy + \int_{a_1}^{a_2} \int_{b_1}^{b_2} \frac{\partial N(x, y)}{\partial y} \, dy \, dx \right\}$$

$$= L \int_{b_1}^{b_2} \int_{a_1}^{a_2} \left(\frac{\partial M}{\partial x} + \frac{\partial N}{\partial y} \right) dx \, dy. \tag{2.66}$$

Equating (2.61) with (2.66) gives

$$\int_{b_1}^{b_2} \int_{a_1}^{a_2} \left(\frac{\partial M}{\partial x} + \frac{\partial N}{\partial y} + \frac{p_1 - p_2}{L} \right) dx \, dy = 0. \tag{2.67}$$

But the rectangle in Fig. 2.6 is completely arbitrary. It could have been placed in any of the other quadrants or expanded or shrunk to any size. Provided that the integrand in (2.67) is continuous, which we shall assume, this can only be so if the integrand is identically zero. Hence

$$\frac{\partial M}{\partial x} + \frac{\partial N}{\partial y} + \frac{p_1 - p_2}{L} = 0. \tag{2.68}$$

For suppose that (2.68) were false at (x_0, y_0) because the left-hand side had, say, a strictly positive value. Then, by continuity, it would remain positive throughout some small region of the form (2.60) surrounding (x_0, y_0); so that its integral over that region would also be positive, contradicting (2.67). To cover the possibility that the left-hand side of (2.68) might be negative at (x_0, y_0), simply multiply by -1.

On using (2.57) and (2.59), and assuming that fluid density, ρ, is constant, we obtain

$$\mu \left(\frac{\partial^2 w}{\partial x^2} + \frac{\partial^2 w}{\partial y^2} \right) = \frac{p_1 - p_2}{L}. \tag{2.69}$$

This equation, which is usually known as Poisson's equation, is our first example of a *partial differential equation*, or PDE for short. Further examples of PDEs will emerge in Chapters 9 and 11. You can verify by direct substitution that a solution of (2.69) satisfying the boundary condition (2.55) is

$$w(x, y) = \frac{p_2 - p_1}{2\mu L} \frac{a^2 b^2}{a^2 + b^2} \left(1 - \frac{x^2}{a^2} - \frac{y^2}{b^2} \right). \tag{2.70}$$

Moreover, it can be shown that (2.70) is the only solution of (2.69) that satisfies the boundary condition. Hence (2.70) describes the flow of water through our elliptical pipe.[6]

The expression for velocity, namely, $w(x, y)$, represents the distance travelled per unit of time, perpendicular to the cross section of the pipe,

---

[6]See Supplementary Note 2.2.

by the particle of fluid with coordinates (x, y). Hence the volume of water traversing the cross section per unit of time is the integral, over the section, of the distances travelled by all such particles. Denote this flux of water through the pipe by Q. Then (Exercise 2.14)

$$Q = \iint\limits_{\frac{x^2}{a^2} + \frac{y^2}{b^2} \leq 1} w(x, y) dx dy = \frac{\pi(p_2 - p_1)a^3 b^3}{4\mu L(a^2 + b^2)}. \tag{2.71}$$

By setting $a = b = r$ in (2.71), we deduce that the flux of water through a circular pipe of radius r is given by

$$Q_c = \frac{\pi(p_2 - p_1)r^4}{8\mu L}, \tag{2.72}$$

which is known in fluid dynamics as Poiseuille's formula. From (2.53), a circular pipe with the same cross-sectional area as the ellipse (2.54) must have radius $\sqrt{ab}$. Hence the flux reduction caused by deforming the circle into an ellipse with the same area is

$$1 - \frac{Q}{Q_c} = \frac{(a - b)^2}{a^2 + b^2}. \tag{2.73}$$

This is always nonnegative and is clearly minimized by taking $a = b$.

But why should the circle have this optimal property? To understand this, observe from (2.70) that velocity is constant over any ellipse, say E, which is similar to the wall. Now, as discussed above, there are two entirely equivalent interpretations of equilibrium flow in a pipe. We can say either that the pressure drop is exactly balanced by an outward momentum flux or that the pressure drop is exactly balanced by a frictional (or viscous) drag. Thus, recalling our earlier discussion, the momentum flux across E represents the viscous drag of the water outside E on the water inside it. In particular, the momentum flux into the wall itself represents the drag of that wall on the entire fluid. Thus the circle, being the curve of minimum length for a given area, minimizes that outward flux at the wall—and hence the pipe's resistance to the flow. The higher the value of μ in (2.57)–(2.59), the greater the resistance; or, which is the same thing, the greater the amount of momentum that is transported toward the wall. Thus the analogy between (2.45) and (2.57) is a perfect one, provided only that we interpret μ not as the viscosity, or stickiness, of the fluid but as a conductivity of momentum.

Expression (2.72) for the flux of water through a circular pipe can be obtained much more readily by exploiting the pipe's symmetry. See Exercise 2.15.

2.7 Equilibrium Shifts

Let's return to the lake we considered in Section 2.1 and suppose that the level of contaminant has already reached its steady state. Thus pollution outflow rx exactly balances pollution inflow P_{in}, so that $x = P_{in}/r$, as in (2.3).

Now suppose that the lake's outflow is suddenly increased, at $t = 0$, to $r + \Delta r$. Then, from (2.4), the pollution subsequently evolves according to

$$\frac{d}{dt}\{Vx(t)\} = P_{in} - (r + \Delta r)x(t), \qquad x(0) = \frac{P_{in}}{r}; \qquad (2.74)$$

whence

$$x(t) = \frac{P_{in}}{r + \Delta r} + \left(\frac{P_{in}}{r} - \frac{P_{in}}{r + \Delta r}\right) e^{-(r+\Delta r)t/V}. \qquad (2.75)$$

In the course of time, the exponential decay term will become so small that the pollution level will have shifted from $x^* = P_{in}/r$ to the new equilibrium,

$$x^*_{new} = \frac{P_{in}}{r + \Delta r}, \qquad (2.76)$$

which is lower than x^*. Thus increasing r while P_{in} remains fixed shifts the equilibrium pollution level downward. A similar analysis would show that increasing P_{in} while r remains fixed shifts the equilibrium level upward. In other words x^*, though a constant as far as its dependence on time is concerned, is nevertheless a function of the two parameters r and P_{in}. We therefore write

$$x^*(P_{in}, r) = \frac{P_{in}}{r}. \qquad (2.77)$$

Partial differentiation reveals that

$$\frac{\partial x^*}{\partial P_{in}} = \frac{1}{r} > 0 \qquad \text{and} \qquad \frac{\partial x^*}{\partial r} = -\frac{P_{in}}{r^2} < 0, \qquad (2.78)$$

in agreement with our conclusions above.

If we were using the direct method then we would obtain (2.77) from the right-hand side of (2.4) in the form

$$0 = P_{in} - rx^*(P_{in}, r). \qquad (2.79)$$

Now suppose we were unable to solve (2.79) to obtain (2.77)—just suppose! Then we could differentiate (2.79) instead to obtain

$$0 = 1 - r\frac{\partial x^*}{\partial P_{in}}, \qquad 0 = -r\frac{\partial x^*}{\partial r} - x^*; \qquad (2.80)$$

whence

$$\frac{\partial x^*}{\partial P_{in}} = \frac{1}{r} > 0, \qquad \frac{\partial x^*}{\partial r} = -\frac{x^*}{r} < 0, \qquad (2.81)$$

as before.

In (2.75), (2.78), and (2.80), we have three different ways to determine how shifting a parameter will shift an equilibrium of dynamical system (2.4):

1. indirectly, from an expression for $x(t)$;

    2. directly, from an *explicit* expression for x^*; (2.82)

    3. directly, from an *implicit* expression for x^*.

More generally, let $\alpha = (\alpha_1, \alpha_2, \ldots, \alpha_m)^T$ be an m-dimensional vector of parameters, and consider the dynamical system[7]

$$\frac{dx_i}{dt} = f_i(\mathbf{x}, \boldsymbol{\alpha}) = f_i(x_i, x_2, \ldots, x_n, \alpha_1, \alpha_2, \ldots, \alpha_m), \qquad i = 1, 2, \ldots, n. \quad (2.83)$$

Let $\mathbf{x} = \mathbf{x}^*(\alpha)$ be an equilibrium of this system. Then the effect on this equilibrium of a parameter shift from α_j to $\alpha_j + \Delta\alpha_j$ can also be determined in one of three ways:

    1. indirectly, from an expression for $\mathbf{x}(t)$

    2. directly, from an *explicit* expression for $\mathbf{x}^*$ (2.84)

    3. directly, from an *implicit* expression for $\mathbf{x}^*$.

But the lake model that led to (2.82) was a very simple one. We have already seen, in Sections 1.4 and 1.5, that an explicit expression $\mathbf{x}(t) = \xi(t, \alpha)$ for solutions to (2.83) may simply not be available. This is the rule, rather than the exception, and so (1) is rarely of use.

    The equilibrium $\mathbf{x} = \mathbf{x}^*(\alpha)$ must satisfy $f_i(\mathbf{x}, \alpha) = 0$, $i = 1, 2, \ldots, n$; or, in vector form,

$$\mathbf{f}(\mathbf{x}^*(\alpha), \alpha) = \mathbf{0}. \quad (2.85)$$

Now, solutions $\mathbf{x} = \mathbf{x}^*(\alpha)$ of these equations exist in principle whenever the conditions of the implicit function theorem are satisfied.[8] But (2.85) may be impossible to solve analytically. Then (2) is rendered useless, too, and only (3) remains.

    We shall dub this alternative the *equilibrium shift technique*. Mathematically, it amounts to no more than implicit differentiation of the equilibrium equations. Applying the chain rule to (2.85), we obtain

$$\sum_{k=1}^{n} \frac{\partial f_i}{\partial x_k}\frac{\partial x_k}{\partial \alpha_j} + \frac{\partial f_i}{\partial \alpha_j} = 0, \qquad 1 \le i \le n, 1 \le k \le m. \quad (2.86)$$

It is just possible that these nm equations might be simple enough to allow determination of the sign of $\partial x_i/\partial \alpha_k$ for some i and k. If so, the information, though qualitative, might be very useful. Economists are frequent users of the equilibrium shift technique because it is often helpful to know, for example, whether a certain tax rate should be increased or decreased to raise, for example, the gross national product (it would be even more helpful, of course, to know by how much).

---

[7] This is the special case of (1.24) for which f_i is independent of t. The dynamical system is said to be *autonomous*.

[8] See Supplementary Note 2.3.

We pause to emphasize that, when applied to an equilibrium of the differential dynamical system (2.83), the equilibrium shift technique is meaningful only if the equilibrium is stable. To appreciate this, let's apply the shift technique to the ecosystem model

$$\frac{1}{x}\frac{dx}{dt} = a_1 - b_1 y \qquad \frac{1}{y}\frac{dy}{dt} = a_2 - b_2 x, \qquad (2.87)$$

first introduced in Section 1.5 and further considered in Exercise 2.3 (ii). This system has the equilibrium

$$x^*(\alpha) = \frac{a_2}{b_2}, \qquad y^*(\alpha) = \frac{a_1}{b_1}, \qquad (2.88)$$

where $\alpha = (a_1, b_1, a_2, b_2)^T$. You can see directly that

$$\frac{\partial x^*}{\partial a_2} = \frac{1}{b_2} > 0. \qquad (2.89)$$

This suggests that an increase in the parameter a_2 will shift x^* upward. Now let's see what really happens. Suppose that the ecosystem described by (2.87) is in equilibrium. Recall the remark in Section 1.5 that the phase-plane diagram Fig. 1.4 retains its topology if the parameters are changed. The effect of an increase in a_2 is merely to shift the equilibrium point x^* to the right. This shift in topology is sketched in Fig. 2.8. The old trajectories are dashed, the new ones solid, and the dot denotes the *old* equilibrium x^*. Notice that, in the new topology, the initial point x^* lies on a trajectory that veers off to the northwest as time increases. Thus, far from shifting x upward, an increase in a_2 will ultimately reduce it all the way to zero! The discrepancy between what (2.89) suggests and what actually happens is due to the equilibrium's instability. It illustrates that caution is necessary in applying the shift technique to systems of the form (2.83).

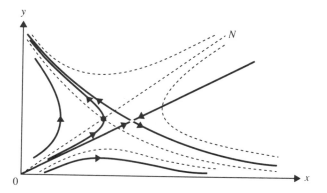

Fig. 2.8 Phase-plane topology shift for (1.22) when a_2 is increased.

As we saw in Section 1.6, however, not every dynamical system is a differential dynamical system. Indeed the equilibrium shift technique is probably most useful when applied to nondifferential ones. Consider, for example, the economic growth model of Section 1.6, in which capital at time t is denoted by $K(t)$, labor by $L(t)$, and output by $Q(t)$. The economy is not in equilibrium, because K, L, and Q do not persist at fixed levels; rather, they increase with time. On the other hand, according to the model, they increase in such a way that, at any given time,

$$Q(t) = aL(t)^\gamma K(t)^{1-\gamma}, \tag{2.90}$$

where a and γ are constants. Thus, although K, L, and Q keep changing, the relationship (2.90) between Q and the quantity $aL\gamma K^{1-\gamma}$ persists. Now, we associate persistence with equilibrium; and it therefore seems that the economy of Section 1.6 has at least some of the qualities of equilibrium, even though it is not truly in equilibrium in the sense of this chapter. We will recognize this state of affairs by saying that the nondifferential dynamical system (2.90) represents a *quasi-equilibrium* between the exogenous variables, namely $K(t)$ and $L(t)$, and the endogenous variable, namely, $Q(t)$.

A more general *nondifferential dynamical system* would consist of n relations

$$f_i(x_1(t), x_2(t), \ldots, x_n(t), \alpha_1(t), \alpha_2(t), \ldots, \alpha_m(t)) = 0, \qquad i = 1, \ldots, n, \tag{2.91}$$

which we may regard as representing a quasi-equilibrium between the n endogenous functions $x_1(t), x_2(t), \ldots, x_n(t)$ and the m exogenous functions $\alpha_1(t), \alpha_2(t), \ldots, \alpha_m(t)$. We can use the chain rule to differentiate (2.91) with respect to $\alpha_j(t)$. By analogy with (2.86), we obtain

$$\sum_{k=1}^{n} \frac{\partial f_i}{\partial x_k(t)} \frac{\partial x_k(t)}{\partial \alpha_j(t)} + \frac{\partial f_i}{\partial \alpha_j(t)} = 0, \qquad 1 \le i \le n, 1 \le j \le m. \tag{2.92}$$

It is just possible that these nm equations might be simple enough to allow determination of the sign of $\partial x_i(t)/\partial \alpha_k(t)$ for some i and k. We would then know the effect of a change in the exogenous function $\alpha_k(t)$ on the endogenous function $x_i(t)$, assuming the quasi-equilibrium (2.91) to be maintained at all times. As a simple example of this idea, observe that (2.90) yields

$$\frac{\partial Q(t)}{\partial K(t)} = a(1 - \gamma) \left(\frac{L(t)}{K(t)} \right)^\gamma > 0, \tag{2.93}$$

expressing the fact that an increase in capital will always result in an increase in output (if the quasi-equilibrium (2.90) persists throughout).

Further application of the shift technique to nondifferential dynamical systems is deferred until Chapter 10, i.e., until after we have had an opportunity to appreciate the usefulness of such models. Meanwhile, to appreciate the limitations of this technique when applied to systems of the form (2.83), try Exercise 2.16.

2.8 How Quickly Must Drivers React to Preserve an Equilibrium?

Consider a platoon of cars, each of length L, driving in equilibrium through a tunnel.[9] All cars are moving at the same speed, v. The gap between cars is a constant, d. The cars will remain in equilibrium only if they all continue to travel at exactly the same speed and hence remain a distance d apart. Now, suppose that the leading driver slows down for a moment, for whatever reason, before accelerating back to speed v. Thus she subjects the equilibrium of the platoon to a small perturbation. Following this perturbation, will the drivers regain their equilibrium? In other words, is the equilibrium stable? Intuition suggests that the drivers will regain equilibrium if they react sufficiently quickly to the first driver's braking, but if they react too slowly, then two or more of the cars will collide. But how quickly is quickly enough? In an attempt to answer that question, we will develop the model for a platoon of N cars that we introduced in Section 1.13. For the sake of definiteness, we will assume in this section that there are just ten cars in the platoon, but with obvious modifications, the analysis would apply just as well to any other number of cars.

Let x be the traffic density in this equilibrium; i.e., let $x = 1/(d + L)$. Then, on using (2.28) and assuming the tunnel to be at least so congested that $x > x_c$, we find that the speed of the platoon in the equilibrium is

$$v = v_{max}\frac{\ln(x_{max}/x)}{\ln(x_{max}/x_c)}, \tag{2.94}$$

where v_{max} is the speed a car would have if the tunnel were empty. For reasons that will emerge in Section 3.5, and again for the sake of definiteness, we shall assume that the traffic density in equilibrium is

$$\frac{x_{max}}{e} = 0.368x_{max}, \tag{2.95}$$

where $e = \exp(1)$. Then, on using (2.94), we find that the speed of the platoon is

$$v = \frac{v_{max}}{\ln(x_{max}/x_c)}. \tag{2.96}$$

Suppose that the cars have remained in equilibrium throughout the interval $t < 0$ and that the first car of the platoon crosses station $z = 0$ at time $t = 0$. If this equilibrium persists then, for $t \geq 0$, the displacement of the jth car in the platoon is given by

$$\zeta_j(t) = vt - (j - 1)(d + L)$$
$$= vt - \frac{(j - 1)}{x}$$

---

[9]You may wish to review Section 2.3 before proceeding.

$$= vt - \frac{e(j-1)}{x_{\max}}, \qquad j = 1, 2, \ldots, 10. \qquad (2.97)$$

We will refer to $\zeta_j(t)$ as the *equilibrium displacement* of the jth car. This does not, of course, mean that the displacement is constant (it increases linearly with time); rather, the speed $d\zeta_j/dt = v$ of the car is constant, and $\zeta_j(t)$ is simply the displacement associated with that equilibrium speed.

Suppose that the equilibrium of the platoon is now perturbed, because the leading driver ($j = 1$) suddenly brakes. Let $z_j(t)$ denote the subsequent displacement of the jth car, where $2 \le j \le 10$. Then (Exercise 2.17), on using (2.26) and (2.27) we get

$$\frac{dz_j(t+\tau)}{dt} = v \ln |z_j(t) - z_{j-1}(t)| + \alpha_j, \qquad 2 \le j \le 10. \qquad (2.98)$$

Here τ is the driver response time (assumed the same for all drivers) and α_j a constant. We will assume that for all j the constant α_j takes the value implied by the analysis of Section 2.3, namely, (verify) $\alpha = v \ln(x_{\max})$. Then (Exercise 2.17), because $z_{j-1}(t) > z_j(t)$, the subsequent displacements of cars 2 to 10 satisfy

$$\frac{dz_j(t+\tau)}{dt} = v \ln \left\{ x_{\max}(z_{j-1}(t) - z_j(t)) \right\}, \qquad 2 \le j \le 10. \qquad (2.99)$$

Now suppose that at $t = 0$ the first driver brakes for exactly one second, reducing her car's speed to $v(1 - \Delta)$ before accelerating back to the equilibrium speed v. Then a reasonable description of the car's velocity for $t \ge 0$ would be (Fig. 2.9)

$$\frac{dz_1}{dt} = v(1 - \Delta t e^{1-t}), \qquad (2.100)$$

whence (verify)

$$z_1(t) = v \left[t + \Delta \left\{ (t+1)e^{-t} - 1 \right\} e \right]. \qquad (2.101)$$

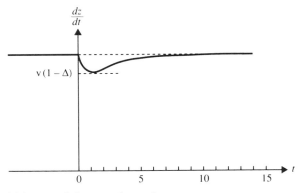

Fig. 2.9 Speed history of first car in a platoon.

From (2.97), with $j = 1$, this displacement differs from the equilibrium displacement $\zeta_1(t)$ by an amount $z_j(t) - \zeta_j(t) = v\Delta\{(t+1)e^{-t} - 1\}\,e$, when $t \geq 0$. Let us define the *perturbation displacement* of the leading car at time t, $\phi_1(t)$, by $\phi_1(t) \equiv z_1(t) - \zeta_1(t)$. Then

$$\phi_1(t) = \begin{cases} v\Delta\{(t+1)e^{-t} - 1\}\,e, & \text{if } t \geq 0; \\ 0 & \text{if } t < 0. \end{cases} \tag{2.102}$$

Note that $\phi_1(t) < 0$ when $t > 0$ because as a result of braking, the leading driver is nearer to $z = 0$ at time t than she would have been otherwise.

Cars 2 to 10 will react to this perturbation by also braking. Let the perturbation displacement of the jth car be denoted by $\phi_j(t) \equiv z_j(t) - \zeta_j(t)$, $2 \leq j \leq 10$. Then, on using (2.97) we have

$$\phi_j(t) = \begin{cases} z_j(t) - vt + \dfrac{e(j-1)}{x_{\max}}, & \text{if } t \geq 0; \\ 0 & \text{if } t < 0. \end{cases} \tag{2.103}$$

On substituting $z_j(t) = \zeta_j(t) + \phi_j(t)$ into (2.99), we obtain (Exercise 2.17)

$$v + \frac{d\phi_j(t + \tau)}{dt} = v\ln\left[x_{\max}\left\{\frac{e}{x_{\max}} + \phi_{j-1}(t) - \phi_j(t)\right\}\right]. \tag{2.104}$$

Replacing t by $t + \tau$, we thus obtain (verify)

$$\frac{d\phi_j}{dt} = v\ln\left[1 + \frac{x_{\max}}{e}\left\{\phi_{j-1}(t - \tau) - \phi_j(t - \tau)\right\}\right]. \tag{2.105}$$

In this equation, ϕ_j is identically zero whenever its argument is negative. Thus $\phi_j(t - \tau) = 0$ whenever $\tau > t$.

The equilibrium described by (2.97) is stable if, following the perturbation depicted in Fig. 2.9, the cars all regain speed v without colliding. Now, two cars will collide if the relative displacement of their front bumpers falls below a car length, L; i.e., if $z_{j-1}(t) - z_j(t) < L$, at any time $t(\geq 0)$ and for any j such that $2 \leq j \leq 10$. From the definition of ϕ_j, the condition $z_{j-1}(t) - z_j(t) < L$ is equivalent to $\phi_{j-1}(t) - \phi_j(t) + \zeta_{j-1}(t) - \zeta_j(t) < L$ (verify). But from (2.97), $\zeta_{j-1}(t) - \zeta_j(t) = d + L$. Hence two cars will collide if, for any j satisfying $2 \leq j \leq 10$ and for any $t \geq 0$,

$$\phi_j(t) - \phi_{j-1}(t) > d. \tag{2.106}$$

For the sake of definiteness, let's choose a particular value for d. Now, as we will discuss in Section 3.3, an appropriate traffic density for New York's tunnels is about 65 cars per mile, and an appropriate equilibrium traffic speed is about 20 miles per hour. Working in feet and seconds, therefore, it seems reasonable to take $v = 20 \times 5280 \div 3600 = 88/3$ and $x = x_{\max}/e = 65/5280 = 13/1056$. Then $d + L = 1056/13$. Thus, if we choose $d = 65$, then the length of a car is a little more than sixteen feet, which does not seem unreasonable. Let's adopt these values (if you don't like them, then you can easily work through the analysis again using different ones). Then the collision criterion (2.106) becomes

$$-\phi_{j-1}(t) + \phi_j(t) > 65, \tag{2.107}$$

and on using our chosen values in (2.102)–(2.105), we find (verify)

$$\phi_j(t) = 0, \qquad\qquad \text{if } t < 0, 1 \le j \le 10;$$

$$\phi_1(t) = \frac{88}{3}\Delta\left\{(t+1)e^{-t} - 1\right\}e, \qquad\qquad \text{if } t \ge 0;$$

$$\frac{d\phi_j}{dt} = \frac{88}{3}\ln\left[1 + \frac{13}{1056}\left\{\phi_{j-1}(t-\tau) - \phi_j(t-\tau)\right\}\right], \text{ if } t \ge 0, 2 \le j \le 10. \tag{2.108}$$

These differential-delay equations are not especially easy to solve. We will therefore approximate them by recurrence relations. Now, the quantity $d\phi_j/dt$ is approximately equal to

$$\frac{\phi_j(t + \epsilon) - \phi_j(t)}{\epsilon} \tag{2.109}$$

if ϵ is small.[10] It will be convenient to suppose that $\epsilon = 1/M$, where M is a large enough integer, and measure time in units of one Mth of a second. Thus n units of new time correspond to $t = n/M$ seconds. Let $\Phi_j(n)$ denote the perturbation displacement of driver j after n units of new time; i.e., $\Phi_j(n) \equiv \phi_j(n/M)$. Then (2.109) becomes $M\{\Phi_j(n+1) - \Phi_j(n)\}$, and for large enough M we can approximate the last equation of (2.108) by (verify)

$$M\left\{\Phi_j(n+1) - \Phi_j(n)\right\} = \frac{88}{3}\ln\left[1 + \frac{13}{1056}\left\{\Phi_{j-1}(n-\sigma) - \Phi_j(n-\sigma)\right\}\right], \tag{2.110}$$

where σ is the reaction time measured in new units; i.e., in old units, the reaction time is $\tau = \sigma/M$ seconds.

How large should M be? Clearly, the larger the value of M we choose, the smaller the error in approximating $d\phi_j/dt$ by (2.109). On the other hand, you must perform $60M$ recursions of (2.110) for each minute of time during which you would like to know what happens to the platoon; and these recursions will be time-consuming if M is very large. I experimented with different values of M and decided that $M = 8$ gave an acceptable trade-off between accuracy and computing cost. Hence we use $M = 8$ in the remainder of this section. (You may wish to experiment with larger values, but you will find that they do not qualitatively alter our conclusions from the analysis.) With this value, and on replacing n by $n - 1$ in (2.110),

---

[10]See Supplementary Note 2.4.

(2.108) is approximated (verify) by

$$\Phi_j(n) = 0 \qquad \text{if } n < 0, 1 \le j \le 10;$$

$$\Phi_1(n) = \frac{11}{3}\Delta\left\{(n+8)e^{-n/8} - 8\right\}e \qquad \text{if } n \ge 0;$$

$$\Phi_j(n) = \Phi_j(n-1) + \frac{11}{3}\ln\left[1 + \frac{13}{1056}\left\{\Phi_{j-1}(n-\sigma-1) - \Phi_j(n-\sigma-1)\right\}\right]$$

$$\text{if } n \ge 0, 2 \le j \le 10.$$

$$(2.111)$$

These recurrence relations are easily solved by calculator or computer. The results are shown in Fig. 2.10 for $\Delta = 0.15$ and $\tau = 1.5$ seconds ($\sigma = 12$). In this particular case, the leading driver reduces speed from 20 m.p.h. to 17 m.p.h. and then accelerates back to 20 m.p.h. again, ac-

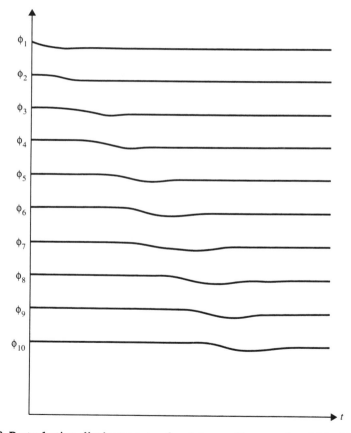

Fig. 2.10 Perturbation displacements of a platoon of ten cars in which drivers take 1.5 seconds to respond when the first car slows temporarily to 17 m.p.h., according to Fig. 2.9, from the equilibrium speed of 20 m.p.h. The horizontal scale is 1 inch to 14.3 seconds, the vertical scale 1 inch to 186 feet.

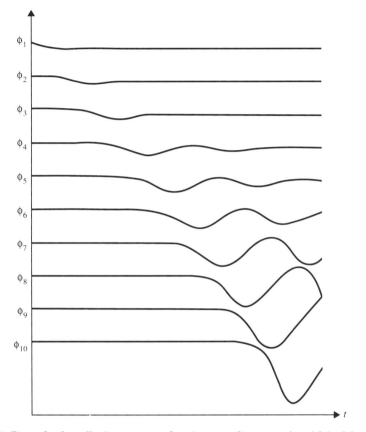

Fig. 2.11 Perturbation displacements of a platoon of ten cars in which drivers take 2.25 seconds to respond when the first car slows temporarily to 17 m.p.h., according to Fig. 2.9, from the equilibrium speed of 20 m.p.h. The horizontal scale is 1 inch to 14.3 seconds, the vertical scale 1 inch to 186 feet.

cording to Fig. 2.9. The corresponding perturbation (2.102) is the uppermost graph in Fig. 2.10. The remaining graphs are (approximations to) the perturbation displacements $\{\phi_j(t): t \geq 0, 2 \leq j \leq 10\}$ of the following cars, plotted 0.35 inches apart, which corresponds to a real distance of 65 feet. Notice that equilibrium is restored within three-quarters of a minute, with all cars once again moving at 20 m.p.h., but (taking the limit of (2.102) as $t \to \infty$) in such a way that all cars are a fixed distance $v\Delta e \approx 12$ feet behind where they would have been if the first driver had not braked. Notice also that each of the following cars takes longer to adjust to the new equilibrium than the car in front of it. Its phase of adjustment is represented by an oscillation of decreasing amplitude about the new equilibrium, but the amplitude of this oscillation increases with j as the initial perturbation

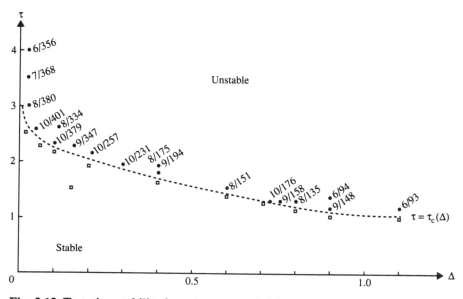

Fig. 2.12 Tentative stability boundary $\tau = \tau_c(\Delta)$ in the Δ-τ plane.

propagates backward through the platoon of cars. (From the point of view of pure mathematics, of course, the equilibrium is never restored. Rather, the cars will move toward the equilibrium forevermore, but will ultimately be an imperceptible distance away from it.)

For the same perturbation but the slower response time of $\tau = 2.25$ seconds ($\sigma = 18$), the graphs of $\phi_j(t)$ are plotted in Fig. 2.11. Notice that (2.107) is violated after 347 units of time (43.4 seconds), when the ninth car collides with the car in front of it. Thus, in the $\Delta - \tau$ plane, (0.15, 1.5) is a stable point for a platoon of ten cars; whereas (0.15, 2.25) is unstable.

The same analysis can be repeated for any other value of τ that is a multiple of an eighth of a second and for any other value of Δ (between 0 and 1). Thus we discover, for example, that (0.4, 1.625) is a stable point, but that (0.4, 1.75) is unstable, because the ninth car collides with the car in front of it after 24.25 seconds. A selection of such points is plotted in Fig. 2.12. A square denotes stability and a dot instability, with the label j/n recording that the jth car collided with the one in front after n units of time ($n/8$ seconds). By continuing in this fashion, we see that there is a boundary curve $\tau = \tau_c(\Delta)$ in the Δ-τ plane, below which the platoon is stable and above which it is unstable. On the unstable side of this boundary, for given Δ, a higher value of τ (slower response time) means a sooner collision between lower numbered cars; and, for given τ, a higher value of Δ (larger perturbation) means the very same thing. The boundary drawn (dashed) in Fig. 2.12 is a tentative one. To locate it precisely, you would have to solve

(2.108) numerically and use a significant amount of computer time. You would find, however, a similar picture to the one in Fig. 2.12.

The analysis in this section is easily reproduced for a platoon that contains an arbitrary number of cars. Simply replace 10, wherever it appears, by N. Increasing N (higher number of cars in the platoon) will shift the $\tau_c(\Delta)$ curve downward and to the left; whereas decreasing N will shift it upward and to the right (as indicated by the labels j/n in Fig. 2.12).

But do you consider that Figs. 2.10–2.12 yield a reasonable picture? Perhaps you would like to mull this over, and then try Exercise 2.18.

Exercises

2.1 Verify (2.20).

2.2 Verify that ζ, as well as η, satisfies the differential equation (2.16).

2.3 By finding the roots of (2.19), classify the static equilibria encountered in
(i) Section 1.4,
(ii) Section 1.5,
(iii) Exercise 1.5.

*__2.4__ What would be the form of $v(x)$ in Section 2.3 if, in Section 1.13, we had assumed that braking drivers do not respond to the size of a gap but only to the rate at which it's closing?

*__2.5__ Suppose that we stand at a fixed point z_0 on a road and count the number of vehicles that pass that point in ϵ units of time, say during the interval $t_0 - \epsilon/2 < t \leq t_0 + \epsilon/2$, then divide our answer by ϵ. We call the quantity thus estimated the *flux* of traffic (at station z_0, at time t_0) and denote it by $F(z_0, t_0)$. By repeating this experiment throughout the day, we obtain a function of time, $F(z_0, t)$, that records variations in the flux of traffic at that particular point on the road. The interval length ϵ must be neither so long that important variations in flux are not recorded (recall Fig. 1.6) nor so short that $F(z_0, t)$ would cease to be a smooth function of time (in the extreme case, it would take the values 0 or 1 in an apparently random sequence!). For a busy road, an hour is too long, a second too short, a minute about right; for a quiet road, we would not obtain a smooth function.

Once traffic flux F and traffic density x have been defined, we may define $v \equiv F/x$ to be the velocity of traffic at station z at time t. Thus, in general:

Number of cars passing per unit of time		Number of cars per unit length of road		Length of road covered per unit of time
$F(z,t)$	$=$	$x(z,t)$	$\times$	$v(z,t)$.

But it is sometimes reasonable to assume that F is some function $f(x)$ of the traffic density, so that $F(z, t) = f(x(z, t))$. Then, in equilibrium, we have

$$
\begin{array}{ccccc}
\text{Number of cars passing} \\
\text{per unit of time}
\end{array}
=
\begin{array}{c}
\text{Number of cars} \\
\text{per unit length} \\
\text{of road}
\end{array}
\times
\begin{array}{c}
\text{Length of road} \\
\text{covered per unit} \\
\text{of time}
\end{array}
$$

$$
f(x) \qquad = \qquad x \qquad \times \qquad v(x),
$$

where v is the velocity of traffic. The graph of f as a function of x is known as the fundamental diagram of traffic flow. Its shape will be similar to that in Fig. E2.1. In particular, $f(0) = 0 = f(x_{max})$ and f has a global maximum at $x = x^*$. We say traffic is *light* if $x < x^*$ and *heavy* if $x > x^*$, and $f(x^*)$ is the *capacity* of the road.

Find the capacity of the road when $v(x)$ is given
(i) by the model of Exercise 2.4;
(ii) by the model of Section 2.3.

2.6 Verify (2.38).

2.7 If r lies to the right of the cusp in Fig. 2.3 then the arguments of the absolute value function in (2.40) are negative. For such r, $\Delta(r) = (1 - ry(r)) \cdot (2r - ry(r) - 1)$. Hence $\Delta(r_2) = 1$ is equivalent to $f(r_2) = 0$, where $f(r) \equiv r^2 y(r) \cdot (y(r) - 2) + 2(r - 1)$. Find r_2 by using the Newton-Raphson iterative technique to solve $f(r) = 0$.
Hint: You will need a computer. First differentiate $g(w) = 2 - w$ implicitly with respect to r and arrange to obtain

$$
y'(r) = \frac{y(r) \cdot (y(r) - 2) \cdot (1 - y(r))}{2 - 2ry(r) + ry(r)^2}.
$$

This will enable you to express $f'(r)$ in terms of r, $y(r)$, and $y'(r)$. Estimate r_2 from the graph. This will be your first value in the iterative sequence. For each subsequent value, you will need a separate subroutine to supply the value of $y(r)$. The Newton–Raphson technique will be required within this subroutine as well.

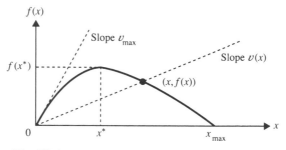

Fig. E2.1

2.8 By solving (2.36) numerically, show that there is a period-4 limit cycle $\{v_1, v_2, v_3, v_4\}$ when $r = 2.56$. To four significant figures, the values of $\{v_k\}$ are, in sequence, 0.2101, 1.587, 0.3531, and 1.850. Show similarly that the steady state when $r = 2.67$ is the 8-period limit cycle $\{1.924, 0.1631, 1.524, 0.3763, 1.990, 0.1417, 1.401, 0.4798\}$ and that the steady state when $r = 2.69$ is the period-16 limit cycle $\{2.000, 0.1357, 1.388, 0.4890, 1.933, 0.1570, 1.516, 0.3781, 2.014, 0.1315, 1.360, 0.5160, 1.897, 0.1699, 1.585, 0.3288\}$.

2.9 For $r = 2.56$, the period-2 equilibrium cycle $\{y(r), z(r)\} = \{0.26357, 1.7364\}$ is unstable, because $r > r_2$. By solving (2.36) numerically with either y or z as initial value, show that the equilibrium cycle diverges toward the period-4 limit cycle in Exercise 2.8. (To keep the population in period-2 equilibrium, the exact value of y or z would be needed. Even if $y(= 0.263573742169)$ or $z(= 1.73642625783)$ were correct to 12 significant figures, the rounding error would be sufficient to shift the period-2 cycle toward the period-4 one, though the divergence would be extremely slow.)

*2.10 The wall of a building is in thermal equilibrium if it transfers heat at a constant rate (just as a rowing boat is in mechanical equilibrium if it moves at a constant speed). Let T_1 be the temperature of air inside a building and T_2 the temperature of the outside air. Assume that $T_2 < T_1$ (it's winter). Let F denote the outward heat flux per unit area, i.e., the amount of heat energy per unit of time that flows across unit area of the wall. Let d be the thickness of the wall.
(i) Which kind of wall will lose heat at the greater rate, a thick one or a thin one? Accordingly, how do you think F varies with d?
(ii) When will the rate of heat loss be greater, when $T_2 << T_1$ (much colder outside than inside) or when T_2 is just a little lower than T_1? Accordingly, how do you think F varies with $T_1 - T_2$?
(iii) Can you suggest an expression for F, in terms of T_1, T_2, d, and some arbitrary constant, which embodies (i) and (ii)? Make the simplest hypothesis consistent with your observations.

*2.11 Instead of installing double-glazing, some folks create a "storm window" by using a sheet of transparent plastic to trap a layer of stagnant air between the air in a room and the pane of glass in the existing window. Adapt the model of Section 2.5 to calculate the heat loss prevented by this contraption. Show, in particular, that it is always less than the loss prevented by double-glazing. (You do not need to know the thermal conductivity of the plastic. Why?)

*2.12 The outward heat flux across area A of wall in Section 2.5 is $AF(x)$ only because we have assumed that heat flux and temperature do not vary within any $x = $ constant plane. If heat flux and temperature also varied with respect to y and z, however, where x, y, and z are orthogonal Cartesian coordinates, then the outward heat flux across area A would be

$$\int_A F(x,y,z)dA = \int_A F(x,y,z)dydz = -\int_A k\frac{\partial T(x,y,z)}{\partial x}dydz,$$

on using (2.45). Similarly, the outward heat flux across area A of a sphere of radius x would be

$$\int_A F(x,\theta,\phi)dA = \int_A F(x,\theta,\phi)x^2 \sin(\theta)d\theta d\phi$$

$$= -\int_A k\frac{\partial T(x,\theta,\phi)}{\partial x}x^2 \sin(\theta)d\theta d\phi,$$

where x, θ, and ϕ are spherical polar coordinates. Use this result to solve the following problem.

A spherically symmetric planet is in equilibrium with radius a. It consists of a radioactive core $0 \leq x \leq c$ surrounded by a solid mantle $c \leq x \leq a$, where x is the distance from the center of the planet. The core generates a steady heat flux $4\pi c^2 F$, which is conducted through the mantle to the surface of the planet. If the thermal conductivity of the mantle is k_M, and if the mantle heat flux per unit area is given by (2.45), show that the temperature drop across the mantle is

$$\frac{c(1 - c/a)F}{k_M}.$$

2.13 Using the methods of Section 1.14, verify that kinematic viscosity has dimensions $[\nu] = L^2/T$.

2.14 Establish (2.71). *Hint:* Use the transformation to elliptical cylinder coordinates,

$$x = ra\cos(\theta), \qquad y = br\sin(\theta),$$

to change the region of integration from that given in (2.71) to $0 \leq \theta < 2\pi$, $0 \leq r \leq 1$.

*2.15 Water is flowing in equilibrium with circular symmetry through a cylindrical pipe with radius a. The pressure drops from p_2 to p_1 along length L of the pipe. Consider a cylinder of water with radius r and length L, whose axis of symmetry coincides with that of the pipe.
(i) What is the (outward) flux of momentum *per unit area* across this cylinder?
(ii) What is therefore the momentum flux?
(iii) What is the net pressure force on this cylinder of water?
(iv) Obtain a differential equation for the velocity $w(r)$ and solve it, subject to an appropriate boundary condition.
(v) Hence obtain Poiseuille's formula (2.72).

*2.16 (i) In the model of Exercise 1.17, if x is measured in tonnes and time in years, what are the dimensions of the constant a?
(ii) If the fish are harvested, as in Exercise 1.14, at rate qux, where u is a constant, show that there are two equilibria (other than $x = 0$) provided $u < aK/4q$. Use a graphical method.
(iii) What happens if $u > aK/4q$? We will ignore the possibility that $u = aK/4q$, precisely, on the grounds that the parameters are subject to so much uncertainty. In deterministic models of natural phenomena, only strict inequalities between parameters are realistic.
(iv) Assume that $u < aK/4q$. Let x_L denote the lower equilibrium and x_U the upper one. Use the graph from (ii) to show that $0 < x_L < K/2 < x_U < K$. Using the equilibrium shift technique, show that x_U decreases with u and q but increases with a. Verify your result graphically.
(v) Why may the shift technique not be applied to the equilibrium x_L? Show that it would lead to absurd results.

2.17 (i) Verify (2.98) and (2.99).
(ii) Verify (2.104).

*2.18 (i) According to (1.81), following drivers in Fig. 2.11 will stop braking and start to accelerate 2.25 seconds after they have observed the preceding car begin to move away. Study the graphs in Fig. 2.11 and convince yourself that this is so. What changes geometrically when a driver stops braking and starts to accelerate?
(ii) The restoration of equilibrium in Fig. 2.10 is exceptionally smooth. I would have anticipated larger ripples. What do you think? Is there anything about (1.81) that produces a stabilizing effect that would not be observed in practice? If so, how could (1.81) be improved? *Hint*: Consider how drivers accelerate as the gap between them and the preceding car increases.

*2.19 (You will need a computer for this one.) In 1959, according to Section 3.3, cars were sent through New York's Lincoln tunnel in platoons of 22. Assume that they drove at 20 m.p.h., in equilibrium. Would such a platoon remain in equilibrium if the driver response time were 1.75 seconds and the leading driver slowed temporarily to 16 m.p.h. according to Fig. 2.9? Now that you have written a computer program to solve (2.111), you can reproduce Figs. 2.10–2.12 for yourself, as well as experiment with other values of d and other forms of $\phi_1(t)$.

*2.20 Suppose that the species of fish in Exercise 2.16 is being "overfished," i.e., that $u > aK/4q$. Show that, although x tends to zero as $t \to \infty$, x never actually reaches zero. Thus, according to the productivity law

$$F(x) = ax^2 \left(1 - \frac{x}{K}\right),$$

the population would be certain to recover from any threat of extinction if the harvest were ever stopped. In practice, however, we might expect that the fish would not recover from an extremely low population level, even if that level were greater than zero (for otherwise, why would species ever become extinct?). In other words, our model does not allow for the real possibility of extinction. Can you suggest an improved form of $F(x)$ that would correct this deficiency without destroying any of the other properties which, according to Exercise 1.17, it is desirable for F to have?

Supplementary Notes

2.1 A more general system of differential equations than (2.12) would have the form

$$\frac{dx_i}{dt} = f_i(x_1, x_2, \ldots, x_n), \qquad 1 \le i \le n.$$

An extension of the analysis performed above would lead to the linear system of differential equations $d\eta/dt = G\eta$, where η is an n-dimensional vector of relative displacements and the $n \times n$ matrix G is defined by

$$g_{ij} = \frac{\partial f_i}{\partial x_j}(x_1^*, \ldots, x_n^*), \qquad 1 \le i, j \le n.$$

The matrix G is called the Jacobian matrix of the system, and the static equilibrium x^* is stable if and only if all the eigenvalues of G have negative real parts. The special case of this result obtained in the text is sufficiently general for the purposes of this book.

2.2 For suppose that there are two solutions, say w_1 and w_2, and let $u = w_1 - w_2$. Denote the ellipse (2.54) by ∂A and the region it bounds by A. Then both w_1 and w_2 satisfy (2.69) and (2.55), and so we have $u = 0$ on ∂A and $\partial^2 u/\partial x^2 + \partial^2 u/\partial y^2 = 0$. Put $m = u\partial u/\partial x$ and $n = u\partial u/\partial y$. Then, on using Green's theorem

$$\oint_{\partial A} (m\,dy - n\,dx) = \iint_A \left(\frac{\partial m}{\partial x} + \frac{\partial n}{\partial y} \right) dx\,dy$$

(which has essentially been proved in reaching (2.66)), we have

$$\iint_A \left\{ \left(\frac{\partial u}{\partial x} \right)^2 + \left(\frac{\partial u}{\partial y} \right)^2 \right\} dx\,dy = \iint_A \left(\frac{\partial m}{\partial x} + \frac{\partial n}{\partial y} \right) dx\,dy$$

$$= \oint_{\partial A} u \left(\frac{\partial u}{\partial x} dy - \frac{\partial u}{\partial y} dx \right) = 0.$$

Therefore $\partial u/\partial x = 0 = \partial u/\partial y$, implying that u is a constant. Because $u = 0$ on ∂A, the value of the constant must be zero. Hence $w_1 = w_2$; i.e., the solution is unique.

2.3 Principally, the Jacobian matrix G, defined by

$$g_{ij} = \frac{\partial f_i}{\partial x_j}(x_1^*, \ldots, x_n^*), \qquad 1 \le i, j \le n,$$

must be nonsingular. By even mentioning **G**, of course, we presume that **f** is at least once differentiable with respect to **x**.

2.4 In view of (1.55) and (2.44), you may be wondering why $d\phi_j/dt$ is approximated by (2.109), and not by

$$\frac{M}{2}\left\{\Phi_j(n+1) - \Phi_j(n-1)\right\}.\tag{a}$$

The answer is that using (a) instead of (2.109) would mean approximating a *first order* ordinary differential equation by a *second order* recurrence relation, instead of by a first order one, and would lead to *numerical instability*; i.e., it would lead to spurious oscillations that predict collisions even for very low values of Δ and τ, which are bound to be stable.

To appreciate this phenomenon of numerical instability, consider the simple first order ordinary differential equation

$$\frac{dx}{dt} = -ax,\tag{b}$$

where a (> 0) is a constant. If we approximate the left-hand side of (b) by $x(n+1) - x(n)$, then the result is

$$x(n+1) - (1-a)x(n) = 0.\tag{c}$$

This first order recurrence relation has solution

$$x(n) = x(0) \cdot (1-a)^n.\tag{d}$$

If we approximate the left-hand side of (b) by $\{x(n+1) - x(n-1)\}/2$, then the result is

$$x(n+1) + 2ax(n) - x(n-1) = 0.\tag{e}$$

This second order recurrence has solution

$$x(n) = \frac{(x(1) - \mu x(0))\lambda^n + (\lambda x(0) - x(1))\mu^n}{\lambda - \mu},\tag{f}$$

where

$$\lambda = -a + \sqrt{a^2 + 1}, \qquad \mu = -a - \sqrt{a^2 + 1}.\tag{g}$$

Now the solution of (b) for integer values of time, $t = n$, is

$$x(0) = x(0)e^{-an} = x(0)(e^{-a})^n,\tag{h}$$

so that (d) correctly predicts the qualitative behavior of the solution, namely, that it is decaying exponentially; though it overestimates the decay, because $1 - a < e^{-a}$. Because $|\mu| > 1$, however, (f) introduces an oscillation of increasing amplitude that is completely spurious. For $a = 0.5$, values predicted by (h), (d), and (f) are recorded in the following table.

n	(h)	(d)	(f)
0	1.0000	1.0000	1.000
1	0.6065	0.5000	0.607
2	0.3679	0.2500	0.393
3	0.2231	0.1250	0.213
4	0.1353	0.0625	0.180
5	0.0821	0.0319	0.033
6	0.0498	0.0156	0.148
7	0.0302	0.0078	−0.115
8	0.0183	0.0039	0.263
9	0.0111	0.0020	−0.378
10	0.0067	0.0010	0.641
11	0.0041	0.0005	−1.02
12	0.0025	0.0002	1.66

You can readily see how the spurious oscillation generated by (f) completely overshadows the exponential decay of the true solution after only a handful of time steps. Of course, for such a large value of a as used for illustration, even (d) is not a good approximation to (h); but even when a is close to zero, the spurious oscillation introduced by (f) will ultimately swamp the true solution.

Numerical analysis is an important part of applied mathematics, and if you've never yet studied it, then you might like to do so now. An especially good introductory text is the one by Fröberg (1969), and much can be learned by perusing the article in *Encyclopaedia Britannica*.

3 OPTIMAL CONTROL AND UTILITY

Many of the policies we adopt in our lives are attempts to *optimize*, that is, to maximize gain or to minimize loss. The policy we select depends, of course, on what we are trying to achieve. Suppose, for example, that I'm travelling by car from Florida to Texas. I may wish to minimize fuel consumption. Then I will attempt to drive, as far as is possible, at a constant speed that is probably somewhat lower than the speed limit. The dashed curve in Fig. 3.1(a) indicates how I might decide to control my speed. Or I may wish to reach my destination as quickly as possible without breaking the speed limit. Then I will select the solid curve of Fig. 3.1(a). On the other hand, I may wish to get there as fast as I can, but subject to the constraint that the police don't give me a ticket; and I am invariably constrained by the maximum speed of my car. In these circumstances, I might decide to control my speed according to the curve in Fig. 3.1(b). Finally, I may wish to maximize the pleasure I get from the trip, subject to the constraint of an overnight stop in a hotel on the Texas border. Then I might decide to control my speed according to the curve in Fig. 3.1(c).

Suppose it takes me T days to drive to Texas. T is not fixed, but depends on how I drive, as shown in Fig. 3.1. Let $u(t)$ be my speed at time t. Let J be the quantity I wish to maximize. We will refer to J as my *utility*. If I am trying to minimize loss, then $J = -\{\text{Loss}\}$. Thus, for the first of my four alternatives, $J = -\{\text{fuel consumption}\}$; for the second and third, $J = -\{\text{journey time}\} = -T$, although there are different *constraints* in each case; and for the fourth alternative, $J = $ pleasure. In general, J will depend upon u. Therefore, from all allowable choices of speed $u(t)$ throughout the interval $0 \leq t \leq T$, I will wish to select the one that makes $J(u)$ a maximum. My problem is

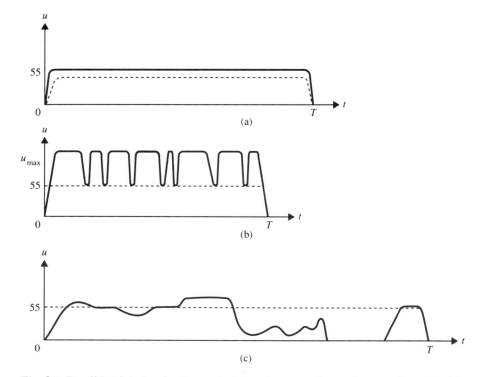

Fig. 3.1 Possible histories for the control function, speed, on a journey from Florida to Texas: (a) minimize fuel consumption (broken line) or minimize travel time, subject to speed limit (solid line); (b) minimize travel time, subject to no tickets (I travel at 55 m.p.h. only when I expect to meet police); (c) maximize pleasure, subject to overnight stay. The speed limit is assumed to be 55 m.p.h. throughout (which it actually was when I first drew this diagram for my students).

thus

To find	$u(t)$ for $0 \leq t \leq T$,	
Which maximizes	$J(u)$	(3.1)
Subject to	any constraints.	

Call u the control function. Then (3.1) is the simplest possible statement of an optimal control problem with a single control function.

Optimal control problems are abundantly found in biology, economics, engineering, management science, and indeed almost every aspect of human endeavor. If there is more than one control function, then u becomes a vector (which we would write in bold typeface as **u**); but otherwise the problem is still (3.1). Even when u is a scalar, however, optimal control problems are usually very difficult to solve—unless, by using intuition, we can argue that the set of potentially optimal controls, i.e., the set of *feasible* controls, is a very simple one. (The most obvious simplification is to assume that u is constant throughout $0 \leq t \leq T$.) Then we might even be able to find the solution graphically, as with the solid curve of Fig. 3.1(a); but it's much more likely that we'll have to use calculus.

The solid graph in Fig. 3.1(a) is indeed the solution to a highly idealized control problem, for which we assume the road to be straight and have zero congestion. It's an example of a *boundary* solution; when optimal, u is always on the constraint boundary (maximum acceleration, maximum speed, maximum deceleration). The dashed curve in Fig. 3.1(a), on the other hand, provides an example of an *interior* solution; when optimal, u is less than maximum speed. An alternative convention that distinguishes the two solutions is to say that the maximum speed constraint is *active* in the first case but *inactive* in the second. Whenever it is appropriate to talk about either, we will always adopt the second of these two conventions.

Now let's discuss some more serious examples.

3.1 How Fast Should a Bird Fly When Migrating?

A bird migrates to be as far away as possible from the unpleasant cold weather of its climate of origin.[1] It starts its journey with a finite source of energy, its body fat, and generally does not feed en route (unlike me, if I drive to Texas). Since this energy is limited, and since every hour in the air uses some of it (even if the bird is just hovering), the bird must not travel too slowly; if it does, the distance travelled will not be considerable. On the other hand, the power required to resist the drag of the air increases so rapidly with speed that the bird must not travel too quickly either; if it does, its energy supply will again be depleted too near to its journey's origin. Then how fast should the bird fly? In this section, we will attempt to answer that question by using a mathematical model.

Let $u(t)$ be the speed of the bird. Then the distance travelled is

$$\int_0^T u(t)dt, \tag{3.2}$$

where T is the flight time. Under benign weather, our bird's flight history is likely to look like the dashed curve in Fig. 3.2. If we allow the speed to be as variable as this, however, then we will not be able to solve our problem with the calculus alone. Therefore, let's make an approximation; let's assume that u is constant, corresponding to the solid line in Fig. 3.2. Our model is thus, in the previous chapter's terminology, an equilibrium model. We ignore the dynamics of ascent and descent in the belief that they affect only a negligible part of the flight. With this assumption, the

---

[1] You might like to review Section 1.12 before proceeding.

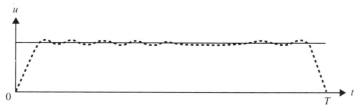

Fig. 3.2 Flight speed of a migrating bird.

distance travelled (3.2) may be approximated by

$$J(u) = u \cdot T(u). \tag{3.3}$$

The bird's objective is to maximize (3.3). The flight time T depends upon u for the reasons given in the opening paragraph.

Let E be the bird's initial energy, stored as fat. We will assume that all of E is used during migration and none is necessary at the end to survive. Then the bird's power output, assumed constant for the entire journey, is E/T, of which only a fraction, say η, can be converted into mechanical energy for the flight. (In the terminology of Section 1.12, $\eta E/T$ is the bird's effective power.) This mechanical energy must do work of two kinds: work against gravity (to keep the bird up) and work against drag (to keep the bird moving along). It seems reasonable to assume that the rate of working against gravity—or *induced power*, I, in the jargon of aerodynamics—does not depend upon the flight speed, u. Moreover, we may borrow from aerodynamics the fact that drag, D, is proportional to the square of the speed and to the surface area that faces the wind, S. This is the same drag law as for the oarsmen of Section 1.12, which is not surprising when you consider that air and water are both fluids. Hence $D = bSu^2$, where b is a constant, and

$$
\begin{aligned}
\eta E/T &= \text{Rate of working against drag} + \text{Induced power} \\
&= \qquad\quad Du \qquad\qquad + \qquad I \\
&= \qquad\quad bSu^3 \qquad\qquad + \qquad I.
\end{aligned}
\tag{3.4}
$$

From (3.3) and (3.4) we deduce that

$$J(u) = \frac{\eta E u}{bSu^3 + I}. \tag{3.5}$$

It is easy to show (see Exercise 3.1) that $J(u)$ attains a maximum on the interval $0 < u < \infty$ at $u = u^*$, where

$$u^* = \left(\frac{I}{2bS}\right)^{1/3} \tag{3.6}$$

and

$$J(u^*) = \frac{2}{3} \frac{\eta E}{(2bSI^2)^{1/3}}. \tag{3.7}$$

Now I, though independent of u, is known from aerodynamics to be proportional to the weight of the bird (a bird that's twice as heavy, for example, has to work twice as hard against gravity). We therefore write

$$I = mgv_i, \tag{3.8}$$

where m is the bird's mass, g is the constant acceleration due to gravity, and v_i, the constant of proportionality, is known in aerodynamics as the *induced velocity*. (You should check that v_i does indeed have the dimensions of velocity, by the method of Section 1.14.) Then, from (3.6)–(3.8), we obtain

$$u^* = \left(\frac{mgv_i}{2bS}\right)^{1/3} \tag{3.9}$$

and

$$J(u^*) = \frac{2}{3} \frac{\eta}{(2bSg^2v_i^2)^{1/3}} \frac{E}{m^{2/3}}. \tag{3.10}$$

In obtaining the optimal speed (3.9), we have overlooked a potential constraint. There is a certain speed, let's call it u_{max}, above which the bird is incapable of flying (however wantonly it squanders fuel). Therefore, instead of seeking the maximum of $J(u)$ on the interval $0 < u < \infty$, we should really have sought its maximum on the interval $0 < u \le u_{max}$. But the constraint $u \le u_{max}$ will not affect our answer (3.9) if $u^* \le u_{max}$, which we have tacitly assumed. In other words, we have tacitly assumed that the constraint $u \le u_{max}$ is inactive. If a constraint is inactive, then we obtain the same result by ignoring it completely. Clearly, if $u^* > u_{max}$, then $J(u)$ would be an increasing function on the interval $0 < u \le u_{max}$ and would take its maximum at u_{max}. In principle, therefore, it is more correct to say that the optimal flight speed is the minimum of u^* and u_{max}, but we shall continue to assume that the constraint $u \le u_{max}$ is inactive, because the "sprinting" speed of any creature always greatly exceeds its "long distance" speed.

How good is this model—does (3.10) tell us how far a bird flies when migrating? Perhaps you would like to think about that; we'll raise the topic again in Chapter 4.

3.2 How Big a Pay Increase Should a Professor Receive?

Lawmakers in a certain state have just granted a pay increase to faculty at the state university. They have done this merely by increasing the budget for professors' salaries, and it is up to the university to decide how these additional funds are to be distributed among faculty. In principle, all the money could be used to raise the salaries of a few outstanding professors.

But this would hurt the morale of, and hardly be fair to, all those less than outstanding professors who still do an excellent job. On the other hand, an "across-the-board" increase (the same percentage for everyone) would hurt the morale of lower paid faculty who have distinguished themselves as outstanding teachers and scholars. A compromise is needed, according to which everyone gets some increase, and more deserving faculty get more. But who is more deserving, and how much more of an increase should they receive? In this section, we will indicate an approach to solving such a problem by using a mathematical model.

To simplify matters, we will begin by assuming that the university administration has already partially solved the problem by allotting certain amounts of the total salary budget increase to each department. These amounts would not, of course, be equal in general. It is perhaps most likely that they would be proportional to the number of faculty in each department (although the university administration may have found a way to increase its allotment to one or two outstanding departments). Let's suppose that, as a consequence, the Mathematics Department now has a budget of B dollars for salary increases. We will also assume that the extra B dollars is available every year from now on, so that the increases will be permanent. How should the Mathematics Department's chairperson distribute this money among the faculty? The method we use to answer this question is also applicable, with rather more work, to the university as a whole (so we have not cheated by concentrating our attention on a single department).

Every year the research, service, and teaching of faculty in the Mathematics Department are subject to peer evaluation. They are classified as outstanding, strong, satisfactory, weak, or inadequate. Service might include not only administrative and pastoral duties within the university but also community service (e.g., judging at a high school science fair). The chairperson has decided to distribute the Mathematics Department's salary increase on the basis of this evaluation. This year, no one has been judged as weak or inadequate. Thus research, service, and teaching are placed in one of three categories for each professor, implying that faculty are placed in one of $3 \times 3 \times 3 = 27$ overall categories. Using O to denote outstanding, G to denote strong, and S to denote satisfactory, the results of this year's peer evaluation are shown in Table 3.1. Notice that only 14 of the 27 possible categories are represented (for example, no one is classified as OGS, i.e., outstanding in research, strong in service, satisfactory in teaching). Notice also, from the final column of the table, that the Department of Mathematics has 30 faculty.

Let us agree to describe research, service, and teaching as factors in the evaluation process. Now, it is generally felt within the department that someone whose factor evaluation is strong has achieved approximately twice as much as someone whose factor evaluation is satisfactory. Moreover, it is generally felt that someone whose factor evaluation is outstanding

Table 3.1 This year's peer evaluation of faculty in the Department of Mathematics.

Category	Research	Service	Teaching	Number of Faculty in Category
1	O	G	O	1
2	O	S	O	1
3	O	S	G	2
4	G	O	G	1
5	G	G	O	1
6	G	G	G	10
7	G	S	G	3
8	G	S	S	1
9	S	O	O	1
10	S	O	G	1
11	S	G	O	1
12	S	G	G	3
13	S	S	G	3
14	S	S	S	1

has achieved twice as much as someone whose factor evaluation is strong. It therefore seems reasonable to say that S corresponds to one unit of factor achievement, G to two units of factor achievement, and O to four units of factor achievement. But faculty in another department or at another university might not agree! Therefore, with a view to making our model as general as possible, we will label the classifications O, G, and S as $i = 1$, 2 and 3, respectively; and we will label the factors research, service, and teaching, respectively, as $j = 1$, 2 and 3. Let m_{ij} denote the number of units of factor j achievement corresponding to the classification i. Then the perceptions of the faculty on this matter can be stored in a 3×3 merit matrix **M**, in which m_{ij} occupies row i and column j. For the particular department we are considering,

$$\mathbf{M} = \begin{bmatrix} 4 & 4 & 4 \\ 2 & 2 & 2 \\ 1 & 1 & 1 \end{bmatrix} \tag{3.11}$$

The columns of **M** are identical in our illustration. But they do not have to be. If, for example, O for teaching were felt to represent five times the achievement of S for teaching, then m_{13} would be 5, not 4. At this juncture, you might like to attempt Exercise 3.2.

The chairperson has decided to solve the allocation problem by paying each faculty member an additional v_j dollars for every unit of factor j achievement. The problem therefore reduces to determining appropriate numbers v_1, v_2, and v_3. We will refer to v_1, v_2, and v_3 as controls. Let $i(k, j)$ denote the factor j label associated with a faculty member in category k,

$1 \leq k \leq 14$. These labels are contained in a 14×3 matrix equivalent to the middle three columns of Table 3.1. For example, $i(5, 2) = 2$, because a professor in category 5 is rated as strong in service, and $i(11, 3) = 1$, because a professor in category 11 is rated outstanding in teaching. Then a faculty member in category k will receive a salary increase of $m_{i(k,1)1} \cdot v_1$ dollars for his research, $m_{i(k,2)2} \cdot v_2$ dollars for his service, and $m_{i(k,3)3} \cdot v_3$ dollars for his teaching. For example, on using (3.11), a professor in category 3 will receive $4v_1$ dollars for research, v_2 dollars for service, and $2v_3$ dollars for teaching. The total salary increase of a professor in category k will therefore be

$$s_k \equiv \sum_{j=1}^{3} m_{i(k,j)j} \cdot v_j. \tag{3.12}$$

For example, the total salary increase of the professor in category 9 will be $v_1 + 4v_2 + 4v_3$ dollars. Let n_k denote the number of faculty in category k. Then the total faculty salary increase for that particular category will be $n_k s_k$. Summing over values of k between 1 and 14, we find the total faculty salary increase. This must equal B, by definition. Hence

$$\sum_{k=1}^{14} n_k s_k = B. \tag{3.13}$$

For (3.11) and Table 3.1, you can easily verify (Exercise 3.3) that the *budget constraint* (3.13) reduces to

$$58v_1 + 55v_2 + 68v_3 = B. \tag{3.14}$$

No matter how our chairperson selects the controls v_1, v_2, and v_3, he must at least ensure that (3.14) is satisfied!

The budget constraint could be satisfied in a variety of ways. The chairperson could, for example, set $v_1 = v_2 = v_3 = B/181$. But paying equal amounts for equal achievements in each of the factors may not be in the department's best interests. Suppose, for example, that the chairperson is concerned about the department's research activity. He may have observed that, whereas only two of the faculty are merely satisfactory in teaching, a third of the faculty is merely satisfactory in research. As an incentive to faculty to concentrate more on research, the chairperson may wish to increase the bonus for a unit of research achievement. The extreme case would be to set $v_1 = B/58$ and $v_2 = v_3 = 0$. This would mean that teaching and service achievements went wholly unrewarded, however, and would not be in the department's best interests either; a successful academic department requires a balance between the three factors. How can the chairperson attain such a balance?

One way would be to insist that the payment for a unit of achievement in one factor can never be more than, say, twice or three times the payment for a unit of achievement in another. We're not sure whether to say twice or three times or even some other ratio; therefore let's simply insist that the control for one factor can never be more than λ times that for another, where the choice of λ is at the chairperson's discretion. In other words, we constrain v_1, v_2, and v_3 so that

$$\frac{1}{\lambda}v_1 \leq v_2 \leq \lambda v_1, \qquad \frac{1}{\lambda}v_1 \leq v_3 \leq \lambda v_1, \qquad \frac{1}{\lambda}v_2 \leq v_3 \leq \lambda v_2, \qquad (3.15)$$

where $\lambda > 1$. For the sake of morale within the department, it is probably also a good idea to set an absolute minimum on the amount of increase that a professor can receive, provided that his work is at least satisfactory in each of the factors. Suppose we agree that such a professor should obtain at least a fraction ϵ of the total faculty increase. Here ϵ is some small number, to be agreed upon by the faculty, and it is clear at the outset that ϵ cannot possibly exceed 1/30 (why?). A further constraint is therefore

$$v_1 + v_2 + v_3 \geq \epsilon B. \qquad (3.16)$$

Let's assume that (3.14)–(3.16) are the only constraints that the chairperson, in consultation with the faculty, finds necessary.

It is convenient at this point to make the controls dimensionless. Because v_j and B have the dimensions of value, we define:

$$u_j \equiv \frac{v_j}{B}, \qquad 1 \leq j \leq 3. \qquad (3.17)$$

Then u_j is dimensionless and the constraints (3.14)–(3.16) take the dimensionless form

$$
\begin{array}{rcrcrcl}
u_1 & - & \lambda u_2 & & & \leq & 0 \\
u_1 & & & - & \lambda u_3 & \leq & 0 \\
-\lambda u_1 & + & u_2 & & & \leq & 0 \\
& & u_2 & - & \lambda u_3 & \leq & 0 \\
-\lambda u_1 & & & + & u_3 & \leq & 0 \\
& & -\lambda u_2 & + & u_3 & \leq & 0 \\
-u_1 & - & u_2 & - & u_3 & \leq & -\epsilon \\
58u_1 & + & 55u_2 & + & 68u_3 & = & 1.
\end{array}
\qquad (3.18)
$$

It will shortly emerge why the seventh of these eight constraints has been written as above, rather than in the more natural form $u_1 + u_2 + u_3 \geq \epsilon$.

We will say that the control vector $\mathbf{u} = (u_1, u_2, u_3)^T$ is *feasible* if u_1, u_2, and u_3 are nonnegative numbers satisfying (3.18). Clearly, the need to satisfy (3.18) gives the chairperson considerably less freedom than would exist if (3.14) were the only constraint. On the other hand, there may still be infinitely many choices of feasible controls! Suppose, for example, that $\lambda = 2$ and $\epsilon = 0.01$. Then you may verify that two feasible choices are $\mathbf{u} = (1/181, 1/181, 1/181)^T$ and $\mathbf{u} = (1/236, 1/118, 1/236)^T$, and it can be shown that there are infinitely many others. Which of all these feasible controls will the chairperson select?

Let's suppose, as earlier, that the chairperson is concerned about the department's research activity and would like to provide an incentive to improve it. Then a rational strategy for him to pursue would be to award the maximum feasible amount to a unit of research achievement. In other words, from all feasible controls, he would choose the one for which u_1 is largest. He is therefore faced with the following control problem:

$$
\begin{aligned}
&\text{Find} && \mathbf{u} = (u_1, u_2, u_3)^T, \\
&\text{Which maximizes} && J(\mathbf{u}) \equiv u_1 \\
&\text{Subject to} && \mathbf{u} \text{ feasible.}
\end{aligned}
\tag{3.19}
$$

This control problem is a special case of the following one:

$$
\begin{aligned}
&\text{Find} && \mathbf{u} = (u_1, u_2, \ldots, u_n)^T, \\
&\text{Which maximizes} && J(\mathbf{u}) \equiv \sum_{k=1}^{n} c_k u_k \\
&\text{Subject to} && \sum_{k=1}^{n} a_{ik} u_k \le b_i, && i = 1, \ldots, q \\
& && \sum_{k=1}^{n} a_{ik} u_k = b_i, && i = q + 1, \ldots, r \\
& && u_j \ge 0, && j = 1, \ldots, n.
\end{aligned}
\tag{3.20}
$$

You can verify that (3.19) is the special case of (3.20) for which $n = 3$, $q = 7$, $r = 8$, $a_{11} = a_{21} = a_{32} = a_{42} = a_{53} = a_{63} = 1.0$, $a_{12} = a_{23} = a_{31} = a_{43} = a_{51} = a_{62} = -\lambda$, $b_1 = b_2 = b_3 = b_4 = b_5 = b_6 = 0$, $b_7 = -\epsilon$, $b_8 = 1.0$, $c_1 = 1.0$, and $c_2 = c_3 = 0$.

The control problem (3.20) is known as a *linear program*. To solve it we must find a set of n nonnegative controls, $\{u_1, u_2, u_3, \ldots, u_n\}$, which maximizes some linear combination of controls subject to r additional constraints, of which q are linear inequality constraints and $r - q$ are linear equality constraints (additional because requiring u_j to be nonnegative is already a constraint). A linear program of the form (3.20) is completely specified once we have specified entries for an $r \times n$ matrix $\mathbf{A}$, an r-dimensional vector $\mathbf{b}$, and an n-dimensional vector $\mathbf{c}$. Efficient computer programs for solving linear programs are found at almost every computer installation and are extremely simple to use. On the other hand, the theory behind these programs would take at least a chapter to expound and would have nothing to do with modelling per se. Therefore, we do not discuss it. The interested reader is referred to one of the many excellent textbooks on optimization, for example, Jeter (1986). From the point of view of modelling, it is important only to know that a feasible solution of (3.20) may not exist and may not be unique; but if it does exist, then a computer package will quickly find it for you (it will also tell you if there is no feasible solution). Moreover, the set of optimal controls is usually unique in practical problems; and, should there be more than one solution, the decisionmaker would use some other criterion to distinguish between them. If there were more than one solution to (3.19), for example, then the chairperson of the Mathematics Department would adopt the most popular one.

Let's now return to (3.19). My students and I used the IMSL routine ZX3LP to solve this problem. The routine is extremely simple to use and fully documented in the IMSL manuals. But other routines exist, even for small personal computers, and they are usually fast, reliable, and easy to use at least for moderate values of r and n. (This state of affairs should not be used as an excuse for ignoring the theory of linear programming completely. It simply makes it out of place in a modelling course.)

To specify A and b for the computer, it is necessary to supply values for λ and ϵ. Suppose, for example, we take $\lambda = 2$ and $\epsilon = 0.01$. Thus one factor payment for a unit of achievement may not exceed twice another, and every professor will receive a pay increase that is at least 1% of the total available funds. You can easily verify that the optimal controls for (3.19) are then given by

$$\mathbf{u}^* = (0.0083682, 0.0041841, 0.0041841). \tag{3.21}$$

This result is quoted with far more accuracy than is useful in practice, but that will enable you to check your own results. If $\lambda = 2$ and $\epsilon = 0.017$, on the other hand, then the optimal controls are given by

$$\mathbf{u}^* = (0.0068421, 0.0067368, 0.0034211). \tag{3.22}$$

On using (3.12), we can now deduce the salary increase that a professor in category k would receive according to (3.21) and (3.22). The results are presented in Table 3.2.

Notice that solutions (3.21) and (3.22) are fundamentally different. When $\epsilon = 0.01$, only the first and second of the inequality constraints in (3.18) are active. If we had known in advance that constraints $i = 3$ to $i = 7$ would be inactive, then we could have overlooked them (but of course we did not know, which is why we included them). In particular, the minimum salary increase constraint (3.16) is inactive. In this case, the value for ϵ set by the chairperson is so low that the lowest salary increase would exceed it anyway. It means that the greatest possible incentive can be given to research by making u_1 exactly twice u_2 or u_3.

When $\epsilon = 0.017$, on the other hand, the value for ϵ set by the chairperson is sufficiently high that the minimum salary constraint (3.16) has become active. At the same time, the first constraint has become inactive;

Table 3.2 Salary increase in dollars awarded to faculty member in category k, $1 \le k \le 14$, for each \$1000 of total salary increase awarded to the Department of Mathematics, for $\lambda = 2$ and for two values of ϵ.

Category	$\epsilon = 0.01$	$\epsilon = 0.017$	Category	$\epsilon = 0.01$	$\epsilon = 0.017$
1	58.58	54.53	8	25.10	23.84
2	54.39	47.79	9	41.84	47.47
3	46.03	40.95	10	33.47	40.63
4	41.84	47.47	11	33.47	34.00
5	41.84	40.84	12	25.10	27.16
6	33.47	34.00	13	20.92	20.42
7	29.29	27.26	14	16.74	17.00

so that, although u_1 is still twice u_3, u_2 is now almost as great as u_1. This is surely not what the chairperson intended by trying to maximize the reward for a unit of research achievement, because one can fare almost as well by concentrating on service. It shows that the choice of ϵ is critical in determining the distribution of salary increases, and it might even be argued that the constraint (3.16) should be eliminated from the problem.

Although we shall not do so here, one can show more generally that for any given $\lambda (> 1)$ there are two critical values of ϵ, namely, $\epsilon_L(\lambda)$ and $\epsilon_U(\lambda)$, defined by

$$\epsilon_L(\lambda) \equiv \frac{\lambda + 2}{58\lambda + 123}, \qquad \epsilon_U(\lambda) \equiv \frac{2\lambda + 1}{113\lambda + 68}, \tag{3.23}$$

such that there is no feasible solution to (3.19) if $\epsilon > \epsilon_U(\lambda)$; however, if $\epsilon < \epsilon_L(\lambda)$, then the optimal control for (3.19) is

$$\mathbf{u}^* = \frac{1}{58\lambda + 123}(\lambda, 1, 1)^T, \tag{3.24}$$

constraint (3.16) being inactive. If $\epsilon_L(\lambda) < \epsilon < \epsilon_U(\lambda)$, on the other hand, then constraint (3.16) is active and the optimal control for (3.19) is

$$\mathbf{u}^* = \frac{1}{3\lambda + 13}(\{1 - 55\epsilon\} \lambda, \{58\lambda + 68\} \epsilon - \lambda - 1, 1 - 55\epsilon)^T. \tag{3.25}$$

Note that $u_1^* = \lambda u_3^*$ in both cases. For $\lambda = 2$, as in Table 3.2, we have $\epsilon_L(2) = 0.0167364$ and $\epsilon_U(2) = 0.0170068$ (so that the maximum value the chairperson can choose for ϵ is then considerably lower than 1/30).

How good is this model—would a university department use it? Notice that, hitherto, we have always made the question "How good is this model?" equivalent to "Does it fit the facts?" Why do we now abandon this equivalence? Perhaps you would like to mull this important matter over; we'll return to it in Chapter 4. Meanwhile, try Exercise 3.5.

3.3   How Many Workers Should Industry Employ?

Suppose that the manufacturing sector of a certain economy seeks to maximize its profits.[2] How many workers should it employ? In this section, we will attempt to provide at least a rough answer to this question.

Let $u(t)$ denote the number of workers employed at time t; this will be our control function. For the sake of simplicity, we will assume that all are full-time workers (though the effect of part-time labor could be included by counting, say, two half-time workers as a single unit of labor). Let $w(t)$ be the average wage per worker per year at time t (more precisely, as in Section 1.6, the instantaneous average annual wage). Then industry's (instantaneous) annual wage bill at time t is $w(t) \cdot u(t)$.

Let $Y(t)$ denote the monetary value of industry's output at time t (again, as in Section 1.6, measured as an instantaneous annual quantity). Let $c(t)$ denote annual costs independent of the size of the labor force. Then, if we assume that all goods produced are also sold, industry's annual profits at time t are $\pi(t) = Y(t) - c(t) - w(t) \cdot u(t)$. On using (1.43), we deduce that $\pi(t) = P(t) \cdot Q(t) - c(t) - w(t) \cdot u(t)$, where Q denotes real output and P the price index. Let's now assume that real output can be modelled adequately by using a Cobb–Douglas production function. Then, from (1.32), we obtain $Q(t) = au(t)^\gamma K(t)^{1-\gamma}$, where K denotes capital stock and $\gamma (0 < \gamma < 1)$ and a are constants. Industry wishes to maximize profits, and so its utility is simply π. Hence it will wish to control $u(t)$ so as to maximize

$$J(u(t)) \equiv P(t)au(t)^\gamma K(t)^{1-\gamma} - c(t) - w(t)u(t). \tag{3.26}$$

Differentiating with respect to $u(t)$, we easily find that

$$\frac{\partial J(u(t))}{\partial u(t)} = \gamma aP(t) \left(\frac{K(t)}{u(t)} \right)^{1-\gamma} - w(t); \tag{3.27}$$

and, because $\gamma < 1$, that $\partial^2 J / \partial (u(t))^2 < 0$. Hence J has a maximum on $0 < u(t) < \infty$ at $u(t) = u^*(t)$, where

$$u^*(t) = \left(\frac{\gamma aP(t)}{w(t)} \right)^{1/(1-\gamma)} K(t). \tag{3.28}$$

Our model predicts that (3.28) is the optimal amount of labor for industry to employ.

How good is this model—does (3.28) agree with the amount of labor *actually* employed? Does it tell us anything about the usefulness of the Cobb-Douglas model of economic growth? Essentially this question will surface again in Chapter 4, as Exercise 4.18. Meanwhile, in deriving (3.28), we have overlooked a possible constraint. What is it? See Exercise 3.6.

---

[2]You might like to review Section 1.6 before proceeding.

3.4   When Should a Forest Be Cut?

A forest is an investment. The trees that grow there are felled for their wood, and the wood has a market value. But if the trees are harvested too soon, then the price they fetch may not even cover the costs of felling them. On the other hand, if the forester waits too long before harvesting, then the doddery old trees may not be growing fast enough to justify their occupation of the soil. Somewhere in between these extremes lies an optimal harvesting time. What is it? In this section, we will attempt to answer that question by using a mathematical model.

The key to solving this problem is to understand that every investment is also an act of abstention. The money a forester spends on planting trees could instead have been deposited in a bank, where its value at time t would have grown at the *instantaneous annual rate of interest*, $v(t)$, defined to be the specific growth rate of the value of money. Thus, if $M(t)$ is the value of money deposited in a bank at time t, we have

$$v(t) \equiv \frac{1}{M(t)} \frac{dM}{dt}. \tag{3.29}$$

For the sake of simplicity, we will assume in this section that v is a constant. Then

$$M(t) = M(0)e^{vt}. \tag{3.30}$$

The annual rate v when interest is compounded instantaneously must be distinguished from the annual rate $\bar{v}$ when interest is compounded only once a year. In the former case, the value of money after one year is $M(0)e^{v}$. In the latter case, the value of money after one year is $M(0) \cdot (1 + \bar{v})$. Hence $v = \ln(1 + \bar{v})$. To distinguish verbally between the two rates, we might call $\bar{v}$ the yearly rate and v the annual one. Moreover, to preserve a distinction that will shortly emerge, we will refer to v as the *nominal* annual interest rate.

By planting trees, the forester hopes that proceeds $\bar{V}(t)$ from the eventual sale of the wood, i.e., revenue minus cutting costs minus planting costs, will exceed whatever the balance of his bank account would have been if he had placed his money there instead. It seems reasonable to call $\bar{V}(t)$ the value of the forest and to expect that the forester would be happy if $\bar{V}$ were made large. Does it, therefore, correspond to his utility in (3.1)?

To understand that the answer is no, consider a purely imaginary event. Suppose that the yearly rate of interest is 8%, i.e., $\bar{v} = 0.08$; and that I have purchased a beaten-up old bicycle for $100, secure in the knowledge that I can clean it, polish it, and sell it at the end of the year for $112. Ignoring riskiness, inflation, and so on (because the event is imaginary), am I right to tell myself that the true worth $\$W$ of my bicycle, at the time of purchase, is $112? No, clearly not; for if I had $112, then I could put it in a bank where, after one year, it would become $112 \cdot (1 + 0.08) = \$120.96$, which is more than I can eventually get for the bicycle. Thus the true worth of my bicycle is less than $112. On the other hand, the true worth of my bicycle is more than the $100 I paid for it, because that amount of money would grow to only $108 if I placed it in a bank. Thus $100 < W < 112$. A few moments thought should now reveal that the true worth of my bicycle is the amount of money that would grow to $112 if deposited in a bank, instead of being used to buy the bicycle. Hence $W(1+0.08) = 112$; whence $W = 112/1.08 = 103.70$. In other words, the bicycle's true worth today is equal to its value one year from now divided by the multiplicative factor by which money would grow in a bank in that time.

For exactly the same reason, the true worth of the forest today is $e^{-\nu}\bar{V}(1)$. By an obvious extension of the same argument, its true worth is also equal to $e^{-2\nu}\bar{V}(2)$, $e^{-3\nu}\bar{V}(3)$, $e^{-4.76\nu}\bar{V}(4.76)$, and so on. Thus, in general, the true worth of a forest with value $\bar{V}(t)$ at time t, *when assessed in the light of alternative investment possibilities*, is the quantity

$$e^{-\nu t}\bar{V}(t);\qquad(3.31)$$

which we shall call the *present value* of the trees in the forest. This is the utility that the forester's harvesting policy will seek to maximize.

Let there be N trees in the forest and let $u(t)$, the control function, be the rate at which trees are cut at time t. To simplify the control problem, let's assume that the entire forest is felled in a single operation, which takes place during the interval $s - \epsilon/2 \le t < s + \epsilon/2$. If the rate of felling within this interval is constant (and therefore equal to N/ϵ), then the graph of u will appear as the dashed curve in Fig. 3.3(a). If ϵ is very small (weeks, say) compared to the time for which a forest grows (several decades), then a useful mathematical approximation is to assume that the trees are all felled at the same instant, i.e., to take the limit of Fig. 3.3(a) as $\epsilon \to 0$ (Fig. 3.3(b)).

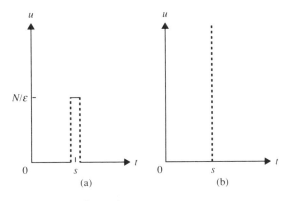

Fig. 3.3 Clearcutting control for a forest.

Then u is an *impulse* control, which can be represented crudely as[3]

$$u(s) = \infty;$$
$$u(t) = 0, \qquad \text{if } t \ne s. \tag{3.32}$$

The utility $J(u)$ has thus been reduced to a function of s, say f. The optimal control problem is therefore to

Find s

Which maximizes $f(t) \equiv e^{-\nu t} \bar{V}(t). \tag{3.33}$

It remains to determine the form of $\bar{V}$.

Assume that the forester always was, and always will be, a forester. Whenever he cuts the forest he replants it again, so that cutting and planting costs occur together at the end of each cycle; or *rotation*, to use the accepted forestry terminology. Let these (combined) costs be denoted by $\bar{c}$; and, at time t, let $\xi(t)$ be the volume of wood in a tree and $\bar{p}(t)$ its selling price per unit volume. Assume, moreover, that all N trees in the forest are identical. Then

$$\bar{V}(t) = N\bar{p}(t)\xi(t) - \bar{c}. \tag{3.34}$$

Prices and costs (which are simply prices of other commodities) will rise over time at the going rate of inflation, r, defined in Section 1.6 by

$$r \equiv \frac{1}{P}\frac{dP}{dt}, \tag{3.35}$$

where $P(t)$ is the price index. Let's assume that the rate of inflation is constant. Then, if c and p denote cost and selling price at the start of the current rotation, we have

$$\bar{p}(t) = pP(t) = pe^{rt}$$
$$\bar{c}(t) = cP(t) = ce^{rt}, \tag{3.36}$$

on recalling that $P(0) = 1$. Exploiting the analogy between (1.43) and (3.36), we say that $\bar{p}$ is the *nominal* selling price and p the *real* selling price; and similarly for $\bar{c}$ and c. Substituting from (3.34) and (3.36) into (3.33), we obtain

$$f(t) = e^{-\nu t}\{Npe^{rt}\xi(t) - ce^{rt}\} = e^{-\delta t}V(t), \tag{3.37}$$

where we have defined the *real* (annual) interest rate

$$\delta \equiv \nu - r \tag{3.38}$$

---

[3]See Supplementary Note 3.1.

to be the difference between the nominal interest rate and the rate of inflation; and where

$$V(t) \equiv Np\xi(t) - c = e^{-rt}\bar{V}(t) \tag{3.39}$$

is the deflated value of the forest at time t, i.e., its value after a correction has been made for the illusory growth that is merely due to inflation of prices (see Section 1.6). Exploiting the analogy between (1.43) and (3.39), we refer to $\bar{V}$ as the nominal value of the forest and V as its *real* value.

Thus, although the words "true" and "real" are virtually synonymous in everyday speech, they have specialized meanings for us in an economic context. An economic asset has real value $V(t)$ at time t if we have corrected for the illusory growth in value, prior to time t, that is merely due to inflation of prices; or, which is exactly the same thing, if we have divided its nominal value $\bar{V}(t)$ by the price index $P(t) = e^{rt}$ (or, which of course is the same, multiplied by e^{-rt}). An economic asset has *true* worth $f(t)$ today if we have corrected its value at time t from now for the illusion that it could not be sold today and the proceeds deposited in a bank; or, which is exactly the same thing, if we have multiplied its value at time t by either $e^{-\nu t}$ or $e^{-\delta t}$, according to whether that value is either nominal or real. You should compare (3.31) with (3.37) to verify that the forest's true worth can indeed be obtained either (i) by always using nominal prices and interest rates or (ii) by always using real prices and interest rates. The second alternative is the norm in economic modelling and, in the absence of a statement to the contrary, we shall assume henceforward that (ii) prevails. Moreover, we will usually refer to $f(t)$ henceforward as present value, rather than true worth, as an insurance against confusing the specialized meanings of real and true. We will call $e^{-\delta t}$ the *discount* factor.

Now a tree is a plant, and a model of plant growth was studied in Section 1.3. Although that model was really intended for annual species of plants, and not for trees, its basic assumption—that growth is proportional to the remaining amount of a limited quantity of nutrient—does not seem unreasonable for trees as well. Thus the model of Section 1.3 suggests that the graph of $\xi(t)$ will be S-shaped, and we will cautiously assume this. Multiplying by the constant Np will not alter the S-shape, and subtracting the constant c merely displaces the graph downward. We therefore expect that the graph of $V(t) = Np\xi(t) - c$ will appear as shown in Fig. 3.4. Thus the forest has no commercial value for $t \leq t_0$, and its value can never exceed V_∞, where the graph has a horizontal asymptote.

It is now (Exercise 3.7) a matter of simple calculus to show that $f(t)$, as defined by (3.37), has a maximum at $t = s$, where

$$\frac{V'(s)}{V(s)} = \delta, \tag{3.40}$$

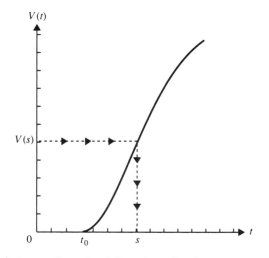

Fig. 3.4 Expected shape of graph of forest's real value.

provided

$$\frac{V''(s)}{V(s)} < \delta^2. \tag{3.41}$$

If $\xi(t)$ does have the logistic form (1.16), then

$$V(t) = \frac{Np\xi_f}{1 + be^{-k\xi_f t}} - c, \tag{3.42}$$

where the constant b is defined by $(1 + b) \cdot \xi(0) = \xi_f$. Thus $V_\infty = Np\xi_f - c$, and $k\xi_f t_0 = \ln(bc/V_\infty)$. Note that t_0 is sure to be a positive number, because $t_0 > 0$ is equivalent to our assumption $V(0) < 0$. It is then easy to deduce from (3.42) and (3.40) that $V(s)$ is the sole positive root of the quadratic equation

$$Np\delta V = k(V + c)(V_\infty - V) \tag{3.43}$$

(see Fig. 3.5); whence, from (3.42), the optimal rotation time is

$$s = \frac{1}{k\xi_f} \ln\left(\frac{b\{V(s) + c\}}{V_\infty - V(s)}\right). \tag{3.44}$$

Referring to Fig. 3.4, therefore, we find the optimal rotation time by moving along the dashed curve in the direction of the arrows.

How good is this model—does (3.44) determine when a forest should be cut? Perhaps you would like to mull this question over; we'll return to it in Chapter 4. Meanwhile, try Exercise 3.9.

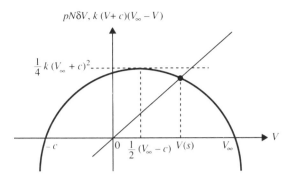

$pN\delta V, k\,(V+c)(V_\infty - V)$

$\frac{1}{4}k\,(V_\infty + c)^2$

$-c$

0 $\frac{1}{2}\,(V_\infty - c)$ $V(s)$ V_∞

V

Fig. 3.5 Graphical determination of $V(s)$ when V is given by (3.42).

3.5 How Dense Should Traffic Be In a Tunnel?

Tunnels are notorious bottlenecks where motorists converge from several directions to compete for the means to cross a river.[4] To reduce these traffic delays, the traffic flux, i.e., number of vehicles using the tunnel per hour, should be maximized. This means solving an optimal control problem for which the utility is traffic flux.

We have already seen in Chapter 2 that traffic flux depends upon traffic density, i.e., number of cars per mile. Accordingly, let the control function, $u(t)$, be the traffic density at time t. To make matters simple, however, we'll assume that u is independent of t. The traffic flow is then in equilibrium, and we'll assume that its velocity is given by (2.28). Thus, from Exercise 2.5, the traffic flux is

$$J(u) = \begin{cases} uv_{max}, & \text{if } 0 \le u \le x_c; \\ Vu\ln(x_{max}/u), & \text{if } x_c < u \le x_{max}; \end{cases} \qquad (3.45)$$

where we have defined

$$V \equiv \frac{v_{max}}{\ln(x_{max}/x_c)} \qquad (3.46)$$

and x_c is the critical traffic density below which cars may travel at the speed limit, v_{max}.

We also saw in Exercise 2.5 that the maximum of (3.45) occurs at $u = x_{max}/e = 0.37x_{max}$ if $x_{max}/e \ge x_c$ and at $u = x_c$ if $x_{max}/e < x_c$. Which of these two inequalities is likely to hold? Let d_{min} denote the distance between cars at maximum density and d_c the distance between cars at the critical density, x_c. Then

$$\frac{x_{max}}{x_c} = \frac{d_c + L}{d_{min} + L}, \qquad (3.47)$$

---

[4]You may wish to review Section 2.3 before proceeding.

where L is the length of a car. Cars move if they are more than a fraction of a carlength apart. Let this fraction be ϵ (to be determined). Then

$$d_{\min} = \epsilon L, \tag{3.48}$$

where $0 < \epsilon < 1$. On the other hand, many state laws declare that for each 10 m.p.h. of speed you should be at least a carlength behind the car in front of you. Let's suppose that this applies to the speed limit. Then $d_c > v_{\max}L/10$. Hence, on using (3.47) and (3.48), the maximum of $J(u)$ depends upon whether

$$\frac{\frac{v_{\max}}{10} + 1}{\epsilon + 1} > e = 2.718. \tag{3.49}$$

Because $\epsilon < 1$, (3.49) is sure to be satisfied if $v_{\max}$ is 45 m.p.h. or greater. In practice, $v_{\max} \geq 45$ is very likely (even if the official speed limit is lower); and ϵ is probably quite close to zero, making (3.49) valid even for much lower values of $v_{\max}$ than 45. Let us therefore assume that $x_{\max}/e > x_c$. Then the optimal density is $u = u^*$, where

$$u^* = \frac{x_{\max}}{e}; \tag{3.50}$$

while the maximum traffic flux, or tunnel capacity, is

$$J(u^*) = Vu^*. \tag{3.51}$$

Thus V, as defined by (3.46), is the optimal speed of traffic through the tunnel.

How would this optimal solution be implemented? In practice, one would control the optimal "throughput" of traffic itself, by placing a stop light at the start of the tunnel that limited influx to $Vu^*/60$ cars per minute. According to data presented by Prigogine and Herman (1971, p. 9), the optimal density for New York's tunnels appears to be about 65 cars per mile, and the optimal speed is about 20 m.p.h. (though it varies slightly from tunnel to tunnel, the newer ones being more efficient.) Thus according to (3.51), the capacity of a New York tunnel is about 1300 cars per hour. This optimal traffic flux can be achieved by using a stop light to allow only about 22 cars (per lane) per minute to enter the tunnel. Gazis (1972, p. 422) reports that, starting in 1959, this very strategy was used to increase the throughput of the Lincoln tunnel. According to Gazis, the strategy worked very well for part of the time, but it was difficult to reset the (optimal) equilibrium after perturbations due to any sudden increase in congestion.

What does this mean? Is this a good model because it sometimes mimics reality well, or a bad model because it sometimes fails to mimic reality at all? You should mull this question over. We'll return to it in Chapter 4.

3.6 How Much Pesticide Should a Crop Grower Use—and When?

Crops are afflicted by insects, and the damage they cause is an economic loss for the grower. Using pesticides helps to reduce this loss, but it is also expensive. Let C_d denote the cost of damage and C_s the cost of spraying. If a grower uses a lot of pesticide, then C_d will be low but C_s will be high. If the grower uses little pesticide, then C_s will be low but C_d will be high. Then how much pesticide should a grower use if he wishes to minimize his total loss, $L = C_s + C_d$? There is also the question of when to apply the pesticide. Too early an application will mean that not enough insects are yet available for slaughter; too late an application will mean that most of the damage has already been done. In both these ways, the grower of a crop is faced with an optimal control problem. In this section, we will try to help him solve it.

Let the concentration of pesticide when a crop is sprayed at time t, denoted by $u(t)$, be the control function. A grower requires a given amount of (diluted) mixture to spray a given area of cropland; therefore the concentration is proportional to the amount of pesticide used. As usual, to make the problem tractable, we will assume that the optimal control is of a particular kind. We'll assume that the grower sprays his crop only once per season, at time s, using pesticide with concentration U; and that any effect on the insects is immediate. We can represent this control mathematically by writing

$$u(t) = \begin{cases} U, & \text{if } t = s; \\ 0, & \text{if } t \neq s, \end{cases} \qquad 0 \leq t \leq T, \qquad (3.52)$$

where the growing season extends from time $t = 0$ to time $t = T$. In practice, of course, the pesticide may take some time to work; but as long as this time is sufficiently short when compared to the length of the growing season, then its effect may be regarded as immediate. According to Plant (1984, p. 5), one application of pesticide is frequently all that is needed in the San Joaquin Valley cotton fields, so that (3.52) is not unreasonable. Then the control $\{u(t): 0 \leq t \leq T\}$ is completely determined once s and U have been determined; i.e., we may write the loss (negative utility) as

$$L(u(t)) = f(s, U), \qquad (3.53)$$

where f is a function of two variables (see (3.63) below). You should compare (3.52) with the form of control assumed in Section 3.4. Here, the optimal policy depends both on the timing of an action (spraying) and its intensity (concentration of pesticide); there, we took the intensity of the action (rate of felling) to be infinite, so that the policy then depended only on the timing.

We will assume that the crop is afflicted by a single pest, that the native population of these insects grows at the constant specific rate R, and that its growth is boosted by immigration at constant rate q per unit of time. Hence if $x(t)$ denotes the magnitude of the insect population at time t, then (Chapter 1, Exercise 1.16):

$$x(t) = \left(x(0) + \frac{q}{R}\right) e^{Rt} - \frac{q}{R}, \qquad 0 \le t < s. \tag{3.54}$$

We will denote the limit of this expression as t tends to s by $x_-(s)$. Thus

$$x_-(s) = x(0)e^{Rs} + \frac{q}{R}\left(e^{Rs} - 1\right). \tag{3.55}$$

Let S denote the proportion of this population that survives the pesticide application at time s. From Exercise 1.13, it seems reasonable to assume that there exists a threshold U_0 such that

$$S(U) = \begin{cases} 1, & \text{if } U < U_0; \\ e^{-\alpha(U-U_0)}, & \text{if } U \ge U_0; \end{cases} \tag{3.56}$$

where α is a constant. Then, because we have assumed that the effect of the pesticide is immediate, the magnitude of the insect population immediately after the crop has been sprayed is

$$x(s) = S(U)x_-(s); \tag{3.57}$$

and the growth of that population for the remainder of the season is given by

$$x(t) = \left(x(s) + \frac{q}{R}\right) e^{R(t-s)} - \frac{q}{R}, \qquad s \le t \le T, \tag{3.58}$$

on using Exercise 1.16 again.

The crop grower's objective, you'll recall, is to minimize his total loss, $L = C_s + C_d$. Let c_0 be the fixed cost of spraying the crop (cost of renting an aircraft, etc.) and c_1 the additional cost of purchasing and applying each unit of pesticide. Then the cost of spraying is

$$C_s = c_0 + c_1 U. \tag{3.59}$$

Let $D(t)$ denote the number of units of crop (say tons of cotton) destroyed by the pest per unit of time. Let p be the price, per unit of crop, that the grower obtains at the end of the season. Then the cost of the damage is the loss of revenue,

$$C_d = p \int_0^T D(t)dt. \tag{3.60}$$

Crop destruction increases with the number of pests. The simplest hypothesis consistent with this observation is that

$$D(t) = bx(t), \tag{3.61}$$

where b is a constant. Hence, on using (3.59)–(3.61), the loss $L = C_s + C_d$ is given by

$$f(s, U) = c_0 + c_1 U + pb \int_0^T x(t)dt, \qquad 0 \leq s \leq T, 0 < U \leq U_{max}, \quad (3.62)$$

where U_{max} is the maximum concentration allowed by law. On using (3.54)–(3.58) we obtain (Exercise 3.10):

$$f(s, U) = c_0 + c_1 U + pb \left[\frac{e^{Rs} - 1}{R} \left(x(0) + \frac{q}{R} \right) \right.$$
$$\left. + \frac{e^{R(T-s)} - 1}{R} \left\{ S(U) \left(x(0) + \frac{q}{R} \right) e^{Rs} + \frac{q}{R}(1 - S(U)) \right\} - \frac{qT}{R} \right],$$
$$0 \leq s \leq T, 0 < U \leq U_{max}. \quad (3.63)$$

If $u = 0$ then there is no fixed cost ($C_s = 0$) and the total cost is simply C_d, the cost of damage to the crop. Hence

$$f(s, 0) = pb \int_0^T x(t)dt = pb \left\{ \frac{e^{RT} - 1}{R} \left(x(0) + \frac{q}{R} \right) - \frac{qT}{R} \right\}; \quad (3.64)$$

naturally, this quantity is independent of s. Now, the pesticide has no effect if $0 < U \leq U_0$, and so the grower will either apply no pesticide (with cost (3.64)) or apply pesticide at time s^* with concentration U^*, where (s^*, U^*) denotes the point in the "rectangle" $0 \leq s \leq T, U_0 < U \leq U_{max}$ where $f(s, U)$ is least. The grower will adopt the second strategy only if

$$f(s^*, U^*) < f(s, 0). \quad (3.65)$$

Finding the minimum of (3.63) in the given rectangle is a calculus problem (Exercise 3.11). Combining its results with the remarks above, we find that if the grower sprays, then the optimal spraying time is given by

$$s^* = \begin{cases} 0, & \text{if } q \leq q_c; \\ \frac{1}{2}T - \frac{1}{2R} \ln \left(1 + \frac{Rx(0)}{q} \right), & \text{if } q \geq q_c; \end{cases} \quad (3.66)$$

where

$$q_c \equiv \frac{Rx(0)}{e^{RT} - 1}. \quad (3.67)$$

There is therefore a critical immigration rate, q_c, below which the native population so dominates the immigrant one that the optimal strategy is to eradicate as much as possible of the native population at the very beginning of the season. Above the critical rate, the effect of the immigrants is strong enough to delay the optimal spraying time, but never by more than half the length of the growing season.

Stating the optimal concentration is somewhat easier if we first define the (continuous) function

$$
\Delta(q) \equiv
\begin{cases}
pb \dfrac{e^{RT} - 1}{R} x(0), & \text{if } q \leq q_c; \\[3ex]
pb \left(\dfrac{\sqrt{q + Rx(0)} e^{RT/2} - \sqrt{q}}{R} \right)^2, & \text{if } q > q_c.
\end{cases}
\tag{3.68}
$$

The quantity Δ is best regarded as an alternative measure of the cost of damage that would occur if no pesticide were applied. Indeed it agrees with (3.64) if $q = 0$. The grower should not spray at all if (3.65) is violated, i.e., if $W \geq 0$, where

$$
W(q) \equiv U_0 + \frac{\alpha}{c_1} \{c_0 - \Delta(q)\} + \ln \left(\frac{\alpha \Delta(q)}{c_1} \right) + 1;
\tag{3.69}
$$

but if (3.65) is satisfied, i.e., if

$$
W(q) < 0,
\tag{3.70}
$$

then the optimal concentration is

$$
U^* =
\begin{cases}
U_0 + \dfrac{1}{\alpha} \ln \left(\dfrac{\alpha \Delta}{c_1} \right), & \text{if } 1 < \dfrac{\alpha \Delta}{c_1} < e^{(U_{max} - U_0)} \\[3ex]
U_{max}, & \text{if } \dfrac{\alpha \Delta}{c_1} \geq e^{(U_{max} - U_0)}.
\end{cases}
\tag{3.71}
$$

Note in particular that the grower should not spray if $\alpha \Delta / c_1 \leq 1$, for which there is a simple economic interpretation. Recall from Exercise 1.13 that α is the fraction of insects killed by a unit increase in concentration. Thus $\alpha \Delta$ is the fraction of the uncontrolled damage avoided by the unit increase. The cost of the unit increase is c_1, and so it is clearly uneconomic to spray unless $c_1 < \alpha \Delta$.

Do (3.66) and (3.71) correspond to what growers actually do? We'll return to that question in Chapter 4. For now, observe that the utility function (negative of the loss function) for the problem we have solved may be written

$$
J(u) = \int_0^T g(x(t), u(t), t) dt,
\tag{3.72}
$$

where

$$
g(x(t), u(t), t) = g(x(t), s, U) = -pbx(t) - \frac{c_0 + c_1 U}{T}.
\tag{3.73}
$$

(Note that g does not depend upon t explicitly here—the dependence on s is part of the dependence on u, not on t—but it does in Section 3.7 below; see (3.80).) The form of utility appearing in (3.72) is extremely common, so common that we give $x(t)$ a name: we call it the *state* of the system at time t. The state of the system for the crop-spraying problem was the magnitude of the insect population. Convince yourself that J really is just a function of u. It appears at first that J should depend on x, but observe from (3.54)–(3.58) that $x(t)$ is completely determined once (3.52) is specified. In other words, $x(t)$ is a function of $u(t)$. Then why do we bother to record explicitly the dependence of J upon x? The answer is that the dependence of $x(t)$ upon $u(t)$ may be known only implicitly. For example, x may depend upon u by being constrained to satisfy a differential equation; and very few differential equations can be solved explicitly, as we have already discovered in Chapter 1.

The following example should help to clarify this point (even though we are going to ensure that the differential equation is solvable in this particular case by arguing that the set of potentially optimal controls is a very limited one).

3.7 How Many Boats in a Fishing Fleet Should Be Operational?

Consider a vast expanse of ocean roamed by a species of edible fish, and let $x(t)$ denote the magnitude of its population at time t. Suppose that all vessels that fish in this territory belong to a single owner, who pays the wages of all the fishermen. Let $u(t)$ be the number of boats at sea at time t, w a fisherman's average wage (per unit of time), c_B the overhead cost per boat (per unit of time), and n the number of crew per boat, assumed the same for each vessel. To fix ideas, you might like to imagine that the sole owner is a communist government; though, as we shall see in due course, the model of this section is also applicable to a capitalist economy. At any rate, the question we pose is this: how many boats should be operational at any given time?

Assume that the owner is a profit maximizer. Let p denote the selling price per unit of fish. Let $h(t)$ denote the harvest rate, i.e., the amount of fish caught per unit of time. Then the owner's revenue per unit of time is $ph(t)$, and the owner's costs per unit of time are $c_B u(t) + wnu(t) = cu(t)$, where

$$c \equiv c_B + nw \tag{3.74}$$

is the owner's cost per boat. Then, at time t, the owner's rate of profit is apparently

$$P(u(t)) = ph(t) - cu(t). \tag{3.75}$$

As explained in Section 3.4, however, the true profit rate at time t from now, when viewed from the present moment and in the light of alternative investment possibilities, is

$$e^{-\delta t}(ph(t) - cu(t)), \tag{3.76}$$

where δ is the interest rate. We shall take it for granted that δ is the real interest rate, as defined in Section 3.4, and that p, w, and c_B are constant (in real terms). Then, with $u(t)$ boats operational at time t, the owner's true return from now until eternity is

$$\int_0^\infty e^{-\delta t}(ph(t) - cu(t))dt. \tag{3.77}$$

By fixing the upper integration limit at eternity, we assume that the owner is very far-sighted. We also make the mathematics easier.

Now suppose that the fish are preyed upon solely by the human species, so that we may regard each vessel as a predator; then the number of active predators is $u(t)$. From Exercise 1.15, it is thus fair to assume that

$$h(t) = qu(t)x(t) \tag{3.78}$$

and that the growth and decay of the fish population is governed by

$$\frac{dx}{dt} = Rx(t)\left(1 - \frac{x(t)}{K}\right) - h(t). \tag{3.79}$$

In the scientific literature, the constant q is known as the *catchability*, and from Section 1.9, we interpret R as the maximum specific growth rate in the absence of harvesting and K as the capacity of the expanse of ocean. On substituting (3.78) into (3.77) and (3.79), the owner's control problem is to find

$$u(t), \qquad 0 \le t < \infty,$$

which maximizes

$$J(u(t)) = \int_0^\infty e^{-\delta t}u(t)\{pqx(t) - c\}\, dt \tag{3.80}$$

subject to the differential equation constraint

$$\frac{dx}{dt} = Rx(t)\left(1 - \frac{qu(t)}{R} - \frac{x(t)}{K}\right). \tag{3.81}$$

Notice that (3.80) is the special case of (3.72) for which $g(x(t), u(t), t) \equiv e^{-\delta t}u(t)\{pqx(t) - c\}$.

To make this control problem tractable, we will further assume that the owner has a special reason for wanting to solve it; namely, that the number of fish has dropped to such a low level that the species might even be in danger of extinction. To be specific, suppose it is somehow known that

$$x(0) = \frac{K}{N}, \tag{3.82}$$

where N is large. The cause of this is quite clear to the owner: the number of boats in operation, $u(0)$—and hence the level of predation—is too high. The owner's vast experience suggests that the best thing to do would be to recall the entire fleet for s units of time, so that the fish have a chance to recuperate, and then maintain the fleet forevermore at the constant level U ($< u(0)$), in such a way that the fish population remains in equilibrium. In other words, it is being assumed that the optimal control is of the form

$$u(t) = \begin{cases} 0, & \text{if } 0 < t \le s; \\ U, & \text{if } s < t < \infty; \end{cases} \tag{3.83}$$

where s and U are to be determined. Under these assumptions, it follows from Exercises 1.10 and 1.15 that (Exercise 3.12)

$$x(t) = \begin{cases} \dfrac{K}{1 + (N-1)e^{-Rt}}, & \text{if } 0 \le t \le s; \\ K\left(1 - \dfrac{qU}{R}\right), & \text{if } s < t < \infty. \end{cases} \tag{3.84}$$

The population varies continuously; therefore letting t tend to s gives

$$\frac{1}{1 + (N-1)e^{-Rs}} = 1 - \frac{qU}{R} (> 0). \tag{3.85}$$

Thus s is a function of U, defined implicitly by (3.85) and explicitly by

$$s(U) = \frac{1}{R} \ln \left\{ (N-1)\left(\frac{R}{qU} - 1\right) \right\}. \tag{3.86}$$

We are therefore entitled to write $J(u(t)) = f(U)$ where, on using (3.83) in (3.80),

$$f(U) \equiv \frac{pqKU}{\delta} e^{-\delta s(U)} \left\{ 1 - \frac{Uq}{R} - b \right\}, \tag{3.87}$$

and where

$$b \equiv \frac{c}{pqK} \tag{3.88}$$

is a dimensionless parameter (Exercise 3.15), which is best regarded as a lower bound on the cost/price ratio of a fishing vessel (being the cost/price ratio that would take effect if the fish population could be maintained artificially at its capacity K despite the harvesting). Clearly $b < 1$; if not, the owner would do better to abandon fishing in favor of some more profitable enterprise. More precisely, the owner will abandon fishing unless $f(U) > 0$ or $1 - b > Uq/R$.

It's a simple matter of the calculus (Exercise 3.13) to show that the maximum of (3.87) on the interval $0 < U < R(1 - b)/q$ occurs at $U = U^*$, where

$$U^* = \frac{R}{4q} \left\{ 3 - b + \frac{\delta}{R} - \sqrt{\left(1 + b - \frac{\delta}{R}\right)^2 + \frac{8b\delta}{R}} \right\}. \tag{3.89}$$

The optimal recovery time $s^* = s(U^*)$ is given by (3.86).

What is the significance of (3.89) from the economic viewpoint? The fact that U^* is the optimal number of operational boats means that $U^* - 1/2$ is too few, and $U^* + 1/2$ too many. Note that, in our context, it is perfectly legitimate to have half a boat, because we were careful enough to define $u(t)$ as the number of boats operational at time t. Thus a boat that spends half its time at sea and half in port would count as an operational half-boat, and similarly for other fractions. Were this not the case, we should hardly be justified in using a noninteger control function.

Let us therefore consider what the effect would be of increasing U from $U = U^* - 1/2$ to $U = U^* + 1/2$. In the short term, harvesting would begin sooner, because $s'(U) < 0$, as you can see directly from (3.86). Thus the owner's profit would be increased by

$$\int_{s(U^* + 1/2)}^{s(U^* - 1/2)} P(u(t))dt, \tag{3.90}$$

where P is defined by (3.75). This expression is approximately equal to

$$P(U^*)\{s(U^* - 1/2) - s(U^* + 1/2)\} \approx -P(U^*)s'(U^*). \tag{3.91}$$

On the other hand, because straightforward algebraic manipulation reveals that $P'(U^*) < 0$ (remember that $b < 1$), the additional boat has the long-term effect of reducing the profit rate by

$$P(U^* - 1/2) - P(U^* + 1/2) \approx -P'(U^*). \tag{3.92}$$

Thus the true worth of the long-term loss of profit is approximately[5]

$$-\int_0^\infty e^{-\delta t}P'(U^*)dt = -\frac{P'(U^*)}{\delta}. \tag{3.93}$$

But by differentiating (3.87) with respect to U we obtain

$$f'(U) = e^{-\delta s(U)}\left\{\frac{P'(U)}{\delta} - P(U)s'(U)\right\}. \tag{3.94}$$

Thus, because $f'(U^*) = 0$, we learn that U^* is the optimal number of operational boats because the short-term gain from slightly increasing that number would be exactly cancelled by the long-term loss, in much the same way as you stop eating candy when the short-term gain in pleasure of taste in the mouth is exactly cancelled by the long-term loss in pleasure of looking at your figure in the mirror.

Arguments akin to the one we have just used are very popular with economists. They provide what is known as a *marginal* interpretation. The importance of these arguments stems from the fact that most people make

---

[5]See Supplementary Note 3.2.

decisions by balancing pros against cons, rather than by consciously maximizing some utility function. Thus, even though the results are equivalent, the marginal derivation may come closer to mimicking what people actually do than the maximization of $J(u)$. In particular, there is an important marginal interpretation of equation (3.40) of Section 3.4 (see Exercise 3.14). There is also an alternative marginal interpretation of (3.89) in terms of the stock level, rather than the number of boats (Exercise 3.17).

How good is our model—would fishermen use it? One technical matter can be disposed of immediately. We assumed above that the optimal control function that maximizes (3.80) subject to the constraint (3.81) and initial condition (3.82) has the form (3.83). How do we know that this is true? The answer is that more advanced mathematics (see Clark, 1990, Chapter 2) can be used to show, *without assuming* (3.83), that the solution of the optimal control problem (3.80)–(3.81) is

$$
\begin{aligned}
u(t) &= U_N, && \text{if } 0 \le t \le s, \\
&= U^*, && \text{if } s < t < \infty,
\end{aligned}
\tag{3.95}
$$

where U^* is defined by (3.89), s is defined as the soonest time for which $x(s) = U^*$ when s is governed by (3.81) and (3.82), and U_N is defined by

$$
U_N \equiv
\begin{cases}
0, & \text{if } N > \left(1 - \dfrac{qU^*}{R}\right)^{-1}, \\[2ex]
U_{\max} & \text{if } N < \left(1 - \dfrac{qU^*}{R}\right)^{-1}.
\end{cases}
\tag{3.96}
$$

Here N is simply $K/x(0)$, as in (3.82), and $U_{\max}$ denotes the maximum number of boats that can be made operational. Thus the correctness of our (mathematical) assumption is assured by the fact that N is large. But the question remains—how good is this model?

We've asked this question so many times that we really ought to provide some answers! Which brings us nicely to Chapter 4.

Exercises

3.1 Verify (3.6) and (3.7).

*3.2 In Section 3.2, suppose that at least one faculty member had a factor classification of weak or inadequate. Would **M** still be a 3 × 3 matrix? Why or why not? Don't jump to an apparently obvious conclusion—it may be a wrong one!

3.3 Obtain (3.14).

3.4 (i) Verify that $\epsilon_U(\lambda)$ and $\epsilon_L(\lambda)$ in (3.23) satisfy $\epsilon_U(\lambda) > \epsilon_L(\lambda)$.
(ii) Verify that (3.25) satisfies $u_j > 0$ for $j = 1, 2, 3$.
(iii) Verify that (3.24) and (3.25) satisfy all other constraints.
(iv) By using a linear programming package such as the IMSL routine ZX3LP to solve (3.19) for various values of ϵ and λ, demonstrate the validity of the optimal control policy implied by (3.24) and (3.25). A rigorous proof of its validity would require the theory of linear programming.

*3.5 (i) Suppose the chairperson of the Mathematics Department in Section 3.2 considered that the faculty's research effort was about right but wished to encourage more devotion to teaching. What would be the chairperson's problem in these circumstances? Use a package such as ZX3LP to find the optimal controls for various values of ϵ and λ. Can you conjecture a more general formula for the optimal policy, corresponding to (3.23)–(3.25)?
(ii) Suppose the chairperson felt his department's problem was that too much effort was devoted to service and wished to discourage this. What would be the chairperson's problem in these circumstances? Use a package such as ZX3LP to find the optimal controls for various values of ϵ and λ. Can you conjecture a more general formula for the optimal policy, corresponding to (3.23)–(3.25)?
(iii) Do you think that a fairer policy would be obtained if the minimum salary constraint (3.16) were eliminated? Why or why not?

*3.6 What constraint have we overlooked in Section 3.3? Modify (3.28) accordingly.

3.7 (i) Verify (3.40), (3.41), (3.43), and (3.44). Show that (3.41) is satisfied when $V(t)$ is given by (3.42).
(ii) Verify that the form of (3.40) would be unchanged if we worked instead with nominal value and interest rate.

3.8 Using the shift technique of Chapter 2, show that the optimal forest rotation time (3.44) is a decreasing function of interest rate. Verify this graphically. The significance of this will be discussed in Chapter 4.

*3.9 The following net stumpage values, i.e., values of $V(t) = pN\xi(t) - c$, for a "typical" stand of British Columbia Douglas fir trees are quoted by Clark (1976, p. 261) as due to P. Pearse; t is measured in years, V in dollars.

t	$V(t)$	t	$V(t)$	t	$V(t)$
30	0	70	497	100	913
40	43	80	650	110	1000
50	143	90	805	120	1075
60	303				

(i) Plot V against t. What do your results tell you about an assumption we made in Section 3.4?

(ii) Using a method suggested by Section 1.9, estimate $V(t)$ and $V'(t)$ for $t = 35, 45, \ldots, 115$.

(iii) Plot θ against t, where θ is defined by

$$\theta(t) \equiv \frac{V(t)}{V'(t)}.$$

Use your graph to estimate the optimal rotation time for Douglas fir trees, according to the model of Section 3.4, when the real interest rate is $\delta = 0.05$.

3.10 Obtain the expression (3.63).

3.11 Obtain the minimum of (3.63) on the open rectangle $0 \le s \le T, U_0 < U \le U_{max}$. Hence verify (3.66) and (3.71).

3.12 Verify (3.84).

3.13 Show that the maximum of (3.87) on the interval $0 < U < R(1 - b)/q$ occurs at (3.89).

*3.14 In this exercise, you will attempt to provide a marginal interpretation of equation (3.40).

(i) A forester cuts his forest at time $t = s$ and immediately invests the proceeds (real value $V(s)$) in a bank. What is his gain after one unit of time if the real interest rate is δ?

(ii) Assuming that δ is small, write down a first approximation to the expression you obtained in (i).

(iii) What additional proceeds would the forester have accrued during the interval from $t = s$ to $t = s + 1$ if he had not cut his forest at $t = s$?

(iv) Write down a first approximation to the expression you obtained in (iii).

(v) Hence show that the optimal cutting time, according to the model of Section 3.4, is when the incremental loss from failing to let the trees grow a little longer is exactly balanced by the incremental gain from alternative investment.

3.15 What are the dimensions of catchability? Verify that the constant b defined by (3.88) is dimensionless.

*3.16 If $\delta = 0$, so that the possibility of alternative investment does not arise, then the integral (3.77) does not converge. In these circumstances, we would expect the owner to maximize the profit rate $P(u(t)) = \{pqx(t) - c\} u(t)$. Throughout this problem, assume that the fish population is maintained in a static equilibrium $x(U)$ by using a constant number U of operational

boats. Thus the profit rate is $P(U) = U\{pqx(U) - c\}$, and the harvest rate is $qUx(U)$. Call $P(U)$ the *sustained profit rate* and $qUx(U)$ the *sustained yield*.
(i) What is the optimal value of U?
(ii) What is then the equilibrium population?
(iii) Suppose that, for political reasons, it became expedient to maximize the sustained yield instead of the sustained profit rate. What would then be the optimal value of U?
(iv) What would then be the equilibrium population?
(v) If you were concerned about the preservation of the species of fish, which objective would you prefer, maximization of sustained profit or maximization of sustained yield?
(vi) Now suppose that $\delta \neq 0$. There are still two possible objectives, maximization of profit (from now until eternity) and maximization of sustained yield. Show that the objective that is more conducive to preserving the species of fish is now determined by the relative magnitudes of b and δ/R.
(vii) Suppose that a government is concerned about preserving the species but has no control over the interest rate that is used to determine the optimal fishing strategy. What should it do?

*3.17 (i) In any fishing equilibrium, whether optimal or not, the fish population is related to the number of boats by $x = X \equiv K(1 - qU/R)$. Hence the optimal control problem of Section 3.7 could have been formulated with X, not U, as the independent variable. Obtain the expression $\phi(X)$ that then replaces (3.87) and show that it has a maximum at $X = X^*$ where

$$X^* = \frac{K}{4}\left\{1 + b - \frac{\delta}{R} + \sqrt{\left(1 + b - \frac{\delta}{R}\right)^2 + \frac{8b\delta}{R}}\right\}. \tag{a}$$

(ii) Show that the cost of harvesting a unit of fish is c/qx when the population is x.
(iii) Hence provide a marginal interpretation for Eq. (a) *Hint*: Write the optimality condition in the form

$$\frac{pR}{\delta}\left(1 - \frac{2X^*}{K} + b\right) = p - \frac{c}{qX^*}$$

and find an expression for the sustained profit rate as a function of X.

*3.18 A dealer has an approximately constant daily demand for one of his products, of which he sells quantity Q per year. He orders the product from a manufacturer in batches of size u, and he wishes to determine the batch size u^* that will minimize his costs. These costs arise from two sources, a set-up cost of c_1 per order and a storage cost of c_2 per unit stock per year.
(i) Show that the average stock held between orders is $u/2$.
(ii) Deduce the total yearly cost.
(iii) Find u^* and show that the minimum total cost is $\sqrt{2Qc_1c_2}$.

Hint: Make life easy for yourself by assuming that the last item is sold around closing time one day and that the new consignment arrives before opening time the following day.

Note: This model is not especially useful because products for which costs arise principally from these two sources are unlikely to have such a regular demand. Probabilistic models for optimal ordering strategies are introduced in Chapter 7.

Supplementary Notes

3.1 Although the crude definition (3.32) will be adequate for the purposes of this section, it is possible to proceed more rigorously. Let us define the function D by

$$D(t-s;\epsilon) \equiv \begin{cases} 0 & \text{if } t-s < -\frac{\epsilon}{2} \\ \frac{1}{\epsilon} & \text{if } -\frac{\epsilon}{2} \le t-s \le \frac{\epsilon}{2} \\ 0 & \text{if } t-s > \frac{\epsilon}{2} \end{cases} = \begin{cases} 0 & \text{if } t < s - \frac{\epsilon}{2} \\ \frac{1}{\epsilon} & \text{if } s - \frac{\epsilon}{2} \le t \le s + \frac{\epsilon}{2} \\ 0 & \text{if } t > s + \frac{\epsilon}{2}. \end{cases}$$

Then $D(t-s;\epsilon) = 0$ unless $-\epsilon/2 \le t \le \epsilon/2$ and $\int_{-\infty}^{\infty} D(t-s;\epsilon)dt = 1$; so that the control depicted in Fig. 3.3(a) is simply $N \cdot D(t-s;\epsilon)$. Moreover, the impulse control that is crudely defined by (3.32) and depicted in Fig. 3.3(b) is simply

$$N \cdot \delta(t-s) \equiv N \lim_{\epsilon \to 0} D(t-s;\epsilon),$$

provided that the limit exists. Strictly, the limit does not exist as an ordinary function, but mathematicians have been able to construct a meaningful calculus of so called *generalized functions*, according to which δ exists as the *Dirac delta function*, which has the property that

$$\delta(x) = 0 \quad \text{if } x \ne 0, \qquad \int_{-\infty}^{\infty} \delta(x)dx = 1.$$

For an alternative approach to the Dirac delta function, see Section 11.5.

3.2 The lower limit of integration on the left-hand side of (3.93) is the time at which harvesting begins. It is legitimate to call it $t = 0$ because (3.90) has not been discounted to the present. If you insist on calling $t = 0$ the present time, then the present value of the long-term loss of profit is approximately

$$-\int_{s(U^*)}^{\infty} e^{-\delta t}P'(U^*)dt = -\frac{P'(U^*)e^{-\delta s(U^*)}}{\delta},$$

and the present value of the owner's short-term gain is $e^{-\delta s(U^*)}$ times (3.91). When the two are equated, of course, the factor $e^{-\delta s(U^*)}$ cancels out.

II | VALIDATING A MODEL

4 VALIDATION: ACCEPT, IMPROVE, OR REJECT

It's a fundamental tenet of the philosophy of science that the truth of a model can never be proved; only disproved. Thus validating a model means proposing a test that the model, if false, would fail to pass. If a model does fail such a test, then we are bound to reject it. If it passes, however, then we propose a more stringent test; if it passes that one, too, then we propose still another one; and so on, until ... when? When do we stop? In theory, we could go on forever. Indeed, broadly speaking, the received wisdom of science consists of mathematical models for which no one has yet invented a test that results in failure. In practice, however, there is always an element of subjectivity about the decision to stop testing. As we'll see in the examples, it depends upon the purpose for which the model was constructed. For the time being, let's simply say that we will accept a model if it explains all the facts that we would like it to explain. Otherwise, we will reject it; or else improve it, then test it again. Later, when we come to discuss the difference between descriptive models and prescriptive ones, we will have to enlarge upon this simple philosophy, but by then we will be better prepared to appreciate the enlargement.

4.1 A Model of U.S. Population Growth

We have seen the cycle of validation at work already, in Section 1.9.[1] There, we proposed $x(t) = 3.9e^{0.3t}$ as a model for the magnitude of the U.S. population, on the strength of supporting data for the years 1790 to 1840. We invented a test for this model, namely, that it should predict the population levels recorded in later years. It failed this test miserably; so we rejected the model and proposed another one, namely, (1.58), in which $x(t) = 3.9e^{0.3t}$

---

[1]You may wish to review Section 1.9 before proceeding.

for $0 \leq t \leq 6.5$ but $x(t) = 31e^{0.24(t-7)}$ for $t > 6.5$. This provided an acceptable description of U.S. population growth for about a century, from 1790 until 1890, but still failed adequately to estimate the magnitude of the twentieth century U.S. population. So we rejected this model, too, and proposed yet another one, namely,

$$x(t) = \frac{198}{1 + 49.4e^{-0.31t}}. \tag{4.1}$$

This provided an acceptable description of U.S. population growth until the middle of the twentieth century; but it failed another test, for it could not predict the 1960 and 1970 population levels. Therefore, in Exercise 1.11, we improved the model to take account of the postwar population surge. A possible improved model is

$$x(t) = \begin{cases} \dfrac{198}{1 + 49.4e^{-0.31t}}, & \text{if } t < 15.5; \\ \dfrac{373}{1 + 1.48e^{-0.28(t-16)}}, & \text{if } t > 15.5. \end{cases} \tag{4.2}$$

This provides an acceptable description of the U.S. population growth from 1790 onward, one that is correct to at least one, mostly two, and occasionally three significant figures.

According to the 1991 *Britannica Yearbook*, the U.S. population was 251.4 million in 1990, i.e., to two significant figures, $x(20) = 250$. This agrees with the prediction of the model (4.2). If, therefore, we require only a rough description of U.S. population growth throughout the history of the nation, then the model (4.2) should be accepted. This does not mean that its prediction (to two significant figures) of 270 million people for the turn of the century is bound to be true; it simply means that we cannot prove it to be false. We'll just have to wait and see.

The model is certainly useful, even if we just regard it as a simplification of the actual data. But the real value of a model is often its ability to generate hypotheses. If (4.2) is true, and if the conceptual derivation of (1.62) is true, then there was a sudden jump around 1945 in the capacity of the land—or at least a jump in the general perception of what that capacity might be. Did the capacity suddenly jump from 200 million to almost twice that much? If so, why did this happen—was it due to improvements in technology, or was it simply due to postwar euphoria? These are questions you might like to discuss. For now, simply observe that an hypothesis emerges with greater impact when suggested by a mathematical model.

4.2 Cleaning Lake Ontario

According to the model of Section 1.1, if all pollution input to a lake of volume V suddenly ceases and if water flows out at rate r, then the existing pollution will take $t_{0.05} \approx 3V/r$ units of time to decay to 5% of its current level.[2] According to Rainey (1967, p. 1242), $V = 1636 \times 10^9$ cubic meters for Lake Ontario and $r = 572,562,432$ cubic meters per day, so that $t_{0.05} \approx 8572$ days or 23.5 years. Can we accept this figure as the cleaning time for Lake Ontario?

As stated at the beginning of this chapter, all we can do is attempt to reject it. This, as it happens, is a simple matter, for the prediction $t_{0.05} \approx 23.5$ is based on the assumption that all pollution inflow to Ontario has ceased. But the water outflow from Lake Erie is known to constitute five-sixths of the water inflow to Lake Ontario. Thus, even if pollution has ceased to enter the Great Lakes as a whole, pollution from Erie will continue to flow into Ontario. In these circumstances, we would expect the 5% cleaning time for Lake Ontario to be considerably longer than 23.5 years and to be an increasing function of the current pollution level for Lake Erie. These expectations can be confirmed by an improved model, which relaxes the assumption of zero pollution input. We develop this model in Section 9.3. For now, simply observe that the model of Section 1.1 has failed a test (the validity of one of its basic assumptions) and must therefore be rejected.

4.3   Plant Growth

It is sometimes necessary to generate a whole chain of predictions before arriving at one that can be tested.[3] Even when this is not absolutely necessary, it is often still convenient. Consider, for example, the plant growth model of Section 1.3. This predicts the S-shaped logistic growth curve of Fig. 1.1. In Section 3.4 we adapted this model to predict that the growth curve $\xi = \xi(t)$ of an individual tree in a forest would also be S-shaped. It wouldn't be difficult to compare this shape with those of tree growth curves that have actually been observed. But suppose that such data were not readily available—would that mean that the model could not be tested?

The answer is no for, as we saw in Section 3.4, the assumption that $\xi = \xi(t)$ is S-shaped leads to the further prediction that the real value $V(t) = Np\xi(t) - c$ of a forest will grow according to a curve like that depicted in Fig. 3.4; and perhaps the shape of this curve can be tested more easily. The graph in Fig. 4.1 depicts the actual growth of net stumpage value for a typical stand of Douglas fir trees in British Columbia. It is based on data given in Exercise 3.9 and ought to remind you of Fig. 3.4. As it happens, the dashed curve in Fig. 4.1 is identical to the solid curve in Fig. 3.4. In other words, Fig. 3.4 is just Fig. 4.1 with the coordinate scales removed. Thus, although we did not have the data to test it directly, the assumption that individual trees grow according to $\xi = \xi(t)$ has led to a prediction that is not inconsistent with actual data for stumpage value. We have proposed

---

[2]You may wish to review Section 1.1 before proceeding.
[3]You may wish to review Sections 1.3 and 3.4 before proceeding.

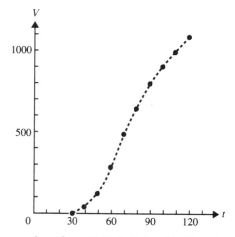

Fig. 4.1 Net stumpage values for a British Columbia Douglas fir. The dots denote Pearse's data points, quoted by Clark (1976, p. 261) and in Exercise 3.9.

a test, and the logistic growth model for trees has passed. We have therefore no reason to reject the model. But this doesn't mean, of course, that we won't be forced to reject it later on other grounds; and we note that many other forms of the function $\xi(t)$ might also have led to a stumpage value curve like that in Fig. 4.1.

4.4   The Speed of a Boat

In Sections 1.12 and 2.2, we constructed a dynamic model of the approach to equilibrium of a rowing boat.[4] The equilibrium speed of the boat was

$$V = 2 \left(\frac{P}{bS} \right)^{1/3},$$ (4.3)

where S denotes wetted surface area, P denotes the effective power supplied by an oarswoman, and b is a constant. In this section, we will test this model by comparing the course times it predicts for two different crews, a lightweight one and a heavyweight one, in a race of length d.

Now, in Exercise 1.19, we found that the rapidity with which the boat approached its equilibrium speed depended upon the dimensionless parameter

$$q \equiv \frac{M}{6bSd},$$ (4.4)

where M is the mass of the boat (plus crew). By setting $u = 0.99V$ in (1.78), for example, we find that the boat is within 1% of its equilibrium speed

---

[4]You may wish to review Sections 1.12 and 2.2 before proceeding.

after a time $8.495qd/V$. Because the course time would be d/V if the crew rowed the entire race at speed V, the results of that exercise show that, provided q is sufficiently small, it will be a good approximation to assume that the boat is in equilibrium throughout the entire race. Whether q is small or not will depend on the values of the parameters M, S, d, and b. The first three of these are either known or easily measured, but a value for b is more difficult to supply. Therefore, we are simply going to assume that a rowing boat reaches equilibrium after a small fraction of the course time has elapsed, on the grounds that this is what seems to happen whenever we watch a boat race. Indeed the best way to estimate the parameter b might well be to determine how long a boat takes to reach, say, 99% of equilibrium speed, deduce the value of q, and hence deduce the value of b.

Let t_L, t_H denote, respectively, the course times of a lightweight crew and a heavyweight one. Let V_L and V_H denote the speeds at which they row. Having made the assumption that q is small, and if our model is correct, then it is at least approximately true that $t_L = d/V_L$ and $t_H = d/V_H$; whence $t_H/t_L = V_L/V_H$. Using S_L to denote the wetted surface area of the lightweight boat, P_L the effective power supplied by a lightweight oarswoman throughout the race, and S_H, P_H to denote the corresponding quantities for a heavyweight crew, we deduce from (4.3) that

$$\frac{t_H}{t_L} = \left(\frac{S_H}{S_L}\right)^{1/3} \left(\frac{P_L}{P_H}\right)^{1/3}. \tag{4.5}$$

Now, according to McMahon (1971, p. 350), the course time of a heavyweight crew is about 5% faster than that for a lightweight one. If our model is correct, then the quantity on the right-hand side of (4.5) should have a value of about 0.95. We thus have a testable prediction. (McMahon's data was obtained for male crews, and here we consider a female crew; but that will not alter the substance of what is to follow.)

Before we can actually test our prediction, however, we require a value for P_L/P_H. Now, we do not have values for P_L and P_H. But McMahon does tell us that the average weight of a heavyweight crew member is about 86 kilograms, and that of a lightweight crew member about 73 kilograms. Let us therefore make the additional assumption that

EFFECTIVE POWER IS PROPORTIONAL TO WEIGHT.   (4.6)

Thus an oarswoman who is $100\alpha\%$ heavier than another oarswoman will supply $1 + \alpha$ times as much effective power. Now $P_L/P_H = 73/86$.

We also do not have values for S_L and S_H. Let us define

$$\lambda \equiv \frac{S_L}{S_H}. \tag{4.7}$$

The heavyweight boat will certainly have a greater wetted surface area than the lightweight one, because extra weight will lower the boat. If the heavyweight boat is larger than the lightweight one, however, then it will also displace a greater volume of water; and the additional upthrust will counterbalance the additional weight, so that S_H, though greater than S_L, will not exceed it by very much. This argument suggests that λ, though less than 1, is also close to 1, and experiment confirms this. On using (4.5)–(4.7), our model predicts that

$$\frac{t_H}{t_L} = \left(\frac{73}{86\lambda}\right)^{1/3} = \frac{0.9468}{\sqrt[3]{\lambda}}. \tag{4.8}$$

Thus, as long as the value of λ is between about 0.975 and 1, the value our model predicts for t_H/t_L will be 0.95 to two significant figures. We have therefore no reason to reject our model; it has passed the test we proposed for it.

Notice, however, that, had it failed, we would have been left to decide by other means whether the fault lay with the assumptions inherent in Sections 1.12 and 2.2 or with the additional assumption contained in (4.6).

4.5   The Extent of Bird Migration

In Section 3.1 we constructed a mathematical model that predicts that birds will migrate a distance

$$d = \frac{2}{3}\frac{\eta}{(2bSg^2 v_i^2)^{1/3}}\frac{E}{m^{2/3}}, \tag{4.9}$$

where E denotes energy stored for the flight and m the mass of the bird, g denotes acceleration due to gravity, η denotes efficiency at converting fat into mechanical energy, v_i denotes induced velocity, S denotes surface area facing the wind, and b is a drag constant.[5] The constant g is certainly the same for all birds. Moreover, it seems reasonable to assume that v_i, b, and S are also the same for all birds, at least for birds of similar species. Then, if subscript Z denotes species Z, our model predicts that

$$d_Z = C \cdot \frac{E_Z}{m_Z^{2/3}}, \tag{4.10}$$

where d_Z is the distance travelled by a member of that species, with initial energy E_Z and mass m_Z; and where C is independent of Z. This is a testable prediction.

According to Tucker (1971, p. 122), a hummingbird (species H) of mass around 5×10^{-3} kg flies 500 miles across the Gulf of Mexico; a Blackpoll warbler (species B) of mass about 20×10^{-3} kg flies 800 miles from New England to Bermuda; and a golden plover (species G) of mass around 0.2 kg flies 2400 miles from the Aleutian to the Hawaiian islands.

---

[5]You may wish to review Section 3.1 before proceeding.

Hence $d_B/d_H = 800/500 = 1.6$, $m_B/m_H = 20/5 = 4$, $d_G/d_B = 2400/800 = 3$, and $m_G/m_B = 0.2/(20 \times 10^{-3}) = 10$. Let's use this data to test our model.

From (4.10), our model predicts that

$$\frac{d_B}{d_H} = \frac{E_B}{E_H} \cdot \left(\frac{m_H}{m_B}\right)^{2/3} \tag{4.11}$$

We have values for d_B/d_H and m_B/m_H but not for E_B/E_H. Let's therefore make the additional assumption that stored fat is proportional to weight. Because weight $= mg$, from (1.74), and because g is the same for all birds, our additional assumption is entirely equivalent to

$$E \text{ is proportional to } m. \tag{4.12}$$

Thus a bird that is α times heavier than another bird will store α times as much energy as fat. Now our model predicts that $E_B/E_H = m_B/m_H$ and hence, on using (4.11), that

$$\frac{d_B}{d_H} = \left(\frac{m_B}{m_H}\right)^{1/3}. \tag{4.13}$$

According to our data, the left-hand side of (4.13) is 1.6 and the right-hand side is $4^{1/3} = 1.587$. This agreement is as good as we could wish for. On the other hand, when we use (4.10), our model also predicts that

$$\frac{d_G}{d_B} = \left(\frac{m_G}{m_B}\right)^{1/3}, \tag{4.14}$$

but our data do not support this prediction, because the cube root of 10 is about 2.15 and certainly not 3. A test of our model has resulted in failure. Which of the assumptions responsible for it is false?

Zoologists seem to be quite content with (4.12). Indeed Tucker (1971, p. 119) says that 25% of body mass is a reasonable estimate of the amount of fat that birds reserve as fuel for long flights. The problem therefore appears to rest with the assumptions we made in Section 3.1. Because $2.15 < 3$, a golden plover flies further than the model predicts. Is there, then, an effect we ignored in Section 3.1 that makes a bird into a more efficient traveller? There's an obvious answer to this question. We have neglected the effect of a following wind, and we must improve our control model to allow for this.

Let u, the control, now be the speed of a bird *relative to the air*, and let V be the following wind. Then distance travelled in time δt is

$$\delta z = (u + V)\delta t. \tag{4.15}$$

This result agrees with Section 3.1 when $V = 0$. From Section 1.12, resisting the drag force D for this distance requires mechanical energy $D(u+V)\delta t$. From aerodynamics, the drag on a bird is proportional to the surface area facing the wind and to the square of its velocity *relative to the wind*. Hence

$D = bSu^2$, as in Section 3.1; but u has a different meaning now (which, however, coincides with that of Section 3.1 when $V = 0$). The mechanical energy required to resist the drag in moving distance δz is therefore $bSu^2(u + V)\delta t$; of which $bSu^2 V\delta t$ is supplied by the wind and the remaining $bSu^3 \delta t$ must be supplied by the bird. The induced power, or rate at which mechanical energy must be supplied to keep the bird airborne, remains unchanged. From (3.8), it's $I = mgv_i$. Hence the rate at which the bird supplies energy, i.e., the effective power, must equal $bSu^3 + mgv_i$. The effective power is simply η times the rate at which the bird actually burns up its energy, and so we have

$$\eta \frac{E}{T} = bSu^3 + mgv_i, \tag{4.16}$$

where

$$T = \frac{d}{u + V} \tag{4.17}$$

is the flight time. The bird still wishes to maximize d; therefore we deduce from (4.16) and (4.17) that its optimal speed relative to the air will be the value of u that maximizes

$$J(u) \equiv \frac{\eta E(u + V)}{bSu^3 + mgv_i}. \tag{4.18}$$

This expression agrees with (3.5) when $V = 0$.

The mathematics of optimization will be simpler if we scale both the flight speed and the following wind with respect to the speed

$$u_0^* \equiv \left(\frac{mgv_i}{2bS} \right)^{1/3} \tag{4.19}$$

that would be optimal if the wind were zero; (4.19) is merely (3.9) with a subscript 0 added to denote zero wind speed. Let $\bar{u}$ denote dimensionless flight speed (relative to the wind) and ϵ dimensionless wind; i.e., define

$$\bar{u} = \frac{u}{u_0^*} \tag{4.20}$$

and

$$\epsilon = \frac{V}{u_0^*}. \tag{4.21}$$

Then we can replace (4.18) by

$$J(\bar{u}) = \left(\frac{\eta E}{bSu_0^{*2}} \right) \cdot \left(\frac{\bar{u} + \epsilon}{\bar{u}^3 + 2} \right). \tag{4.22}$$

It is a simple exercise (Exercise 4.1) to show that (4.22) has a maximum at $\bar{u} = \bar{u}^*$, where $\bar{u}^*(\epsilon)$ is the only positive solution of the equation

$$\bar{u}^2(2\bar{u} + 3\epsilon) = 2, \qquad (4.23)$$

the value of the maximum being $J(\bar{u}^*(\epsilon))$. Note that $\epsilon = 0$ and $\bar{u}^*(0) = 1$ correspond to Section 3.1. Because (4.23) can be solved numerically (by the Newton–Raphson technique) for any value of ϵ, the graph of $\bar{u}^*(\epsilon)$ is easily found and is sketched in Fig. 4.2, for $0 \le \epsilon \le 1$. Note that $\bar{u}^*(\epsilon) \le 1$; i.e., a following wind reduces the optimal flight speed. Can you prove this without solving (4.23) numerically? See Exercise 4.2.

The fact that a Blackpoll warbler migrates four times as far as a hummingbird has already been explained by our model. Let us therefore assume that no wind follows a hummingbird or a Blackpoll warbler, but that a golden plover is followed by a (dimensionless) wind ϵ_G. Then, assuming that (4.12) still holds but that d_G and d_B are given by our improved model, we deduce from (4.22) that

$$\frac{d_G}{d_B} = \left(\frac{m_G}{m_B}\right)^{1/3} \frac{\Delta(\epsilon_G)}{\Delta(0)} = 3\left(\frac{\Delta(\epsilon_G)}{0.4642}\right), \qquad (4.24)$$

where we have defined

$$\Delta(\epsilon) \equiv \frac{\bar{u}^*(\epsilon) + \epsilon}{\bar{u}^*(\epsilon)^3 + 2} \qquad (4.25)$$

and used the facts that $m_G/m_B = 10$ and $\Delta(0) = 1/3$. See Exercise 4.3.

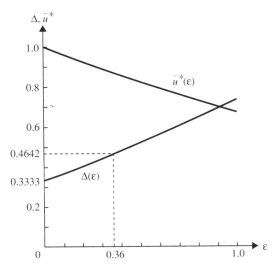

Fig. 4.2 Migrating birds travel slower (relative to the air) and further with a wind behind them. The diagram shows how the optimal airspeed and distance travelled vary with (dimensionless) wind speed.

The graph of $\Delta(\epsilon)$ is sketched in Fig. 4.2. Notice that $\Delta(\epsilon) \geq 1/3$. This is the mathematical equivalent of saying that a bird migrates further with a following wind. The function takes the value 0.4642 where ϵ is about 0.36. Thus, because $\bar{u}^*(0.36) \approx 0.85$, the model of this section would pass its test if golden plovers were habitually followed by a wind of just over 42% of their actual flight speed (relative to that wind). According to Tucker (1971, p. 122), small birds have been observed to migrate at 10 to 15 meters per second, with a tail wind of at least 5 meters per second; so that, at the very least, our model appears to be along the right lines.

But does it *actually* pass its test? To answer that question, you would have to acquire the wind-speed data. Perhaps this is a matter you would like to pursue. But if the data match the model's prediction, it does not mean that the model is true; it means only that we have no good reason to doubt its truth and should therefore accept it, at least until we have adequate data for a more stringent test. And if the data do not match the model's prediction, it does not necessarily mean that our model should be rejected completely; it may only mean that our model must be further improved to take account of spatial or temporal variations of wind speed, or indeed any other factor. See, for example, Exercises 4.20 and 4.21.

4.6   The Speed of Cars in a Tunnel

In Section 2.3, we used the conceptual approach to derive a model for the speed, $v(x)$ miles per hour, of cars through a tunnel when the traffic density is x cars per mile.[6] The model we obtained was

$$v(x) = \begin{cases} v_{max}, & \text{if } 0 \leq x \leq x_c; \\ v_{max}\dfrac{\ln(x_{max}/x)}{\ln(x_{max}/x_c)}, & \text{if } x_c < x \leq x_{max}. \end{cases} \tag{4.26}$$

Here x_c is the highest density at which drivers are free to move at the speed limit v_{max}, and x_{max} is the density that will just bring traffic to a halt.

We have already explained, in Chapter 2, how density is measured on an actual road. Data points obtained in this way from New York's Holland tunnel are depicted in Fig. 4.3. Think of the data point (x, v) as recording the average value of all speeds observed when the density was measured and found to be x. The dashed curve in the diagram is (4.26), with v_{max} about 55 m.p.h., x_c about 10 cars per mile, and x_{max} about 175 cars per mile. Observe that this curve fits the data quite well in the range $x_c < x \leq x_{max}$. (We do not have data for $0 \leq x \leq x_c$. The effect of this is considered toward the end of the section.)

We have raised the point, in Section 4.3, that a model may yield several successive generations of predictions, and that any of these may be tested.

---

[6]You may wish to review Sections 2.3 and 3.5 before proceeding.

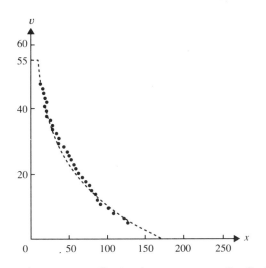

Fig. 4.3 Speed, v m.p.h., versus traffic density, x cars per mile, for New York's Holland tunnel. The 32 data points were obtained from a sample of over 20,000 cars. The dashed curve is (4.26). The sketch is adapted by permission of the publisher from Prigogine and Herman, *Kinetic Theory of Vehicular Traffic*, p. 6. (Copyright 1971 by Elsevier Science Publishing Co., Inc.)

In Section 4.3, for example, a test was applied to a second generation prediction, the growth in value of a forest. Viewed in this light, (4.26) is a first generation prediction, which, by virtue of Fig. 4.3, has passed a first generation test. But the same model has already yielded, in Section 3.5, the second generation prediction that the optimal speed for cars through a tunnel is $v_{max} \ln(x_{max}/x_c)$, the optimal flux being $x_{max}/e = 0.368 x_{max}$ times that speed. We already know from Chapter 3 that these predictions are compatible with data from the Lincoln tunnel. Thus the model has also passed a second generation test. Moreover, the fact that the model, as mentioned in Chapter 3, fails to mimic reality in the immediate aftermath of perturbations, due to any sudden congestion, is of no concern to us, because we cannot expect an equilibrium model to describe nonequilibrium conditions! We therefore conclude that the model provides an acceptable description of equilibrium traffic flow, the purpose for which it was designed.

For $x_c < x \le x_{max}$, the dashed curve in Fig. 4.3 is the curve of the form (4.26) which best fits the data, in the sense of Section 8.4. It is found that $x_{max} \approx 175$ and

$$\frac{v_{max}}{\ln(x_{max}/x_c)} \approx 19, \tag{4.27}$$

if x_{max} and the parameter on the left-hand side of (4.27) are the two parameters that are varied, to give the dashed curve its best fit property. In view of (3.46) and (3.51), it is also found that the optimal speed for cars through New York's Holland tunnel is about 19 m.p.h.; see Prigogine and Herman (1971, p. 9).

Because no data are available for $0 \leq x \leq x_c$, however, the appropriate values of v_{max} and x_c cannot be determined accurately. All we know is that they are constrained, so that (4.27) is satisfied and the horizontal section of the dashed curve lies above the leftmost data point; but within these constraints they are arbitrary. Thus the horizontal section of curve in Fig. 4.3 is purely speculative and should, perhaps, be lower. As we have already seen in Chapter 3, however, this is a matter of minor importance when using the model to determine optimal flux.

4.7 The Stability of Cars in a Tunnel

Sometimes, even without a conscious effort to compare your model with data, it will dawn on you that one of your predictions is somewhat dubious.[7] Improvements that emerge as the result of such doubts are the subject of this and the following section.

In Section 2.8 we developed a model for instability of traffic in a tunnel. We found that for any given amplitude of the initial disturbance Δ there is a critical value, $\tau_c(\Delta)$, which driver response time must not exceed if a platoon of cars is to stay in equilibrium. We found, moreover, that $\partial \tau_c / \partial \Delta < 0$ (see Fig. 2.12). This picture is surely in qualitative agreement with everyone's highway experience. But it seems that the return to equilibrium in Fig. 2.10 is almost too smooth to be true. If this is right, then something is wrong with the assumptions that led to Fig. 2.10; and hence with equation (1.81), i.e.,

$$\frac{d^2 z_j(t+\tau)}{dt^2} = -\lambda \frac{z_j'(t) - z_{j-1}'(t)}{|z_j(t) - z_{j-1}(t)|}. \tag{4.28}$$

Recall that the car with index j is following the car with index $j-1$. Thus the equation says two things. First, if two cars are moving toward one another then the closer they are, the harder the second one brakes. With this I agree. Second, if two cars are moving away from one another then the further away they are, the more gradually the second one accelerates. I suspect that this isn't true (though the follower will certainly accelerate if the leader is moving away).

A possible improvement of (4.28) would be to combine it with the model of Exercise 2.4 in the following way:

$$\frac{d^2 z_j(t+\tau)}{dt^2} = \begin{cases} -\lambda \dfrac{z_j'(t) - z_{j-1}'(t)}{|z_j(t) - z_{j-1}(t)|}, & \text{if } z_j'(t) > z_{j-1}'(t); \quad \text{(4.29a)} \\[3mm] -\mu \left\{ z_j'(t) - z_{j-1}'(t) \right\}, & \text{if } z_j'(t) < z_{j-1}'(t); \quad \text{(4.29b)} \end{cases}$$

---

[7]You may wish to review Section 2.8 before proceeding.

where λ and μ are constants. Thus, if a gap between cars is closing, the following driver responds both to the size of the gap and the rate at which it is closing; but if a gap between cars is widening, then the following driver responds only to the rate at which it is widening. Integrating, and choosing the constants of integration so that the equilibrium solution of (4.29) will satisfy $v(x_{max}) = 0$, we find (verify) that

$$\frac{dz_j(t+\tau)}{dt} = \begin{cases} \lambda \ln\left(x_{max}(z_{j-1}(t) - z_j(t))\right), & \text{if } z'_j(t) > z'_{j-1}(t); \\ \mu\left(z_{j-1}(t) - z_j(t) - \dfrac{1}{x_{max}}\right), & \text{if } z'_j(t) < z'_{j-1}(t). \end{cases} \tag{4.30}$$

Because (4.29a) and (4.29b) must yield the same equilibrium solution, we have

$$\lambda \ln\left(\frac{x_{max}}{x}\right) = v(x) = \mu\left(\frac{1}{x} - \frac{1}{x_{max}}\right). \tag{4.31}$$

Assuming, as in Section 2.8, that $x = x_{max}/e$ (the optimal density for maximum traffic flow) and denoting $v(x_{max}/e)$ simply by v, we determine the constants λ and μ to be

$$\lambda = v, \qquad \mu = \frac{vx_{max}}{e-1}, \tag{4.32}$$

where $e = \exp(1) = 2.718$. On writing $z_j(t)$ as the sum of the equilibrium displacement $\zeta_j(t)$, given by (2.97), and a perturbation displacement $\phi_j(t)$, substituting into (4.30), and using (4.32), we find (Exercise 4.4) that the differential-delay equations for the perturbation displacements are

$$\frac{d\phi_j(t+\tau)}{dt} = \begin{cases} v \ln\left\{1 + \dfrac{x_{max}}{e}(\phi_{j-1}(t) - \phi_j(t))\right\}, & \text{if } \phi'_j(t) > \phi'_{j-1}(t); \\ \dfrac{x_{max}v}{e-1}\left\{\phi_{j-1}(t) - \phi_j(t)\right\}, & \text{if } \phi'_j(t) < \phi'_{j-1}(t). \end{cases} \begin{matrix} (4.33) \\[12pt] (4.34) \end{matrix}$$

In view of Exercise 4.4(ii), the expression for $d\phi_j(t+\tau)/dt$ given by (4.33) exceeds that given by (4.34) if

$$-0.632 < \frac{x_{max}(\phi_{j-1} - \phi_j)}{e} < 0. \tag{4.35}$$

Equations (4.33)–(4.34) were discretized and solved for a platoon of 10 cars, as described in Section 2.8 for equations (2.105). I used the same perturbation displacement for the leading car as in Fig. 2.9 and the same parameter values $d = 65$, $v = 88/3$, and $x_{max}/e = 13/1056$ as were used in the earlier model. Thus (4.35) became

$$-51.3 < \phi_{j-1} - \phi_j < 0. \tag{4.36}$$

If you completed Exercise 2.19, then a simple modification of your computer program will enable you to do this for yourself. The results for $\Delta = 0.15$ and $\tau = 0.15$ are shown in Fig. 4.4 as the solid graphs. The

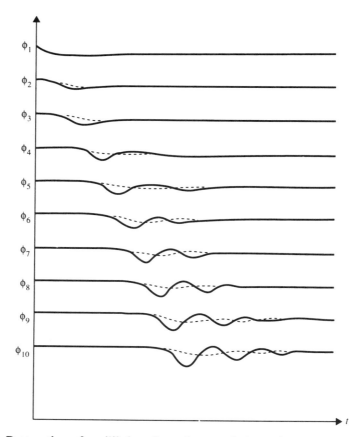

Fig. 4.4 Restoration of equilibrium for a platoon of 10 cars in a tunnel. The dashed curves are identical to those of Fig. 2.10. The solid curves are predicted by the model of (4.29). Scales are as in Fig. 2.10.

dashed ones, for comparison, are those of Fig. 2.10. The strong upward curvature of the solid curves during the accelerative phase is a symptom of (4.34), i.e., of acceleration being independent of spacing between cars. Notice that the solid curves always dip below the dashed ones when they first depart from them. The reason is that when driver j first reacts to driver $j-1$, $\phi_{j-1} < 0$ and $\phi_j = 0$, in such a way that (4.36) is satisfied. Then (4.33) yields a less negative slope than (4.34).

The results for $\Delta = 0.15$ and $\tau = 2.25$ are shown in Fig. 4.5 as the solid graphs. The dashed ones, for comparison, are those of Fig. 2.11. The eighth car collides with the seventh one after only 26.25 seconds (210 time units), much sooner than a collision occurs in Fig. 2.11. The effect introduced by (4.29b) is clearly a destabilizing one.

A selection of points is labeled in Fig. 4.6, according to whether the points are stable or unstable. As in Section 2.8, a square denotes stability,

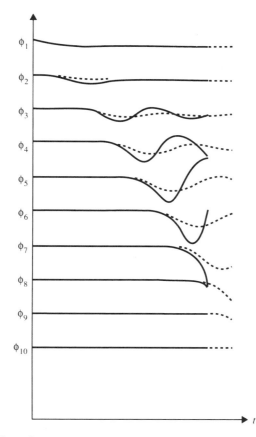

Fig. 4.5 Instability of a platoon of 10 cars when $\Delta = 0.15$ and $\tau = 2.25$, for first 30 seconds following perturbation. The dashed curves are identical to those of Fig. 2.11. No collision has yet occurred. The solid curves are the predictions of (4.29). The eighth car collides with the seventh after 26.25 seconds. Scales are as in Fig. 2.11.

a dot instability, and j/n indicates that the jth car collided with the car in front of it after $n/8$ seconds. A tentative stability boundary $\tau = \tau_c(\Delta)$ is drawn as a solid curve. The dashed curve, for comparison, is that of Fig. 2.12. Once again, the effect of (4.29b) is seen to be a destabilizing one.

But is this a genuine improvement? How do drivers actually drive—according to (4.28), according to (4.29), or according to neither of these? In the absence of hard data, the best we can do is to compare the predictions of (4.28), namely, Figs. 2.10–2.12, with those of (4.29), namely, Figs. 4.4–4.6, and use our personal experience to guess which picture is a fairer one of reality. Perhaps you would like to pursue this matter and obtain some data (though this will be far from easy). It is even possible that one model describes drivers in, say, Boston, and the other describes drivers in New

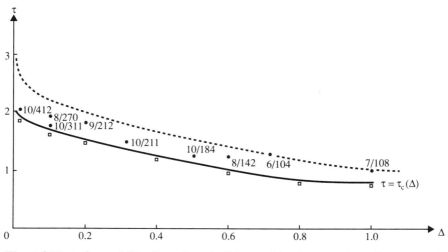

Fig. 4.6 Tentative stability boundary (solid curve) in the Δ-τ plane for the model of (4.29). The dashed curve is identical to that of Fig. 2.12.

York. But we now at least have a rational basis for accepting one model and rejecting the other (or rejecting both), even though, without the data, we cannot actually make that choice. Therefore, as good scientists, we can conclude only that the matter remains open to question.

On a real highway, of course, both τ and L would be functions of j, but such complexity could not usefully be embodied in a deterministic model. We would expect, however, that such variations make a real highway even more unstable than our models predict, and that a collision is likely if any driver (except the first) has a slower reaction time than $\tau_c(\Delta)$. Individual values of τ are utterly beyond the control of traffic police, and so they must endeavor instead to keep Δ as close as possible to zero, i.e., to ensure that the stability region in Fig. 2.12 or Fig. 4.6 is as high as possible. It's no wonder that police are so much in evidence in New York's tunnels, urging traffic to maintain equilibrium!

4.8   The Forest Rotation Time

In Exercise 3.14, we considered the effect of reducing a forest rotation period by a unit of time, from $t = s + 1$ to $t = s$.[8] We found that the forester's loss from unrealized tree growth would be approximately $V'(s)$; whereas her gain from alternative investment would be approximately $\delta V(s)$. Thus, according to Section 3.4, it is too soon to harvest if

---

[8]Before proceeding, you may wish to review your solutions to Exercise 3.9 and 3.14.

$V'(s) > \delta V(s)$, and too late to harvest if $\delta V(s) > V'(s)$, but if

$$V'(s) = \delta V(s) \tag{4.37}$$

then the harvest time, s, is exactly right. In this way, we have derived an intuitive marginal interpretation of the optimality condition, namely, that the best time to harvest is when the incremental loss from denying the trees a little more growth is exactly cancelled by the incremental gain from alternative investment (money deposited in a bank). The fact that this marginal analysis is far from rigorous is without consequence, because rigorous derivation of the optimality condition (4.37) is provided by the analysis of Section 3.4. (An alternative verbal expression of (4.37) is that the specific growth rate of forest value is equal to the specific growth rate of alternative investment value, and that the forest should be cut as soon as the former specific growth rate ceases to exceed the latter. Although this latter interpretation requires no approximation for its derivation, it is much less useful than the one we have chosen to use.)

But a moment's reflection reveals that (4.37) is inconsistent with our assumption that the forester always was, and always will be, a forester. By comparing $V'(s)$ with $\delta V(s)$, we balance an incremental loss from one generation of trees against an incremental gain from the same generation of trees—and forget about all future generations! If the forester's land were to become worthless after the harvest, then (4.37) would indeed be correct, but the land does not become worthless, because the forester plants new trees. Then the new seedlings will have acquired value, say $\epsilon(s)$, within the interval from $t = s$ to $t = s + 1$, and the forester should add this incremental gain to $\delta V(s)$ in her marginal analysis of the decision to harvest. Thus s is too soon if $V'(s) > \delta V(s) + \epsilon(s)$ and too late if $V'(s) < \delta V(s) + \epsilon(s)$, but s is just right if

$$V'(s) - \delta V(s) = \epsilon(s). \tag{4.38}$$

It is easy to verify that the optimal rotation time predicted by (4.38) is shorter than that predicted by (4.37); see Exercise 4.5.

To determine ϵ precisely, suppose that our forester harvests her forest every s years, i.e., at times $t = js$, $j = 1, 2, \ldots$, and replants, of course, immediately. Then between $t = (j - 1)s$ and $t = js$ the forest will acquire (real) value $V(s)$, independently of the value of j, if we assume that (real) costs and prices are constant and the biological growth mechanism is independent of time. Hence the present value of the jth harvest is $e^{-\delta js}V(s)$, and the present value of all harvests, from now until eternity, is

$$f(s) \equiv \sum_{j=1}^{\infty} e^{-\delta js} V(s) = V(s) \sum_{j=1}^{\infty} (e^{-\delta s})^j = \frac{V(s)}{e^{\delta s} - 1}, \tag{4.39}$$

on summing the geometric series. The quantity $f(s)$ is known by foresters as the value of the land. A more complete description would be the value of the land *for forestry*, because the possibility of using the land for alternative ends is not entertained. Notice, however, that the attractiveness of such alternative investment (following the next harvest) increases rapidly with δ, since the ratio of the value of the first cutting to that of the land (for forestry) is $e^{-\delta s}V(s)/f(s) = 1 - e^{-\delta s}$.

It is straightforward (Exercise 4.6) to show that (4.39) has a maximum at $s = s^*$, where

$$V'(s^*) = \delta V(s^*) + \delta f(s^*), \qquad (4.40)$$

provided $V''(s^*) < \delta V'(s^*)$, which is true in particular if $V'(s^*) > 0$ and $V''(s^*) < 0$. Comparing (4.38) with (4.40), we have $\epsilon(s) = \delta f(s)$. Moreover, the marginal interpretation of the opening paragraph of this section can be made valid again if we interpret the incremental gain from alternative investment as including not only the interest that would accrue from selling the wood but also the interest that would accrue from selling the land itself (for forestry).

In applying our new optimality condition, it is convenient to define

$$\theta(t) \equiv \frac{V(t)}{V'(t)}. \qquad (4.41)$$

Then (4.40) becomes

$$\theta(s^*) = \frac{1 - e^{-\delta s^*}}{\delta}; \qquad (4.42)$$

whereas the condition for a maximum, i.e., $V''(s^*) < \delta V'(s^*)$, becomes

$$\theta'(s^*) + \delta\theta(s^*) > 1. \qquad (4.43)$$

For a typical stand of Douglas fir trees in British Columbia, values of V were given in Exercise 3.9. In that exercise, by assuming that $(V(t + 5) + V(t - 5))/2 \approx V(t)$ and $(V(t + 5) - V(t - 5))/10 \approx V'(t)$, we obtained the estimates of V, V', and θ given in Table 4.1. Notice that $V''(t)$ appears to change sign between $t = 65$ and $t = 85$. This doesn't seem quite right, and we suspect that more refined methods are needed to estimate adequately $V'(75)$. Therefore, when joining the corresponding points (t, θ) in the t-θ plane by a dashed curve, we have ignored the data point at $(75, 37.5)$; see Fig. 4.7. The solid curve in Fig. 4.7 is the graph of $(1 - e^{-\delta t})/\delta$, for $\delta = 0.05$. This meets the dashed curve where $t = s^* \approx 63$. It appears from the diagram that $\theta(s^*) > 18$ and that $\theta'(s^*) \approx 0.8$. Thus $\theta'(s^*) + \delta\theta(s^*) \approx 1.7$, implying that (4.43) is satisfied. We now deduce from (4.42) that, when the real interest rate is 5%, the optimal rotation period is about 63 years. As remarked above, this is shorter (by about two years) than the value obtained in Chapter 3. Moreover, our estimation of s^* is clearly unaffected

Table 4.1 Values of V, V' and θ, defined by (4.41), as used to determine optimal rotation period.

t	$V(t)$	$V'(t)$	$\theta(t)$	t	$V(t)$	$V'(t)$	$\theta(t)$
35	21.5	4.3	5.0	85	727.5	15.5	46.9
45	93	10	9.3	95	859	10.8	79.5
55	223	16	13.9	105	956.5	8.7	110
65	400	19.4	20.6	115	1037.5	7.5	138
75	573.5	15.3	37.5				

by the ignored data point. Let us therefore expend no further energy on our suspicion that $V'(75)$ has been poorly estimated. The same method can readily be used to determine s^* for any other value of δ.

The optimality condition (4.40) is certainly an improvement upon (4.37), but the question posed in Chapter 3 has not been answered. Does (4.40) determine when a forest should be cut? How can we test our model? This raises a most important question concerning what is testable and what is not, and two remarks are in order.

First, the choice of utility function in an economic problem is a task for moral philosophers. We did not use data to decide that a forester should maximize present value, as opposed to some other quantity; we simply

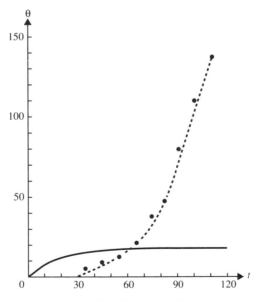

Fig. 4.7 Graphical determination of optimal rotation period for a stand of Douglas fir trees.

accepted the prevailing opinion among practicing economists (you may challenge this, but not within the framework described at the beginning of this chapter). Second, as we stated long ago in our ABC of modelling, mathematical deductions from given assumptions are not (or should not be) open to question. Mathematics does not lie. Thus, if real prices are constant, if all trees are identical, if a forest is felled and replanted in an instant, and if all other effects which our model neglects are truly negligible, then there is no question that (4.40) provides the optimal rotation period. In brief, we cannot use data to test the truth of either the utility function or the solution to the pure mathematical problem.

We must use data, however, to test the truth of our assumptions that prices are constant, that trees are identical, that harvesting and replanting are instantaneous, and that other effects may be ignored; and it will not surprise you to learn that all four assumptions are doubtful for most forests. Indeed the forest rotation problem is both more controversial and more complicated mathematically than befits further discussion in a general work on mathematical modelling. We do relax the assumption that all trees are identical in Sections 5.8 and 9.2, but the reader who wishes to study these matters in depth is referred to the literature.

4.9   Crop Spraying

According to Plant (1985, p. 5), a cotton grower in the San Joaquin Valley adopts the following strategy in attempting to control the spider mite.[9] Every time she measures the pest population, the grower compares the cost of applying pesticide at the maximum level allowed by law with the projected cost of crop damage if no pesticide is applied. If the latter is greater than the former, then the grower applies the maximum allowable amount of pesticide. Can we use this information to test the model we proposed in Section 3.6?

Let's make the assumption that, by virtue of experience (i.e., trial and error), what growers do is also what is optimal. Then, if the strategy they adopt agrees with the prediction of our model, we will accept our model, at least for the time being, on the grounds that we have as yet no reason to reject it. If the strategy they adopt disagrees with our model's predictions, however, then we will have to decide, by other means, whether the fault lies with the assumption that growers do what is optimal, with (3.54), (3.56), or (3.61), or even with some other assumption.

Now, in the notation of Section 3.6, the growers' strategy is

$$\text{Spray if} \qquad f(s^*, 0) > c_0 + c_1 U_{\max} \qquad (4.44)$$

$$\text{Don't spray if} \qquad f(s^*, 0) < c_0 + c_1 U_{\max}. \qquad (4.45)$$

---

[9] You may wish to review Section 3.6 before proceeding.

If (4.44) is satisfied then the grower sprays with concentration U_{max}. We will denote the time of spraying by s^*, on the assumption that when growers spray is also when it is optimal (but remember that $f(s, 0)$ is independent of s).

Because $f(s^*, U_{max})$ exceeds $c_0 + c_1 U_{max}$, (4.44) may be satisfied in one of two ways: either

$$c_0 + c_1 U_{max} < f(s^*, U_{max}) < f(s^*, 0) \qquad (4.46)$$

or

$$c_0 + c_1 U_{max} < f(s^*, 0) < f(s^*, U_{max}). \qquad (4.47)$$

If (4.46) holds then (3.65) is satisfied with $U^* = U_{max}$; whence, from (3.70) and (3.71), the growers' strategy is the same as our model's if

$$\frac{\alpha \Delta}{c_1} \geq e^{(U_{max} - U_0)}. \qquad (4.48)$$

Let's assume that (4.48) holds. Then, because $U^* = U_{max}$, the right-hand side of (4.37) is less than $f(s^*, U^*)$. Therefore, from (3.44), (4.45) agrees with our model also. Our model is thus in complete agreement if

1. (4.48) is true and
2. (4.48) implies that (4.47) is impossible. (4.49)

To show that (4.47) cannot hold we must show that $W(q)$ is negative; i.e., that $\xi \geq \exp(U_{max} - U_0)$ implies that

$$\Psi(\xi) \equiv U_0 + \alpha \frac{c_0}{c_1} + 1 + \ln \xi - \xi \qquad (4.50)$$

is negative. Because Ψ is a decreasing function for $\xi > 1$ (and $\xi \geq \exp(U_{max} - U_0)$ implies $\xi > 1$), $\Psi(\xi)$ will certainly be negative if $\Psi(\exp\{U_{max} - U_0\})$ is negative; or, upon rearrangement,

$$U_{max} + \alpha \frac{c_0}{c_1} + 1 < e^{(U_{max} - U_0)} \leq \frac{\alpha \Delta}{c_1}. \qquad (4.51)$$

More generally, if we define

$$\lambda \equiv \frac{\alpha \Delta}{c_1} e^{-(U_{max} - U_0)}, \qquad (4.52)$$

then (4.49) is satisfied if

$$U_{max} + \frac{\alpha c_0}{c_1} + 1 + \ln \lambda < \lambda e^{(U_{max} - U_0)}, \qquad \lambda > 1. \qquad (4.53)$$

The greater the value of λ—i.e., in view of (4.37), the more valuable the crop or the more destructive the pest—the more easily that (4.53) is satisfied.

In summary, if (4.53) holds, then the only piece of data we have is consistent with our model. If (4.53) holds, then we have no reason to reject the model, at least until more extensive data enable us to apply a more stringent test. But does (4.53) hold? My guess is that pb is probably large enough to ensure that it does; but I don't actually know, and perhaps this is a matter you would like to pursue.

4.10   How Right Was Poiseuille?

In Section 2.6, we constructed a model for the flow of water through a pipe with an arbitrary elliptical cross section.[10] Because the circle is a special case of the ellipse, this model should at least be able to explain the flow of water through a circular pipe. Let's use this fact to test our model.

Let $Oxyz$ be an orthogonal Cartesian coordinate system, with Oz along the symmetry axis of the pipe. Then recall, from (2.70), that our model predicts the flow velocity in the direction of Oz, at station (x, y), to be

$$w(x, y) = \frac{p_2 - p_1}{2\mu L} \frac{a^2 b^2}{a^2 + b^2} \left(1 - \frac{x^2}{a^2} - \frac{y^2}{b^2} \right), \tag{4.54}$$

where a b, are the semi-axes of the ellipse and

$$-\frac{\partial p}{\partial z} = \frac{p_2 - p_1}{L} > 0 \tag{4.55}$$

is the constant (adverse) pressure gradient that drives the flow. Let our circular pipe have radius a, and let $r = \sqrt{x^2 + y^2}$. Then, on setting $b = a$ in (4.54) and using (4.55), we find that the flow velocity at radius r within the pipe is given by

$$w(r) = -\frac{a^2}{4\mu} \frac{\partial p}{\partial z} \left(1 - \frac{r^2}{a^2} \right), \qquad 0 \le r \le a. \tag{4.56}$$

The maximum velocity

$$w_{\max} = -\frac{a^2}{4\mu} \frac{\partial p}{\partial z} \tag{4.57}$$

occurs at the center of the pipe; whereas the average velocity is

$$w_{av} = \frac{1}{\pi a^2} \int_0^{2\pi} \int_0^a w(r) r \, dr \, d\theta = \frac{1}{2} w_{\max}. \tag{4.58}$$

Hence, according to the model of Section 2.6, we have

$$\frac{w(r)}{w_{\max}} = 1 - \frac{r^2}{a^2}; \tag{4.59}$$

---

[10]You may wish to review Section 2.6 before proceeding.

and the flux of water along the pipe is

$$Q_c = -\frac{\pi a^4}{8\mu}\frac{\partial p}{\partial z} = \pi a^2 w_{av}, \qquad (4.60)$$

on using (2.72). As we remarked in Chapter 2, this last result is known as Poiseuille's formula (by some, and the Hagen–Poiseuille law by others).

Experimental values of $w(r)$ at ten different radii are presented by Hansen (1967, p. 309) and plotted in Fig. 4.8. If you plot (4.59) on the same graph then you will find that the agreement between theory and practice is excellent—the model fits the facts. Moreover, experiments by Hagen and Poiseuille themselves established that Q/a^4 was remarkably constant for water flowing through pipes of different radii, under the same adverse pressure gradient; see, for example, Batchelor (1967, p. 181). In all of this there is no reason at all to reject the model of Section 2.6.

We must accept it, however, subject to (at least) two limitations. First, if $z = 0$ is the entrance to the pipe, then there is a region $0 < z < z_c$, known as the entry region, in which (4.59) does not apply. To understand why, imagine that the pipe is inserted into fluid flowing with constant velocity w_{av}, parallel to the pipe. Then at $z = 0$ we have

$$w = w_{av}, \qquad 0 \le r < a; \qquad (4.61)$$

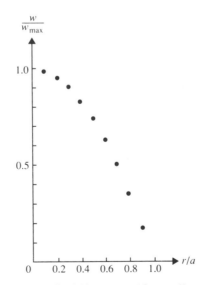

Fig. 4.8 Experimental values of $w(r)/w_{max}$ at 10 equally spaced radii for flow in a circular pipe. The sketch, based on a Ph.D. thesis by V. E. Senecal, is adapted from Arthur G. Hansen, *Fluid Mechanics*, p. 309 by kind permission of the author (who now holds the copyright).

whereas for $z \geq z_c$ we have w given by (4.59). Between $z = 0$ and $z = z_c$, the fluid velocity profile must make a transition from the "flat" profile of (4.61) to the "parabolic" profile of (4.59). This transition (by a mechanism known as boundary layer thickening) is indicated schematically in Fig. 4.9. The flow in the entry region is difficult to analyze but the length of the region may be estimated (Hansen, p. 309) as

$$z_c = 0.232a^2 w_{av}\frac{\rho}{\mu}, \tag{4.62}$$

where ρ is the density and μ the shear viscosity of the fluid. The wider the pipe or the faster the flow, the longer is the entry region.

Second, and more fundamentally, we must recall our assumption in Section 2.6 that the flow in our pipe is in equilibrium. Now in Section 2.8 we saw an example of a dynamical system, namely, a platoon of cars, that was stable for certain values of the system parameters but not for others. Instability set in when a certain parameter exceeded a critical value. This also happens to flow in a pipe. For traffic flow, the relevant parameter was τ, the driver response time, and its critical value was $\tau_c(\Delta)$. For flow in a pipe, the relevant parameter is

$$R \equiv \frac{2a\rho w_{av}}{\mu}, \tag{4.63}$$

and its critical value, R_c, is about 2000. R, which is dimensionless, is known in fluid dynamics as the Reynolds number. Fig. 4.8 was obtained for a Reynolds number of 1094, well within the stable region; but for $R > R_c$ the equilibrium (4.59) cannot be sustained, just as (2.96) could not be sustained for $\tau > \tau_c$.

If you wish to pursue the study of flow at Reynolds numbers above R_c, i.e., at *supercritical* Reynolds numbers, then you ought to begin with an introductory text like Batchelor (1967) or Hansen (1967). We don't have space to pursue it here. We have raised the matter merely to illustrate that a model that is excellent for certain conditions may be quite inadequate for others, because (4.59) provides an excellent description of *subcritical* $(R < R_c)$ pipe flow away from the entrance $(z > z_c)$ but is inapplicable either to the entry region $0 < z < z_c$ or to supercritical flow. To put things another way, it is rare that an assumption on which a model is based is valid without exception, and we should expect that the model may cease

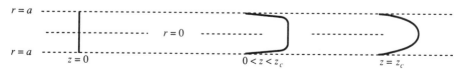

Fig. 4.9 Schematic representation of transition to Poiseuille flow in the entry region of a circular pipe.

to provide an adequate description of reality as soon as the assumption ceases to be justifiable. In the case of the entry region, for example, the assumption that ceases to be justifiable is that w is independent of z, as is clear from Fig. 4.9.

4.11   Competing Species

We have good grounds for believing that if an ecosystem consists solely of two species competing for the *same* food supply then, in the long run, only one species will survive.[11] The other species will become extinct (or else migrate, in search of an alternative food supply). For example, *The Economist* of April 26, 1986 reports, in a lively and informative article, "The rows behind ecology," that three kinds of parasitic wasps were once introduced to Hawaii to control fruit flies. The first established itself and brought the flies under control. When the second species was introduced, it drove the first one to extinction and assumed for itself the job of controlling the fruit flies. Later, the third species in turn drove out the second one. Indeed the idea that only one of two competing species can survive has become enshrined in theoretical ecology as "Gause's principle of competitive exclusion," even though it was apparently never enunciated by Gause; see Maynard Smith (1974, pp. 61 and 103).

Let $x(t)$ and $y(t)$ denote the population magnitudes, at time t, of the two competing species. Then it appears at first sight that (1.22), i.e.,

$$\frac{1}{x}\frac{dx}{dt} = a_1 - b_1 y, \qquad \frac{1}{y}\frac{dy}{dt} = a_2 - b_2 x, \qquad (4.64)$$

where $a_1, a_2, b_1, b_2 > 0$, might be an acceptable model of a competing-species ecosystem, because, from Fig. 1.4, it predicts that species x will prevail if $(x(0), y(0))$ lies below ON, but species y will prevail if $(x(0), y(0))$ lies above ON. But (4.64) also predicts that the surviving species will become infinitely numerous. Suppose, for example, that species x survives and species y has been driven to extinction. Then $y(t) = 0$, so that (4.64) yields $(1/x)dx/dt = a_1$, whence $x(t) = x(0)\exp(a_1 t)$ grows without bound. In other words, our model assumes that one species will grow exponentially in the absence of the other.

Now, although exponential growth may be a reasonable assumption in the short term, it clearly cannot last forever. In the long run, the growth of species x must be limited, because its food supply is distinctly finite. Let us therefore make the fresh assumption that, in the absence of the other, each species would grow not exponentially but according to a logistic growth

---

[11] You may wish to review Section 1.5 before proceeding.

law, i.e., that, for some positive parameters c_1 and c_2,

$$\frac{1}{x}\frac{dx}{dt} = a_1 - c_1 x, \qquad \frac{1}{y}\frac{dy}{dt} = a_2 - c_2 y. \tag{4.65}$$

Thus, if $i(= 1, 2)$ represents the surviving species, then it cannot become more numerous than a_i/c_i. Adding the interaction terms to (4.65), we are thus led to propose that (4.64) should be replaced by the model

$$\frac{dx}{dt} = x(a_1 - c_1 x - b_1 y),$$
$$\frac{dy}{dt} = y(a_2 - b_2 x - c_2 y). \tag{4.66}$$

This is a dynamical system of the form (2.12), with $u(x, y) \equiv x(a_1 - c_1 x - b_1 y)$ and $v(x, y) \equiv y(a_2 - b_2 x - c_2 y)$. It has equilibria where u and v both vanish, i.e., at points (x^*, y^*) such that $u(x^*, y^*) = 0 = v(x^*, y^*)$. We exclude the unstable equilibrium $x^* = 0 = y^*$ on the grounds that at least one species is presumed to exist, and so there are three potential equilibria:

$$x^* = 0, \qquad y^* = \frac{a_2}{c_2}, \tag{4.67}$$

$$x^* = \frac{a_1}{c_1}, \qquad y^* = 0, \tag{4.68}$$

or (x^*, y^*) is the intersection of the line pair

$$L_1: \quad c_1 x + b_1 y = a_1, \tag{4.69a}$$
$$L_2: \quad b_2 x + c_2 y = a_2. \tag{4.69b}$$

Because a_1, a_2, b_1, b_2, c_1, and c_2 are all positive, the lines L_1 and L_2 in (4.69) must have positive intercepts on both axes in the $x - y$ plane. Thus there are four possible configurations, as depicted in Fig. 4.10, in which dots denote equilibria.

In (a) and (b) of Fig. 4.10, the lines L_1, L_2 do not intersect, so that the only equilibria are at (4.67) and (4.68). In (c) and (d) the lines do intersect, and there is therefore an additional equilibrium, where

$$x^* = \frac{a_1 c_2 - a_2 b_1}{c_1 c_2 - b_2 b_1}, \qquad y^* = \frac{a_1 b_2 - a_2 c_1}{b_1 b_2 - c_2 c_1}. \tag{4.70}$$

The arrows in the diagrams are a schematic representation of the direction of the vector dx/dt—in the various regions, in the first quadrant of the $x - y$ plane, which are separated by L_1, L_2 and the coordinate axes. These are easily deduced from the observation that dx/dt is positive below L_1, zero on it, and negative above it; whereas dy/dt is positive below L_2, zero on it, and negative above it. Thus solution curves always cross L_1 vertically and L_2 horizontally. The curve marked S in Fig. 4.10 (d) is a separatrix, like ON in Fig. 1.4, which solution curves may never cross. If you sketch a

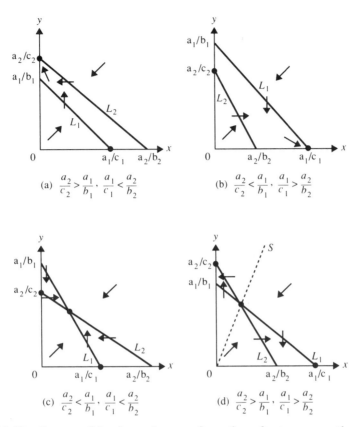

Fig. 4.10 The four possible phase-plane configurations for two competing species.

few trajectories by using a computer routine such as DVERK (Section 1.4), then you will discover that Fig. 4.10(d) is closer to Fig. 1.4 than any of the remaining three diagrams. Indeed Fig. 1.4 could be regarded as the special case of Fig. 4.10(d) in which the equilibria at $(a_1/c_1, 0)$ and $(0, a_2/c_2)$ have retreated along their respective axes to infinity.

A few minutes spent pondering these diagrams will convince you that, for arbitrary initial population magnitudes $x(0)$ and $y(0)$, the ecosystem approaches one of the three potential equilibria. The identity of this steady state is determined by the signs of the numbers I_x and I_y, where

$$I_x \equiv \frac{a_1}{c_1} - \frac{a_2}{b_2}, \qquad I_y \equiv \frac{a_2}{c_2} - \frac{a_1}{b_1}. \tag{4.71}$$

The dependence of the species' destiny on I_x and I_y is recorded in Table 4.2. You can easily confirm, by the method of Section (2.2), that the attracting equilibrium is stable, and the other(s) unstable (Exercise 4.7).

Table 4.2   Destiny of Competing Species as a Function of Interference.

Diagram	Sign of I_x	Sign of I_y	Steady state that attracts all solutions
(b)	+	−	$(a_1/c_1, 0)$
(a)	−	+	$(0, a_2/c_2)$
(c)	−	−	(4.70)
(d)	+	+	$(a_1/c_1, 0)$ if $(x(0), y(0))$ lies below S
			$(0, a_2/c_2)$ if $(x(0), y(0))$ lies above S

How do we interpret the quantities defined by (4.71)? It is clear from (4.65) that a_1/c_1 is the capacity of the first species (as defined in Section 1.9); and Section 1.5 shows that a_2/b_2 is the equilibrium population when the species is assumed to have infinite capacity. Although that equilibrium is unstable and would never be observed, it nevertheless represents the only level at which species x could coexist with species y if both were assumed to have infinite capacity. If this level is less than a_1/c_1 then coexistence with species y would prevent species x from reaching capacity; whereas if a_2/b_2 exceeds a_1/c_1 then the growth of species x is more restricted in isolation than it would be if it were obliged to coexist with species y. Thus I_x is an *interference* factor, being negative or positive according to whether species x interferes more with itself than with species y, or vice versa. We interpret y similarly.

Now, if both I_x and I_y are negative, then the two species interfere less with each other than they would with themselves; and can hardly be said to compete for a common food supply. Possibility (c) in Table 4.2 must therefore be ruled out. In all other circumstances, i.e., when at least one species interferes with the other, (4.66) predicts that only one species will survive, and Table 4.2 specifies which. You might describe (a) and (b) as unfair competition, in which case (d) would represent a fair fight, although its outcome would still be determined by the initial strengths of the armies. All this is perfectly in accordance with Gause's principle. We have therefore no reason to reject (4.66), and until we are able to apply a more stringent test, we tentatively accept the improved version of our model.

4.12 Predator–Prey Oscillations

We have two good grounds for believing that if a carnivorous predator feeds solely on an herbivorous prey, then both populations may oscillate forever.[12] The first ground is conceptual. If there are many prey but few predators then the predators will have an abundant food supply and their

---

[12] You may wish to review Section 1.4 before proceeding.

numbers will grow, at the expense, of course, of the prey. The consequent decline in the numbers of prey will choke off the food supply of the more abundant predators, many of whom will die of hunger. The smaller predator population that results will have a smaller demand for food, giving the prey population an opportunity to recover, but now there are many prey again, which is where we began. The second ground is empirical. Such predator–prey oscillations have actually been observed in nature. The most famous example concerns the lynx and the snowshoe hare of America's northern coniferous forests, whose populations fluctuate with a period of about ten years; see Messenger (1980, p. 1046).

If $x(t)$ and $y(t)$ denote, respectively, the magnitudes of the prey and predator populations at time t, then such a cycle of growth and decay would appear as depicted in Figs. 1.2 and 1.3. At first sight, therefore, it seems that (1.20), i.e.,

$$\frac{1}{x}\frac{dx}{dt} = a_1 - b_1 y, \qquad \frac{1}{y}\frac{dy}{dt} = -a_2 + b_2 x, \qquad (4.72)$$

where $a_1, a_2, b_1, b_2 > 0$, might be an acceptable model of a predator–prey ecosystem. But, like (4.64) above, the model (4.72) embodies the assumption that species x (here the prey) would grow exponentially in the absence of the predator ($y = 0$), and we have just agreed that this is not a sound assumption. No matter how much vegetation there may be to eat, crowding and disease are bound to ensure an upper limit to the capacity of the land, no matter how large that limit may be. Let's call it $1/\epsilon$, where we have in mind that ϵ is a small number. Before we consider any further whether (4.72) is acceptable or not, we must at least correct this one defect. Let's therefore assume that, in the absence of predators, the growth of the prey population would obey the logistic equation of Section 1.9. Thus

$$\frac{1}{x}\frac{dx}{dt} = a_1(1 - \epsilon x). \qquad (4.73)$$

This is just (1.63) with $R = a_1$ and $K = 1/\epsilon$. We therefore replace (4.72) by the dynamical system

$$\frac{dx}{dt} = x\{a_1(1 - \epsilon x) - b_1 y\},$$
$$\frac{dy}{dt} = y\{-a_2 + b_2 x\}. \qquad (4.74)$$

We propose this as an improvement to our model. Note that, because the prey capacity $1/\epsilon$ must exceed *all* actual prey population levels, it must in particular exceed the level a_2/b_2 at which, according to our old model (4.72), the prey could coexist with the predator in static equilibrium. Thus

$$\epsilon < \frac{b_2}{a_2}. \qquad (4.75)$$

A typical solution of this pair of ordinary differential equations, obtained as in Sections 1.4 and 1.5 by numerical approximation, is represented in the $x - y$ plane by the spiral shown in Fig. 4.11. The curve is actually drawn for $x(0) = 1 = y(0)$ and parameter values $a_1 = 4$, $a_2 = 3$, $b_1 = 2$, $b_2 = 1$, and $\epsilon = 1/30$; but for any other initial population levels $x(0) > 0$ and $y(0) > 0$, and for any other values of $a_1, a_2, b_1, b_2,$ and ϵ satisfying (4.75), the solution curve would still spiral in toward the stable focus

$$\mathbf{x}^* = (x^*, y^*)^T = \left\{ \frac{a_2}{b_2}, \frac{a_1}{b_1} \left(1 - \frac{\epsilon a_2}{b_2} \right) \right\}. \tag{4.76}$$

Where are the cycles of Fig. 1.3? They have disappeared!

Herbivorous species have been observed to coexist with carnivorous ones in apparently static equilibrium. Examples concern the Rocky Mountain lion and mule deer, and the wolf and caribou in Alaska (May, 1976c, p. 55). Such simple ecosystems might well be described accurately by the model (4.74). But (4.74) cannot represent the kind of interaction that determines the population cycles of the lynx and hare in Canada; i.e., (4.74) is not an acceptable model of that particular predator–prey system. Moreover, the cycles of Fig. 1.3 are absent from (4.74) for any positive value of ϵ, *however small*. In other words, the addition of a small quantity, namely, $-\epsilon a_1 x^2$, to the right-hand side of one of the equations (4.72) has a drastic effect on the topology of the phase-plane, because the equilibrium $\mathbf{x}^*$ erupts (or "bifurcates") from a metastable center into a stable focus as ϵ becomes positive. This is easily deduced by the methods of Section 2.2, because the quadratic equation (2.19) for the system (4.74) is found (Exercise 4.7) to be

$$\omega^2 + \frac{\epsilon a_1 a_2}{b_2}\omega + a_1 a_2 \left(1 - \frac{\epsilon a_2}{b_2} \right) = 0. \tag{4.77}$$

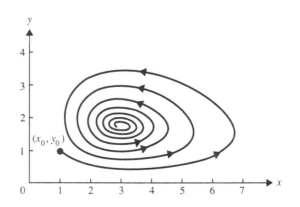

Fig. 4.11 Form of solutions to (4.74).

For $\epsilon = 0$, the roots of (4.77) are purely imaginary, so that x^* is a center, by (2.21); whereas for small positive ϵ the roots are complex, with negative real parts, so that x^* is a stable focus, again by (2.21).

Whenever the phase-plane topology of the dynamical system

$$\frac{dx}{dt} = u(x, y), \qquad \frac{dy}{dt} = v(x, y) \tag{4.78}$$

is different from that of the dynamical system

$$\frac{dx}{dt} = u(x, y) + \epsilon a(x, y), \qquad \frac{dy}{dt} = v(x, y) + \epsilon b(x, y) \tag{4.79}$$

for arbitrarily small ϵ, we say that (4.78) is *structurally unstable*. In terms of Chapter 2's mythical land of Phaseplane, it means that a slight change in the mechanism that determines the wind pattern, *while the wind is still blowing*, will drastically change that pattern. Thus the concept of structural stability is fundamentally different from those we discussed in Chapter 2. It applies to the mechanism of a dynamical system; whereas earlier concepts applied to the destinies of the entities which that mechanism governed. Structural stability is also much more difficult to treat mathematically and is well beyond the scope of this book. Fortunately, however, the only result we need to proceed is that dynamical systems that lead to metastable cycles are always structurally unstable; whereas those that lead to limit cycles are not. Biological systems are constantly subjected to changes in the mechanism that governs their evolution and so any population cycle that is actually observed must surely be a limit cycle, because it would not persist if it were not associated with a structurally stable dynamical system.

Our predator–prey model is unacceptable. But we have also stumbled upon a test that any acceptable model is obliged to pass: it must give rise to a limit cycle. We will discuss such a model in Chapter 10. Meanwhile, try Exercise 4.14.

4.13 Sockeye Swings, Paradigms, and Complexity

We now consider circumstances in which, as an interim measure, we are obliged to relax our criteria and consider models that cannot be tested.[13] We said at the beginning, and we'll say it again, that the *ultimate* test of a model's worth is often its ability to predict observable data. But nature is complex, with so many variables, that *all* essential characteristics of many natural phenomena can be captured only by a sophisticated model, one that is synthesized from many components. We do not expect that some of the components will explain observable data without the others, for that would imply that the others were redundant anyway; but without

---

[13] You may wish to review Sections 1.8 and 2.4 before proceeding.

the components, the fully integrated model will never be constructed, and hence never tested. We are thus obliged to develop models that cannot themselves be tested, on the understanding that the final product will be tested eventually. If it fails this test, we must then decide which components are faulty and how to replace them. This would be a poor policy for an engineer to adopt; but nature does not allow controlled experiments in her laboratory and so it is frequently the best that a mathematical modeller of natural phenomena can do. This shouldn't take you completely by surprise, for we broached the idea earlier, in Sections 4.3 and 4.6, when we spoke of several generations of predictions. Moreover, in certain circumstances, we are even obliged to develop models that will never be tested at all, as in the controversy over the nuclear arms race. At least, we fervently hope that such models will never be tested.

The salmon model of Sections 1.8 and 2.4 can be regarded as a model component that cannot, itself, be tested, because all available data will embody the age structure, environmental variability, and interactions with other species (including man!) that the model simply ignores. Nevertheless, there are indications in the scientific literature that our model component may prove to be a useful one. According to Ricker (1954, p. 592), the appropriate unit of time in the metered model of Section 1.8 is about four years for a certain lineage of sockeye salmon on the Fraser river in Canada. These salmon are supposed to have been much less plentiful in 1909 than in either 1905 or 1913. Could this have corresponded to a period-2 cycle such as the one suggested by the third column of Table 1.2? The evidence in favor includes the fact that the percentage exploitation of those salmon by the fishery was low at the time. From (1.53) and Section 2.4, a period-2 limit cycle could occur only if the percentage exploitation were low enough to keep the parameter $r = \ln(\gamma\alpha)$ above 2, where γ denotes the fraction of young salmon surviving to breed and α the number of eggs laid by a female for each existing member of the population. Low percentage exploitation would have made it possible for γ to exceed $7.389/\alpha$ during the early part of this century. During the middle of the century, however, exploitation was high enough to keep r below 2 (or γ below $7.389/\alpha$) and ensure that no swings would be observed. The latter is a liberal but nevertheless accurate translation of Ricker's statement (p. 592) that "under present conditions the fishery normally takes enough to keep the spawning stock near the dome of the curve." (If you want to know which curve, read the paper.)

But the salmon model may also be regarded as a *paradigm*, i.e., a simple model that isolates and explores the consequences of a particular idea. Viewed in this way, the model's simplicity is its greatest strength. Suppose, for example, you had the idea that the best way to preserve a species was to hunt it in small quantities. How could you test this idea? One way would be to apply the model of Section 2.4. According to this model, if exploitation is high enough to ensure $r < 2$, then the stock of the species will remain in equilibrium at a moderate level. In particular, it will never reach a low level. But if the species is hunted only in small quantities then r may be much larger than 2, implying that the stock will swing back and forth between low levels and high levels. Now, suppose that the species is afflicted one year by pathological doses of pollution, disease, or natural catastrophe. Under which circumstances is the species most likely to survive—if the stock level is guaranteed to be at least moderate, or if the stock level is guaranteed to be at least low? Because we cannot arrange for catastrophes to occur when the stock is abundant, I'm sure you'll agree that the species is better off if $r < 2$. Low levels of exploitation do not *necessarily* help to preserve the species (it would depend upon whether r fell below 2 even without human intervention). Thus the truth of your idea, as captivating as it sounds, is seen to be in doubt. Do you believe that you would have realized this without the help of a paradigm?

It is no coincidence that the salmon model may be regarded either as a component or as a paradigm, for the two are closely related. A model component is an attempt to capture one (or more) essential characteristics of a complex phenomenon; a paradigm is an attempt to isolate one (or more) particular idea; and the foundations of even the most sophisticated models are ideas. When the idea is testable, we call it an assumption. When, as here, it is merely explorable, it simply remains an idea.

A good idea now might be to tackle Exercise 4.8.

4.14 Optimal Fleet Size and Higher Paradigms

The dual view, that a nontestable model may be regarded either as a paradigm or as a component of a later, testable model, applies in particular to Section 3.7.[14] The number of components in a mathematical model depends, of course, on how you look at it; but a possible view is that the fleet size model has two components, a biological one and an economic one. In the hierarchy of complexity, therefore, it enjoys a higher station than the model discussed in the previous section, which is purely biological. Stretching the dual relationship to its very limits now, we might expect the fleet size model to be the kind of paradigm that would let us explore not a single idea but the interrelationship between two—a higher paradigm, as it were.

One idea, which some people believe, is that you should live for today, on the grounds that tomorrow may never come. If it does come, however, then you might wish that you had prepared for it, which is why some

---

[14]You may wish to review Section 3.7 before proceeding.

people say that you should live for tomorrow. Who is right? As we discussed toward the end of Section 4.8, a mathematical model could never answer such a question. What it can do, however, is to provide a conceptual framework for assessing the implications of two conflicting viewpoints, and, more importantly, of the whole spectrum of viewpoints that lie between. So consider the woman who lives for today. How would you convince her to put her money in a bank? Not by offering her 10% interest, or 20% for that matter, because she is far too worried that she won't be around to enjoy the income. The only rate that might tempt her to put her money in a bank would be ∞%. Now consider the woman who lives for the end of the next millennium. She's so sure she will still be around that she would accept a (real) interest rate of 0.001% or so, because she can still nearly triple her money. Indeed if she felt that it was worth paying so much a year for the safekeeping, then she would probably accept an interest rate of zero. Here are the extreme viewpoints. The woman who lives for the far distant future is modelled by $\delta = 0$. The woman who lives for the here and now is modelled by $\delta \to \infty$. Most of us mortals lie somewhere in between, in the spectrum $0 < \delta < \infty$, the great irony being that δ probably decreases with age. (When Hurricane Kate swept through Tallahassee in November, 1985, she destroyed the oak tree of an 84-year-old neighbor of mine. Next day he removed it and planted a new one.) In this section, we will refer to δ as the *discount* rate. The new terminology merely reflects a change of emphasis. When δ was used to measure the lucrativeness of alternative investment, we called it the (real) interest rate, but when using δ to measure the degree to which the future is ignored (or "discounted"), we shall call it the discount rate.

Let's now assess the implications of concern about the future for the fishing fleet of Section 3.7. The optimal number of boats, as a function of δ, is given by (3.89):

$$U^*(\delta) = \frac{R}{4q}\left\{3 - b + \frac{\delta}{R} - \sqrt{\left(1 + b - \frac{\delta}{R}\right)^2 + \frac{8b\delta}{R}}\right\}. \tag{4.80}$$

Thus, because $b < 1$,

$$U^{*\prime}(\delta) = \frac{1}{4q}\left\{1 - \frac{\frac{\delta}{R} - 1 + 3b}{\sqrt{\left(\frac{\delta}{R} - 1 + 3b\right)^2 + 8b(1 - b)}}\right\} > 0; \tag{4.81}$$

also

$$U^*(0) = \frac{R(1 - b)}{2q}, \tag{4.82}$$

and[15]

$$U^*(\infty) = \frac{R(1 - b)}{q}.\tag{4.83}$$

You can see immediately from the profit expression (3.87), i.e., from

$$f(U) \equiv \frac{pqKU}{\delta} e^{-\delta s(U)} \left\{ 1 - \frac{Uq}{R} - b \right\},\tag{4.84}$$

that the sole owner of a fishery who lived purely for today would drive it to a state of zero profit (*after* her huge salary had already been included as part of costs (3.74)!); whereas the sole owner who lived purely for her great grandchildren's great grandchildren would maximize sustained profit, ensure that the fish stocks remained above the level of maximum sustainable yield, and thus be very popular with the conservationists in Exercise 3.16 (where sustained profit and sustained yield are defined). Because $U^{*\prime}(\delta) > 0$, moreover, the more that the owner cared about the future, the fewer boats she would use and the higher the stock level that would be maintained.

Even in Western democracies, fisheries can be viewed as having a sole owner, namely society. If we could only agree on a suitable value for δ then models like this one might help to determine a system of licensing laws that would keep exploitation at the optimal level. Naturally, many components would need to be added to the model, because the oceans are roamed by many boats in search of many species, with many breeding cycles and as many selling prices; and all are vulnerable to unexpected environmental shocks. The problem of developing a theory of conservation of natural resources is such a difficult one that, at the time of writing, relatively few components have been added by anyone to the model of Section 3.7 (see Clark, 1990). Yet as difficult as the modelling sounds, the really tough problem is still that of getting society to agree!

4.15   On the Advantages of Flexibility In Prescriptive Models

We close this chapter by addressing a question we posed at the end of Section 3.2, where we were trying to decide how big a pay increase a professor should receive. Instead of asking whether our linear programming model fitted the facts, we wondered whether anyone might use it. Why did we suddenly change our criterion of relevance?

---

[15]Use the fact that

$$\lim_{x \to \infty} \left\{ x - \sqrt{(C - x)^2 + Ax} \right\} = \lim_{x \to \infty} \frac{(2C - A)x - C^2}{x + \sqrt{(C - x)^2 + Ax}} = C - \frac{A}{2}$$

for any constants A and C.

The answer lies in the purpose behind the model. The models we encountered in Chapters 1 and 2 were descriptions of natural or social phenomena, i.e., attempts to say what things or people *do* do, in the active sense of the verb. By contrast, the flight model of Section 3.1 was an attempt to say what a bird *should* do, in the modal sense of the verb. If we believe that what birds do is also what is optimal by virtue of trial and error, then our flight model is essentially also an attempt to *describe* a phenomenon. But birds have migrated for thousands of years. They've had all that time to solve their problem, and no one has ever pressed them for an answer. Administrators, on the other hand, have generally had little experience of the problems they currently face, yet are expected to *prescribe* solutions within a relatively short period of time. They must therefore short-circuit the process of trial and error by imagining what people *would* do, in the subjunctive sense of the verb, if certain actions were taken, then consciously select from among those actions the one that seems most desirable.

Suppose, for example, that our chairperson is worried about her department's research rating. She may imagine that her faculty would improve the quality of their research if she took the action of raising the payment for a unit of research achievement. From many such actions, she may consider it most desirable to select the one that maximizes that payment, subject to certain constraints. Suppose that her action fails to achieve its intended goal, perhaps because faculty are still devoting too much time to service. What will she do? She will try to imagine what different actions would achieve, then select from among those different actions the one that seems most desirable. She may, for example, consider lowering the payment for a unit of service achievement. From many such actions, she may now consider it most desirable to select the one that minimizes that payment, subject to the same constraints as before. Moreover, she may not have the time or energy to think through the entire problem, whenever it appears that a change of policy is required. Then what she needs is a *flexible* model, one easily altered to suit the needs of the moment. The greater her model's flexibility, the greater its usefulness.

If you did Exercise 3.5, then you will probably agree that the model we developed in Section 3.2 is indeed a flexible one. Provided they remain linear, the constraints and utility function in the linear program (3.20) may be changed at almost a moment's notice, simply by redefining the matrices A, b, and c. Moreover, a moment later, ZX3LP will be just as happy to reveal the new optimal control (or tell you that no control is feasible).

We have said already that, in managing society and its resources, the choice of utility function lies outside the scope of the philosophy set forth at the beginning of this chapter. In a democratic nation or institution, that choice is made by political debate, whether at the national level or at the grass-roots level. In a dictatorship, of course, it is made by decree. In these circumstances, constraints are likewise political choices. Our chairperson's budget, for example, is (at least indirectly) the result of a vote in the state legislature. Whenever a model is prescriptive rather than descriptive, i.e., whenever a model attempts to predict what should be done rather than what is actually done, ability to explain facts is no longer a suitable criterion of worth. There are no facts to test such a model! A descriptive model is concerned with the past or present, a prescriptive model with the future; and, as Benjamin Franklin once wryly remarked, the only certainties about the future are death and taxes. Without certainty, there are no facts.

We should emphasize here that all prescriptive models have descriptive elements. For example, the manager of our fishing fleet assumed that the logistic law would describe future growth of fish stocks in the absence of harvesting. But even these descriptive elements are untestable, except by virtue of waiting and using hindsight. A decision maker must use foresight, however; she cannot wait for the future to happen.

If fitting the facts is no longer relevant, how do we measure the worth of a prescriptive model? That raises a controversial issue. I think it fair to say that there is no widely accepted criterion of worth for prescriptive models. I propose, however, that flexibility should be a candidate. Even if it is not the only one, flexibility is at least desirable in a decision maker's model, and we shall try to retain as much as possible. In practice, as we shall see in Part III and beyond, this usually means keeping parameters free.

It is also desirable that the descriptive elements of a decision making model should explicitly recognize that the future is uncertain, which brings us nicely to Chapter 5.

Exercises

4.1 Show that (4.22) has a maximum where (4.23) is satisfied.

4.2 Show graphically that (4.23) has a unique solution $\bar{u}^*(\epsilon)$, which is lower than the optimal flight speed in zero wind.

4.3 (i) Verify (4.24).
(ii) The graph of $\bar{u}^*(\epsilon)$ in Fig. 4.2 seems almost linear but has a slight upward curvature. Can you prove this by using the shift technique of Chapter 2?

4.4 (i) Derive (4.33) and (4.34).
(ii) Verify (4.35).
Hint: Show that $\ln(1 + w) > bw$ whenever $-0.632 < w < 0$, where $b = e/(e - 1) \approx 1.582$.

4.5 Show that the effect of future generations of trees in (4.38) is to make the optimal rotation time shorter than predicted by Section 3.4. *Hint:* You need only the fact that ϵ is positive.

4.6 Show that $f(t) = V(t)/(e^{\delta t} - 1)$ has a maximum at $t = s^*$, where s^* is defined by (4.40), provided $V''(s^*) < \delta V'(s^*)$.

4.7 (i) Verify that the attracting equilibria in Fig. 4.10 are stable but any others are unstable, by using the methods of Section 2.2.
(ii) Obtain (4.77).

*4.8 Student A thinks that in all of Section 4.13 there is no good reason to reject the model (1.54), so why not accept it and forget all this business about components and paradigms? Student B thinks the problem with (1.54) is that it will explain almost any kind of behavior for some value of a. Student C thinks the problem with (1.54) is that so many real effects are incorporated into a that it's almost impossible to infer from data what its value should be. What do you think?

*4.9 Student A says that the logistic growth model is a paradigm, because it was used in Section 3.7 to capture some qualitative features of a fish population's growth and is too simple to provide a realistic description of any population. Student B says that the logistic growth model is not a paradigm but a predictive model, because it was used in Sections 1.9 and 4.1 to predict the growth of the U.S. population. What do you think?

*4.10 Student C says that the plant growth model of Section 1.3 was a paradigm until it predicted the growth in value of a stand of Douglas firs, then it became a predictive model. Student A says once a paradigm, always a paradigm. Student B says that the plant growth model can't be a predictive model because it's used as a component in Section 3.2, whose model is not predictive. What do you think?

*4.11 Student A says Section 2.8 provides a better model than Section 4.7. Student B says Section 4.7 provides a better model than Section 2.8. Student C says that it doesn't matter, because neither of them is any good. What do you think?

*4.12 (i) Let $x(t)$ denote the number of species of birds on an island at time t (not the number of individuals—thus if an island's only inhabitants are 6 million gulls and 2 turtle doves, then $x = 2$). Let E denote the number of species becoming extinct per unit of time; let I denote the number of species immigrating per unit of time. Write down a relation between dx/dt, E, and I.

(ii) Will the number of species immigrating per unit of time decrease or increase with x? Why? What is the simplest dependence of I upon x which incorporates your answer? Interpret any parameters you introduce.

(iii) Will the number of species becoming extinct per unit of time decrease or increase with x? What is the value of E if $x = 0$? What is the simplest dependence of E upon x which incorporates your answer? Interpret any parameters you introduce.

(iv) The biota of the island of Krakatoa was completely destroyed by a volcanic eruption in 1883 (Diamond and May, 1976, p. 175). Suppose that birds subsequently began to immigrate at the rate of ν species per unit of time. Identify ν in your analysis.

(v) The number of bird species on Krakatoa returned to equilibrium quite quickly (see Diamond and May, op. cit.). Use the direct method to find this equilibrium, according to your model. Show graphically that the equilibrium is stable.

(vi) Verify that this equilibrium is stable by solving the ordinary differential equation in (i). What does the fact that the equilibrium was reached quickly tell you about some of your parameters? According to Diamond and May, the number of plant species is still rising toward equilibrium. What would this tell you about your parameters, if your model were used for plants instead of birds?

(vii) Criticize your assumptions, both the ones that you've made and the ones we've made here. What is wrong with the form of $E(x)$ for small x? Attempt to improve your model. How does any new assumption affect the equilibrium?

Note: Even if the *number* of species on an island remains static, their *identities* will still change, because species are continually immigrating and becoming extinct.

4.13 In the model of Section 4.11, we tacitly assumed that the parameters b_1, b_2, c_1, c_2 did not satisfy $b_1 b_2 - c_1 c_2 = 0$, as is clear from (4.70). There are good reasons for this: because of environmental fluctuations, only inequalities between system parameters can be expected to have any validity. In a laboratory, however, where the environment is kept constant, equalities between parameters may be realistic. Therefore, in this exercise, we will discover the dynamical behavior of the system

$$\frac{dx}{dt} = x\left(a_1 - c_1 x - \frac{c_1 c_2}{b_2} y\right),$$

$$\frac{dy}{dt} = y(a_2 - b_2 x - c_2 y).$$

(i) Show that, if $a_1 b_2 - a_2 c_1 \neq 0$, then the only equilibria are at $(0, a_2/c_2)$ and $(a_1/c_1, 0)$. Show graphically that $(x(t), y(t))$ approaches $(a_1/c_1, 0)$ if $a_1/c_1 > a_2/b_2$ and approaches $(0, a_2/c_2)$ if $a_2/b_2 > a_1/c_1$.

(ii) Show that, if $a_1 b_2 - a_2 c_1 = 0$, then every point on the straight line between $(0, a_2/c_2)$ and $(a_1/c_1, 0)$ is an equilibrium. Show graphically that these equilibria are metastable.

*4.14 If the model (4.74) of Section 4.12 does not give rise to a limit cycle, then at least one of the assumptions embodied in it must be false. Which assumption concerning predators is questionable, and why? Can you suggest a better one? (*Hint*: Do Canadian lynx eat only snowshoe hare?)

*4.15 The net specific growth rate of a human population is the difference between the specific birth rate b and the specific death rate d, i.e.,

$$\frac{1}{x}\frac{dx}{dt} = b - d,$$

where $x(t)$ denotes the population at time t in suitably large units (of a million people, say) and b and d are, respectively, the numbers of births and deaths per unit of time, per unit of existing population. A very special kind of population, which demographers call a *birth cohort*, consists of a large number of individuals of a certain sex (female, say) *who were all born at a certain time*, say $t = 0$. The large number in question is usually taken to be 100,000 individuals, which we shall designate henceforward as the unit of population. Hence $x(0) = 1$. The population in question is so exclusive that, by definition, you cannot enter it unless you were already in it at $t = 0$. Thus $b = 0$. Demographers call d the force of mortality and prefer to denote it by the Greek letter μ. Hence

$$\frac{1}{x}\frac{dx}{dt} = -\mu, \qquad x(0) = 1.$$

Now, a person who is born at time $t = 0$ will be of age z at time z, so that time and age are identical for our birth cohort. Thus we can use age z instead of time t as our independent variable. The advantage of this is that the mortality, i.e., the specific death rate of the birth cohort, usually depends much more strongly on age than it does on time. If we ignore the dependence of μ on time completely, then we have either of the following two equivalent forms:

$$\frac{1}{x}\frac{dx}{dz} = -\mu(z), \qquad x(0) = 1,$$

or

$$x(z) = \exp\left\{-\int_0^z \mu(\xi)d\xi\right\}.$$

(i) Why should μ depend much more strongly upon age than upon time? When $\mu = \mu(z)$, demographers say that mortality is *fixed*—i.e., fixed at birth—though mortality would generally be lower at each age z for a cohort born in, say, 1950 than for one born in, say, 1890. How could exceptions to this arise? *Hint*: You may find the first of the two data sets in Table E4.1 helpful.

(ii) (You may wish to read Appendix 1 before attempting this part.) If you used Table E4.1 in answering (i) then you realized that decreasing mortality is related to increasing expectation of life at birth. We will now derive this relationship explicitly. Because $x(z)$ is the fraction of the cohort alive at age z, the fraction of the cohort dying between the ages of z and $z + \delta z$, where

Table E4.1   Some life expectancies.

Life expectancy at birth (in years) of Swedish females born in a given year		Life expectancy in 1967 of a Swedish female with the given age in years	
Year	Expectancy e_0	Age x	Expectancy
1880	49.53	0	76.58
1885	51.13	5	72.59
1890	52.44	10	67.68
1895	54.09	15	62.76
1900	54.26	20	57.85
1905	56.77	25	52.98
1910	58.95	30	48.13
1915	59.58	35	43.32
1920	58.71	40	38.54
1925	63.77	45	33.84
1930	64.11	50	29.26
1935	66.15	55	24.78
1940	68.42	60	20.46
1945	70.16	65	16.35
1950	72.76	70	12.61
1955	74.18	75	9.36
1960	75.22	80	6.86
1965	76.14	85	4.89

Calculated on the assumption that specific death rates prevailing at birth remain constant throughout a lifetime (adapted from Keyfitz and Flieger, 1971, pp. 102 and 464).

δz is infinitesimally small, is

$$x(z) - x(z + \delta z) = -x'(z)\delta z + o(\delta z),$$

where we use the "little oh" notation defined in Chapter 1. Hence, if the random variable L denotes length of life and f its probability density function, then we can write

$$f(z)\delta z + o(\delta z) = \text{Prob}(z < L \leq z + \delta z) = -x'(z)\delta z + o(\delta z).$$

The expectation of life at birth, e_0, is simply the expected value of the random variable L. Hence show that

$$e_0 \equiv E[L] \equiv \int_0^\infty zf(z)dz = \int_0^\infty x(z)dz.$$

If $\mu(z)$ were known as a function of z, then e_0 could be calculated from the above formula as

$$e_0 = \int_0^\infty \exp\left\{-\int_0^z \mu(\xi)d\xi\right\} dz.$$

(iii) One of the first attempts to determine the form of $\mu(z)$ was made by Benjamin Gompertz (1825). He reasoned that mortality is partly caused by humankind's "deterioration, or an increased inability to withstand destruction" and that "at the end of equal infinitely small intervals of time, he lost equal portions of his remaining power to oppose destruction," which is another way of saying that "power to avoid destruction" decreases at a constant specific rate. If "power to avoid destruction" is inversely proportional to mortality, what form of $\mu(z)$ would be implied by Gompertz's reasoning, if that were the sole cause of death? This form of $\mu(z)$ has become known as the Gompertz law of mortality.

(iv) Demographers and actuaries use $l(z)$ to denote the probability that an individual chosen randomly from a certain population will survive to reach age z (or beyond). Identify $l(z)$ with a quantity defined above. (*Note*: Strictly, your answer is what demographers call "$l(z)$ on radix 1." Radix denotes an arbitrary scaling factor, here 1 but usually 100,000, by which values of $l(z)$ are multiplied.)

(v) Tables of values of $l(z)$ are readily available as part of so-called *life tables*, which actuaries and demographers regularly compile and update. We will indicate how such tables are compiled in Section 8.6. Use the following excerpt from the 1966 life table for U.S. females (Keyfitz and Flieger, 1971, p. 354) to validate the Gompertz law of mortality.

Age z	$l(z)$	Age z	$l(z)$
0	1.00000	45	0.93486
1	0.98003	50	0.91519
5	0.97672	55	0.88698
10	0.97493	60	0.84706
15	0.97345	65	0.79115
20	0.97058	70	0.70671
25	0.96711	75	0.59260
30	0.96285	80	0.44656
35	0.95677	85	0.27641
40	0.94805		

Say whether you believe the law to be valid for all ages, for some ages, or not at all. *Hint*: Review Section 1.9 and your solution to Exercise 1.10.

(vi) Gompertz never believed, however, that "deterioration" was the sole cause of death. Rather, he believed that there were "two generally coexisting causes," the second of which was "chance, without previous disposition to death or deterioration." We might prefer to regard the second cause as constant exposure to the risk of accident. What form would $\mu(z)$ take if this constant exposure to accident were the sole cause of death? Deduce the form of $l(z)$. Validate this mortality law by using the data above for U.S. females in 1966. Is there any range of values of z in which this law explains the data better than Gompertz's law? Under what conditions could the second cause be the dominant cause of death?

(vii) Suggest a form for $\mu(z)$ when Gompertz's two causes are allowed to coexist. Find the corresponding form of $l(z)$.

*4.16 There is a certain age, usually denoted as ω by demographers, beyond which no one can survive. Fries and Crapo (1981) call this the *maximum life potential*. As in the previous exercise, let $l(z)$ denote the probability of surviving to age z (or more). Then $l(z) = 0$ for $z \geq \omega$; and the specific death rate, $\mu(z)$, must tend to infinity as $z \to \omega$ to ensure that no one survives past age ω. A form of $\mu(z)$ consistent with this property is

$$\mu(z) = \frac{\epsilon}{\omega - z}, \qquad 0 \leq z \leq w,$$

where ϵ (> 0) is a constant.

(i) With this form of $\mu(z)$, what is the probability of surviving to age z? (See Exercise 4.15) What is the expectation of life at birth, e_0?

(ii) Using the 1966 survival probabilities for U.S. females, quoted in Exercise 4.15, validate the mortality law above for the particular choices $\omega = 90$, $\omega = 100$, and $\omega = 110$ (years). In each case, is the law valid for all z? If not, is there any indication that it might be valid for some z, i.e., that there exist z_1 and z_2 such that

$$\mu(z) = \frac{\epsilon}{\omega - z}, \qquad z_1 < z < z_2?$$

(iii) Do your results surprise you in any way? What is the conceptual difficulty in using this form of $\mu(z)$ for large values of z? Elaborate, comparing your analysis with that of Exercise 4.15.

(iv) Can you suggest a way of resolving this difficulty and, at the same time, suggest a "hybrid" mortality law that gives a better fit to the life table data for U.S. females in 1966, in the range $30 < z < \omega$, than either the Gompertz law or the one considered here?

(v) Criticize this hybrid mortality law. Does its better fit to data grant sufficient grounds for accepting it? If not, do you have sufficient grounds for rejecting it?

*4.17 (i) For the U.S. population in 1970 (male and female), the probabilities of survival at age z (or more) are given by Fries and Crapo (1981, p. 146) as follows:

Age z	$l(z)$	Age z	$l(z)$
69	0.64077	89	0.11387
74	0.52308	94	0.03927
79	0.38279	99	0.00921
84	0.23805	100	0.00656

The same authors estimate maximum life potential to be $\omega = 115$. Use the above data to validate the mortality model

$$\mu(z) = \frac{\epsilon}{\omega - z},$$

where ϵ is a constant, for ages beyond 70. *Hint:* Use the expression for l derived in Exercise 4.16. For the range of z for which you consider the model valid, estimate ϵ graphically.

(ii) The same authors give probabilities of survival to age z (or more) in the range $29 \leq z \leq 64$ as follows:

Age z	$l(z)$	Age z	$l(z)$
29	0.95454	49	0.89559
34	0.94675	54	0.85969
39	0.93600	59	0.80730
44	0.91995	60	0.73560

Use these data and those of (i) to validate the Gompertz model $\mu(z) = ae^{cz}$ for ages beyond 29. For the range of z in which you consider the model valid, estimate a and c graphically.

(iii) Using your answers to (i) and (ii), can you suggest a continuous function that models mortality for all ages beyond about 35?

*4.18 In Section 1.6 and again in Section 3.3, we used the Cobb–Douglas production function $Q = AL^{\gamma}K^{1-\gamma}$, where a, γ are constants satisfying $a > 0$ and $0 < \gamma < 1$, to model the relationship between (real) output $Q(t)$, capital stock $K(t)$, and labor $L(t)$ at time t, in the manufacturing sector of an economy. But the output was expressed in dollar terms as $Y(t)/P(t)$, where Y denotes nominal value of goods produced and P denotes price index. Because everything is expressed in dollar terms, there is no reason to confine this model to a manufacturing sector. It can be applied just as easily to the produce grown by an agricultural sector or the services provided by office workers. Hence the model can be applied to an entire economy.

In Section 3.3 we predicted that the optimal amount of labor for such an economy to employ would be

$$L^*(t) = \left(\frac{\gamma aP(t)}{w(t)} \right)^{1/(1-\gamma)} K(t),$$

where $w(t)$ is the annual wage paid to a unit of labor. As when validating the crop-spraying model of Section 3.6, let's assume that what businesswomen do is also what is optimal, by virtue of trial and error. Having made this assumption, if the Cobb–Douglas model is a good one, then the wages paid to labor in the entire economy should equal $w(t)L^*(t)$.

Denison (1974, p. 260) has estimated $w(t)L^*(t)$ as a fraction of nominal output $Y(t)$ for the entire economy of the United States (including Alaska and Hawaii) for most of the 1960s. Denison's estimates for the years $\{196t: 0 \leq t \leq 8\}$, are as follows:

t	$\dfrac{w(t)L^*(t)}{Y(t)}$	t	$\dfrac{w(t)L^*(t)}{Y(t)}$	t	$\dfrac{w(t)L^*(t)}{Y(t)}$
0	0.8222	3	0.8047	6	0.7814
1	0.8226	4	0.7980	7	0.7983
2	0.8087	5	0.7833	8	0.7984

Based on these estimates, how well did the Cobb–Douglas production function model the relationship between capital, labor, and output in the decade of the 1960s?

*4.19 How good are the chemical dynamic models of Sections 1.10 and 1.11? First decide how you could test them, then ask a friendly chemistry professor about where you might find some data.

*4.20 In Section 4.5, investigate the possibility that the false assumption is not that there is no following wind, but rather that S is the same for all birds.

*4.21 In Section 3.1, even if there were no following wind, (3.10) would underestimate how far a bird can migrate because we have assumed that mass is constant throughout. In reality, however, mass decreases during flight from m to λm, where λ is significantly less than 1. For example, $\lambda = 5/8$ when a yellow wagtail migrates across the Sahara desert (Alexander, 1990, p. 447).

How much further than (3.10) predicts can a bird migrate? How much does this result affect the analysis in Section 4.5?

III THE PROBABILISTIC VIEW

5 BIRTH AND DEATH. PROBABILISTIC DYNAMICS

The dynamical models we met in Chapter 1 are all deterministic.[1] They predict that the state of some system at time t—perhaps the speed of a boat or the weight of a plant—is certain to be, say, x. If the model is correct, then there is no possibility that the state in question might instead turn out to be y or z. Now, this approach is quite acceptable if the system being modelled is protected from unexpected shocks; for example, if the plant is in a laboratory, where unexpected drought will not affect its growth, or if the boat is rowed on a calm day, so that no gusts of wind will affect its speed. But an element of uncertainty is an intrinsic part of the world in which we live, and we cannot ignore it indefinitely. Accordingly, in this chapter, we develop models that recognize uncertainty explicitly. They do not predict that the state of a system is certain to be x at time t; rather, they predict the *probability* that the state will be x at time t.

For reasons that will soon emerge, we begin our study of probabilistic models by taking a fresh look at a deterministic one, namely, Section 1.9's logistic model for U.S. population growth. According to this model, the state of the population at time t, namely, $x(t)$ units, is the number of *millions* of people in the United States at time t. Because the addition or removal of one individual, to or from the population, corresponded to addition or removal of such a tiny fraction of the unit in which the population was measured in Section 1.9, we could imagine that x varied continuously. So we made $x(t)$ a real-valued function. If the number of individuals in an entire population is small, however, then one individual corresponds to a significant fraction of it, and it is no longer legitimate to let the state of the population be a real-valued function. In this chapter, because we are interested in the dynamics of small populations, we will de-

---

[1] You may wish to study Appendix 1 before proceeding.

part from the approach of Section 1.9 by making the state of a system an integer. Because we have already departed from the approach of Chapter 1 by choosing probabilistic models, the state of a system at time t must be an *integer-valued random variable*. We will denote it by $X(t)$ and allow it to take values $i = 0, 1, 2, \ldots, N$. Thus the dynamical systems we describe in this chapter, by using probabilistic models, have $N + 1$ possible states. But N may be as large as we please, and it will sometimes be convenient to regard it as infinite.

For a fixed value of t, $X(t)$ is a single random variable. But t may take any value satisfying $0 \leq t < \infty$. Hence we deal, in effect, with an infinite set of random variables, namely, $S_t \equiv \{X(t): 0 \leq t < \infty\}$. We give S_t a name: we call it a *continuous-time stochastic process*. All random variables in this stochastic process have the same sample space, namely, the integers between 0 and N. But they do not in general have the same distribution. Rather, their distributions depend on the value of the parameter t. (In the special case where all distributions are indeed identical, the stochastic process is said to be stationary, as we shall see in the following chapter.)

Although we depart from the approach of Section 1.9 in two respects, by making the state a random variable and by making it an integer, one important property of the logistic model is retained in our probabilistic models. Suppose that a population that is growing logistically, with maximum specific growth rate R and capacity K, is known to have magnitude I at time $t = s$. Then its size at time $t = s + \tau$ is obtained by integrating the differential equation

$$\frac{1}{x}\frac{dx}{dt} = R\left(1 - \frac{x}{K}\right) \tag{5.1}$$

forward in time from $t = s$ to $t = s + \tau$, subject to the initial condition $x(s) = I$. We obtain

$$x(s + \tau) = \frac{IK}{I + (K - I)e^{-R\tau}}. \tag{5.2}$$

Notice that this expression is completely independent of s. Thus, τ units of time after it had magnitude I, the population will have magnitude (5.2), no matter where in the course of history it had magnitude I. In other words, if the model is correct, and if $t = s$ is regarded as the present time, then you can foretell the population's future by knowing only its current magnitude I. You do not need to know anything about its past. In this sense, the logistic growth model has no memory, and it is this memorylessness which our probabilistic models will retain.

The property that the future depends only on the present, and not on the past, is known in the literature on probability as the *Markov* property. Thus the models of this chapter are Markov models. Because these models are probabilistic, the mathematical formulation of the Markov property is not quite as simple as saying that $x(s + \tau)$ depends on τ and $x(s)$, but not on s itself. Rather, we say that, if $X(s) = i$, then the probability that $X(s+\tau) = j$ depends on τ and i, but not on s itself. In terms of conditional probabilities, $\text{Prob}(X(s+\tau) = j \mid X(s) = i)$ is independent of s. Because it's independent of s, we might as well set s equal to zero, and that's why the Markov property is usually stated as

$$\text{Prob}(X(s+t) = j \mid X(s) = i) = \text{Prob}(X(t) = j \mid X(0) = i), \quad (5.3)$$

where we have replaced τ by t.

What gives rise to the Markov property? We can understand best by returning to the logistic growth model, where the net specific growth rate $\mu = b - d$ is given by

$$\mu(x) = R\left(1 - \frac{x}{K}\right), \quad (5.4)$$

and imagining instead that, for some strange reason which need not concern us, we had another population for which the net specific growth rate was

$$\mu = \mu(x,t) = Rt\left(1 - \frac{x}{K}\right). \quad (5.5)$$

Replacing the right-hand side of (5.1) by (5.5) and integrating from $t = s$ to $t = s + \tau$, subject to $x(s) = I$, yields

$$x(s+\tau) = \frac{IK}{I + (K-I)e^{-R\tau(\tau+2s)}}. \quad (5.6)$$

Notice that this expression is no longer independent of s. Our strange population has a sense of history! Comparing (5.5) and (5.6) with (5.1) and (5.2), we see that the Markov property of the logistic growth model arises from the assumption that the specific birth rate b, the specific death rate d, and, as a consequence, the net specific growth rate μ depend only on the state; i.e., $b = b(x)$ and $d = d(x)$, implying $\mu = \mu(x)$. An *explicit* dependence of μ on t, as in (5.5), would destroy the system's memorylessness (there is always, of course, an *implicit* dependence of μ on t, because x depends upon t).

In the logistic growth model, the number of units added to the population in the infinitesimal interval $[t, t+\delta t)$, *per unit of existing population*, is $b(x)\delta t + o(\delta t)$. Similarly, the number of units removed in the interval $[t, t+\delta t)$, per unit of population, is $d(x)\delta t + o(\delta t)$. Because the logistic growth model is deterministic, and because x varies continuously, we can be certain that these amounts are added and removed in the infinitesimal interval. In a probabilistic model, however, we cannot be sure that anyone was born or died in the interval $[t, t+\delta t)$. We can do no more than assign some probability to a birth or death. The longer the interval $[t, t+\delta t)$, the more likely it is that someone is born or dies. It therefore seems reasonable to assume that the probability of a birth or death is proportional to δt. On the other hand, if the interval is short enough, then the probability of two or more births, or two or more deaths, or a birth and a death, is negligible. We can make these ideas precise by saying that the probabilities of the latter events are $o(\delta t)$; whereas the probability of a birth when the state of the population is i is $b(i)\delta t + o(\delta t)$. It is customary to denote functional dependence upon an integer by using a subscript, however, and so we prefer to write the probability of a birth as

$$b_i \delta t + o(\delta t). \qquad (5.7)$$

Similarly, we let the probability of a death when the state of the population is i be $d(i)\delta t + o(\delta t)$ or, in subscript notation,

$$d_i \delta t + o(\delta t). \qquad (5.8)$$

The previous paragraph suggests that our models should then retain the Markov property; and indeed they do, as can be verified from the examples below. It will be convenient to go on calling b_i the birth rate and d_i the death rate, but these should now be interpreted as probabilities per unit of time, rather than specific rates of addition or removal.

In Chapter 1, we introduced a generic term for growth and decay; we called it dynamics. Similarly, we shall find it useful to have a generic term for the events that increase or decrease the state of a system by 1, i.e., for births and deaths; we shall call them *transitions*. On the other hand, birth and death are not always the most suggestive labels for these two transitions; and, where appropriate, we shall replace the labels birth and death by arrival and departure, entrance and exit, rise and fall, or even repair and malfunction. But the transitions will still be described by (5.7) and (5.8). Only the label will have changed, to conform to everyday usage.

The goal of our models, as stated in the opening paragraph, will be to assign probability to the event that the state of a system is i at time t. We will therefore attempt to assign numerical values to $\pi_i(t)$, where we define

$$\pi_i(t) \equiv \text{Prob}(X(t) = i), \qquad i = 0, \ldots, N. \qquad (5.9)$$

The $N + 1$ values $\{\pi_i(t): i = 0, 1, \ldots, N\}$ constitute the probability distribution of the discrete random variable $X(t)$ and must therefore sum to 1:

$$\sum_{i=0}^{N} \pi_i(t) = 1. \qquad (5.10)$$

It will be convenient to record these values as the $(N + 1)$-dimensional column vector

$$\boldsymbol{\pi}(t) = (\pi_0(t), \pi_1(t), \ldots, \pi_N(t))^T, \qquad (5.11)$$

where a superscript T denotes transpose; $\boldsymbol{\pi}(t)$ will be called the system distribution vector. It will also be convenient to denote (5.3) by $r_{ij}(t)$, i.e., to define

$$r_{ij}(t) = \text{Prob}(X(t) = j \mid X(0) = i), \qquad 0 \le i, j \le N. \qquad (5.12)$$

For fixed i, $\{r_{ij}(t): 0 \le j \le N\}$ defines a conditional probability distribution whose terms must sum to 1; i.e.,

$$\sum_{j=0}^{N} r_{ij}(t) = 1, \qquad 0 \le i \le N. \qquad (5.13)$$

For $0 \le i \le N$ we have $N+1$ conditional distributions, and it is convenient to store them as the rows of the $(N+1) \times (N+1)$ matrix

$$\mathbf{R}(t) = [r_{ij}(t)]. \qquad (5.14)$$

We note in passing that any square matrix with rows summing to 1 is called a *stochastic matrix*. Hence $\mathbf{R}$ is stochastic.

Now, if the events U_0, U_1, ..., U_N exhaust the sample space of a random variable, and if U is any event in the same sample space, then

$$\text{Prob}(U) = \sum_{k=0}^{N} \text{Prob}(U \mid U_k) \cdot \text{Prob}(U_k); \qquad (5.15)$$

see (A.13). Note that the event U does not have to be associated with the same random variable as the events U_0, U_1, ..., U_N. In particular, U_k may be the event that the random variable $X(0)$ takes the value k, and U may be the event that $X(t)$ takes the value i. Then $X(0)$ and $X(t)$ are different random variables, but they have the same sample space and belong to the same stochastic process. Hence:

$$\text{Prob}(X(t) = i) = \sum_{k=0}^{N} \text{Prob}(X(t) = i \mid X(0) = k) \cdot \text{Prob}(X(0) = k). \qquad (5.16)$$

On using (5.12):

$$\pi_i(t) = \sum_{k=0}^{N} r_{ki}(t) \cdot \pi_k(0), \qquad 0 \le i \le N, \qquad (5.17)$$

or in matrix notation,

$$\pi(t)^T = \pi(0)^T \mathbf{R}(t). \qquad (5.18)$$

For given $\pi(0)$, $\pi(t)$ can be deduced from (5.9) as soon as $\mathbf{R}(t)$ is known. For this reason, our major preoccupation in this chapter will be with the matrix $\mathbf{R}(t)$.

The first of our examples is the simplest one possible, a probabilistic model for a population that may contain at most one individual.

5.1   When Will an Old Man Die? The Exponential Distribution

Suppose that a population consists of just one old man. Sooner or later he will die; then there will be no one. Let the random variable $X(t)$ denote the number alive in this population at time t. Then X has sample space $\{0, 1\}$; i.e., the two possible states of the population are $i = 0$ and $i = 1$, and the matrix (5.14) becomes

$$\mathbf{R}(t) = \begin{bmatrix} r_{00}(t) & r_{01}(t) \\ r_{10}(t) & r_{11}(t) \end{bmatrix}.$$

We will now proceed to derive explicit expressions for the entries of this matrix.

If the man were dead at time $t = 0$ then he would have to remain so. Hence $\text{Prob}(X(t) = 0 \mid X(0) = 0) = 1$, $\text{Prob}(X(t) = 1 \mid X(0) = 0) = 0$. Thus, by (5.12), $r_{00}(t) = 1$ and $r_{01}(t) = 0$. But the man is alive at time $t = 0$, so that we would not expect r_{00} and r_{01} to have any bearing on his future; and this is confirmed by (5.20) below. By (5.13) with $i = 1 = N$, i.e., because the last row of the matrix $\mathbf{R}$ must sum to 1, we have $r_{10}(t) + r_{11}(t) = 1$. Hence

$$\mathbf{R}(t) = \begin{bmatrix} 1 & 0 \\ 1 - r_{11}(t) & r_{11}(t) \end{bmatrix}. \tag{5.19}$$

The man is alive initially, and so we have $\pi(0) = (0, 1)^T$. Hence on using (5.18) the system distribution vector $\pi(t)$ is given by

$$\pi(t)^T = (0, 1)\mathbf{R}(t) = (1 - r_{11}(t), r_{11}(t)), \tag{5.20}$$

and π is completely determined if we know $r_{11}(t)$.

We shall obtain $r_{11}(t)$ below from a differential equation, which requires an initial condition. A suitable one emerges from the observation that if the man is alive at time $t = 0$, then he cannot also be dead at $t = 0$, though he could conceivably be dead at time $t = \epsilon$ for any value of $\epsilon > 0$, however small. Hence

$$r_{11}(0) = \text{Prob}(X(0) = 1 \mid X(0) = 1) = 1. \tag{5.21}$$

Let U_1 be the event that the man is alive at time t, i.e., that $X(t) = 1$. Let U_0 be the event that he is dead at time t, i.e., that $X(t) = 0$. Let U be the event that the man is alive at the later time $t + \delta t$, i.e., that $X(t + \delta t) = 1$, where δt is infinitesimally small. Thus U_0 and U_1 are associated with the random variable $X(t)$, and U with the random variable $X(t + \delta t)$; but both random variables have the same sample space $\{0, 1\}$ and belong to the same stochastic process. The man must be either alive or dead at time t, and so U_0 and U_1 exhaust the possibilities then arising. Hence, from (5.15) with $N = 1$, we have

$$\text{Prob}[U] = \text{Prob}[U \mid U_0] \cdot \text{Prob}[U_0] + \text{Prob}[U \mid U_1] \cdot \text{Prob}[U_1]. \tag{5.22}$$

But $\text{Prob}[U \mid U_0] = 0$, because the man cannot be alive at time $t + \delta t$ if he is dead at time t. Also, from (5.9) and (5.20), we have $\text{Prob}[U] = \text{Prob}(X(t + \delta t) = 1) = r_{11}(t + \delta t)$; and $\text{Prob}[U_1] = \text{Prob}(X(t) = 1) = r_{11}(t)$. Moreover, because $\text{Prob}[U \mid U_1]$ is the probability that the man survives from t to $t + \delta t$, $1 - \text{Prob}[U \mid U_1]$ must be the probability that he dies within the interval $[t, t + \delta t)$, which, from (5.8), equals $d_1 \delta t + o(\delta t)$. Hence (5.22) reduces to

$$r_{11}(t + \delta t) = (1 - d_1 \delta t + o(\delta t)) r_{11}(t). \tag{5.23}$$

Rearranging and dividing by δt, we get

$$\frac{r_{11}(t + \delta t) - r_{11}(t)}{\delta t} = -d_1 r_{11}(t) + \frac{o(\delta t)}{\delta t}. \tag{5.24}$$

Taking the limit as $\delta t \to 0$, we have

$$\frac{d}{dt}\{r_{11}(t)\} = -d_1 r_{11}(t). \tag{5.25}$$

Solving this ordinary differential equation and using (5.20) and (5.21), we get

$$\pi_1(t) = r_{11}(t) = r_{11}(0)e^{-d_1 t} = e^{-d_1 t} \tag{5.26}$$

$$\pi_0(t) = r_{10}(t) = 1 - r_{11}(t) = 1 - e^{-d_1 t}. \tag{5.27}$$

Thus the probability that the population remains in state 1—i.e., the probability that the old man does not die—decays to zero exponentially as time increases toward infinity. This is the probabilistic analogue of the natural decay process of Chapter 1.

You can verify directly that this model satisfies the Markov property. Let U_1 be the event that the man is alive at time s, U_0 the event that he is dead at time s, and U the event that he is alive at time $t + s$. Because the man must be either alive or dead at time s, U_0 and U_1 exhaust the possibilities then arising, so that

$$\text{Prob}[U] = \text{Prob}[U \mid U_0] \cdot \text{Prob}[U_0] + \text{Prob}\, U \mid U_1] \cdot \text{Prob}[U_1]. \tag{5.28}$$

But $\text{Prob}[U \mid U_0] = 0$, because the man cannot be alive at time $t + s$ if he is dead at time s. Moreover, $\text{Prob}[U] = \text{Prob}(X(t + s) = 1) = \pi_1(t + s)$, and $\text{Prob}[U_1] = \text{Prob}(X(s) = 1) = \pi_1(s)$. Hence (5.28) implies

$$\text{Prob}(X(t + s) = 1 \mid X(s) = 1) = \text{Prob}[U \mid U_1] = \frac{\pi_1(t + s)}{\pi_1(s)} = \frac{e^{-d_1(t+s)}}{e^{-d_1 s}}$$

$$= e^{-d_1 t} = \text{Prob}(X(t) = 1 \mid X(0) = 1); \tag{5.29}$$

i.e., (5.3) is satisfied.

But there's another way to look at this. Let the random variable G denote the time until death in the old man's life or, with a view to generalization, the *time until transition*. Then the cumulative distribution function of G is given by

$$\text{Prob}(G > t) = e^{-d_1 t} \tag{5.30a}$$

or

$$\text{Prob}(G \le t) = 1 - e^{-d_1 t}. \tag{5.30b}$$

This important distribution, called the *exponential distribution*, is the only distribution with the property that

$$\text{Prob}(G > t + s \mid G > s) = \text{Prob}(G > t) \tag{5.31}$$

for all $t, s \ge 0$. Indeed (5.3) and (5.31) are equivalent statements of the Markov property of memorylessness. Thus, if the model is correct, the old man's chances of living to age 97, having just reached 95, are the same as his chances of living to be 87 were when he had just turned 85.

The cumulative distribution function $\text{Prob}(G \le t)$ is plotted in Fig. 5.1 (we will need this diagram in Section 8.7). The probability density function of the exponential distribution is given by

$$f(t) = \frac{d}{dt} \{\text{Prob}(G \le t)\} = d_1 e^{-d_1 t}, \qquad 0 \le t < \infty; \tag{5.32}$$

whence the mean μ and variance σ^2 of the distribution are given by

$$\mu = \int_0^\infty t f(t) dt = \frac{1}{d_1}, \tag{5.33a}$$

$$\sigma^2 = \int_0^\infty (t - \mu)^2 f(t) dt = \frac{1}{d_1^2}. \tag{5.33b}$$

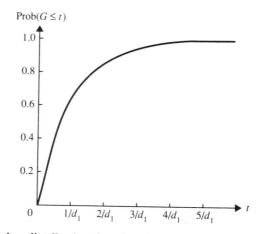

Fig. 5.1 Cumulative distribution function (5.30b) of the exponential distribution.

From (5.33a), we can interpret the death rate as

$$d_1 = \frac{1}{\text{MEAN TIME UNTIL TRANSITION}}. \tag{5.34}$$

5.2   When Will *N* Men Die? A Pure Death Process

The model we have just discussed is a very simple example of a pure *death process*. We will now consider a more general death process, in which a population consists of N old men, dead or alive. Let the random variable $X(t)$ denote the number alive at time t. Then X has sample space $\{0, 1, \ldots, N\}$; i.e., the $N + 1$ possible states of the population are $i = 0, 1, \ldots, N$.

If i men are alive at time 0, then there cannot be more than i men alive when $t > 0$. Hence

$$r_{ij}(t) = \text{Prob}(X(t) = j \mid X(0) = i) = 0 \qquad \text{if } i < j. \tag{5.35}$$

In other words, the matrix $\mathbf{R}(t)$, defined by (5.14), is lower triangular. Moreover, if i men are alive at time 0 then there cannot also be j men alive at time 0 unless $i = j$. It is convenient to express this idea by introducing some new notation. Let us define the *Kronecker delta* by

$$\delta_{ij} = \begin{cases} 0, & \text{if } i \neq j, \\ 1, & \text{if } i = j, \end{cases} \quad i, j = 0, 1, \ldots, N. \tag{5.36}$$

Then

$$r_{ij}(0) = \delta_{ij}, \qquad 0 \leq i, j \leq N. \tag{5.37}$$

Now suppose that i men are indeed alive at time 0; i.e., restrict the sample space of $X(0)$ to the event that $X(0) = i$. Let U denote the event that j men are alive at time $t+\delta t$, where δt is infinitesimally small. Let U_1 denote the event that j men are alive at time t, U_2 the event that $j + 1$ men are alive at time t, and U_3 the event that more than $j+1$ men are alive at time t. Because U_1, U_2, and U_3 exhaust the possibilities then arising, we have

$$\text{Prob}[U] = \text{Prob}[U \mid U_1] \cdot \text{Prob}[U_1] + \text{Prob}[U \mid U_2] \cdot \text{Prob}[U_2]$$
$$+ \text{Prob}[U \mid U_3] \cdot \text{Prob}[U_3]. \tag{5.38}$$

On using (5.8), the probability of a death in the interval $[t, t+\delta t)$, if $X(t) = j+1$, is $d_{j+1}\delta t + o(\delta t) = \text{Prob}[U \mid U_2]$. Similarly, the probability of no death in the interval $[t, t + \delta t)$, if $X(t) = j$, is $1 - d_j\delta t + o(\delta t) = \text{Prob}[U \mid U_1]$; whereas $\text{Prob}[U \mid U_3] = o(\delta t)$. Because $X(0) = i$, $\text{Prob}[U_1] = \text{Prob}(X(t) = j) = \text{Prob}(X(t) = j \mid X(0) = i) = r_{ij}(t)$, on using (5.12). Similarly, $\text{Prob}[U_2] = r_{i,j+1}(t)$ and $\text{Prob}[U] = r_{ij}(t + \delta t)$. Substituting into (5.38), we have

$$r_{ij}(t + \delta t) = \left\{1 - d_j\delta t + o(\delta t)\right\} r_{ij}(t) + \left\{d_{j+1}\delta t + o(\delta t)\right\} r_{i,j+1}(t)$$
$$+ o(\delta t) \cdot \text{Prob}[U_3].$$

Hence

$$\frac{r_{ij}(t+\delta t) - r_{ij}(t)}{\delta t} = -d_j r_{ij}(t) + d_{j+1} r_{i,j+1}(t) + \frac{o(\delta t)}{\delta t}.$$

Note that an expression for $\text{Prob}[U_3]$ is not required. Taking the limit as $\delta t \to 0$ now yields the ordinary differential equation

$$\frac{d}{dt}\{r_{ij}(t)\} = -d_j r_{ij}(t) + d_{j+1} r_{i,j+1}(t), \tag{5.39}$$

with (Exercise 5.1) solution

$$r_{ij}(t) = e^{-d_j t} r_{ij}(0) + d_{j+1} \int_0^t e^{-d_j(t-s)} r_{i,j+1}(s)ds. \tag{5.40}$$

Because $r_{i,i+1}(s) = 0$ by (5.35), we observe, on setting $j = i$ in (5.40) and using (5.37), that $r_{ii}(t) = r_{ii}(0)\exp(-d_i t) = \exp(-d_i t)$. Now let $X(s) = i$; and let the random variable G_i denote time that elapses until the following death or, in more general terms, time until the next transition. Then $\text{Prob}(G_i > t) = \text{Prob}(X(t+s) = i \mid X(s) = i) = \text{Prob}(X(t) = i \mid X(0) = i) = r_{ii}(t)$ on using the Markov property (5.3). Thus the time until transition is exponentially distributed with mean $1/d_i$; or

$$\text{Prob}(G_i > t) = e^{-d_i t}, \tag{5.41a}$$

$$d_i = \frac{1}{\text{MEAN TIME UNTIL TRANSITION}}. \tag{5.41b}$$

Expression (5.40) is valid for all i and j though, in view of (5.35), we do not need it for $i < j$. Thus the probabilities $r_{ij}(t)$ can be found recursively, starting with the diagonal elements of the matrix $\mathbf{R}(t)$ and moving leftward along the rows. For example, suppose that $d_i = D = $ constant (i.e., independent of i) for $i \geq 1$; we must, of course, take $d_0 = 0$, because the death of the entire population precludes further deaths of individuals. Then for $N = 3$ we obtain

$$\mathbf{R}(t) = \begin{bmatrix} 1 & 0 & 0 & 0 \\ 1 - e^{-Dt} & e^{-Dt} & 0 & 0 \\ 1 - e^{-Dt} - Dte^{-Dt} & Dte^{-Dt} & e^{-Dt} & 0 \\ 1 - e^{-Dt}(1 + Dt + \frac{1}{2}D^2t^2) & \frac{1}{2}D^2t^2 e^{-Dt} & Dte^{-Dt} & e^{-Dt} \end{bmatrix}.$$

Notice that this is indeed a stochastic matrix. Notice also the pattern of entries, from which the structure of the general $(N+1) \times (N+1)$ matrix $\mathbf{R}(t)$ is readily apparent. Clearly, the probability that N old men alive today will all have died t years from now is

$$r_{N0}(t) = 1 - e^{-Dt} \sum_{k=0}^{N-1} \frac{(Dt)^k}{k!}, \tag{5.42}$$

where $1/D$ is a man's life expectancy. We should bear in mind, however, that $d_i = D$ is not realistic—why?

5.3 Forming a Queue. A Pure Birth Process

Suppose that a queue is forming at a checkout counter. Let the random variable $X(t)$ denote the number of individuals in the queue at time t, and let N be the maximum number of individuals for which the queue has room. Then X has sample space $\{0, 1, \ldots, N\}$, and we can think of arrivals as births in a population with $N + 1$ possible states, namely, $i = 0, 1, \ldots,$ N. We will assume that there are no departures from the queue or, as it were, no "deaths" in the population. The process we are about to describe is therefore a pure birth process.

Suppose that there are i individuals in the queue at time 0; i.e., $X(0) = i$. Because there are no departures,

$$r_{ij}(t) = \text{Prob}(X(t) = j \mid X(0) = i) = 0, \qquad \text{if } i > j; \qquad (5.43)$$

i.e., the matrix $\mathbf{R}(t)$ is upper triangular. Let U denote the event that there are j individuals in the queue at time $t + \delta t$, where δt is infinitesimally small. Let U_1 denote the event that there are j individuals in the queue at time t, U_2 the event that there are $j - 1$ individuals in the queue at time t, and U_3 the event that there are less than $j - 1$ individuals in the queue at time t. Because U_1, U_2, and U_3 exhaust the possibilities then arising, i.e., exhaust the sample space of $X(t)$, (A.13) implies that

$$\text{Prob}[U] = \text{Prob}[U \mid U_1] \cdot \text{Prob}[U_1] + \text{Prob}[U \mid U_2] \cdot \text{Prob}[U_2]$$
$$+ \text{Prob}[U \mid U_3] \cdot \text{Prob}[U_3]. \qquad (5.44)$$

On using (5.7), the probability of a birth in the interval $[t, t + \delta t)$ if the state of the population is $j - 1$ is $b_{j-1}\delta t + o(\delta t) = \text{Prob}[U \mid U_2]$. Similarly, the probability of no birth in the interval $[t, t + \delta t)$ if the state of the population is j is $1 - b_j\delta t + o(\delta t) = \text{Prob}[U \mid U_1]$; whereas $\text{Prob}[U \mid U_3] = o(\delta t)$. Because $X(0) = i$, $\text{Prob}[U_1] = \text{Prob}(X(t) = j) = \text{Prob}(X(t) = j \mid X(0) = i) = r_{ij}(t)$, on using (5.12). Similarly, $\text{Prob}[U_2] = r_{i,j-1}(t)$ and $\text{Prob}[U] = r_{ij}(t + \delta t)$. Substituting into (5.44), we have

$$r_{ij}(t+\delta t) = \left\{1 - b_j\delta t + o(\delta t)\right\} r_{ij}(t) + \left\{b_{j-1}\delta t + o(\delta t)\right\} r_{i,j-1}(t) + o(\delta t) \cdot \text{Prob}[U_3].$$

Rearranging, as in the previous section, and taking the limit as $\delta t \to 0$ now yields the ordinary differential equation

$$\frac{d}{dt}\left\{r_{ij}(t)\right\} = -b_j r_{ij}(t) + b_{j-1} r_{i,j-1}(t), \qquad (5.45)$$

with (Exercise 5.2) solution

$$r_{ij}(t) = e^{-b_j t}\delta_{ij} + b_{j-1}\int_0^t e^{-b_j(t-s)} r_{i,j-1}(s)ds, \qquad (5.46)$$

because $r_{ij}(0) = \delta_{ij}$, as in the previous section. Because $r_{i,i-1}(s) = 0$ by (5.43), we observe on setting $j = i$ in (5.46) that $r_{ii}(t) = \exp(-b_i t)$. Now let $X(s) = i$, and let the random variable G_i denote the time that elapses until the following birth, or, in more general terms, the time until the next transition. Then $\text{Prob}(G_i > t) = \text{Prob}(X(t+s) = i \mid X(s) = i) = \text{Prob}(X(t) = i \mid X(0) = i) = r_{ii}(t)$, on using the Markov property (5.3). Thus the time until transition is exponentially distributed with mean $1/b_i$; or

$$\text{Prob}(G_i > t) = e^{-b_i t}, \tag{5.47a}$$

$$b_i = \frac{1}{\text{MEAN TIME UNTIL TRANSITION}}. \tag{5.47b}$$

Expression (5.46) is valid for all i and j though, in view of (5.43), we do not need it for $i > j$. Thus the probabilities $r_{ij}(t)$ can be found recursively, starting with the diagonal elements of the matrix $\mathbf{R}(t)$ and moving to the right along the rows. For example, suppose $b_i = B =$ constant for $0 \le i \le N - 1$; we must, of course, take $b_N = 0$, because there is space for only N individuals. Then for $N = 3$ we obtain

$$\mathbf{R}(t) = \begin{bmatrix} e^{-Bt} & Bte^{-Bt} & \frac{1}{2}B^2t^2e^{-Bt} & 1 - e^{-Bt}(1 + Bt + \frac{1}{2}B^2t^2) \\ 0 & e^{-Bt} & Bte^{-Bt} & 1 - e^{-Bt}(1 + Bt) \\ 0 & 0 & e^{-Bt} & 1 - e^{-Bt} \\ 0 & 0 & 0 & 1 \end{bmatrix}.$$

Notice that this is indeed a stochastic matrix. Notice also the pattern of entries, from which the structure of the general $(N + 1) \times (N + 1)$ matrix $\mathbf{R}(t)$ is readily apparent. Clearly, the probability that a queue that began with i people will have j people at time t is given by

$$r_{ij}(t) = \begin{cases} \dfrac{(Bt)^{j-i}}{(j-i)!}e^{-Bt}, & \text{if } 0 \le j \le N - 1; \\[2ex] 1 - e^{-Bt}\displaystyle\sum_{k=0}^{N-i-1}\dfrac{(Bt)^k}{k!}, & \text{if } 0 \le i \le N - 1, j = N; \\[2ex] 1, & \text{if } i = j = N. \end{cases} \tag{5.48}$$

Though N will always be finite for the queue at a checkout, there are other queues for which N may be regarded as infinite. Indeed the pure birth model with constant birth rate $b_i = B$ and infinitely many states occurs so frequently in mathematical modelling that we give it a special name; we call it the *Poisson process*. Let the random variable $A(t)$ denote the number of arrivals by time t in a Poisson process. Then, setting $j = k$, $i = 0$, and letting $N \to \infty$ in the first part of (5.48), the distribution of A is given by

$$\text{Prob}(A(t) = k) = \text{Prob}(X(t) = k \mid X(0) = 0) = r_{0k}(t) = e^{-Bt}\frac{(Bt)^k}{k!}. \tag{5.49}$$

We call this distribution the Poisson distribution. You can easily verify (Exercise 5.2) that its mean is Bt; i.e., an average of Bt individuals arrive in any interval of length t.

The Poisson distribution is closely related to the exponential distribution. To see this, let the ith individual arrive at time s, and let the random variable G_i denote the time that elapses before the next arrival. Then, because of the Markov property, the random variables G_1, G_2, G_3, ..., are independent. We shall call them the *interarrival* times. On using (5.3) and observing that every entry on the diagonal of $\mathbf{R}(t)$ is e^{-Bt}, we have

$$\text{Prob}(G_i > t) = \text{Prob}(X(t + s) = i \mid X(s) = i) = r_{ii}(t) = e^{-Bt}. \qquad (5.50)$$

In other words, the interarrival times in a Poisson process have identical exponential distributions, with mean $1/B$. We shall use this fact in the following section.

Having now described both a pure birth model and a pure death model, the logical next step might be to describe a model in which births (or arrivals) and deaths (or departures) can both occur. We prefer, however, to address this matter later, in Section 5.5. Meanwhile, let's appreciate the power and usefulness of even a simple probabilistic model.

5.4 How Busy Must a Road Be to Require a Pedestrian Crossing Control?

Let λ denote the average number of cars per second that pass a point on a road where pedestrians cross. We will assume that these cars all travel in the same direction, as in each lane of a two-lane road with a median. If traffic is flowing freely, then the passing of a car may be regarded as an arrival in an infinite queue. This suggests using the Poisson process (5.49) as a model for the traffic flow, with interarrival time G_j as the time gap between the $(j - 1)$th car and the jth. But if a second is the unit of time then, from (5.49), the average number of cars per second is

$$\sum_{k=1}^{\infty} k \cdot \text{Prob}(A(1) = k) = \sum_{k=1}^{\infty} k \cdot e^{-B} \frac{B^k}{k!} = Be^{-B} \sum_{k=1}^{\infty} \frac{B^{k-1}}{(k-1)!} = Be^{-B}e^{B} = B.$$
$$(5.51)$$

Thus $\lambda = B$, and gaps between cars have the exponential distribution

$$\text{Prob}(G_j > t) = e^{-\lambda t}. \qquad (5.52)$$

Let T be the time it takes a typical pedestrian to cross the road. Then the probability that a typical pedestrian crosses the road during the jth time gap, i.e., between the $(j - 1)$th car and the jth, is

$$\gamma_j \equiv \text{Prob}(G_0 \leq T, G_1 \leq T, G_2 \leq T, \ldots, G_{j-2} \leq T, G_{j-1} > T)$$
$$= \text{Prob}(G_0 \leq T) \cdot \text{Prob}(G_1 \leq T) \ldots \text{Prob}(G_{j-2} \leq T) \cdot \text{Prob}(G_{j-1} > T)$$
$$= (1 - e^{-\lambda T})^{j-1} \cdot e^{-\lambda T}, \qquad (5.53)$$

where we have used (A.57) and the fact that interarrival times are independent. Thus the mean number of gaps that a typical pedestrian requires to cross the road (including the gap during which he crosses) is (Exercise 5.4)

$$\sum_{j=0}^{\infty} j\gamma_j = e^{\lambda T}. \tag{5.54}$$

Let α denote the average length of the gaps that occur while a typical pedestrian is waiting to cross. Then a typical pedestrian's average waiting time is $(e^{\lambda T} - 1)\alpha$.

Now, let τ be the longest time for which, on average, a typical pedestrian can reasonably be expected to wait. Then a crossing control is needed unless

$$(e^{\lambda T} - 1)\alpha \le \tau. \tag{5.55}$$

Whatever the value of α, it cannot exceed T, for then at least one time gap would have exceeded T and the pedestrian would have been able to cross the road. A sufficient condition for (5.55) is therefore $(e^{\lambda T} - 1)T \le \tau$, or

$$\lambda \le \frac{1}{T} \ln \left(1 + \frac{\tau}{T}\right). \tag{5.56}$$

We can regard the right-hand side of (5.56) as expressing the maximum rate at which cars should be allowed to pass if pedestrians are not to be delayed unreasonably, or, because long delays encourage pedestrians to take risks, as the maximum safe traffic flow.

It is convenient to convert this quantity into number of cars per hour and express it as a function of the width of the road. Suppose that τ is 60 seconds. Let d be the width of the road in feet, and assume that a pedestrian walks at 3.5 feet per second.[2] Then $T = 2d/7$ and, on using (5.56), the maximum safe traffic flow in cars per hour is

$$\phi(d) = \frac{210}{d} \ln \left(1 + \frac{210}{d}\right). \tag{5.57}$$

The graph of ϕ is sketched in Fig. 5.2. Points in the d-ϕ plane below this curve represent safe traffic flows, and points above the curve represent unsafe ones. Notice how sharply the permissible traffic flow decreases with increasing road width. For example, a traffic flow of 1500 cars per hour would be judged safe if the road were 20 feet wide, but a flow of only 875 cars per hour would be deemed unsafe if the road were 30 feet wide. Thus crossing controls for pedestrians are essential on principal highways, where, because traffic is heavy and d is large, it is almost certain that traffic flow exceeds $\phi(d)$.

How good is this model? Does its descriptive component fit the facts for freely flowing traffic? We should not accept this model until we have shown that passing cars may reasonably be regarded as arrivals in a Poisson

---

[2]See Supplementary Note 1.

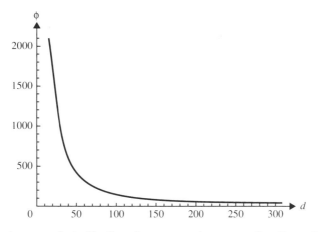

Fig. 5.2 Maximum safe traffic flow in cars per hour as a function of road width in feet, according to (5.57).

process. Therefore, in Section 8.7, we will analyze some traffic data and verify that our assumption is indeed a reasonable one.

5.5 The Rise and Fall of the Company Executive

In this section, we will develop a probabilistic dynamic model in which "births" and "deaths" are both allowed, though the probability of more than one such transition in the interval $[t, t + \delta t)$ is $o(\delta t)$. We will refer to this model as a *birth and death process*. We will introduce the model by describing the movement of an elevator between floors of a company building, but the model has several other applications.

Let the company building have $N+1$ floors, with 0 corresponding to the lowest floor and N to the uppermost, and let a "birth" on floor i correspond to a transition between floor i and the floor above. Let the random variable $X(t)$ denote the floor the elevator occupies at time t if it is at rest, or the floor to which it is headed at time t if it is moving. By allowing X to be the floor of destination as well as of occupancy, the travel time between floor i and floor $i + 1$ can be regarded as part of the elevator's occupancy of floor $i + 1$. Thus a birth takes place instantly, at the moment when a company executive at floor i (or anyone else who wishes to use the elevator) presses the button for floor $i + 1$. Similarly, a "death" on floor i corresponds to a transition between floor i and the floor below, regarded as happening at the instant when an executive at floor i presses the button for floor $i - 1$.

To ensure that a birth and death model provides an adequate description, we will assume that the elevator stops at every floor on its way up or down. We might imagine that the elevator has only two buttons, one marked "up" and the other marked "down," a transition occurring whenever a button is pressed. Thus the elevator cannot rise *directly* from the ground floor to, say, the third. Rather, this journey would require a stop at the second floor en route, so that two births would be involved. This assumption is an obvious imperfection of our birth and death model, but we will be able to eliminate it later. The following section will take a step in this direction, and further developments will be facilitated by Chapter 10.

Because we assume the Markov property, the future must never depend on the past. But suppose that an executive has caught the elevator at the third floor and is on his way to the ground floor when, during the elevator's compulsory stop at the second floor, he meets another executive who wishes to climb to the tenth floor. The second executive decides to ride with the first executive to the ground floor, rather than wait for the elevator to return, so that his place in the elevator is assured. In these circumstances, the next birth would depend not just on the elevator's ground floor position but also on the fact that it had acquired a high-minded executive at the previous stop. In other words, the Markov property would be violated. We exclude this possibility by assuming that executives are always too harassed to wait for an elevator. If it's already there—and either empty or going in the right direction—then they will use it. Otherwise, they will take the stairs.

Having dispensed with these preliminaries, we now proceed to develop our model. Suppose that the elevator is on floor i at time 0, i.e., that $X(0) = i$. Let U be the event that the elevator is on floor j at time $t + \delta t$, where δt is infinitesimally small, and where $1 \leq j \leq N - 1$; $j = 0$ and $j = N$ will be considered later. Then $\text{Prob}[U] = \text{Prob}(X(t + \delta t) = j) = \text{Prob}(X(t + \delta t) = j \mid X(0) = i) = r_{ij}(t + \delta t)$. At time t, the following four events exhaust all possibilities:

> U_1 : the elevator is on floor j,
>
> U_2 : the elevator is on floor $j + 1$,
>
> U_3 : the elevator is on floor $j - 1$,   (5.58)
>
> U_4 : the elevator is above floor $j + 1$ or below floor $j - 1$.

Hence, from (A.13):

$$\text{Prob}[U] = \sum_{k=1}^{4} \text{Prob}[U \mid U_k] \cdot \text{Prob}[U_k]. \tag{5.59}$$

But on using (5.7) and (5.8), we have $\text{Prob}[U \mid U_2] = d_{j+1}\delta t + o(\delta t)$ and $\text{Prob}[U \mid U_3] = b_{j-1}\delta t + o(\delta t)$; whereas $\text{Prob}[U \mid U_4] = o(\delta t)$, being the probability of a birth *and* a death in the interval $[t, t + \delta t]$. The probability of a birth *or* a death, on the other hand, is $b_j\delta t + d_j\delta t + o(\delta t)$ when the elevator is on floor j; and, given U_1, the event U will take place only if neither a birth nor a death occurs in the interval $[t, t + \delta t)$. Thus $\text{Prob}[U \mid U_1] = 1 - b_j\delta t - d_j\delta t + o(\delta t)$. Substituting these results into (5.59), and using (5.12), we thus obtain

$$r_{ij}(t + \delta t) = (1 - b_j\delta t - d_j\delta t + o(\delta t))r_{ij}(t) + (d_{j+1}\delta t + o(\delta t))r_{i,j+1}(t)$$
$$+ (b_{j-1}\delta t + o(\delta t))r_{i,j-1}(t) + o(\delta t) \cdot \text{Prob}[U_4]. \quad (5.60)$$

Rearranging, we get

$$\frac{r_{ij}(t + \delta t) - r_{ij}(t)}{\delta t} = -(b_j + d_j)r_{ij}(t) + d_{j+1}r_{i,j+1}(t) + b_{j-1}r_{i,j-1}(t) + \frac{o(\delta t)}{\delta t}.$$

Taking the limit as $\delta t \to 0$, we obtain the differential equation

$$\frac{d}{dt}r_{ij} = b_{j-1}r_{i,j-1} - (b_j + d_j)r_{ij} + d_{j+1}r_{i,j+1}, \qquad 1 \le j \le N - 1. \quad (5.61)$$

For $j = 0$ the event U_3 is impossible, and for $j = N$ we must exclude the event U_2. Then, by reasoning similar to that which yielded (5.61):

$$\frac{dr_{i0}}{dt} = d_1 r_{i1} - b_0 r_{i0} \qquad (5.62)$$

$$\frac{dr_{iN}}{dt} = b_{N-1}r_{i,N-1} - d_N r_{iN}. \qquad (5.63)$$

Here we have used the fact that $d_0 = 0 = b_N$, because the elevator can neither descend below floor 0 nor rise above floor N. Equations (5.61)–(5.63), together with the initial conditions $r_{ij}(0) = \delta_{ij}$, $i, j = 0, 1, \ldots, N$ are sufficient to determine $r_{ij}(t)$ on $[0, \infty)$ for all i and j. Naturally, the constraint (5.13) must be satisfied by $r_{ij}(t)$ for every value of i, and you should note that the sum of equations (5.61)–(5.63) is just the derivative of this constraint with respect to t. Also, be sure you appreciate that $r_{ij}(0) = \delta_{ij}$ is an initial condition only in the mathematical sense. The initial state of the system itself is the random variable $X(0)$, with probability distribution $\pi(0)$.

To keep the mathematics simple, however, we shall confine our attention for the remainder of this section to the special case $N = 2$. Thus we will model a company building that consists of three stories, with a single elevator connecting the floors. In such a building (for example, the old Florida Education Association/United building in Tallahassee), can we predict where the elevator will be at the end of the day, if it either always starts on the ground floor ($i = 0$) or else is equally likely to start the day on any of the three floors?

Setting $N = 2$ in (5.61)–(5.63) gives

$$\frac{dr_{i0}}{dt} = -b_0 r_{i0} + d_1 r_{i1},$$

$$\frac{dr_{i1}}{dt} = b_0 r_{i0} - (b_1 + d_1) r_{i1} + d_2 r_{i2}, \qquad (5.64)$$

$$\frac{dr_{i2}}{dt} = b_1 r_{i1} - d_2 r_{i2}.$$

It isn't difficult (Exercise 5.5) to solve this set of linear equations for any values of $b_0, b_1, d_1,$ and d_2, subject of course to $r_{ij}(0) = \delta_{ij}, i, j = 0, 1, 2$; but the algebra is messy. We will therefore make the further assumption that executives on the top floor approach the elevator/staircase area throughout the day at the same average rate as executives on the ground floor, so that we may take $b_0 = d_2 = \alpha$, say. Then you can verify by substitution that the solution to (5.64) for $i = 0, 1, 2$ is given by the 3×3 matrix

$$\mathbf{R}(t) = \frac{1}{\alpha + b_1 + d_1} \begin{bmatrix} d_1 & \alpha & b_1 \\ d_1 & \alpha & b_1 \\ d_1 & \alpha & b_1 \end{bmatrix} + \frac{e^{-\alpha t}}{b_1 + d_1} \begin{bmatrix} b_1 & 0 & -b_1 \\ 0 & 0 & 0 \\ -d_1 & 0 & d_1 \end{bmatrix}$$

$$+ \frac{e^{-(\alpha + b_1 + d_1)t}}{(b_1 + d_1)(\alpha + b_1 + d_1)}$$

$$\times \begin{bmatrix} \alpha d_1 & -\alpha(b_1 + d_1) & \alpha b_1 \\ -d_1(b_1 + d_1) & (b_1 + d_1)^2 & -b_1(b_1 + d_1) \\ \alpha d_1 & -\alpha(b_1 + d_1) & \alpha b_1 \end{bmatrix}. \qquad (5.65)$$

If the elevator always starts the day on the ground floor then, from (5.18) and (5.65), $\pi(t)^T = (1, 0, 0)\mathbf{R}(t) =$

$$\frac{1}{\alpha + b_1 + d_1}(d_1, \alpha, b_1) + \frac{b_1 e^{-\alpha t}}{b_1 + d_1}(1, 0, -1)$$

$$- \frac{\alpha e^{-(\alpha + b_1 + d_1)t}}{(b_1 + d_1)(\alpha + b_1 + d_1)}(-d_1, b_1 + d_1, -b_1); \qquad (5.66)$$

whereas if it's equally likely to start on any of the floors, $\pi(t)^T = (1/3, 1/3, 1/3)\mathbf{R}(t) =$

$$\frac{1}{\alpha + b_1 + d_1}(d_1, \alpha, b_1) + \frac{b_1 - d_1}{b_1 + d_1}e^{-\alpha t}(1, 0, -1)$$

$$+ \frac{(b_1 + d_1 - 2\alpha)}{3(b_1 + d_1)(\alpha + b_1 + d_1)}e^{-(\alpha + b_1 + d_1)t}(-d_1, b_1 + d_1, -b_1). \qquad (5.67)$$

We do not have values for α, b_1, and d_1, but it seems reasonable to assume values of several per hour—at the very least, $\alpha > 1$, $b_1 > 1$, and $d_1 > 1$. At the end of the nine-hour working day, therefore, the exponential terms in (5.66) and (5.67) are completely negligible, because $e^{-9} \approx 10^{-4}$, and e^{-27} is even smaller. Thus, regardless of the initial location of the elevator, we have

$$\pi(9)^T \approx \frac{1}{\alpha + b_1 + d_1}(d_1, \alpha, b_1) \tag{5.68}$$

at the end of the day.

Intuitively, this is a sound result. It says that the elevator will ultimately spend a fraction $d_1/(\alpha + b_1 + d_1)$ of each hour on the ground floor. This fraction increases with d_1, i.e., with an increase in the average rate of departure from the middle to the ground floor. You should similarly interpret $\pi_1(9)$ and $\pi_2(9)$.

If executives on the middle floor approach the elevator/staircase area at the same rate as on the other floors, and if any executive's ultimate destination is equally likely to be either of the other two floors, then

$$b_1 = \frac{1}{2}\left\{\text{rate of ground floor departure}\right\} + \frac{1}{2}\left\{\text{rate of approach on middle floor}\right\}$$
$$= \frac{\alpha}{2} + \frac{\alpha}{2} = \alpha,$$

the first contribution being due to executives who are riding all the way from the ground floor to the top; and similarly,

$$d_1 = \frac{1}{2}\left\{\text{rate of top floor departure}\right\} + \frac{1}{2}\left\{\text{rate of approach on middle floor}\right\}$$
$$= \frac{\alpha}{2} + \frac{\alpha}{2} = \alpha,$$

too. In these circumstances, (5.68) produces the obvious answer $\pi(9)^T \approx$ (1/3, 1/3, 1/3), irrespective, of course, of how the elevator starts its day.

Now try Exercises 5.6 and 5.7.

5.6   Discrete Models of a Day in the Life of an Elevator

In Sections 1.1 and 1.7 we constructed a model of a contaminated lake. We found that if we wished to know the state of the lake at all times t, $0 \leq t < \infty$, then we had to use a differential equation to describe the decay of pollution. If we were satisfied to know the state of the lake only at discrete instants $t = 0, 1, 2, \ldots, l, \ldots$, however, then a recurrence relation could be used instead, provided a certain condition was satisfied. We called the resulting model a metered model; and the condition for using it was that the system in question behave qualitatively the same within each unit of time. Because of this qualitative periodicity, the state of the system at time $t = l + 1$ could be predicted if only the state of the system at time $t = l$ were known. What happened prior to time $t = l$ was irrelevant. But this is just the Markov property of memorylessness, and suggests that birth and death models can also be metered if we replace the continuous-time stochastic process $S_t = \{X(t) : 0 \leq t < \infty\}$ by a *discrete-time stochastic process*, namely, the set $S_l \equiv \{X(l) : l = 0, 1, 2, \ldots\}$. We will illustrate this idea by constructing a metered model of the company elevator described in the previous section.

Let an hour be the unit of time. Thus, if $t = 0$ were 8 A.M., then $t = 9$ would be 5 P.M. We will construct a model that predicts the elevator's distribution vector $\pi(t)$ at the discrete instants $t = 0, 1, 2, \ldots, 9$. Let U be the event that the elevator is on floor j at time $l + 1$, and let U_i be the event that the elevator is on floor i at time l, for $0 \leq i \leq 2$. Then, because the disjoint events U_0, U_1, and U_2 exhaust the possibilities arising at time l, we have

$$\text{Prob}[U] = \sum_{i=0}^{2} \text{Prob}[U \mid U_i] \cdot \text{Prob}[U_i]. \qquad (5.69)$$

But from the Markov property, i.e., on setting $t = 0$ and $s = l$ in (5.3), we have

$$\text{Prob}(U \mid U_i) = \text{Prob}(X(l + 1) = j \mid X(l) = i)$$
$$= \text{Prob}(X(1) = j \mid X(0) = i) = r_{ij}(1).$$

Moreover, $\text{Prob}(U) = \text{Prob}(X(l + 1) = j) = \pi_j(l + 1)$, and $\text{Prob}(U_i) = \text{Prob}(X(l) = i) = \pi_i(l)$. Hence (5.69) yields a set of three coupled recurrence relations:

$$\pi_j(l + 1) = \sum_{i=0}^{2} r_{ij}(1)\pi_i(l). \qquad (5.70)$$

Let us define the 3×3 matrix $\mathbf{S}$ by

$$\mathbf{S} \equiv \mathbf{R}(1). \qquad (5.71)$$

Then (5.70) may be written more succinctly in matrix form as

$$\pi(l + 1)^T = \pi(l)^T \mathbf{S}. \qquad (5.72)$$

This matrix recurrence relation is a probabilistic analogue of (1.48), and you can verify by substitution that its general solution is given by

$$\pi(l)^T = \pi(0)^T \mathbf{S}^l. \qquad (5.73)$$

Once $\mathbf{S}$ and $\pi(0)$ are known, $\pi(l)$ can be calculated from (5.73) for any value of l, by recursion.

But $\mathbf{S}$ is known from the previous section. To keep the mathematics simple, let's assume, as at the very end of Section 5.5, that executives on all floors of the building approach the elevator/staircase area at the same rate α and that the other two floors are equally likely to be their final destinations. Then, from (5.65) with $b_1 = d_1 = \alpha$:

$$\mathbf{R}(t) = \frac{1}{3}\begin{bmatrix} 1 & 1 & 1 \\ 1 & 1 & 1 \\ 1 & 1 & 1 \end{bmatrix} + \frac{1}{2}e^{-\alpha t}\begin{bmatrix} 1 & 0 & -1 \\ 0 & 0 & 0 \\ -1 & 0 & 1 \end{bmatrix} + \frac{1}{6}e^{-3\alpha t}\begin{bmatrix} 1 & -2 & 1 \\ -2 & 4 & -2 \\ 1 & -2 & 1 \end{bmatrix}. \tag{5.74}$$

In particular,

$$\mathbf{S} = \mathbf{R}(1) = \frac{1}{3}\begin{bmatrix} 1 & 1 & 1 \\ 1 & 1 & 1 \\ 1 & 1 & 1 \end{bmatrix} + \frac{1}{2}e^{-\alpha}\begin{bmatrix} 1 & 0 & -1 \\ 0 & 0 & 0 \\ -1 & 0 & 1 \end{bmatrix} + \frac{1}{6}e^{-3\alpha}\begin{bmatrix} 1 & -2 & 1 \\ -2 & 4 & -2 \\ 1 & -2 & 1 \end{bmatrix}. \tag{5.75}$$

For the sake of illustration, let's take $\alpha = 1.5$.[3] Then (5.75) yields

$$\mathbf{S} = \begin{bmatrix} 0.4467499 & 0.3296303 & 0.2236198 \\ 0.3296303 & 0.3407393 & 0.3296303 \\ 0.2236198 & 0.3296303 & 0.4467499 \end{bmatrix}. \tag{5.76}$$

You can easily verify (though you'll need a computer) that

$$\mathbf{S}^9 = \begin{bmatrix} 0.3333340 & 0.3333333 & 0.3333326 \\ 0.3333333 & 0.3333333 & 0.3333333 \\ 0.3333326 & 0.3333333 & 0.3333340 \end{bmatrix}. \tag{5.77}$$

Thus if the elevator is certain to start the day on the ground floor, then

$$\pi(9)^T = (1,0,0)\mathbf{S}^9 = (0.3333340, 0.3333333, 0.3333326); \tag{5.78}$$

whereas if it is equally likely to start its day on any floor, then

$$\pi(9)^T = \left(\frac{1}{3}, \frac{1}{3}, \frac{1}{3}\right)\mathbf{S}^9 = \left(\frac{1}{3}, \frac{1}{3}, \frac{1}{3}\right). \tag{5.79}$$

As in Section 5.5, we have $\pi(9)^T \approx (1/3, 1/3, 1/3)$, irrespective of where the elevator began its day. We can interpret these equal probabilities by saying that the elevator will ultimately spend an equal proportion of time on each floor.

A very different result is obtained if we choose to measure time not in hours but in terms of the number of transitions (births or deaths). Let $t = 0$ correspond, as before, to the start of the elevator's day. Now, however, let $t = 1$ correspond to the first transition, $t = 2$ to the second transition, and, in general, $t = n$ to the nth transition. Let U denote the event that the elevator is on floor j after $n+1$ transitions, and let U_i denote the event that

---

[3]The idea for this model came from observing the FEA/United building in Talla-hassee. There, the appropriate value of α was certainly greater than 2, and probably 4 or 5; but I had to choose α sufficiently small that (5.73) would still be converging after 9 time steps (to seven significant figures, see (5.77)).

the elevator is on floor i after n transitions, $0 \leq i \leq 2$. The disjoint events U_0, U_1, and U_2 exhaust the possibilities arising at $t = n$, and so we have

$$\text{Prob}(U) = \sum_{i=0}^{2} \text{Prob}(U \mid U_i) \cdot \text{Prob}(U_i). \tag{5.80}$$

Let us denote $\text{Prob}(U \mid U_i)$ by p_{ij}. Then from the Markov property, i.e., setting $t = 0$ and $s = n$ in (5.3), we have

$$p_{ij} = \text{Prob}(X(n+1) = j \mid X(n) = i) = \text{Prob}(X(1) = j \mid X(0) = i). \tag{5.81}$$

But $\text{Prob}(U) = \text{Prob}(X(n+1) = j) = \pi_j(n+1)$, and $\text{Prob}(U_i) = \text{Prob}(X(n) = i) = \pi_i(n)$. Hence (5.80) becomes

$$\pi_j(n+1) = \sum_{i=0}^{2} p_{ij} \pi_i(n), \tag{5.82}$$

or in matrix terms,

$$\pi(n+1)^T = \pi(n)^T \mathbf{P}, \tag{5.83}$$

where $\mathbf{P}$ is the 3×3 matrix whose (i,j)th entry is given by (5.81). You can verify by substitution that the general solution of this matrix recurrence relation is given by

$$\pi(n)^T = \pi(0)^T \mathbf{P}^n. \tag{5.84}$$

Once $\mathbf{P}$ and $\pi(0)$ are known, $\pi(n)$ can be calculated from (5.84) for any value of n, by recursion.

We may refer to $\mathbf{P}$ as the one-step transition matrix, or simply the *transition matrix*, because from (5.81), it gives the probability of going from floor i to floor j in precisely one transition. The matrix $\mathbf{P}^n$ may be called the n-step transition matrix, because it gives the probability of going from floor i to floor j in precisely n transitions. Like $\mathbf{R}(t)$, $\mathbf{P}$ and $\mathbf{P}^n$ are both stochastic matrices.

It remains to calculate $\mathbf{P}$. It is impossible to go directly from the top to the bottom or the bottom to the top in precisely one transition; therefore we have $p_{20} = p_{02} = 0$. But it is also impossible to return to the same floor in one transition, whence $p_{00} = p_{11} = p_{22} = 0$. If the elevator is on the ground floor or the top floor, then its next transition must be to the middle floor, whence $p_{01} = p_{21} = 1$. It now remains only to calculate p_{10} and p_{12}.

In the limit as $\delta t \to 0$, in continuous time t, the probability that the elevator rises from floor 1 to floor 2 is $b_1 \delta t$, and the probability that the elevator descends from floor 1 to floor 0 is $d_1 \delta t$. The ratio of these two probabilities is b_1/d_1. This is true for any value of continuous time. Hence it must be true, in particular, at the instant of transition (even though we do not know when that occurs in continuous time). Hence

$$\frac{p_{12}}{p_{10}} = \frac{b_1}{d_1}. \tag{5.85}$$

But $p_{10} + p_{12} = 1$. Solving these equations yields $p_{10}(b_1 + d_1) = d_1$, $p_{12}(b_1 + d_1) = b_1$. Thus the transition matrix is

$$\mathbf{P} = \begin{bmatrix} 0 & 1 & 0 \\ \frac{d_1}{b_1+d_1} & 0 & \frac{b_1}{b_1+d_1} \\ 0 & 1 & 0 \end{bmatrix}. \tag{5.86}$$

You can easily deduce from this that for $k = 0, 1, 2, \ldots$,

$$\mathbf{P}^{2k+1} = \begin{bmatrix} 0 & 1 & 0 \\ \frac{d_1}{b_1+d_1} & 0 & \frac{b_1}{b_1+d_1} \\ 0 & 1 & 0 \end{bmatrix}, \quad \mathbf{P}^{2k+2} = \begin{bmatrix} \frac{d_1}{b_1+d_1} & 0 & \frac{b_1}{b_1+d_1} \\ 0 & 1 & 0 \\ \frac{d_1}{b_1+d_1} & 0 & \frac{b_1}{b_1+d_1} \end{bmatrix}. \tag{5.87}$$

Thus, if the elevator is certain to start the day on the ground floor, then $\pi(2k + 1)^T = (1, 0, 0)\mathbf{P}^{2k+1} = (0, 1, 0)$ and

$$\pi(2k + 2)^T = (1, 0, 0)\mathbf{P}^{2k+2} = \left(\frac{d_1}{b_1 + d_1}, 0, \frac{b_1}{b_1 + d_1} \right); \tag{5.88}$$

i.e., the elevator is certain to be on the middle floor after an odd number of transitions, but cannot possibly be on the middle floor after an even number.

Although this result is obvious intuitively, it is quite at odds with the result we obtained in (5.78). It shows that, despite mathematical similarities, the two discrete-time elevator models discussed in this section are fundamentally different. In the first model, time is measured *linearly*: each unit of time corresponds to one hour. In the second model, time is measured *nonlinearly*: on any particular day, two transitions could be as close together as a fraction of a minute or as far apart as an hour or two. We anticipated this distinction by using l to denote linear time and n to denote nonlinear time. In the first model, $t = 9$ means that precisely 9 hours have elapsed, but the number of transitions is unknown. In the second model, $t = 9$ means that precisely 9 transitions have occurred, but in an unknown number of hours. In the first model, even though time has been treated as a discrete variable, the elevator's history up to time l includes a continuum of l hours, throughout most of which the elevator is waiting at one floor or another. In the second model, the elevator's waiting time is completely ignored. It is the latter omission which precludes the second model from predicting that the elevator will ultimately spend a fixed proportion of time at each floor.

A discrete-time Markov model in which time is measured in this nonlinear fashion is known in the literature on probability as a *Markov chain*.[4] Although we restricted our attention above to a Markov chain with only three states, namely, $i = 0, 1, 2$, the extension of the model to $N + 1$ states is immediate. Indeed (5.83), (5.84), and the definition of **P** in (5.81) are valid for any value of N, and we need only replace 2 by N in (5.82) to make that valid, too.

The distinction between a Markov chain and the discrete version of a linear-time Markov model is often blurred in the literature, because the passage of a unit of (linear) time can be defined to correspond to a transition. Both models are described as Markov chains, and **S** is called a transition matrix. In these last two respects, we will follow convention. Nevertheless, we will attempt to preserve the theoretical distinction between the two models by using n for time and **P** for transition matrix when transitions correspond to physical events, but l for time and **S** for transition matrix when transitions are artificially defined as the passage of so many units of linear time.

The periodic behavior that appears in (5.87) is caused by the fact that all entries on the leading diagonal of **P** are zero. But most of the Markov chains that are useful in practice, and all of those introduced henceforward in the main body of the text, have at least one nonzero on the leading diagonal of the transition matrix—which is enough to prevent such periodicity. That single fact is virtually all that we need to know about periodicity in Markov chains, and we shall therefore not dwell on the topic; though we do discuss it further in the Exercises and in Section 6.1.

The following example of a Markov chain involves five states; thus $N = 4$.

5.7   Birds in a Cage. A Birth and Death Chain

A cage in a pet shop contains four budgerigars and is divided into two sections by a trapeze in the middle. For convenience, we will label these sections North and South. At each end of the cage there is a food supply. Thus the cage's horizontal section appears as drawn in Fig. 5.3.

The birds spend almost all of their time in the vicinity of a food supply, either eating or resting. Every so often, however, one of the birds will fly up to the trapeze and swing on it for a while, before flying down again to rejoin its precious food supply. After swinging on the trapeze, a bird forgets where it came from and is equally likely to fly down to either end of the cage. Moreover, the birds act independently of one another. Each

---

[4]The continuous-time model considered in the previous section is a special case of a *Markov jump process*, for which more general equations are developed in Exercise 10.8.

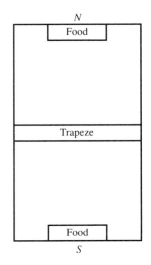

Fig. 5.3 Plan of bird cage.

is equally likely to be the next performing artist. In these circumstances, can we predict where the birds will be after, say, 20 trapeze acts, if initially they were all at the South end of the cage?

We will model this system as a Markov chain. The resulting model, if somewhat fanciful, will be extremely useful in Chapter 10 for the purpose of introducing semi-Markov processes. Let the random variable $X(n)$ denote the number of birds at the North end of the cage at time n. Because there are four birds, its sample space is $\{0, 1, 2, 3, 4\}$; i.e., the possible states of the system are $i = 0, 1, 2, 3, 4$. Let each trapeze act correspond to a transition. Then after each transition, the number of birds at the North end of the cage will either increase by one, decrease by one, or stay the same. Hence the transition matrix satisfies $p_{ij} = 0$ unless $i - 1 \leq j \leq i + 1$. A Markov chain with this property is known as a *birth and death chain*. All Markov chains described in this chapter belong to this category. More general chains are introduced in Chapter 6.

Suppose that there is just one bird at the North end of the cage, i.e., that $X(n) = 1$. Let U be the event that this bird is next to perform; hence $\bar{U}$ is the complementary event that one of the other birds is next to perform. Then, by assumption, $\text{Prob}(U) = 1/4$ and $\text{Prob}(\bar{U}) = 1 - \text{Prob}(U) = 3/4$. Let V be the event that a bird flies down to the opposite side of the cage after trapezing; hence $\bar{V}$ is the complementary event that it returns whence it came. Then by assumption, V and $\bar{V}$ are independent of U and $\bar{U}$, and $\text{Prob}(V) = \text{Prob}(\bar{V}) = 1/2$. Thus, on using (A.49), $p_{10} = \text{Prob}(V \cap U) = \text{Prob}(V) \cdot \text{Prob}(U) = 1/2 \cdot 1/4 = 1/8$. Similarly, $p_{12} = \text{Prob}(V \cap \bar{U}) = \text{Prob}(V) \cdot \text{Prob}(\bar{U}) = 1/2 \cdot 3/4 = 3/8$. The cage will remain in state 1 if either the bird from the North or a bird from the South returns; i.e., $p_{11} = \text{Prob}(\bar{V})\cdot\text{Prob}(U)+\text{Prob}(\bar{V})\cdot\text{Prob}(\bar{U}) = \text{Prob}(\bar{V}) = 1/2$. In this way, we have completely determined the second row of the transition matrix $\mathbf{P}$. Continuing in this fashion, you can easily check that

$$\mathbf{P} = \begin{bmatrix} \frac{1}{2} & \frac{1}{2} & 0 & 0 & 0 \\ \frac{1}{8} & \frac{1}{2} & \frac{3}{8} & 0 & 0 \\ 0 & \frac{1}{4} & \frac{1}{2} & \frac{1}{4} & 0 \\ 0 & 0 & \frac{3}{8} & \frac{1}{2} & \frac{1}{8} \\ 0 & 0 & 0 & \frac{1}{2} & \frac{1}{2} \end{bmatrix}. \tag{5.89}$$

Notice that all terms on the leading diagonal are nonzero, so that there will be no periodicity in powers of $\mathbf{P}$ (see the remarks at the end of the previous section). With $\pi(0)^T = (1,0,0,0,0)$, the cage distribution after 20 trapeze acts is given by $\pi(20)^T = (1,0,0,0,0)\mathbf{P}^{20} =$

$$(0.0632932, 0.2515856, 0.3749993, 0.2484144, 0.0617076), \tag{5.90}$$

on using (5.84). Notice that the expected number of birds in the North end of the cage at that time is

$$\sum_{i=0}^{4} i\pi_i(20) \approx 1.994, \tag{5.91}$$

which is close to 2. Notice also that

$$\pi(20)^T \approx \left(\frac{1}{16}, \frac{1}{4}, \frac{3}{8}, \frac{1}{4}, \frac{1}{16}\right). \tag{5.92}$$

You've probably guessed that, by analogy with the previous sections, (5.92) is true irrespective of where the birds began; and after a large number of trapeze acts, the expected number of birds at either end of the cage would be indistinguishable from 2. In other words, the cage ultimately approaches a *stationary distribution*, the probabilistic equivalent of an equilibrium.

But already we're talking about stationary distributions, and these are the subject of the following chapter.

5.8   Trees in a Forest. An Absorbing Birth and Death Chain

According to Usher (1967, p. 359), on which this section is based, the Scots pine trees in the forest plantations at Corrour, in Scotland, are grouped into classes according to girth. There are six classes in all, which we shall label $i = 0, 1, \ldots, 5$. The thinnest trees belong to class 0 and the fattest to class 5, with increasing i corresponding to increasing girth. A census of trees is effectively taken once every six years. Let this be the unit of time, and let the passage of every sixth year be regarded as a transition. Then a Markov chain with $N = 5$ can be used to model the dynamics of the forest's class structure, if we define $X(l)$ to be the class at time l of a randomly chosen tree. The components of the resulting distribution vector $\pi(l)$ will be interpreted as the proportions of trees in each class at time l. Note that by defining a transition to take place once every unit of time, we make linear and nonlinear time equivalent (see the remarks at the end of the previous section).

Let us assume that all trees reach full stature, i.e., enter class 5, and are then felled. We will regard this sixth class as consisting both of felled and unfelled trees (of full stature). Once a tree has been felled, it remains forever felled. Thus $i = 5$ ultimately "absorbs" all trees. In other words, the last row of the transition matrix $\mathbf{S}$ is given by $s_{5j} = 0$ for $0 \leq j \leq 4$ and $s_{55} = 1$. For this reason, $i = 5$ is called an *absorbing state* of the Markov chain; and the Markov chain itself is described as an *absorbing chain*.

Now, the remaining classes are defined in such a way that, at the end of every six-year period, a tree will either remain in the same class or have sufficient girth to move into the class immediately above, but never sufficient girth to skip a class. For $0 \leq i \leq 4$, let a_i denote the probability that a tree from class i remains in class i at the end of a census period. Then $1 - a_i$ is the probability that it grows sufficiently to move into class $i + 1$. Thus $s_{ii} = a_i$, $s_{i,i+1} = 1 - a_i$ and $s_{ij} = 0$ if $j \geq i + 2$. Moreover, because no tree ever moves down a class, we have $s_{ij} = 0$ for $i > j$; i.e., the transition matrix $\mathbf{S}$ is upper triangular. Combining these results, we find that

$$\mathbf{S} = \begin{bmatrix} a_0 & 1-a_0 & 0 & 0 & 0 & 0 \\ 0 & a_1 & 1-a_1 & 0 & 0 & 0 \\ 0 & 0 & a_2 & 1-a_2 & 0 & 0 \\ 0 & 0 & 0 & a_3 & 1-a_3 & 0 \\ 0 & 0 & 0 & 0 & a_4 & 1-a_4 \\ 0 & 0 & 0 & 0 & 0 & 1 \end{bmatrix}. \tag{5.93}$$

Values for a_0, a_1, a_2, a_3, and a_4 are estimated in Exercise 8.6 to be 0.72, 0.69, 0.75, 0.77, and 0.63. If we knew the forest's initial class structure, then its structure at the end of each six-year census period could now be determined from (5.84). Suppose, for example, that the structure were initially uniform; i.e., $\pi(0)^T = (1/6, 1/6, 1/6, 1/6, 1/6, 1/6)$. Then, because

$$\mathbf{S}^5 = \begin{bmatrix} 0.1935 & 0.3462 & 0.3243 & 0.1165 & 0.0178 & 0.0018 \\ 0.0 & 0.1564 & 0.4180 & 0.3103 & 0.0900 & 0.0253 \\ 0.0 & 0.0 & 0.2373 & 0.4172 & 0.2128 & 0.1327 \\ 0.0 & 0.0 & 0.0 & 0.2707 & 0.2816 & 0.4477 \\ 0.0 & 0.0 & 0.0 & 0.0 & 0.0992 & 0.9008 \\ 0.0 & 0.0 & 0.0 & 0.0 & 0.0 & 1.0 \end{bmatrix}, \tag{5.94}$$

the class structure after 30 years would be given by $\pi(5)^T = \pi(0)^T \mathbf{S}^5 = (0.032, 0.084, 0.163, 0.186, 0.117, 0.418)$. Thus, after 30 years, the percentage of trees in class 2 would be almost as high as originally, and over 40% of the trees would have reached full stature.

This model is of limited value in its present form—why? How does a real forest differ from the one we have just discussed? Perhaps you would like to mull this question over. We'll return to it in the following chapter, and again in Chapter 9.

Exercises

5.1 Deduce (5.40) from (5.39). Verify the expression for $\mathbf{R}(t)$ when $N = 3$.

5.2 Deduce (5.46) from (5.45). Verify the expression for $\mathbf{R}(t)$ when $N = 3$. Show that the Poisson distribution (5.49) has mean Bt; i.e., verify that

$$\sum_{k=1}^{\infty} k \cdot \text{Prob}(A(t) = k) = Bt.$$

5.3 The birth rate in a pure birth model need not be constant (independent of i). It may, for example, increase linearly with i. Show that if

$$b_i = Bi, \qquad 1 \le i < \infty$$

then the entries of the upper triangular matrix $\mathbf{R}(t)$ are given by

$$r_{ij}(t) = \frac{(j-1)!}{(i-1)!(j-i)!} e^{-Bit} \left(1 - e^{-Bt}\right)^{j-i}, \qquad j \ge i \ge 1.$$

Here $K!$ is defined as the product of all integers between 1 and K; i.e., $K! = 1 \cdot 2 \cdot 3 \cdots (K-1) \cdot K$, with $0! = 1$. Note that the lowest state of the population is $i = 1$, because $b_0 = 0$ would mean that the population could never exist. In other words, the first row and column of the matrix $\mathbf{R}(t)$, as defined in Chapter 5, must be deleted.
Hint: The result can be deduced from (5.46) by fixing i and performing mathematical induction on j.

5.4 Obtain (5.54). *Hint:* For $|x| < 1$, the Taylor series

$$\frac{1}{1-x} = \sum_{j=0}^{\infty} x^j$$

converges absolutely and may be differentiated term by term. Choose suitable x.

5.5 Solve equations (5.64) and deduce (5.65) as a special case of your answer.

*5.6 In the three-storey building considered in Section 5.5, executives approach the elevator/staircase area at the same rate on each floor, but ground floor executives always want to get to the top, middle floor executives are equally likely to go up or down, and three-quarters of top floor executives who use the elevator are leaving the building. Show that the elevator is least likely to end its day on the middle floor and most likely to finish on top.

*5.7 The time until failure of a machine in a factory has exponential distribution with mean $1/D$. The repair time has exponential distribution with mean $1/B$. If the machine is working at the beginning of a nine-hour day, show that

$$\frac{B + De^{-9(B+D)}}{B + D}$$

is the probability that it will still be working at the end of it.

*5.8 Suppose that the birds in Section 5.7 have excellent memories, always remember where they came from and, after swinging on the trapeze in the center of the cage, always fly down to the side opposite whence they came. Write down the transition matrix $\mathbf{P}$ that describes these new circumstances.

5.9 In Exercise 5.8, what do you notice about (i) odd powers of $\mathbf{P}$, (ii) even powers of $\mathbf{P}$? You should see a pattern emerging after you have computed six or seven powers of $\mathbf{P}$.

*5.10 An extremely enthusiastic individual decides to found a new religion. He persuades some friends to join, then all new members try to enlist as many new members as possible. Do you think that the early growth of this religion could be modelled adequately by treating new members as arrivals in a Poisson process? Why or why not? If not, which other model encountered in this chapter might prove adequate?

*5.11 A U.S. female born in 1966 had a 98% chance of surviving the first year of life. If she survived the first year of life, then she had a 99.48% chance of surviving until her tenth birthday. The conditional probabilities of surviving various other "milestone" birthdays are given by the following table.

Milestone birthday	Probability of reaching this birthday alive, given that the previous birthday has been reached
1	0.9800
10	0.9948
20	0.9955
30	0.9920
45	0.9709
60	0.9061
70	0.8343
85	0.3911
115	0.0000

Use this data to construct a Markov chain model, with ten states, that describes the passage through life of a female born in 1966 in the United States. Identify the life table probabilities $l(1)$, $l(10)$, $l(20)$, $l(30)$, $l(45)$, $l(60)$,

$l(70)$ and $l(85)$ in Exercise 4.15 with entries of the n-step transition matrix $\mathbf{P}^n$. In each case, give appropriate values of i, j and n. (The values will differ slightly from those presented in Exercise 4.15 because of rounding error). Which special features does your Markov chain have—which model in Chapter 5 does it most resemble?

**5.12 A 4-state Markov chain was used by Liu and Yao (1977) to model traffic flow at a T-junction. Obtain a copy of Liu and Yao's paper and read it. Do you think that the states of their Markov chain have been well defined? Criticize their model, paying particular attention to the circumstances in which a car is able to enter the minor road. Would you accept their model, would you reject their model, or can you improve their model?

*5.13 Fig. E5.1 is plan of a T-junction. There is no left-turn lane on the major road, which runs north and south. A driver who wishes to turn left, i.e., west, from the major road must remain in a waiting zone, marked "wait here" in the diagram, until the "danger zone" in the southbound lane is clear. Naturally, if southbound traffic is heavy and many drivers wish to go west, then northbound traffic may be seriously interrupted. In this exercise, we will attempt to use a 4-state, nonlinear-time Markov chain to model the severity of such interruptions.

Let us make the following definitions:

a: Probability that a northbound driver, having reached the waiting zone, wishes to turn left. (Imagine that the decision whether to turn or not is not made until after the junction has been reached.)

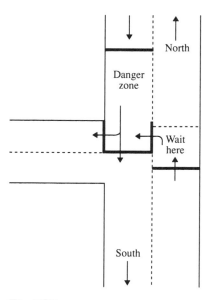

Fig. E5.1

b: Probability that a left-turning driver will be able to enter the minor road immediately after the next southbound car has left the danger zone (given that a driver is waiting to turn left and that at least one car is blocking the danger zone).

λ: Mean southbound flow, in numbers of cars per unit of time.

μ: Mean northbound flow, in number of cars per unit of time.

p: Probability that a northbound car enters the waiting zone before a southbound car enters the danger zone, given that the waiting zone and danger zone are both clear. Assume that at most one northbound car can wait in the waiting zone (whereas more than one southbound car may occupy the danger zone).

q: Probability that a northbound car enters the waiting zone, either to turn or to go forward, before the southbound car approaching the junction can leave the danger zone, given that the danger zone is blocked but the waiting zone clear.

T: Minimum time gap in southbound flow that a left-turning driver requires to cross it safely. Assume that the danger zone is defined in terms of this time rather than in terms of a distance. Thus the danger zone is clear if and only if the next southbound car is at least T seconds away from the junction.

Let there be four possible states in which the T-junction can be at any given instant, and let the random variable $X(t)$ have sample space $\{0, 1, 2, 3\}$. Let $X(t) = 0$ if the danger zone and waiting zone are both occupied. Let $X(t) = 3$ if the danger zone and waiting zone are both clear. Let transitions between states occur as indicated in Fig. E5.1, by arrows over solid bars. For example, a transition occurs if a southbound car turns right or goes forward at the junction.

(i) In terms of b, what is $p_{00} = \text{Prob}(X(n + 1) = 0 \mid X(n) = 0)$? Write down the values of $p_{03} = \text{Prob}(X(n + 1) = 3 \mid X(n) = 0)$ and $p_{33} = \text{Prob}(X(n + 1) = 3 \mid X(n) = 3)$.

(ii) Define states 1 and 2. Hence write down a transition matrix $\mathbf{P}$ for a Markov chain, in terms of a, b, p, and q.

(iii) Suppose that southbound cars arrive according to a Poisson process. Express b as a function of λ and T.

(iv) Suppose that northbound cars arrive according to a Poisson process when the northbound lane is clear, and consider the following argument. Let $X(t) = 3$. Let random variables N and S denote the times that elapse before the next northbound and southbound cars, respectively, arrive. Because interarrival times in a Poisson process are exponentially distributed,

$$\text{Prob}(N > t) = e^{-\mu t}, \qquad \text{(a)}$$
$$\text{Prob}(S > \tau) = e^{-\lambda \tau}. \qquad \text{(b)}$$

Hence the joint probability density function of S and N is given by

$$f(\tau, t) = \lambda \mu e^{-(\lambda \tau + \mu t)}, \qquad 0 \le \tau, t < \infty. \qquad \text{(c)}$$

Thus

$$p = \text{Prob}(N < S) = \int_0^\infty \int_0^\tau f(\tau, t) \, dt \, d\tau = \frac{\mu}{\lambda + \mu}. \qquad \text{(d)}$$

Because b is known as a function of λ and T from part (iii), and because q can be found as a function of μ and T by a calculation similar to (d), the transition matrix $\mathbf{P}$ is effectively expressed as a function of a, λ, μ, and T. Hence the probability that the northbound lane will be blocked at the T-junction can be determined for arbitrary values of a, λ, μ and T.

Do you accept or reject this argument? *Note*: There is nothing wrong with (c), and the integral in (d) has been evaluated correctly.

*5.14 Measles broke out on the campuses of Baylor University, Indiana University, and Florida State University in, respectively, September of 1982, February of 1983, and January of 1986. Each of these outbreaks was controlled by mass immunization of the entire campus, which took place after the disease had been detected. Data pertaining to the three epidemics are contained in the following table, which was obtained from Scime (1986):

University	Number of days from first day of communicability to first day of vaccinations	Total number of infections	Total population
Baylor	26	102	10,320
Indiana	11	67	32,000
F.S.U.	7	4	23,000

According to Scime, the FSU and Baylor outbreaks originated from a single infected person; whereas the Indiana outbreak appeared to originate from several. Moreover, about 3.2% of college-age people are currently "susceptible" to measles, i.e., can contract the disease upon contact with an infected person. But an infected person is "infective," i.e., can transmit the disease to a susceptible person, for a period of only ten days. Thereafter, he recovers and becomes immune (as well as noninfective).

Can you construct a simple model of a measles epidemic that is consistent with these data? Can you estimate how many infections, say, FSU would have had if they had waited as long as Baylor to vaccinate? *Hint*: You will obviously have to make some fairly heroic assumptions, so don't be shy.

Supplementary Notes

5.1 The material in this section is based on Noble (1967, Section 15.3) and Gerlough and Barnes (1971, p. 60–63), whose walking speed we adopt. Nevertheless, the derivation of the maximum safe traffic flow is entirely original and avoids a shortcoming in the analyses of Noble and Gerlough and Barnes. By using different arguments, the earlier authors essentially find that

$$\phi(d) = \frac{210}{d} \ln \left(\frac{210}{d} \right).$$

This is negative for roads wider than 210 feet—an absurd result! Although this absurdity may have no practical consequence (because relevant values of d are much less than 210), it still seems preferable to dispense with it.

6 STATIONARY DISTRIBUTIONS

In Chapters 1 and 2 we met several kinds of dynamical systems. In particular, we met continuous-time systems described by differential equations of the form

$$\frac{d\mathbf{x}}{dt} = \mathbf{f}(\mathbf{x}, \alpha), \qquad 0 \le t < \infty, \tag{6.1}$$

where $\mathbf{x}(t)$ was an n-dimensional vector, α an m-dimensional constant vector, and $\mathbf{f}$ an n-dimensional vector function; and we met discrete-time systems described by recurrence relations of the form $x_{n+1} = F(x_n)$, $n = 0$, 1, 2, 3, ... A more general discrete-time dynamical system would have the form

$$\mathbf{x}_{n+1} = \mathbf{F}(\mathbf{x}_n), \qquad n = 0, 1, 2, 3, \ldots, \tag{6.2}$$

where $\mathbf{x}_n$ is an N-dimensional vector and $\mathbf{F}$ an N-dimensional vector function.

We found that such dynamical systems could have equilibria, states that would persist forever if undisturbed. We agreed, moreover, that any equilibrium observed in nature would be a stable one, to which, after a small perturbation, the system would spend the rest of its life returning. Potentially stable equilibria $\mathbf{x} = \mathbf{x}^*(\alpha)$ of (6.1) could be identified as the solutions of the vector equation

$$\mathbf{f}(\mathbf{x}, \alpha) = 0. \tag{6.3}$$

The stability of these equilibria could then be verified by the method described at the beginning of Chapter 2. Alternatively, we could identify a stable equilibrium indirectly by solving (6.1) and watching the solution converge. Similarly, a stable equilibrium $\mathbf{x}_n = \mathbf{x}^*$ of (6.2) would satisfy

$$\mathbf{x}^* = \mathbf{F}(\mathbf{x}^*). \tag{6.4}$$

It could be identified indirectly by choosing $\mathbf{x}_0$, solving (6.2), and watching the solution converge.

Both continuous-time and discrete-time equilibria have their probabilistic analogues. But the state in a probabilistic model ($X(t)$ or $X(n)$) is a random variable, which cannot be restricted to a single value for the remainder of time—for then it would no longer be random! In this sense, a stochastic process could never converge to an equilibrium. In another sense, however, a stochastic process can converge and often does. We say that a stochastic process has a *stationary distribution* $\pi^* = (\pi_0^*, \pi_1^*, \ldots, \pi_n^*)^T$ when, if π^* is the initial distribution vector, then the distribution vector experiences no tendency to change. In other words, the continuous-time stochastic process $S_t = \{X(t) : 0 \le t < \infty\}$ has a stationary distribution when, if $\pi_i(0) \equiv \text{Prob}(X(0) = i) = \pi_i^*$, $0 \le i \le N$, then for all t satisfying $0 \le t < \infty$, $\pi_i(t) \equiv \text{Prob}(X(t) = i) = \pi_i^*$, $0 \le i \le N$. Again, the discrete-time stochastic process $S_k = \{X(k): k = 0, 1, 2, \ldots\}$ has a stationary distribution when, if $\pi_i(0) \equiv \text{Prob}(X(0) = i) = \pi_i^*$, $0 \le i \le N$, then for all $k = 0, 1, 2, \ldots$, $\pi_i(k) \equiv \text{Prob}(X(k) = i) = \pi_i^*$, $0 \le i \le N$. Here, k can stand for either l or n; i.e., time can be measured either linearly or nonlinearly.

We shall assume henceforward that if a stationary distribution π^* exists, then it is also unique, and, furthermore, that $\pi(t)$ (or $\pi(k)$) converges to π^* as $t \to \infty$ (or $k \to \infty$) from any value of $\pi(0)$. Convergence to stationarity is the probabilistic analogue of stability of equilibrium. By analogy with Chapter 2, the stationary distribution can be identified in either of two ways: by the direct method, which involves solving equations whose solution can only be a stationary distribution; and by the indirect method, which involves finding the limit as $t \to \infty$ of the matrix $\mathbf{R}(t)$, the limit as $l \to \infty$ of the matrix $\mathbf{S}^l$ or the limit as $n \to \infty$ of the matrix $\mathbf{P}^n$. Because, by assumption, the stationary distribution must be reached for any $\pi(0)$, it is legitimate to assume that $X(0) = i$ for any value of i; or, which is exactly the same thing, *all* rows of the matrix $\mathbf{R}(t)$, $\mathbf{S}^l$ or $\mathbf{P}^n$ must converge to $(\pi^*)^T$. Note that assuming $X(0) = i$ is equivalent to assuming that $\pi(0)^T = (0, \ldots, 0, 1, 0, \ldots, 0)$, where the 1 falls in row i of $\pi(0)$ and all other entries are zero.

Not all stochastic processes are as well behaved as we have just assumed. Some do not converge to their stationary distribution. Some have more than one stationary distribution. But our assumptions are valid for many applications, including almost all of the examples considered in the remainder of this text. The only two exceptions are introduced (purely for illustration) at the end of the following section, where we state sufficient conditions for our assumptions to be valid for a Markov chain.

6.1 The Certainty of Death

Recall from Section 5.2 that a matrix $R(t)$ for our 4-state pure death process is given by

$$R(t) = \begin{bmatrix} 1 & 0 & 0 & 0 \\ 1 - e^{-Dt} & e^{-Dt} & 0 & 0 \\ 1 - e^{-Dt} - Dte^{-Dt} & Dte^{-Dt} & e^{-Dt} & 0 \\ 1 - e^{-Dt}(1 + Dt + \frac{1}{2}D^2t^2) & \frac{1}{2}D^2t^2e^{-Dt} & Dte^{-Dt} & e^{-Dt} \end{bmatrix}. \quad (6.5)$$

To find the stationary distribution of this pure death process by the indirect method, we must simply take the limit as $t \to \infty$ and observe that

$$R(\infty) = \begin{bmatrix} 1 & 0 & 0 & 0 \\ 1 & 0 & 0 & 0 \\ 1 & 0 & 0 & 0 \\ 1 & 0 & 0 & 0 \end{bmatrix}. \quad (6.6)$$

Notice that all rows of this matrix R converge to $(\pi^*)^T = (1, 0, 0, 0)$. Thus, irrespective of the value of $X(0)$, the population is certain to end in state 0. In other words, no matter how many of three old men are alive to begin with, it is certain that ultimately all will be dead. To verify this result by the direct method, observe from (6.5) that if $\pi(0) = (1, 0, 0, 0)^T$, then $\pi(t)^T = \pi(0)^T R(t) = (1, 0, 0, 0)$ for all $t, 0 \le t < \infty$. More generally, you can easily verify that $\pi^* = (1, 0, 0, \ldots, 0)^T$ is a stationary distribution of the $N + 1$ state pure death process discussed at the end of Section 5.2, confirming the rather obvious result that for any value of N, N old men will ultimately die.

As a second example of the certainty of death, recall that in Section 5.8 we used the transition matrix

$$S = \begin{bmatrix} 0.72 & 0.28 & 0 & 0 & 0 & 0 \\ 0 & 0.69 & 0.31 & 0 & 0 & 0 \\ 0 & 0 & 0.75 & 0.25 & 0 & 0 \\ 0 & 0 & 0 & 0.77 & 0.23 & 0 \\ 0 & 0 & 0 & 0 & 0.63 & 0.37 \\ 0 & 0 & 0 & 0 & 0 & 1 \end{bmatrix} \quad (6.7)$$

to describe the growth of Scots pine trees in a forest. By computing successive powers of S, you can easily verify that the limit of S^l as $l \to \infty$ is the 6×6 matrix whose every row is the vector $(\pi^*)^T = (0, 0, 0, 0, 0, 1)$. Thus, by the indirect method, $(\pi^*) = (0, 0, 0, 0, 0, 1)^T$ is a stationary distribution for trees in the forest. No matter which class a tree initially occupies, it will ultimately enter class 5 and remain there (recall our assumption that all trees would reach maturity before being felled).

To obtain this result by the direct method, observe that if π^* is a stationary distribution, then $\pi(l) = \pi^*$ for all values of l. But $(\pi(l + 1))^T =$

$(\pi(l))^T S$. Hence

$$(\pi^*)^T = (\pi^*)^T S. \tag{6.8}$$

This is a set of six linear equations, only five of which are independent, the solution being $(\pi^*)^T = (0,0,0,0,0,c)$ for any value of c. But the solution is made unique by observing that

$$\sum_{i=0}^{5} \pi_i^* = 1. \tag{6.9}$$

As we said at the end of Chapter 5, our forest chain is of limited value. A forester doesn't need a mathematical model to know that all her trees will eventually reach maturity if not removed at an earlier stage! In its present form, our forest model ignores the fact that a forester replants trees as well as fells them; for note from the first column of (6.7) that it is impossible for a tree to enter state 0. Our model must therefore be extended before it can describe a real forest, and we will return to this matter in Chapter 9.

For a more general Markov chain with $N + 1$ states, the direct method of finding π^* still reduces to solving equations (6.8) and (6.9), but with 5 replaced by N, and, where necessary, S by P. Thus, if time is measured nonlinearly, the stationary distribution (assumed unique) is determined by

$$(\pi^*)^T = (\pi^*)^T P \tag{6.10a}$$

$$\sum_{j=0}^{N} \pi_j^* = 1. \tag{6.10b}$$

One of the $N + 1$ equations (6.10a) is always redundant because those equations add up to (6.10b). In the terminology of linear algebra, π^* is the left eigenvector of the matrix P with eigenvalue 1, normalized so that the sum of entries is 1.

Before proceeding to the following section, we illustrate that a Markov chain may not possess a unique stationary distribution; or, even if it does, may not converge to it. We also state sufficient conditions on P (or S) for uniqueness and convergence assumptions to be valid. By solving (6.10), you can easily show that $\pi^* = (a, b, a, b, a)^T$ is a stationary distribution of the Markov chain with transition matrix

$$P = \begin{bmatrix} \frac{1}{3} & 0 & \frac{1}{3} & 0 & \frac{1}{3} \\ 0 & \frac{1}{2} & 0 & \frac{1}{2} & 0 \\ \frac{1}{3} & 0 & \frac{1}{3} & 0 & \frac{1}{3} \\ 0 & \frac{1}{2} & 0 & \frac{1}{2} & 0 \\ \frac{1}{3} & 0 & \frac{1}{3} & 0 & \frac{1}{3} \end{bmatrix} \tag{6.11}$$

for all $0 \le a, b \le 1$ satisfying $3a + 2b = 1$. Hence there are infinitely many stationary distributions. To which distribution will the Markov chain converge? Because, as is readily verified, $\mathbf{P}^n = \mathbf{P}$ for all values of n, the Markov chain will converge to the stationary distribution π^* given by $(\pi^*)^T = \pi(0)^T \mathbf{P}$. If $\pi(0) = (1/2, 1/2, 0, 0, 0)^T$, for example, then the stationary distribution will be $\pi^* = (1/6, 1/4, 1/6, 1/4, 1/6)^T$; whereas if $\pi(0) = (1/3, 1/3, 1/3, 0, 0)^T$, then the stationary distribution will be $\pi^* = (2/9, 1/6, 2/9, 1/6, 2/9)^T$. Why does this behavior arise? Directly from the transition matrix, you can see that the states can be partitioned into two mutually disjoint sets, namely, $\{0, 2, 4\}$ and $\{1, 3\}$, such that it is impossible to move from one of these sets to the other. The Markov chain could therefore be decomposed into two smaller Markov chains, one having 3×3 transition matrix with all entries equal to $1/3$ and the other having 2×2 transition matrix with all entries equal to $1/2$. The Markov chain is said to be *reducible*. States 0, 2, and 4 are said to be *inaccessible* from states 1 and 3, and vice versa; and 0, 2, 4 are mutually accessible, and similarly for 1, 3. This suggests that for our uniqueness and convergence assumptions to be valid, a Markov chain should not be reducible, i.e., should be *irreducible*. You can check for irreducibility—i.e., that it's possible to go from any state to any other after a finite number of transitions—merely by inspecting $\mathbf{P}$.

But we do not assume only that a stationary distribution is unique. We also assume that $\mathbf{P}^n$ converges. Consider the Markov chain with transition matrix (5.86):

$$\mathbf{P} = \begin{bmatrix} 0 & 1 & 0 \\ \frac{d_1}{b_1 + d_1} & 0 & \frac{b_1}{b_1 + d_1} \\ 0 & 1 & 0 \end{bmatrix}.$$

You can easily verify by solving (6.10) that this Markov chain has the unique stationary distribution

$$\pi^* = \left(\frac{d_1}{2(b_1 + d_1)}, \frac{1}{2}, \frac{b_1}{2(b_1 + d_1)} \right)^T. \tag{6.12}$$

But we know from (5.87) that this Markov chain exhibits periodic behavior, so that $\mathbf{P}^n$ does not converge. Thus a Markov chain need not converge to its stationary distribution. But there will be no periodic behavior—i.e., the chain will be *aperiodic*—if $p_{ii} > 0$ for even one value of i, $0 \le i \le N$, and you can easily check for this.

All this suggests that sufficient conditions for uniqueness and convergence are that a Markov chain be irreducible and aperiodic. This is not quite enough, however; the chain must also be *positive recurrent*. For finite N this simply means that if $X(0) = i$, $0 \le i \le N$, then state i will be visited infinitely often if the process goes on forever. A Markov chain that is positive recurrent and aperiodic is called *ergodic*, and it can be proved that an irreducible ergodic chain has a unique stationary distribution, to which $\pi(n)$ must converge for any value of $\pi(0)$. See Ross (1983, p. 109) for a proof of this theorem and for further elaboration of the concepts of periodicity, recurrence, and reducibility.

6.2   Elevator Stationarity. The Stationary Birth and Death Process

The 3-state birth and death process used in Section 5.5 to describe an elevator has stationary distribution[1]

$$\pi^* = \lim_{t \to \infty} \pi(t) = \frac{1}{d_1 + \alpha + b_1}(d_1, \alpha, b_1)^T. \tag{6.13}$$

We virtually obtained this result in Section 5.5 by the indirect method, as we showed that $\pi(t)$ was approximately (6.13) after only nine hours. We indicated in (5.68) that the stationary distribution was independent of the initial distribution, and you can check for yourself that all rows of $\mathbf{R}(t)$ in (5.65) do indeed converge to (6.13) as $t \to \infty$.

We can also obtain this distribution by the direct method. Recall from Chapter 5 that the matrix $\mathbf{R}(t)$ for the continuous-time birth and death model with $N + 1$ states is obtained from the solution of the following system of $N + 1$ differential equations:

$$\frac{dr_{i0}}{dt} = d_1 r_{i1} - b_0 r_{i0}; \tag{6.14}$$

$$\frac{d}{dt} r_{ij} = b_{j-1} r_{i,j-1} - (b_j + d_j) r_{ij} + d_{j+1} r_{i,j+1}, \qquad 1 \le j \le N - 1; \tag{6.15}$$

$$\frac{dr_{iN}}{dt} = b_{N-1} r_{i,N-1} - d_N r_{iN}. \tag{6.16}$$

Now, if the distribution of $X(t)$ is stationary, then $\mathbf{R}(t)$ is constant and equal to $\mathbf{R}(\infty)$. Hence we have

$$\frac{dr_{ij}}{dt} = 0, \qquad 0 \le i, j \le N. \tag{6.17}$$

The elevator model of Section 5.5 is the special case of this in which $N = 2$ and $b_0 = d_2 = \alpha$. Hence from (6.14)–(6.17) we have

$$-\alpha r_{i0} + d_1 r_{i1} = 0,$$
$$\alpha r_{i0} - (b_1 + d_1) r_{i1} + \alpha r_{i2} = 0, \tag{6.18}$$
$$b_1 r_{i1} - \alpha r_{i2} = 0.$$

---

[1]You may wish to review Section 5.5 before proceeding.

These equations are linearly dependent; (6.18) sums to $0 = 0$. Their solution is made unique by appending (5.13), namely, $r_{i0} + r_{i1} + r_{i2} = 1$. Hence

$$(r_{i0}, r_{i1}, r_{i2}) = \frac{1}{d_1 + \alpha + b_1}(d_1, \alpha, b_1), \tag{6.19}$$

confirming that (6.13) must be the limit of each row of the matrix $\mathbf{R}(t)$.

Similarly, in Section 5.6, we virtually established by the indirect method that the metered model of the same elevator, with matrix $\mathbf{S}$ defined by (5.76) and $b_1 = d_1 = \alpha$, had stationary distribution

$$\boldsymbol{\pi}^* = \lim_{l \to \infty} \boldsymbol{\pi}(l) = \left(\frac{1}{3}, \frac{1}{3}, \frac{1}{3}\right)^T. \tag{6.20}$$

We indicated that this limit was independent of $\boldsymbol{\pi}(0)$ and virtually showed, through (5.77), that the limit of the matrix $\mathbf{S}^n$ as $n \to \infty$ was the matrix whose every row is equal to (6.20). This result can also be obtained by using (6.10); see Exercise 6.1. Again, we virtually established by the indirect method that the bird chain of Section 5.7 converges to a stationary distribution. This may also be found directly; see Exercise 6.2.

In Chapter 2 we found that by assuming a system had already reached a steady state, we were able (directly) to determine equilibria that could not easily be found by the indirect method, because of the complexity of the dynamical equations. In the same way, by assuming that a stochastic process has reached stationarity, we are able to determine stationary distributions that could not easily be found by the indirect method. The rows of $\mathbf{R}(t)$ are identical in a stationary state, and so we have $r_{ij}(t) = (\boldsymbol{\pi}^*)_j$, independently of the value of i. Henceforward, we will analyze stationary distributions solely by the direct method. It is therefore convenient to drop the asterisk and denote the (stationary) distribution by the $N + 1$ dimensional vector $\boldsymbol{\pi} = (\pi_0, \pi_1, \ldots, \pi_N)^T$. Then, for all i, we write $r_{ij}(t) = \pi_j$, and on using (6.14)–(6.17) we obtain the following equations for a *stationary birth and death process*:

$$-b_0\pi_0 + d_1\pi_1 = 0, \tag{6.21}$$

$$b_{j-1}\pi_{j-1} - (b_j + d_j)\pi_j + d_{j+1}\pi_{j+1} = 0, \qquad 1 \le j \le N - 1, \tag{6.22}$$

$$b_{N-1}\pi_{N-1} - d_N\pi_N = 0; \tag{6.23}$$

$$\sum_{j=0}^{N} \pi_j = 1. \tag{6.24}$$

Only $N + 1$ of these $N + 2$ linear equations are independent, because the left-hand sides of (6.21)–(6.23) sum to zero. By substituting (6.21) into (6.22) for $j = 1$, and proceeding recursively for higher values of j, it is readily shown that $b_{j-1}\pi_{j-1} = d_j\pi_j$, for all $1 \le j \le N$. Hence, on using

(6.24) and defining

$$C_j = \frac{b_{j-1}b_{j-2}b_{j-3}\ldots b_1 b_0}{d_j d_{j-1}d_{j-2}\ldots d_2 d_1}, \qquad j \geq 1, \tag{6.25}$$

we find that

$$\pi_0 = \frac{1}{1 + \sum_{j=1}^{N} C_j}, \tag{6.26a}$$

$$\pi_j = C_j \pi_0, \qquad 1 \leq j \leq N. \tag{6.26b}$$

It is often convenient to let N be infinite. This is legitimate whenever C_j tends to zero sufficiently rapidly, as $j \to \infty$, for the resulting infinite series in the denominator of π_0 to converge.

We now proceed to an application of the stationary birth and death process.

6.3   How Long Is the Queue at the Checkout? A First Look

Suppose that a supermarket checkout has only one cashier, whose serving time is exponentially distributed with mean s minutes. Let's assume that customers arrive independently, the time between arrivals being exponentially distributed with mean c minutes. Then we can model the queue at the checkout as a birth and death process, with arrivals corresponding to births and customer departures to deaths.

Let the random variable $X(t)$, taking integer values $j = 0, 1, 2, \ldots, N$, denote the number of individuals in the queue, including the one being served. Then, from expressions (5.41b) and (5.47b) for the reciprocal of the mean time until transition, we have

$$b_j = \frac{1}{c}, \qquad d_j = \frac{1}{s} \tag{6.27}$$

for all relevant j (or, if you prefer, $d_0 = 0 = b_N$).

Now assume that

$$s < c. \tag{6.28}$$

Thus, on average, the cashier works fast enough to restrict the queue to an acceptable size. Although the pattern of customer arrival will change during the day, it does not seem unreasonable to suppose it constant over a fairly short interval, a lunch hour, perhaps. Let us therefore assume that the queue is stationary, i.e., that the distribution of $X(t)$ is independent of time. Then $\text{Prob}(X(t) = j) = \pi_j$ is independent of time, and from (6.25)–

(6.27) we have

$$\pi_0 = \frac{1}{1 + \sum_{j=1}^{N} \left(\frac{s}{c}\right)^j} = \frac{1 - \frac{s}{c}}{1 - \left(\frac{s}{c}\right)^{N+1}},$$

$$\pi_j = \left(\frac{s}{c}\right)^j \pi_0, \qquad 1 \le j \le N. \tag{6.29}$$

Let's define

$$\rho = \frac{s}{c}. \tag{6.30}$$

The parameter ρ is a measure of business intensity; note from (6.28) that $0 < \rho < 1$. Then (Exercise 6.4) the expected value of the length of the queue is

$$E[X] = \sum_{j=1}^{N} j\pi_j = \frac{N\rho^{N+2} - (N+1)\rho^{N+1} + \rho}{(1-\rho)(1-\rho^{N+1})}. \tag{6.31}$$

This is a rather unwieldy expression. We therefore assume that N is so large as to be effectively infinite. Then, taking the limit of (6.31) as $N \to \infty$, the expected number of individuals in the queue is

$$L_1(\rho) = \frac{\rho}{1-\rho}. \tag{6.32}$$

This function is plotted in Fig. 6.1. Notice how rapidly L_1 increases as ρ approaches 1. It jumps from 3 to 10 as ρ increases from 0.75 to 0.91. Mean queue lengths of up to 3 are probably quite acceptable; whereas mean queue lengths of up to 10 are certainly not, and the manager would have to open another checkout. Thus our checkout model, albeit a very simple one,

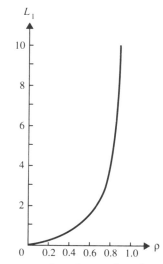

Fig. 6.1 Single-server mean queue length as a function of business intensity.

has already played an important role. It has highlighted the importance of the parameter ρ and told us that s and c are important times, which a successful manager must measure and know.

6.4 How Long Is the Queue at the Checkout? A Second Look

We saw in the previous section that if ρ, the business intensity, is close to 1 then the queue, on average, will be too long and a second checkout will be opened. It is now permissible for s to exceed c, but $s/2$ must still be less than c if the two servers together are to restrict the queue to an acceptable size in the long run. Hence $\rho < 2$. To make matters simple, we will assume that the customers form a single queue whose first member advances to the next available server. This is already what happens in banks and post offices and may one day happen in supermarkets, too.

It is still true that

$$b_j = \frac{1}{c} \tag{6.33}$$

because customers arrive, on average, every c minutes. But it's no longer true that $d_j = 1/s$ because the chance of a "death," or customer departure, increases with the number of servers. It seems reasonable to assume that the new (two-server) death rate will be twice the old one. But we must also remember that one server will be idle if $j = 1$, because the only customer then enjoys the attention of the other server. Hence

$$d_j = \begin{cases} \dfrac{2}{s}, & \text{if } 2 \le j \le N; \\[2mm] \dfrac{1}{s}, & \text{if } j = 1. \end{cases} \tag{6.34}$$

In these circumstances (Exercise 6.5), it is straightforward to show that the expected value of the length of the queue is

$$\sum_{j=1}^{N} j\pi_j = \pi_0 \rho \, \frac{N\left(\frac{\rho}{2}\right)^{N+1} - (N+1)\left(\frac{\rho}{2}\right)^N + 1}{\left(1 - \frac{\rho}{2}\right)^2}, \tag{6.35}$$

where

$$\pi_0 = \frac{1 - \frac{\rho}{2}}{1 + \frac{\rho}{2} - \rho\left(\frac{\rho}{2}\right)^N}. \tag{6.36}$$

These expressions are somewhat unwieldy. Therefore, as in the previous section, we assume that N is so large as to be effectively infinite. Then the mean queue length with two servers is

$$L_2(\rho) = \frac{4\rho}{4 - \rho^2}. \tag{6.37}$$

This function is plotted in Fig. 6.2. You should verify that $L_2(\rho) < L_1(\rho)$ when $\rho < 1$ (when $\rho \geq 1$, of course, L_1 is undefined). The dashed horizontal line represents a mean queue length of three. Suppose the manager believes that this value should not be exceeded. Then, from Fig. 6.2, a single checkout will suffice if $0 < \rho \leq 0.75$, and two will suffice if $0.75 < \rho \leq 1.44$. If ρ significantly exceeds 1.44, however, then at least three servers are needed.

It isn't difficult to see how the results of this section can be generalized to an arbitrary number of servers, k. We will assume, of course, that these k servers restrict the queue to an acceptable size in the long run. Thus $s/k < c$ or

$$\rho < k. \tag{6.38}$$

If all are busy, then the death rate is k times that for a single server, but if at least one is idle ($j < k$), then the death rate is j times that for a single server. Hence (6.34) is replaced by

$$d_j = \begin{cases} \dfrac{k}{s}, & \text{if } j \geq k; \\[2mm] \dfrac{j}{s}, & \text{if } 1 \leq j < k. \end{cases} \tag{6.39}$$

Assume that N is so large as to be effectively infinite. Then (Exercise 6.6) the expected value of the k-server queue length is

$$L_k(\rho) = \frac{\sum_{j=1}^{k} \frac{\rho^j}{(j-1)!} + \frac{k^k}{k!}\left(\frac{\rho}{k}\right)^{k+1} \frac{k-\rho+1}{(1-\rho/k)^2}}{1 + \sum_{j=1}^{k} \frac{\rho^j}{j!} + \frac{\rho^{k+1}}{k!(k-\rho)}}. \tag{6.40}$$

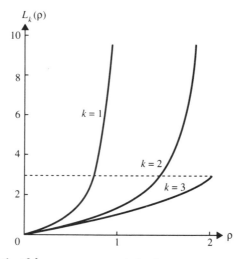

Fig. 6.2 Mean length of k-server queue at checkout.

Note that (6.32) and (6.37) are the special cases of (6.40) for which $k = 1$ and $k = 2$, respectively. The graph of

$$L_3(\rho) = \frac{\rho(18 + 6\rho - \rho^2)}{(3 - \rho)(6 + 4\rho + \rho^2)} \tag{6.41}$$

is also sketched in Fig. 6.2, from which you can see that three servers would be adequate if $\rho \leq 2.0$. But a fourth server would be needed if ρ were significantly greater than 2, and so on, ad infinitum. The notion that three is the critical queue length is entirely arbitrary, of course; but it is clear how Fig. 6.2 may be extended to higher values of k (up to the maximum number of checkouts available), so that a manager can determine the appropriate number of servers for any business intensity.

   If the choice of 3 for a critical queue length is purely arbitrary, however, then how do we decide which length is optimal? Perhaps you would like to think about that—it's the kind of question we shall try to answer in the following chapter. Meanwhile, now might be a good time to attempt Exercises 6.7 and 6.8.

6.5   How Long Must Someone Wait at the Checkout? Another View

Naturally, the length of the queue at a checkout does not tell the whole story about waiting. A queue of ten people might be quite acceptable if each took 30 seconds to pass the checkout, but a queue of two might be unacceptable if you were the second and had to wait half an hour. For certain purposes, therefore, the waiting time is a more reliable indicator of the tedium of queuing than the number of folks in the queue, but, as we shall now discover, the expected value of the waiting time in a stationary birth and death process can always be deduced from the expected length of the queue.

   To make matters simple, we'll begin with the case of a single server and assume that N is infinite. Then from (6.29) and (6.30), we have

$$\pi_j = \rho^j(1 - \rho), \qquad j \geq 0. \tag{6.42}$$

Suppose that there are $i - 1$ people in the queue when an individual arrives. Then, before departing, she will have to wait a total of i service times. Now, from the point of view of the world at large, these service times are associated with a stationary birth and death process. But our individual does not care about the waiting time of people who join the queue behind her. These future births are of no concern. Thus, from her point of view, the birth and death process converts to a pure death process with $X(0) = i$, in which she will witness i transitions before departing. Let the times between these transitions, i.e., the service times, be denoted by the random variables

$G_1, G_2, \ldots, G_i$. Then from (5.41a) and (6.27), we have

$$\text{Prob}(G_i > t) = e^{-t/s}. \tag{6.43}$$

Because the exponential distribution has no memory, (6.43) is valid even if the first person in the queue has already been receiving the server's attention when our individual arrives.

What really interests our individual, however, is the total waiting time, including the time during which she is actually served. Let this be denoted by the continuous random variable Y. Let us also define the random variable V_i by

$$V_i \equiv G_1 + G_2 + G_3 + \cdots + G_{i-1} + G_i. \tag{6.44}$$

Now, *if* our individual knew for certain that there would be, say, $j-1$ people in the queue when she arrived, then her total waiting time would be $Y = V_j$, and the thing she wanted to know would be $\text{Prob}(Y \leq t) = \text{Prob}(V_j \leq t)$. This is readily obtained from Section 5.2. For if $X(t)$ denotes the number of individuals in a pure death process at time t, then

$$\text{Prob}(V_j \leq t) = \text{Prob}(X(t) = 0 \mid X(0) = j) = r_{j0}(t), \tag{6.45}$$

and explicit expressions for all entries of the matrix $\mathbf{R}(t)$ are already known. But our individual does *not* know how many people there will be in the queue on her arrival, so that $\text{Prob}(Y \leq t)$ is not given by (6.45) and further analysis is required. Let U be the event that $Y \leq t$, and let U_j be the event that there are j people in the queue when our individual arrives, for $0 \leq j < \infty$. Then, because the set $\{U_j : 0 \leq j < \infty\}$ includes all possibilities, we have

$$\text{Prob}(U) = \sum_{j=0}^{\infty} \text{Prob}(U \mid U_j) \cdot \text{Prob}(U_j). \tag{6.46}$$

But $\text{Prob}(U \mid U_j) = \text{Prob}(Y \leq t \mid j \text{ in queue on arrival}) = \text{Prob}(V_{j+1} \leq t)$, and $\text{Prob}[U_j] = \pi_j$. Hence (6.45) and (6.46) yield:

$$\text{Prob}(Y \leq t) = \sum_{j=0}^{\infty} r_{j+1,0}(t)\pi_j. \tag{6.47}$$

Now, π_j is given by (6.42), and there are two equivalent expressions for $r_{j+1,0}(t)$:

$$r_{j+1,0}(t) = \text{Prob}(V_{j+1} \leq t) = 1 - e^{-t/s} \sum_{k=0}^{j} \frac{\left(\frac{t}{s}\right)^k}{k!}, \tag{6.48a}$$

$$r_{j+1,0}(t) = \text{Prob}(V_{j+1} \leq t) = \frac{1}{s} \int_0^t \frac{(\zeta/s)^j e^{-\zeta/s}}{j!} d\zeta. \tag{6.48b}$$

The first of these expressions follows directly from (6.27) and (5.42) on replacing N by $j + 1$. The second expression, which is the one we shall use, can be obtained from (5.40) by recursive integration, but it can be obtained just as easily (Exercise 6.9) by observing that the integrand in (6.48b) is the derivative of (6.48a). Hence, on using (6.27), (6.42), (6.47), and (6.48b) we have

$$
\begin{aligned}
\text{Prob}(Y \leq t) &= \sum_{j=0}^{\infty} \rho^j (1 - \rho) \frac{1}{s} \int_0^t \frac{(\zeta/s)^j e^{-\zeta/s}}{j!} d\zeta \\
&= \int_0^t \frac{1 - \rho}{s} e^{-\zeta/s} \sum_{j=0}^{\infty} \frac{(\rho\zeta/s)^j}{j!} d\zeta \\
&= \int_0^t \frac{1 - \rho}{s} e^{-\zeta/s} e^{\rho\zeta/s} d\zeta = \int_0^t e^{-(1 - \rho)\zeta/s} d((1 - \rho)\zeta/s) \\
&= 1 - e^{-(1 - \rho)t/s}.
\end{aligned}
\tag{6.49}
$$

In other words, on comparing (6.49) with (5.30b), our individual's total waiting time is exponentially distributed with transition rate $1/s - 1/c$, the difference between the death rate (serving rate) and birth rate (customer arrival rate). It follows immediately that the expected value of the waiting time, which we will denote by W, is given by

$$
W \equiv E[Y] = \frac{1}{\frac{1}{s} - \frac{1}{c}} = \frac{s}{1 - \rho}.
\tag{6.50}
$$

But from Section 6.3, the expected value of the length of a single server queue is $L = \rho/(1 - \rho) = s/c(1 - \rho)$. Hence W and L are related by

$$
W = cL = \frac{L}{\text{BIRTH RATE}}.
\tag{6.51}
$$

This result is intuitively sound. It says that, under stationary conditions, AVERAGE QUEUE LENGTH = AVERAGE WAITING TIME × AVERAGE RATE OF ARRIVAL. Moreover, it is merely a special case of the more general result that

$$
W = \frac{L}{\bar{b}},
\tag{6.52}
$$

where

$$
\bar{b} = \sum_{j=0}^{\infty} b_j \pi_j
\tag{6.53}
$$

for any stationary queuing process, whether service times are exponentially distributed or not, and for an arbitrary number of servers. It is also true that

$$
W_q = \frac{L_q}{\bar{b}},
\tag{6.54}
$$

where W_q is the expected value of the waiting time *excluding* service time, and L_q is the expected queue length *excluding* the k people being served.

Because L_q and L are related by

$$L_q = \sum_{j=k}^{\infty}(j-k)\pi_j = \sum_{j=1}^{\infty}j\pi_j - \sum_{j=1}^{k-1}j\pi_j - k\left\{\sum_{j=0}^{\infty}\pi_j - \sum_{j=0}^{k-1}\pi_j\right\}$$

$$= L - \sum_{j=1}^{k-1}j\pi_j - k\left\{1 - \sum_{j=0}^{k-1}\pi_j\right\},$$

(6.55)

all four quantities L, W, L_q, and W_q are known as soon as any one of them is known.

These are just a few of the results that are known collectively as *queuing theory*. This vast but quite recent discipline generalizes the stationary birth and death process to deal with queuing problems where the interarrival and service times do not have exponential distributions. Why might this generalization be necessary?

Let's attempt to answer this question by comparing the exponential distribution with mean μ, and hence cumulative frequency distribution

$$\text{Prob}(G \le t) = 1 - e^{-t/\mu},$$

(6.56)

with a frequently used alternative, namely, the so-called normal distribution with mean μ and variance σ^2 (and hence standard deviation σ). The latter has probability density function

$$f(t) = \frac{1}{\sigma\sqrt{2\pi}}e^{-(t-\mu)^2/2\sigma^2}$$

(6.57)

and cumulative distribution function

$$\text{Prob}(G \le t) = \frac{1}{\sigma\sqrt{2\pi}}\int_{-\infty}^{t}e^{-(\zeta-\mu)^2/2\sigma^2}d\zeta.$$

(6.58)

Both functions are defined for $-\infty < t < \infty$ and are sketched in Fig. 6.3.

Notice that the normal distribution is symmetric about its mean, attaching the same probability to the event $\mu - \tau < G \le \mu$ as to the event $\mu < G \le \mu + \tau$, for any value of τ. Moreover, approximately 68% of observations taken from a normal distribution lie within one standard deviation of the mean, approximately 95% within two standard deviations, and almost all (99.7%) within three. Thus, provided that μ exceeds 3σ, (6.57) assigns negligible probability to negative values of a random variable. It can therefore be used to approximate the distribution of random variables that are intrinsically positive.

On the other hand, the exponential distribution is not symmetric about its mean. Indeed it follows immediately from (6.56) that

$$\frac{\text{Prob}(\mu < G \le \mu + \tau)}{\text{Prob}(\mu - \tau < G \le \mu)} = \frac{\text{Prob}(G \le \mu + \tau) - \text{Prob}(G \le \mu)}{\text{Prob}(G \le \mu) - \text{Prob}(G \le \mu - \tau)} = e^{-\tau/\mu} < 1.$$

(6.59)

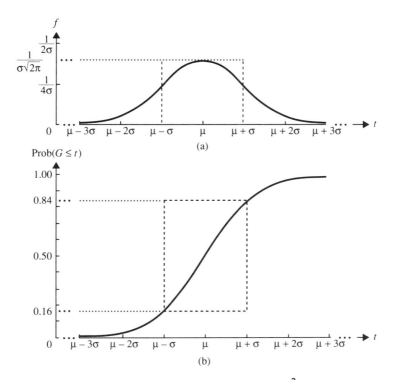

Fig. 6.3 Normal distribution with mean μ and variance σ^2.

(a) Probability density function (6.57). Notice that 68% of the area under the graph falls in the interval $\sigma - \mu < t \le \sigma + \mu$, i.e., within one standard deviation of the mean. More than 95% of the area falls within two standard deviations of the mean and almost all (99.7%) within three standard deviations.

(b) Cumulative distribution function (6.58). For each value of t, this records the area between the horizontal axis, $-\infty < \zeta \le t$, and the graph in (a). Again, notice that 68% of normally distributed values fall within one standard deviation of the mean and almost all (99.7%) fall within three. We will use this observation in deriving the normal approximation to the Poisson distribution; see Section 7.2.

Only $100 \times e^{-1} \approx 37\%$ of exponentially distributed values lie above the mean; the corresponding figure for the normal distribution, of course, is 50%. Yet despite that, an exponentially distributed random variable is more than twice as likely as a normally distributed one to exceed the mean by two standard deviations.

These qualitative considerations enable us to reach some tentative conclusions about inter-transition times in queues. If real service times fall within a narrow band and are as likely to exceed the mean as fall below it, then the normal is a more suitable service distribution than the exponential. This would happen if each customer received the same service in theory but human frailty made service time a random variable in practice—in a soup kitchen, for example, where everyone gets soup and a roll; or at a movie theater, where cashiers appear able to launch tickets and change at a family of seven as rapidly as at a dating couple. In such circumstances, long service times would be exceedingly rare. But customers in a supermarket may be buying anything from a six-pack of beer to a gourmet meal that is fit for a queen. Long service times must therefore be assigned higher probability than according to the normal distribution. Moreover, isn't it true that the majority of supermarket service times are lower than average? Don't checkout queues usually move pretty quickly until someone can't find her check book, or insists on paying with a Bank of Mars card? This all suggests that the exponential is a more suitable service distribution for a supermarket checkout than the normal; and the validity of this hypothesis can be tested by a method to be described in Section 8.7. On the whole, you don't have to worry so much about customer interarrival times—they tend to be exponentially distributed, even when service times don't. But the value of c might vary throughout the day. Thus our model, which assumes c constant, is valid only for fairly short intervals, and different versions of it would have to be used at different times during the day.

Unfortunately, although queues with nonexponentially distributed inter-transition times are important in practice, the technical difficulties associated with studying them far outweigh the additional contribution they make to an understanding of the modelling process. Therefore, we shall not pursue them further in this text. But if you would like to pursue the matter yourself, then you might begin with Chapter 10 of Hillier and Liebermann (1980), followed by Chapter 8 of Ross (1980). Indeed our formal study of queuing theory could have ended with the previous section if it weren't that two of this section's results will be needed later. The first is (6.49), which we shall need in Chapter 7. The second is the probability density function of the random variable V_i, defined by (6.44). Directly from (6.48b), we see that the cumulative distribution function of V_i is given by

$$\text{Prob}(V_i \leq t) = \frac{1}{s}\int_0^t \frac{(\zeta/s)^{i-1}e^{-\zeta/s}}{(i-1)!}\,d\zeta. \tag{6.60}$$

Hence, on using (A.28), its probability density function is given by

$$f_i(t) = \frac{d}{dt}\{\text{Prob}(V_i \leq t)\} = \frac{t^{i-1}e^{-t/s}}{s^i(i-1)!}. \tag{6.61}$$

We say that V_i has a Gamma distribution with parameter i, on $0 \leq t < \infty$. The Gamma distribution is sketched in Fig. 6.4 for several values of i. Note that, because V_i is the sum of i independent random variables, which are identically distributed with mean s, it follows from (A.58) that the mean and variance of the Gamma distribution are $\mu = is$ and $\sigma^2 = is^2$. We shall need (6.61) in Section 6.7.

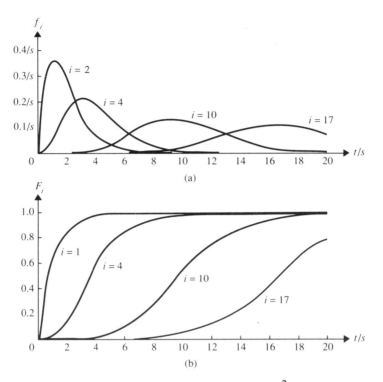

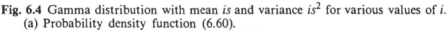

Fig. 6.4 Gamma distribution with mean is and variance is^2 for various values of i.
(a) Probability density function (6.60).
(b) Cumulative distribution function (6.61).

6.6 The Structure of the Work Force

Markov chains have proved extremely useful in the social sciences, where a natural unit of nonlinear time is one generation. In 1953 or thereabouts, David Glass and John Hall made use of this fact to construct the following model of social mobility in Britain.

Let seven classes of occupation, $i = 0, 1, 2, \ldots, 6$, be defined as in Table 6.1. Let the random variable $X(n)$ denote the occupational class of a randomly chosen individual from the nth generation of the British work force, and suppose that her children's chances of belonging to a certain occupational class do not depend on that of her parents. In other words, suppose that

$$\text{Prob}(X(n+1) = j \mid X(n) = i) = \text{Prob}(X(n) = j \mid X(n-1) = i) = p_{ij}. \quad (6.62)$$

Then the seven occupational classes may be regarded as the seven states of a Markov chain, with p_{ij} denoting the probability that a son or daughter of a member of class i will become a member of class j.

Table 6.1   Occupational classes defined by Glass and Hall

State	Class of occupation
0	Professional and higher administrative
1	Managerial and executive
2	Higher grade supervisory and nonmanual
3	Lower grade supervisory and nonmanual
4	Skilled manual and routine nonmanual
5	Semi-skilled manual
6	Unskilled manual

Based on this assumption, and using data collected in 1949, Glass and Hall estimated a transition matrix for England and Wales to be

$$\mathbf{P} = \begin{bmatrix} .388 & .146 & .202 & .062 & .140 & .047 & .015 \\ .107 & .267 & .277 & .120 & .206 & .053 & .020 \\ .035 & .101 & .188 & .191 & .357 & .067 & .061 \\ .021 & .039 & .112 & .212 & .430 & .124 & .062 \\ .009 & .024 & .075 & .123 & .473 & .171 & .125 \\ .0 & .013 & .041 & .088 & .391 & .312 & .155 \\ .0 & .008 & .036 & .083 & .364 & .235 & .274 \end{bmatrix}. \tag{6.63}$$

This says, for example, that about 1.5% of children of professional and higher administrative people grew up to become unskilled manual workers. (The data were actually based on 3500 pairs of fathers and sons and excluded the female work force; but we will ignore this point in the present section, where we are simply concerned with the idea behind the model.)

It is straightforward to show that the Markov chain associated with (6.63) has a unique stationary distribution, π. One way to verify this would be to use a computer to calculate successive powers of $\mathbf{P}$. As $n \to \infty$, all rows of $\mathbf{P}^n$ would converge to π^T. A second way would be to inspect the transition matrix, verifying that the conditions laid down at the end of Section 6.1 are all satisfied; π would then emerge from (6.10). By either method, the unique stationary distribution is found to be given by

$$\pi^T = (0.023, 0.042, 0.088, 0.127, 0.409, 0.182, 0.129). \tag{6.64}$$

This enables us to predict the structure of the British work force several generations from now. We might, for example, predict that about 4% of workers will be managers or executives. Naturally, the correctness of any predictions will strongly depend on the correctness of the model's major assumption, namely, (6.62).

Do workers really migrate in fixed proportions from one occupational class to another? We will return to this question in Chapter 8, where we describe how the transition matrix (6.63) was estimated.

6.7 When Does a T-junction Require a Left-turn Lane?

Consider a T-junction on a narrow two-lane road.[2] As drawn in Fig. 6.5, the major road runs from north to south, with the minor road leaving to the west. There is no left-turn lane on the major road at the junction. Thus if southbound traffic is heavy, then cars turning left may have difficulty entering the minor road and may cause delays to northbound traffic. When is this problem serious enough to warrant building a left-turn lane? In this section, we will attempt to answer that question.

Let λ denote the mean traffic flow from north to south, expressed as number of cars per second. Let μ denote the corresponding flow from south to north. Let α denote the probability that one of these northward bound cars will turn left at the junction; hence $1 - \alpha$ is the probability that it will continue forward. We will suppose that northbound drivers first cross the bar marked "Decision" in Fig. 6.5, then decide to turn left (with probability α) or continue forward (with probability $1 - \alpha$). We will call this bar the *decision bar*. In practice, the decision bar might be 20 or 30 yards

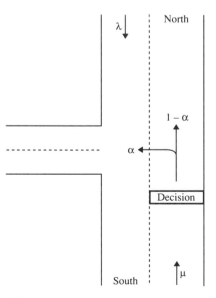

Fig. 6.5 Plan of T-junction.

---

[2]You may wish to review Section 5.4 before proceeding.

south of the junction; but it will be convenient to imagine that drivers make this decision at the last possible moment, so that the decision bar is about a car-length away from the turning point (this is, anyway, how almost everyone in Tallahassee seems to drive). Let T be the minimum time gap in the southbound traffic flow that will allow a north-bound car to turn left (assumed the same for all cars). Because T is a time, as opposed to a distance, it is immaterial whether a southbound car turns right or goes forward at the junction. We will assume throughout that drivers joining the main road at the junction (from the west) interrupt neither the northbound nor southbound flow, but rather enter the main stream of traffic whenever it is completely clear for them to do so. We will also assume that drivers act responsibly, so that southbound cars are not impeded by drivers turning left when such a move would be dangerous. It is then reasonable to regard the approach of southbound cars to the junction as unrestricted; and hence, as for the crossing control problem of Section 5.4, to model it as a Poisson process, with an average of λt arrivals in an interval of length t seconds. Let the random variable S denote the time gap in seconds between southbound cars. Then, from (5.52), the distribution of S is given by

$$\text{Prob}(S > t) = e^{-\lambda t}, \qquad 0 \le t < \infty. \tag{6.65}$$

Similarly, if the random variable N denotes the time gap in seconds between northbound cars whenever northbound traffic is moving freely, then it seems reasonable to assume that

$$\text{Prob}(N > t) = e^{-\mu t}, \qquad 0 \le t < \infty, \tag{6.66}$$

with associated probability density function

$$f_N(t) = \frac{d}{dt}\{\text{Prob}(N \le t)\} = \mu e^{-\mu t}, \qquad 0 \le t < \infty. \tag{6.67}$$

Let us define a *tailback* to be a group of one or more northbound cars that have been brought to a halt behind a car turning left at the junction. Because a tailback may form, northbound traffic need not move freely at the junction itself, but it will still move freely behind the tailback! Thus (6.66) governs times between northbound arrivals at the tailback, if there is a tailback; or at the junction, if not. From our comments above, it follows that the first car of any tailback will be waiting just behind the decision bar.

Now suppose that a northbound car has reached the T-junction and is waiting to turn left. In Fig. 6.5, it has just crossed the decision bar. Let the random variables S_1, S_2, S_3, ... denote the sequence of time gaps in the southbound flow which confronts this driver; and let U_j be the event that she enters the minor road during the jth such gap. Then, because the gaps are independent, her predicament is identical to that of a crossing

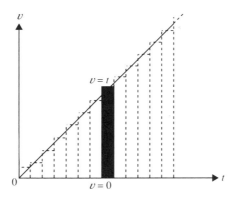

Fig. 6.6 The region in the t-v plane for which $t < v$. This infinite wedge is sliced into infinitesimally thin strips parallel to the v-axis which extend from $v = 0$ to $v = t$. For fixed t, the inner integration with respect to v in (6.70) is performed along such a strip, as indicated schematically in the diagram. The outer integration with respect to t then sums over all such strips.

pedestrian, and we deduce immediately from (5.53) that

$$\text{Prob}(U_j) = \text{Prob}(S_1 \leq T, S_2 \leq T, \ldots, S_{j-1} \leq T, S_j > T)$$

$$= \left(1 - e^{-\lambda T}\right)^{j-1} e^{-\lambda T}. \tag{6.68}$$

Let W denote the event that no tailback forms behind our waiting driver, and let the random variable V_j be defined by $V_j = S_1 + S_2 + S_3 + \cdots + S_j$. Then it follows immediately from (6.61) that V_j has a Gamma distribution with probability density function

$$f_j(v) = \frac{\lambda e^{-\lambda v}(\lambda v)^{j-1}}{(j-1)!}. \tag{6.69}$$

The random variables N and V_j are independent, because northbound and southbound flows of traffic are independent. Thus the joint probability density function of the random variables N and V_j, which we shall denote by f, is simply the product of (6.67) and (6.69); i.e., $f(t, v) = f_N(t) \cdot f_j(v)$. See Appendix 1, in particular remarks following (A.55). Moreover, on using (A.51), the probability that $V_j < N$ is simply the integral of f over the region in the t-v plane, depicted in Fig. 6.6, for which $v < t$. Hence

$$\text{Prob}(V_j < N) = \int_{v < t} f(t, v)\,dv\,dt$$

$$= \int_0^\infty \int_0^t f_N(t) \cdot f_j(v)\,dv\,dt \tag{6.70}$$

$$= \left(\frac{\lambda}{\mu + \lambda}\right)^j,$$

on evaluating the double integral in Exercise 6.10.

Now, *given* that our driver enters the minor road during gap j, the probability that no tailback forms behind her is simply the probability that the $(j-1)$th southbound car passes her before the next northbound car arrives at the junction. In other words,

$$\text{Prob}(W \mid U_j) = \text{Prob}(V_{j-1} < N) = \left(\frac{\lambda}{\lambda + \mu}\right)^{j-1}, \qquad (6.71)$$

on using (6.70). But the infinite sequence of events $\{U_1, U_2, U_3, \ldots\}$ exhausts all possibilities for entering the minor road. Hence from (A.13),

$$\text{Prob}(W) = \sum_{j=1}^{\infty} \text{Prob}(W \mid U_j) \cdot \text{Prob}(U_j) \qquad (6.72)$$

$$= e^{-\lambda T} \sum_{j=1}^{\infty} \left\{\frac{\lambda(1 - e^{-\lambda T})}{\mu + \lambda}\right\}^{j-1}, \qquad (6.73)$$

on using (6.68) and (6.71). Let θ denote the probability that a tailback does form behind a left-turning driver. Then (Exercise 6.10)

$$\theta = 1 - \text{Prob}(W) = \frac{\mu(1 - e^{-\lambda T})}{\mu + \lambda e^{-\lambda T}}. \qquad (6.74)$$

Note that because the formation of a tailback is governed by two memoryless Poisson processes, θ is also the probability that a tailback containing two or more cars will form behind a tailback containing precisely one. Let us now define four traffic configurations, labelled $i = 0, 1, 2, 3$, as shown in Fig. 6.7:

Configuration	Last car to drive over decision bar	Status of northbound lane
$i = 0$	has continued forward	no tailback
$i = 1$	is waiting to turn left	no tailback
$i = 2$	has continued forward	tailback
$i = 3$	is waiting to turn left	tailback

$$(6.75)$$

Because transitions between these configurations are governed by two memoryless Poisson processes, we may regard them as the states of a Markov chain. Notice that each transition corresponds to a physical event. States 0 or 1 may be entered only when a car crosses the decision bar,

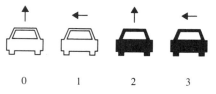

| 0 | 1 | 2 | 3 |

Fig. 6.7 Possible traffic configurations at a T-junction.

states 2 or 3 either when a car crosses the decision bar or when one joins the tailback. Thus time is measured nonlinearly. Therefore, let $\mathbf{P} = [p_{ij}]$ be the transition matrix; i.e., let p_{ij} be the probability that j is the next configuration if the current one is i, $0 \leq i, j \leq 3$. Then, because state 2 cannot be entered from states 0 or 1, and because state 3 cannot be entered from state 0, we have $p_{02} = p_{03} = p_{12} = 0$.

Suppose that the junction is in state 0. Then northbound traffic is moving freely, the last car over the decision bar has gone forward, and the next state will be 0 or 1 according to whether the next car goes forward or turns left, after passing the decision bar. Hence $p_{00} = 1 - \alpha$, $p_{01} = \alpha$.

Suppose that the junction is in state 1. Thus a car is waiting to turn, and traffic behind it is still moving freely. If it fails to turn before the arrival of the next northbound car, however, then a tailback will form and the junction will enter state 3. Hence $p_{13} = \theta$. Otherwise, the car succeeds in turning before the next arrival, no tailback forms and the next state will be 0 or 1 according to whether the next northbound car goes forward or turns. We have thus determined p_{1j}, $0 \leq j \leq 3$.

Suppose that the junction is in either state 2 or state 3. Then a tailback exists; i.e., at east one northbound car has been brought to a halt behind the decision bar. Let r be the probability that this queue contains precisely one car; hence $1 - r$ is the probability that it contains more than one car. But we have observed already that if θ denotes the probability that a tailback will form behind a left-turning driver, then θ is also the probability that a tailback containing more than one car will develop from a tailback containing precisely one car. Thus $1 - r = \theta$, or

$$r = 1 - \theta. \tag{6.76}$$

Now suppose that the junction is in state 2 for sure. Then a car has just gone forward, and the junction will enter state 0 or state 1 with probability r, state 2 or state 3 with probability $1 - r$. If it enters state 0 or state 1, then it will enter state 1 with probability α and state 0 with probability $1 - \alpha$. Similarly for state 2 or state 3. This is sufficient to determine p_{2j}, $0 \leq j \leq 3$.

Suppose, finally, that the junction is in state 3. Then it will enter state 0 or state 1 only if there is precisely one car in the tailback and the waiting car succeeds in turning before the next arrival. Hence $p_{30} + p_{31} = r(1 - \theta)$. If the junction enters state 0 or state 1, then it will enter 1 with probability α and 0 with probability $1 - \alpha$. Otherwise it will enter state 2 or state 3. It will enter state 2 if the waiting car succeeds in turning before the next arrival, *and* the tailback contains more than one car, *and* the next car goes forward. Thus

$$p_{32} = (1 - \theta)(1 - r)(1 - \alpha). \tag{6.77}$$

It will remain in state 3 if *either* the waiting car fails to turn before a new northbound car arrives, regardless of how many cars are in the tailback; *or* if the tailback contains more than one car, *and* the waiting car turns before an arrival, *and* the next car then waits to turn left. Thus

$$p_{33} = \theta + (1-r)(1-\theta)\alpha. \tag{6.78}$$

Combining our results, we find that the transition matrix for the Markov chain is

$$\mathbf{P} = \begin{bmatrix} 1-\alpha & \alpha & 0 & 0 \\ (1-\theta)(1-\alpha) & (1-\theta)\alpha & 0 & \theta \\ r(1-\alpha) & r\alpha & (1-r)(1-\alpha) & (1-r)\alpha \\ r(1-\theta)(1-\alpha) & r(1-\theta)\alpha & (1-r)(1-\theta)(1-\alpha) & \theta+(1-r)(1-\theta)\alpha \end{bmatrix}. \tag{6.79}$$

Inspection of (6.79) reveals that it is the transition matrix of an irreducible ergodic Markov chain; see the end of Section 6.1. Hence, irrespective of the value of $\pi(0)$, the T-junction's distribution vector $\pi(n)$ will converge to a unique stationary distribution, $\pi(\infty)$. On using (6.10) and (6.76), this stationary distribution is found to be (Exercise 6.10)

$$\pi(\infty) = \frac{1}{\Gamma} \begin{bmatrix} (1-\theta)^2(1-\alpha) \\ \alpha(1-\theta)^2 \\ \alpha\theta^2(1-\theta)(1-\alpha) \\ \alpha\theta\{1-\theta(1-\alpha)\} \end{bmatrix}, \tag{6.80}$$

where

$$\Gamma \equiv (1-\theta)^2 + \alpha\theta(1-\theta^2(1-\alpha)). \tag{6.81}$$

Note that the numerator of $\pi_2(\infty)$ is the product of five numbers with magnitude less than one; whereas for $j = 0, 1, 3$, the numerator of $\pi_j(\infty)$ is the product of only three numbers with magnitude less than one. Thus 2 is always the least probable configuration, as confirmed by numerical results.

From (6.80) we can readily deduce the value of $\pi_2(\infty) + \pi_3(\infty)$, i.e., the stationary proportion of configurations in which a tailback exists in the northbound lane. This is a function of θ and α. But from (6.74), θ is a function of λ, μ and T. Hence the proportion $\pi_2(\infty) + \pi_3(\infty)$ will be a function of λ, μ, T, and α; we will denote it by $\Delta(\lambda, \mu, T, \alpha)$. Then on using (6.74) and (6.80), we easily obtain:

$$\Delta(\lambda, \mu, \alpha, T) = 1 \Big/$$

$$\left\{ 1 + \frac{(\lambda+\mu)^2(\lambda+\mu e^{\lambda T})}{\alpha\mu(e^{\lambda T}-1)[\alpha\mu^2 e^{2\lambda T} + 2\mu e^{\lambda T}(\lambda+(1-\alpha)\mu) + \lambda^2 - (1-\alpha)\mu^2]} \right\}. \tag{6.82}$$

Let δ_c denote the maximum proportion of traffic configurations in which a tailback is deemed acceptable. Of course, defining δ_c may require a value judgment on the part of a traffic engineer or community activist, depending on the purpose for which the model is being used, but we shall suppose that somehow δ_c is known. Then a left-turn lane is necessary if

$$\Delta(\lambda, \mu, \alpha, T) > \delta_c. \tag{6.83}$$

If $\Delta > \delta_c$ but there is no left-turn lane, then drivers may try to reduce Δ to δ_c by lowering their perception of the value of T, thus increasing the risk of collisions at the junction.

The junction of High Road with Hartsfield Road in Tallahassee was a T-junction like the one we have described above. High Road, the main road, runs north and south; Hartsfield Road, the minor road, runs east and west. High Road is not used excessively by large vehicles, and those that do use it almost never turn left into Hartsfield. Thus assuming T to be the same for all cars is a better approximation to reality than it might be at many other junctions. Unfortunately, the other major source of such variation, namely, the personality of the driver, cannot be so easily eliminated.

On April 10, 1986, a normal working day, the City of Tallahassee Traffic Engineering Department enumerated peak-hour traffic movements at this junction. In the morning, 437 vehicles traversed the junction in the southbound direction and 154 in the northbound direction, of which 41 turned left into Hartsfield. For the morning rush hour, it therefore seems reasonable to take $\lambda = 437/3600 = 0.121$, $\mu = 154/3600 = 0.0428$ and $\alpha = 41/154 = 0.266$. In the evening, 383 vehicles traversed the junction in the southbound direction and 337 in the northbound direction, of which 133 turned left into Hartsfield. For the evening rush hour, it therefore seems reasonable to take $\lambda = 383/3600 = 0.106$, $\mu = 337/3600 = 0.0936$ and $\alpha = 133/337 = 0.395$. For these morning and evening rush hours, Δ has been plotted in Fig. 6.8 as a function of T. For $T = 6$, for example, the model predicts that the northbound lane would have been brought to a halt behind cars turning left for only 8.4% of traffic configurations in the morning rush hour, but for as many as 18.2% of traffic configurations in the evening rush hour. Unfortunately, data for T were not available, though it seems likely that T would be less than 10 seconds (which is why the graph has been truncated at that value).

The T-junction model we developed in this section represents a considerable improvement upon Liu and Yao (1977); see Exercise 5.12. Our model allows arbitrary values of the four most relevant traffic parameters, namely, southbound traffic flow λ, northbound traffic flow μ, left turn probability α, and minimum crossing time T. By contrast, Liu and Yao's model allows arbitrary values of only two parameters, namely, the left turn probability α, and the probability that a southbound vehicle will turn right. The latter is one of the least relevant parameters. More fundamentally, however, Liu and Yao chose traffic configurations in which a northbound vehicle can turn left only after a southbound vehicle has turned right. This is clearly unrealistic, because left-turning vehicles could not take advantage of breaks

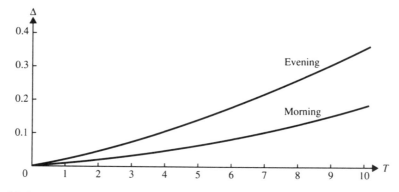

Fig. 6.8 Proportion of traffic configurations, $\Delta(\lambda, \mu, \alpha, T)$, as a function of minimum crossing time T, in which flow along a narrow two-lane major road develops a tailback in one direction behind cars turning left onto a minor road at a T-junction. The values of the parameters λ, μ, and α are fixed for the morning and evening rush hours at a T-junction in Tallahassee, Florida.

in the oncoming flow and would block the northbound lane permanently if all southbound traffic were continuing forward! (In the morning rush hour described above, only 53 of the 437 southbound cars, i.e., only 12%, turned right.)

   I devised this model toward the end of 1986. The junction of High with Hartsfield in Tallahassee has since been redesigned.

Exercises

6.1 By solving (6.8) and (6.9), with 2 replacing 5 in (6.9) and S given by (5.76), verify directly that (6.20) is the unique stationary distribution of the linear time elevator model introduced in Section 5.6.

6.2 Verify directly that (5.92) is the unique stationary distribution of the bird chain considered in Section 5.7. Confirm that, when the distribution is stationary, the mean number of birds in the North end of the cage is 2, as suggested by (5.91).

6.3 Show that the modified bird chain in Exercise 5.8 has the same stationary distribution as in Exercise 6.2. This further illustrates that stationary distributions exist which cannot be found by the indirect method, at least as described at the beginning of this chapter. The relationship of this stationary distribution to the pattern discovered in Exercise 5.9 is explained, for example, by Hoel et. al. (1972, pp. 78–79). It is interesting mathematically but has little to do with the subject of this book.

6.4 Verify (6.31). *Hint:* First differentiate

$$\sum_{j=0}^{N} \rho^j = \frac{1 - \rho^{N+1}}{1 - \rho}$$

with respect to ρ.

6.5 Verify (6.35) and (6.36). *Hint:* Adapt Exercise 6.4.

6.6 Derive (6.40).

6.7 (i) Let the random variables A and I denote, respectively, the number of active servers and the number of idle servers in a stationary supermarket. Then

$$I = \begin{cases} 0, & \text{if } A \geq k; \\ k - A, & \text{if } A \leq k - 1. \end{cases}$$

Show that the expected number of idle servers is $E[I] = k - \rho$.
(ii) Deduce that the expected number of active servers agrees with intuition. *Hint:* Assume N infinite and use the fact that $jC_j = \rho C_{j-1}$ if $j \leq k - 1$, whereas $kC_j = \rho C_{j-1}$ if $j \geq k$.

*** 6.8** A single worker is responsible for the maintenance of N machines. The time she takes to repair a machine is exponentially distributed with mean s, and the time until failure of a machine is exponentially distributed with mean μ. Model this as a stationary birth and death process in which the birth rate b_j actually depends upon j. Show that the expected number of machines in need of repair is

$$L = N - (1 - \pi_0)\frac{\mu}{s},$$

where

$$\pi_0 = \frac{1}{\displaystyle\sum_{j=0}^{N} \frac{N!}{(N-j)!} \left(\frac{s}{\mu}\right)^j}.$$

Hints: It may be easier to think of the birth rate as being a function of $N - j$, the number of machines in working condition. You may find the mathematics easier if you observe that $C_j = s(N - j + 1)C_{j-1}/\mu$. Note that the model would apply as well to a team of repair workers if every member were needed to fix each machine.

6.9 (i) Obtain (6.48b) by successive application of (5.40). *Hint:* Remember that (6.27) is valid only for $j \geq 1$; $d_0 = 0$.
(ii) Verify that the integrand in (6.48b) is the derivative of (6.48a).

6.10 (i) Verify (6.70) by showing that

$$\int_0^\infty \int_0^t \mu e^{-\mu t} \frac{\lambda e^{-\lambda v}(\lambda v)^{j-1}}{(j-1)!} \, dv \, dt = \left(\frac{\lambda}{\mu+\lambda}\right)^j.$$

(ii) Verify (6.74).
(iii) Verify (6.80).

**6.11 Use the model of Section 6.7 to compare several T-junctions in your vicinity. At least in North Florida, there are still many such junctions in rural and outer suburban areas. Moreover, from time to time a left-turn lane appears. Why did the traffic engineers put it there and not at the junction further down the road? Was their decision a good one? Which of the remaining T-junctions is most in need of a turn lane?

7 OPTIMAL DECISION AND REWARD

Toward the end of Chapter 3, on the subject of optimal control, we introduced the state variable $x(t)$ and saw that it was generally a function of the control, $u(t)$. We also saw that utility J is generally a function of both x and u in the first instance; although, because x is a function of u, it is permissible to regard J as a function of u alone.

The control function is always deterministic, but in many circumstances, particularly where human action is concerned, uncertain knowledge of future events makes it more appropriate to regard the state as a random variable, $X(t)$. Because utility depends on state, it must also be a random variable and can no longer itself be maximized. Nevertheless, we can still maximize an associated statistic, e.g., its expected value. To distinguish these circumstances from those we have met in Chapter 3, we shall call the control a decision variable whenever the state is a random variable, and the statistic to be maximized will be called the *reward* (so "reward" is given a special meaning, which departs from common usage). As we shall discover below, however, matters of optimal decision and reward are identical in all other respects to matters of optimal control and utility. The optimal decision is always the one that maximizes reward.

Throughout the following examples, we shall assume that reward is indeed the expected value of utility. We will denote it by J, so that $J(u)$ continues to be the function that we maximize. Rewards other than expected value are considered in Exercise 7.13 and in Section 8.8.

7.1 How Much Should a Buyer Buy? A First Look

It is customary for retail stores and other buyers to order merchandise at regular intervals. Let such an interval correspond to a unit of time. Let u denote the amount of some product that a buyer buys at the beginning of the interval, and suppose that the merchandise is delivered immediately (for if it took time s to be delivered, we could simply suppose that the buyer placed his order time s in advance). Let Z denote the demand for the product during the interval. Demand is uncertain, so Z is a random variable. Assuming that the distribution of Z is known, how much should the buyer buy? In this section, we will attempt to answer that question. We will assume that the buyer is a profit maximizer, so that his reward is the expected value of his profit. We will first derive an expression for this reward, then find the value u^* that maximizes it.

To make matters simple, we shall suppose that the stock is initially zero and that the buyer cannot reorder before the end of the interval. If demand exceeds supply, then tardy customers are out of luck. Let the demand Z have probability density function $f(z)$. Then $f(z) \geq 0$ on $0 \leq z < \infty$ and

$$\int_0^\infty f(z)dz = 1.$$

Let c_0 denote the cost to the buyer of a unit of merchandise. Let k denote the delivery cost, plus any other administrative costs that are independent of the size of the order (assume that the order is never so big that an extra truck is required to deliver it). Then the cost of placing that order is $k + c_0 u$. During the subsequent interval, one of two things may happen: demand may exceed supply or supply may exceed demand. In the former case, no further costs are incurred and, because all units are sold, the retailer's revenue is simply pu, where p is the unit selling price. But in the latter case, where the retailer's revenue is pZ, further costs—known as *holding costs*—arise from the unsold units.

Included in holding costs are stock insurance and rent for additional storage space, if needed. Most holding costs, however, are *opportunity costs*, i.e., penalties for losing opportunities. For example, the warehouse space wasted by items for which demand is low could have been used more profitably by items for which demand is high. Moreover, the money used to buy the unsold items could instead have been deposited in a bank, and the unearned interest is an opportunity cost. These opportunity costs depend in quite a complicated way upon the circumstances of the retailer and the nature of the commodity being sold. For example, if the retailer is a jeweler and diamonds are the commodity, then insurance is high but the warehouse space that could have been used instead for grandfather clocks is negligible; and vice versa. Moreover, the unit of time for which an order is placed may be as short as 7 days, in which case unearned interest is negligible; or as long as 365 days, in which case it might be considerable. Despite this complexity, one thing is clear: on any given day within the interval $0 \leq t \leq 1$, the more items in stock, the greater the holding costs for that particular day. We incorporate this observation into a model by making the robust assumption that holding costs per unit of time are proportional to the average stock within that interval. Let $\zeta(t)$ be the stock level at time t, $0 \le t \le 1$. Then $\zeta(0) = u$, and the average stock per unit of time is

$$\int_0^1 \zeta(t)dt. \qquad (7.1)$$

Thus holding costs are

$$c_1 \int_0^1 \zeta(t)dt, \qquad (7.2)$$

where c_1 is the constant of proportionality.

Suppose, for example, that the buyer ordered every week and that almost all sales were made on a single day (Saturday, say) at the beginning of the 7-day interval $0 \le t \le 1$. (This is not such an unlikely circumstance. As a high school student in England at a time when stores were forbidden to open on Sunday, I worked in a furniture store and observed directly that items unsold by Saturday night were almost always still unsold by the following Friday.) Then ζ would have the form depicted by the solid curves in Fig. 7.1. The exact expression for (7.2) would be rather unwieldy. But ζ in Fig. 7.1 could reasonably be approximated by the dashed curves. For these curves, (7.1) equals $u - Z$ if $Z < u$ and equals 0 if $Z > u$. Let us define a threshold function H by

$$H(z) \equiv \begin{cases} 0, & \text{if } z \le 0; \\ 1, & \text{if } z > 0. \end{cases} \qquad (7.3)$$

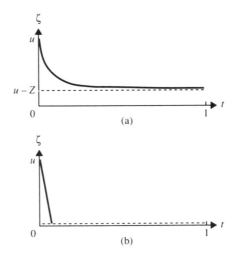

Fig. 7.1 Possible graph of depletion of stock $\zeta(t)$ during the interval $0 < t < 1$ for which order is placed. If $Z < u$ then $\zeta(1) = Z$. (a) Unsold items: $Z < u$; (b) unsatisfied demand: $Z > u$. (Technically, $\zeta(t)$ is a random variable, so the graph should be interpreted as that of its expected value, $E[\zeta(t)]$.)

Then from (7.2), the associated holding costs are $c_1(u - Z)H(u - Z)$. In the present section, we will assume that the holding costs are of this form. An alternative form for $\zeta(t)$, and hence holding costs, is depicted in Fig. 7.2; and we'll pursue it in Section 7.3.

With the assumptions we have now made, total costs per unit of time, i.e., ordering costs plus holding costs, are given by

$$k + c_0 u + c_1(u - Z)H(u - Z); \qquad (7.4)$$

and revenue per unit of time is pZ if $Z < u$ and pu if $Z > u$, i.e.,

$$p\{Z + (u - Z)H(Z - u)\}. \qquad (7.5)$$

Thus profit from the period is the random variable

$$P_u \equiv p\{Z + (u - Z)H(Z - u)\} - c_1(u - Z)H(u - Z) - c_0 u - k. \qquad (7.6)$$

Note that this is a *mixed* random variable (see Appendix 1) because it takes the value $(p - c_0)u - k$ with finite probability

$$\text{Prob}(Z > u) = \int_u^\infty f(z)dz;$$

whereas all other values are continuously distributed between $-\{(c_0+c_1)u+ k\}$ and $(p-c_0)u-k$. Recall that our buyer is a profit maximizer. His reward is therefore the expected value of profit, i.e., $E[P_u]$. The mixed random variable P_u is defined in (7.6) as a function of the purely continuous random variable Z, and so we can calculate $E[P_u]$ directly from the distribution of Z and thus avoid finding P_u's distribution explicitly; see (A.32). Thus

$$\begin{aligned} J(u) = E[P_u] &= pE[Z + (u - Z)H(Z - u)] \\ &\quad - c_1 E[(u - Z)H(u - Z)] - c_0 u - k \\ &= p \int_0^\infty \{z + (u - z)H(z - u)\} f(z)dz \\ &\quad - c_1 \int_0^\infty (u - z)H(u - z)f(z)dz - c_0 u - k. \qquad (7.7) \end{aligned}$$

After some manipulation (Exercise 7.1), we obtain

$$J(u) = (p + c_1)\left\{ \int_0^u zf(z)dz - u \int_0^u f(z)dz \right\} + (p - c_0)u - k. \qquad (7.8)$$

By Leibnitz's rule from the calculus, the first and second derivatives of this with respect to u are

$$J'(u) = -(p + c_1) \int_0^u f(z)dz + p - c_0 \qquad (7.9)$$

and $J''(u) = -(p+c_1)f(u)$, which is negative. Thus subject to a proviso that will be discussed at the end of this section, the optimal decision is to order

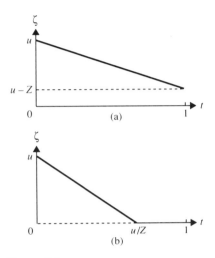

Fig. 7.2 Same as for Figure 7.1.

u^*, where u^* is defined by

$$\int_0^{u^*} f(z)dz = \frac{p - c_0}{p + c_1} \tag{7.10}$$

or, which is exactly the same thing (why?),

$$\int_{u^*}^{\infty} f(z)dz = \frac{c_0 + c_1}{p + c_1}. \tag{7.11}$$

Note that u^* is unique (see Exercise 7.2). Then, on substituting (7.10) into (7.8), we find the maximum reward to be

$$J(u^*) = (p + c_1) \int_0^{u^*} zf(z)dz - k. \tag{7.12}$$

If the probability density function is explicitly known, then it may be possible to determine u^* analytically. We explore this possibility in the Exercises (see, for example, Exercise 7.4). Even if the distribution of Z is known only empirically, however, it is always possible to estimate u^* from the graph of its cumulative distribution function. To illustrate this idea, let's suppose that demand Z has a normal distribution, with mean μ and variance σ^2. Then (Section 6.5):

$$\text{Prob}(Z \le u) = \frac{1}{\sigma\sqrt{2\pi}} \int_{-\infty}^{u} e^{-(z-\mu)^2/2\sigma^2} dz. \tag{7.13}$$

Let Φ be the cumulative distribution function of a normally distributed random variable with mean 0 and variance 1; i.e., define

$$\Phi(t) \equiv \frac{1}{\sqrt{2\pi}} \int_{-\infty}^{t} e^{-\zeta^2/2} d\zeta. \tag{7.14}$$

Then (Exercise 7.2)

$$\text{Prob}(Z \le u) = \Phi\left(\frac{u - \mu}{\sigma}\right). \tag{7.15}$$

The left hand side of (7.10) is just $\text{Prob}(Z \le u^*)$; therefore we find that

$$\Phi\left(\frac{u^* - \mu}{\sigma}\right) = \frac{1 - \frac{c_0}{p}}{1 + \frac{c_1}{p}}. \tag{7.16}$$

The graph of Φ is drawn in Fig. 7.3. Let γ denote the unique value of t for which $\Phi(t)$ equals the right-hand side of (7.16); γ is a function of the two cost/price ratios, c_0/p and c_1/p, and is readily determined from the graph of Φ, as indicated in Fig. 7.3. Thus, for arbitrary μ, σ, c_0/p, and c_1/p, the optimal order quantity is given by

$$u^*(\mu, \sigma, c_0/p, c_1/p) = \mu + \sigma \cdot \gamma(c_0/p, c_1/p). \tag{7.17}$$

Because u^* lies above or below the mean of Z according to whether γ is positive or negative, we deduce from Fig. 7.3 that $u^* > \mu$ or $u^* < \mu$ according to whether the right-hand side of (7.16) is greater or less than $1/2$, i.e., according to whether $p > c_1 + 2c_0$ or $p < c_1 + 2c_0$. In other words, when demand is normally distributed, it is risky to order a greater quantity than $E[Z]$ unless the buyers markup per item, $p - c_0$, exceeds the cost per item plus the unit holding cost.

In all of this we have tacitly assumed that the optimal reward is positive. Now, if k is zero or very small, then $J(u^*)$ is certainly positive. If k is large, however, then the possibility arises that $J(u^*)$ might be negative. On using (7.10) to eliminate $p + c_1$ from (7.12), we find that $J(u^*) < 0$ if

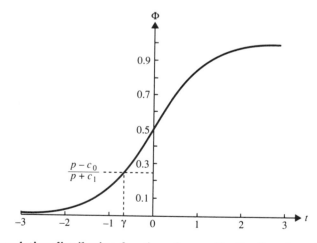

Fig. 7.3 Cumulative distribution function of normally distributed random variable with mean $\mu = 0$ and variance $\sigma^2 = 1$; i.e., $\Phi(t)$, as defined by (7.14).

$k > k_c$, where

$$k_c = (p - c_0)\frac{\int_0^{u^*} zf(z)dz}{\int_0^{u^*} f(z)dz}. \tag{7.18}$$

It would then be better not to order at all; for then there would be no delivery cost k and a profit of zero would be guaranteed. In conclusion, therefore, the buyer should order an amount

$$u^*H(k_c - k), \tag{7.19}$$

where H is defined by (7.3), u^* by either (7.10) or (7.11), and k_c by (7.18).

In practice, the cost of actually shipping the goods from supplier to retailer is often proportional to the amount delivered (particularly if an independent trucking company is used) and is then best regarded as incorporated into c_0. The fixed delivery cost k, due to the order (and not due simply to being in business), is then essentially the cost of promoting the items in question. In these circumstances, a possible interpretation of k_c in (7.18) is that it is the maximum amount the retailer in question should even consider spending on promoting the goods for which the order is placed.

7.2   How Many Roses for Valentine's Day?

Let's attempt to apply our model to a man selling roses by the side of the road. Despite the exhortations of florists to order early for Valentine's Day, these street vendors are still in business. Indeed they seem to function all year long in Tallahassee; but we shall confine our attention to Valentine's Day, when they are extraordinarily busy, serving the needs of those for whom romance cannot be planned. The interval for which the rose-vendor orders, namely, part of a day, is so short that he loses no money by tying up his money in flowers (as opposed to having it in a bank). Because his product is highly perishable, he throws the surplus away at the end of the day, so that there are no storage charges. Hence

$$c_1 = 0, \tag{7.20}$$

and the shape of $\zeta(t)$ in Fig. 7.1 becomes irrelevant. Moreover, because the vendor does not pay to advertise his roses, we can safely assume that $k = 0$ (see the remarks at the end of the previous section).

Let's suppose that the vendor makes all his sales during a period of two or three hours, say the morning rush hours, which therefore correspond to the interval $0 < t < 1$. Because this interval is short, it is reasonable to suppose that customers arrive according to a Poisson process. If the random variable $A(t)$ denotes the number arriving by time t, then from (5.49),

$$\text{Prob}(A(t) = k) = e^{-Bt}\frac{(Bt)^k}{k!} \tag{7.21}$$

for some parameter B. Let's suppose that each customer buys one bunch of roses. Then demand for the period $0 < t < 1$ is given by $Z = A(1)$. Hence on using (7.21), Z has probability distribution

$$\text{Prob}(Z = k) = \frac{B^k}{k!}e^{-B}. \tag{7.22}$$

This is the cumulative distribution of a discrete random variable; whereas the buying model of Section 7.1 assumes that Z has a continuous distribution. Then can we not use the model?

To answer this question, let us define the cumulative distribution function of the discrete random variable (7.22):

$$F(k) \equiv \text{Prob}(Z \le k) = \sum_{j=0}^{k} \frac{B^j}{j!}e^{-B}. \tag{7.23}$$

The function F is plotted in Fig. 7.4 for three values of B, namely, $B = 10$, $B = 30$, and $B = 70$. What do you notice about its shape? First of all, it resembles that of the normal distribution. This suggests that we might approximate the discrete Poisson probability function (7.22) by a continuous normal probability density function (6.57). But which values should we choose for the parameters μ and σ^2? Recall from Exercise 5.2 that Z in (7.22) has mean B, the expected number of arrivals for the entire interval $0 < t < 1$. This suggests we take $\mu = B$. Recall from Fig. 6.3, moreover, that almost all normally distributed values lie within three stan-

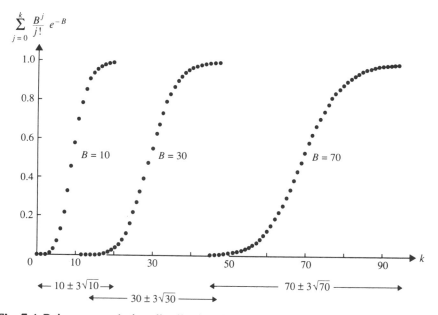

Fig. 7.4 Poisson cumulative distribution function (7.23) for various values of B.

dard deviations of the mean. Then observe from Fig. 7.4 that the interval $B - 3\sqrt{B} \leq k \leq B + 3\sqrt{B}$ contains almost all values distributed according to (7.23). This suggests that we take $\sigma = \sqrt{B}$. In other words, it leads us to conjecture that the discrete Poisson distribution with mean B can be approximated by the continuous normal distribution with mean B and variance B.

Is this conjecture correct? The answer is yes, if B is sufficiently large. It can be established rigorously that the Poisson distribution with mean B approaches the normal distribution with mean and variance B as $B \to \infty$; but the verification is somewhat lengthy, and the reader who is unwilling to accept this result on faith is referred to Fry (1965, p. 251) for the proof. Fry warns against using the approximation far from the mean of the distribution, where the relative error might still be large. His caveat can be ignored for our purposes, however, because u^* will always be close to the mean.

But how large is large? Fig. 7.4 suggests that values of B exceeding about 30 are big enough, and you can verify this numerically; see Exercise 7.6. It would hardly be worth a vendor's while to brave the early morning cold if he couldn't sell at least 30 bunches of roses! Hence the approximation can safely be used, and on using (7.20), (7.17) yields the optimal order quantity

$$u^*(B, \sqrt{B}, c_0/p, c_1/p) = B + \sqrt{B}\gamma(c_0/p, c_1/p). \tag{7.24}$$

Of course, real demand is always an integer, and so the value u^* predicted by (7.24) must be rounded to the nearest one. Fig. 7.3 has been drawn for $c_0/p = 3/4$ and $c_1/p = 0$; which would apply to the rose vendor if he marked up his roses by a third. The graph shows that $\Phi(\gamma) = 0.25$ when γ is just a little larger than -0.7. More precisely, from tables of values of the function Φ (e.g., Lindley and Miller, 1953), we find that $\gamma = -0.6745$. Hence $u^* = \sqrt{B}(\sqrt{B} - 0.6745)$. If, for example, B were 100 then our model would recommend that the vendor acquire 93 bunches of roses.

But should the vendor accept that recommendation? Perhaps you would like to think about this matter, then try Exercise 7.13.

7.3   How Much Should A Buyer Buy? A Second Look

In Section 7.1, we assumed that merchandise sold to customers during the interval $0 < t < 1$ was mostly demanded shortly after $t = 0$. This gave rise to Fig. 7.1, which might be appropriate if the interval in question were about a week. If the interval is much longer (a month or a quarter, say), however, then we might expect demand to occur more uniformly throughout it. To simplify the mathematics, we will assume that the short-term demand, that is the rate at which the stock is depleted within the interval $0 < t < 1$, is absolutely uniform. Then stock $\zeta(t)$ will vary linearly with time, as shown in Fig. 7.2. If supply exceeds demand, then the area under the graph in Fig. 7.2(a) is $u - Z/2$, whence from (7.2), holding costs are $c_1(u - Z/2)$. If demand exceeds supply, then only a proportion u/Z of demand is satisfied. Because short-term demand is uniform, supplies are exhausted after a proportion u/Z of the interval $0 \leq t \leq 1$ has elapsed, whence the holding costs are c_1 times the area in Fig. 7.2(b), or $c_1 u^2/Z$. On using (7.3), the expression for holding costs is therefore

$$c_1 \left\{ (u - \frac{1}{2}Z)H(u - Z) + \frac{u^2}{2Z}H(Z - u) \right\}. \tag{7.25}$$

The expressions for revenue and delivery cost are unchanged from the previous section. Thus the retailer's reward is $J(u) = E[\text{Revenue}] - E[\text{Costs}] =$

$$p \int_0^\infty \{z + (u - z)H(z - u)\} f(z)dz$$

$$- c_1 \int_0^\infty \left\{ \left(u - \frac{1}{2}z\right) H(u - z) + \frac{u^2}{2z}H(z - u) \right\} f(z)dz - c_0 u - k$$

$$= p \left(\int_0^\infty z f(z)dz + \int_u^\infty (u - z)f(z)dz \right)$$

$$- c_1 \left\{ \int_0^u \left(u - \frac{1}{2}z\right) f(z)dz + \int_u^\infty \frac{u^2}{2z}f(z)dz \right\} - c_0 u - k$$

$$= \left(p + \frac{c_1}{2}\right) \int_0^u z f(z)dz - (p + c_1)u \int_0^u f(z)dz$$

$$- \frac{1}{2}c_1 u^2 \int_u^\infty \frac{f(z)}{z}dz + (p - c_0)u - k, \tag{7.26}$$

by algebraic manipulations similar to those that produced (7.8). The first and second derivatives of this expression with respect to u are, on using Leibnitz's rule from the calculus,

$$J'(u) = -(p + c_1) \int_0^u f(z)dz - c_1 u \int_u^\infty \frac{f(z)}{z}dz + p - c_0 \tag{7.27}$$

and

$$J''(u) = -pf(u) - c_1 \int_u^\infty \frac{f(z)}{z}dz, \tag{7.28}$$

which is negative. Thus (for k sufficiently small) the optimal decision is to order u^*, where

$$\int_0^{u^*} f(z)dz + \frac{c_1 u^*}{p + c_1} \int_{u^*}^\infty \frac{f(z)}{z}dz = \frac{p - c_0}{p + c_1} \tag{7.29}$$

or, which is exactly the same thing,

$$\int_{u^*}^\infty \left(1 - \frac{c_1 u^*}{(p + c_1)z}\right) f(z)dz = \frac{c_0 + c_1}{p + c_1}. \tag{7.30}$$

On substituting (7.29) into (7.26) we find the maximum reward to be

$$J(u^*) = \left(p + \frac{1}{2}c_1\right) \int_0^{u^*} zf(z)dz + \frac{1}{2}c_1u^{*2} \int_{u^*}^{\infty} \frac{f(z)}{z}dz - k. \tag{7.31}$$

For $z > u^*$, the (positive) integrand on the left-side of (7.30) is smaller than the one on the left-hand side of (7.11), but the value of the integral must still be the same. We immediately conclude that the value of u^* given by (7.30) is smaller than the value given by (7.11), regardless of the shape of the distribution of Z. Thus the optimal order is always smaller when sales are uniform throughout the interval $0 \leq t \leq 1$ than when sales take place soon after the stock arrives at the store. For an illustration of this result, see Exercise 7.5.

As in Section 7.1, we should allow for the possibility that it might be better not to order at all. You can easily verify that the buyer should then order an amount $u^*H(k_c - k)$, where, from (7.31),

$$k_c = \left(p + \frac{c_1}{2}\right) \int_0^{u^*} zf(z)dz + \frac{1}{2}c_1u^{*2} \int_{u^*}^{\infty} \frac{f(z)}{z}dz. \tag{7.32}$$

This is the analogue of (7.18).

At the end of the previous section, we spoke of circumstances in which k would denote the costs of advertising. There is a critical value, k_c that these costs must not exceed if a positive profit is to be expected, but if $k < k_c$ then, according to both buying models, the reward is simply $k_c - k$. This is a paradoxical result, because it suggests that zero is the optimal amount to spend on advertising; whereas common sense suggests that a new item of merchandise might not sell at all if nothing were spent on advertising it. How can this be so? You should attempt to resolve this paradox before proceeding.

7.4 How Much Should a Retailer Spend on Advertising?

The answer to the question posed at the end of the previous section is that, hitherto, we have tacitly assumed that $f(z)$ is independent of k. In reality, of course, if k denotes advertising costs, then $E[Z]$ increases with k. A model with this property is now discussed.

Consider a bookseller who is promoting a new book by a famous marine biologist, with many colored photographs of deep-sea exploration. Let $b(k)$ denote the number of *potential* buyers of the book when k dollars are spent on advertising it; $b(k)$ is the number of people who are genuinely interested and would like to own a copy (but may not actually buy one because they need to spend the money on something else). Then $b(k)$ is an increasing function of k, and because the book is new, we shall suppose that $b(0) = 0$. (It is possible that a few well informed customers would order the book before the bookseller had spent even a dime on publicity; then $b(0)$ would be positive. But we ignore this possibility, because it complicates the algebra while adding little to the model.) The bookseller intends to promote the book by inviting the author to sign copies one day in the bookstore, and a brochure describing the book's virtues and advertising the author's visit will be mailed to prospective customers. In such circumstances, copies of the book will mostly be sold soon after publication; hence the model of Fig. 7.1 is appropriate. Moreover, in the absence of fuller information about actual buyers, it is not unreasonable to assume that demand is equally likely to fall anywhere between $b(0)$ and $b(k)$. Hence Z has probability density function

$$f(z) = \frac{1}{b(k)} H(b(k) - z), \tag{7.33}$$

where H is defined by (7.3). Thus f depends on k, but we will continue to use the notation $f(z)$. For given k, the optimal order $u^*(k)$ is obtained by substituting (7.33) into (7.10). We obtain

$$u^*(k) = \frac{p - c_0}{p + c_1} b(k). \tag{7.34}$$

From (7.12), the corresponding reward is

$$J(u^*(k)) = \alpha_0 b(k) - k, \tag{7.35}$$

where we define

$$\alpha_0 = \frac{(p - c_0)^2}{2(p + c_1)}. \tag{7.36}$$

For any value of k, (7.35) is the maximum reward, and we can maximize this reward in turn to obtain the particular value of k that is best overall.

Before proceeding to do this, however, we must make some assumptions about the form of $b(k)$. The number of potential buyers created by the kth dollar spent on advertising is measured by $b(k + 1/2) - b(k - 1/2)$. By a slight modification of the argument used to justify (1.55), this is approximately $b'(k)$ if third derivatives are small (and assuming, of course, that b is differentiable). No matter how much is spent on advertising, there exists a limit that the number of potential buyers cannot exceed; call it B. Then $b'(k)$ is decreasing for sufficiently large k and

$$\lim_{k \to \infty} b(k) = B, \qquad \lim_{k \to \infty} b'(k) = 0. \tag{7.37}$$

Suppose it costs \$$\alpha$ to print and mail a brochure, and the setup costs (for printing the brochure) and author's expenses amount to \$$k_0$. Then the first k_0 dollars spent on advertising create no potential buyers: $b(k) = 0$ for $0 \le k \le k_0$. Suppose that the first b_0 people on the bookseller's mailing list are known to have a genuine interest in marine biology (perhaps because they belong to a learned society). Each brochure mailed to one of these people is guaranteed to create a potential buyer; whence each dollar will create $1/\alpha$ potential buyers:

$$b'(k) = \frac{1}{\alpha}, \qquad k_0 \le k \le k_1, \tag{7.38}$$

where

$$k_1 \equiv k_0 + \alpha b_0. \tag{7.39}$$

For $k \ge k_1$, we cannot be sure that a brochure will create a potential buyer; hence $b'(k) \le 1/\alpha$. All that has been assumed so far about the form of $b(k)$ is summarized by the solid curve in Fig. 7.5.

The dashed curve is purely speculative. We have supposed that, for $k > k_1$, the person whose brochure is paid for by the kth dollar is a little less likely to be a potential buyer than the person whose brochure was paid for by the $(k-1)$th ; i.e., we have supposed that the bookseller has sorted his mailing list according to some criterion of decreasing interest in marine biology. Hence $b'(k)$ decreases continuously from $1/\alpha$ to 0 as k increases from k_1 to infinity, in such a way that $b(k)$ tends to B. The simplest function that behaves in this way for $k \ge k_1$ is $b(k) = B(k + k_2)/(k + k_3)$, where k_2 and k_3 are constants chosen to ensure the continuity of $b(k)$ and $b'(k)$ at k_1. We now have an explicit expression for the potential buyer function:

$$b(k) = \begin{cases} 0, & \text{if } 0 \le k \le k_0; \\ \dfrac{k - k_0}{\alpha}, & \text{if } k_0 \le k < k_1 \\ \dfrac{B(k - k_1) + b_0\alpha(B - b_0)}{k - k_1 + \alpha(B - b_0)}, & \text{if } k \ge k_1. \end{cases} \tag{7.40}$$

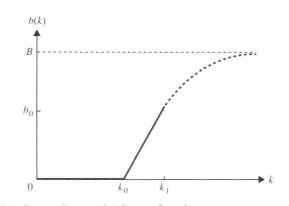

Fig. 7.5 Possible form of potential buyer function.

Hence, from (7.35), we have an explicit expression for the reward:

$$J(u^*(k)) = \begin{cases} -k, & \text{if } 0 \le k \le k_0; \\ \left(\dfrac{\alpha_0}{\alpha} - 1\right) k - k_0 \dfrac{\alpha_0}{\alpha}, & \text{if } k_0 \le k < k_1; \\ \alpha_0 \left\{ \dfrac{B(k - k_1) + b_0 \alpha(B - b_0)}{k - k_1 + \alpha(B - b_0)} \right\} - k, & \text{if } k \ge k_1. \end{cases} \quad (7.41)$$

The maximization of (7.41) is a straightforward application of the calculus (Exercise 7.8). If $\alpha > \alpha_0$ then the maximum is at $k = 0$. Hence there is no reward unless $\alpha < \alpha_0$. We shall, in fact, make a stronger assumption, namely, that the bookseller would not undertake the venture of promoting this book unless he could expect a positive profit solely from sales to his prime targets, the first b_0 customers on his mailing list. Hence $J(u^*(k_1)) > 0$; or $\alpha < \alpha_c$, where we define

$$\alpha_c \equiv \alpha_0 - \frac{k_0}{b_0}. \quad (7.42)$$

Then the reward $J(u^*(k))$ has the form depicted in Fig. 7.6. The maximum occurs at $k = k^*$, where

$$k^* = k_1 + \left(\sqrt{\frac{\alpha_0}{\alpha}} - 1\right) \alpha(B - b_0); \quad (7.43)$$

and the optimal strategy is to order

$$u^*(k^*)H(\alpha_c - \alpha) \quad (7.44)$$

copies of the new book, where H is defined by (7.3), α_c by (7.42), k^* by (7.43), and $u^*(k^*)$ by setting $k = k^*$ in (7.34). Notice from Fig. 7.6 that, as in Sections 7.1 and 7.3, there is a critical value k_c which advertising

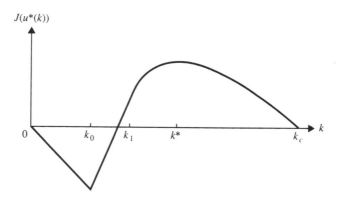

Fig. 7.6 Reward $J(u^*(k))$ plotted against k for $\alpha < \alpha_c$.

expenditure must not exceed; but there is also a second critical value,

$$\frac{k_0 \alpha_0}{\alpha_0 - \alpha},\tag{7.45}$$

below which there is no (positive) reward for advertising expenditure. For an application of the model, see Exercise 7.9.

How good is this model—would a bookseller use it? By definition, $f(z) = 0$ unless $b(0) \le z \le b(k)$, but the bookseller might argue that he has enough information about actual buyers to know that (7.33) is a poor assumption. He might, for example, have learned from experience that the fraction of potential buyers who actually buy the book is ten times more likely to fall between a sixth and a half than anywhere else. Then the distribution can easily be altered; see Exercises 7.10 and 7.11. There is no reason in principle why all the information that a bookseller could possibly have about actual buyers cannot be incorporated into a probability density function with parameter k. His first objection is therefore overruled!

Suppose that the bookseller now objects to the cavalier way in which we continued the potential buyer function past the value $k = k_1$. Then we will ask to see his mailing list of, say, $B - \epsilon$ members. We will use our ingenuity to arrange it in order of decreasing propensity to buy. We might, for example, divide it into N classes, such that the ith class has b_i members, a fraction ξ_i of which are assumed to be potential buyers. Then

$$1 = \xi_0 > \xi_1 > \xi_2 > \cdots > \xi_{N-1} > 0;\tag{7.46}$$

and, for $i = 0, 1, \ldots, N - 1$,

$$b'(k) = \frac{\xi_i}{\alpha} \qquad \text{if } k_i < k < k_{i+1},\tag{7.47}$$

where

$$k_i = k_0 + \alpha \sum_{j=0}^{i-1} b_j,\tag{7.48}$$

and, of course,

$$\sum_{j=0}^{N-1} b_j = B - \epsilon.\tag{7.49}$$

This empirically derived potential buyer function has been sketched in Fig. 7.7. The straight line segments could be approximated by the dashed curve (with continuous derivative). Doesn't this remind you of the conceptually derived potential buyer function in Fig. 7.5? We do not, of course, pretend to know precisely which $\xi_i b_i$ members of class i are the potential buyers (if we did, we'd save postage on the others, who simply hurl their brochure in the trash). We just have a hunch that they are there. We might, for example, believe that a tenth of all people who adore photography will

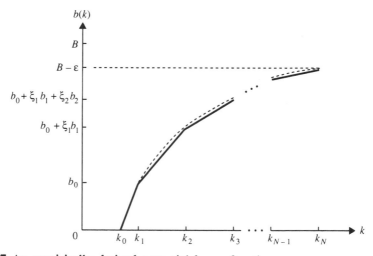

Fig. 7.7 An empirically derived potential buyer function.

buy the book for its lavish color photographs, even though they hate marine biology. Surely the bookseller cannot object to the function in Fig. 7.7!
(Even if some of his advertising is done by telephone, newspaper ads, or
other means, this simply makes α a function of i and creates occasional
horizontal segments in the graph of $b(k)$, corresponding to the additional
setup costs.)

If the distribution of Z is "piecewise uniform," as in Exercise 7.10,
then (7.35) is still valid for some constant α_0 (not, however, defined by
(7.36)); and you can easily show that $J(u^*(k))$ has the piecewise linear form

$$J(u^*(k)) = \begin{cases} -k, & \text{if } 0 \le k < k_0; \\ \alpha_0\left\{\displaystyle\sum_{j=0}^{i-1}\xi_j b_j + \frac{\xi_i}{\alpha}(k - k_i)\right\} - k, & \text{if } k_i \le k \le k_{i+1}, \\ & 0 \le i \le N - 1. \end{cases} \quad (7.50)$$

Its maximum therefore occurs at a corner, specifically at $k = k_{i^*}$,
where i^* is the least value of i such that $\xi_i \le \alpha/\alpha_0$. (If equality
should happen to hold, then the maximum would be achieved through-
out the interval $k_{i^*} \le k \le k_{i^*+1}$, the graph of $J(u^*(k))$ being horizontal there.) This result has a simple economic interpretation. From
(7.35), the additional revenue expected from the potential buyers in
class i is $\alpha_0\xi_i b_i$. The additional cost of turning them into potential
buyers is αb_i. Hence advertising the book to class i is considered
profitable only if $\alpha_0\xi_i b_i > \alpha b_i$. For a numerical example, see Exercise 7.12.

7.5   How Much Should a Buyer Buy? A Third Look

We have assumed until now that our buyer places a single order at $t = 0$, for the period $0 < t < 1$, and that items not sold by $t = 1$ remain unsold forever. In practice, however, our buyer may order regularly, say at times $t = 0, 1, 2, \ldots$. Items not sold by $t = 1$ may be sold during $1 < t < 2$ or later. Our model must be extended to account for this.

Let $t = N$ be the buyer's *horizon*, i.e., the time beyond which demand for the product is either nonexistent or so low as to be effectively zero. Items not sold by $t = N$ are assumed to remain unsold forever. Note that N may be as large as we please, sometimes even infinite (recall the forester's model from Sections 3.4 and 4.8). The buyer therefore places a total of N orders at times $t = 0, 1, \ldots, N - 1$. Let u_j denote the quantity ordered at time $t = j$ for the interval $j < t < j + 1$. Let demand during this period be denoted by the random variable Z_j, with probability density function f_j. For simplicity's sake, we shall assume that the random variables Z_j are all uniformly distributed between a and b, i.e., that

$$f_j(z) = \begin{cases} 0, & \text{if } z < a; \\ \dfrac{1}{b - a}, & \text{if } a \leq z < b; \\ 0, & \text{if } z \geq b, \quad j = 0, 1, \ldots, N - 1. \end{cases} \tag{7.51}$$

No new concepts are needed in extending this section's model to cover circumstances where the probability density functions differ from (7.51) or depend upon j (or both); but the mathematical details are so complicated as virtually to obscure the model's essential insights, and it is more in keeping with the spirit of this book to overlook them.

What distinguishes this problem from that studied previously is that surplus stock at $t = j$ may be sold during $j < t < j + 1$ or later. Let X_j denote this surplus, i.e., the number of items ordered by $t = j - 1$ but not yet sold by $t = j$. Then the single most important point to observe about X_j is that, *from the point of view of the buyer's decision at $t = j - 1$, X_j is a* random variable; whereas, *from the point of view of the buyer's decision at $t = j$, X_j is fully determined* (being the value the random variable actually took). Specifically, if we view the surplus with the benefit of hindsight from $t = j$, then

$$X_j = x_j, \qquad j = 0, 1, \ldots, N - 1, \tag{7.52}$$

where x_j is a number that is known for certain; whereas, if we view the surplus from $t = j - 1$, then X_j is the random variable

$$X_j = \begin{cases} u_{j-1} + x_{j-1} - Z_{j-1}, & \text{if } Z_{j-1} < u_{j-1} + x_{j-1}; \\ 0, & \text{if } Z_{j-1} > u_{j-1} + x_{j-1}. \end{cases} \tag{7.53}$$

These equations are valid for $j = 1, \ldots, N$. The initial surplus is

$$x_0 = 0, \tag{7.54}$$

because nothing was ordered at time $t = -1$.

Now, X_j is a mixed random variable (see Appendix 1), because there is a finite probability that it will take the value zero, whereas all other values are distributed uniformly between 0 and $u_{j-1} + x_{j-1} - a$. Specifically, on using (7.51) and (7.53), we have

$$\text{Prob}(X_j = 0) = \text{Prob}(Z_{j-1} > u_{j-1} + x_{j-1})$$

$$= \int_{u_{j-1}+x_{j-1}}^{\infty} f_{j-1}(z)dz = \frac{b - u_{j-1} - x_{j-1}}{b - a} \tag{7.55}$$

and, for x lying between 0 and $u_{j-1} + x_{j-1} - a$,

$$\text{Prob}\left(x - \frac{1}{2}\delta x < X_j < x + \frac{1}{2}\delta x\right) = \frac{\delta x}{b - a} + o(\delta x). \tag{7.56}$$

For completeness, we append that

$$\text{Prob}(X_j < 0) = 0 = \text{Prob}(X_j > u_{j-1} + x_{j-1} - a). \tag{7.57}$$

You can verify that (7.55)–(7.57) define a probability distribution by observing that

$$\text{Prob}(X_j = 0) + \int_0^{u_{j-1}+x_{j-1}-a} \frac{dx}{b - a} = 1. \tag{7.58}$$

We will assume throughout that costs for each period are assessed according to the formula (7.4), suggested by Fig. 7.1. Because demand is recurrent, it is reasonable to suppose that the product in question is not promoted by the retailer (except insofar as it advertises itself by being displayed). Let shipping costs be absorbed into c_0, as suggested at the end of Section 7.1. Then $k = 0$ in (7.4); i.e., costs from the interval $j < t < j+1$ are

$$
\begin{aligned}
&c_0 u_j + c_1(u_j + x_j - Z_j), &&\text{if } Z_j < u_j + x_j; \\
&c_0 u_j, &&\text{if } Z_j > u_j + x_j.
\end{aligned}
\tag{7.59}
$$

Revenue from the same period is

$$
\begin{aligned}
&pZ_j, &&\text{if } Z_j < u_j + x_j; \\
&p(u_j + x_j), &&\text{if } Z_j > u_j + x_j.
\end{aligned}
\tag{7.60}
$$

Thus reward from the interval $j < t < j + 1$ is the expected value of the difference between (7.60) and (7.59) or

$$\int_0^{u_j+x_j} \left\{ pz - c_0 u_j - c_1(u_j + x_j - z) \right\} f_j(z)dz$$

$$+ \int_{u_j+x_j}^{\infty} \left\{ p(u_j + x_j) - c_0 u_j \right\} f_j(z)dz. \tag{7.61}$$

This quantity depends upon u_j. It also depends upon x_j, but because x_j is not a decision variable, we prefer to denote (7.61) by $I_j(u_j)$. Hence, on substituting (7.51) into (7.61), we have

$$I_j(u_j) = p(u_j + x_j) - c_0 u_j - \frac{(p + c_1)(u_j + x_j - a)^2}{2(b - a)}. \tag{7.62}$$

You may be wondering why we have now used the symbol I for reward when previously we have always used J. The answer is that I_j is not *in general* the quantity a rational buyer would wish to maximize, so it is perfectly logical to denote it by a different symbol. To understand this, we will confine our attention to the short horizon $N = 2$ for the remainder of this section. We will extend our analysis to longer horizons in Section 12.1.

Consider, therefore, the buyer whose horizon is $N = 2$. He makes two decisions, the first at $t = 0$, the second at $t = 1$. His objective at $t = 0$ is to maximize reward from the interval $0 < t < 2$. This reward will depend upon the quantity u_0 he orders at $t = 0$; hence denote it by $J_2(u_0)$, the subscript on J indicating the number of decisions that remain to be made. At $t = 1$, however, the interval $0 < t < 1$ is part of history. The buyer's objective changes from maximization of reward from $0 < t < 1$ to maximization of reward from the remaining period, $1 < t < 2$. This reward will depend upon the quantity u_1 ordered at $t = 1$; hence denote it by $J_1(u_1)$. Now, in this particular case, the buyer's objective function at $t = 1$ coincides with (7.62); i.e., $J_1(u_1) = I_1(u_1)$. But it would be incorrect to say that $J_2(u_0) = I_0(u_0)$, because then the buyer would be basing his decision at $t = 0$ solely on reward from $0 < t < 1$ and ignoring reward from $1 < t < 2$. Does this mean that $J_2(u_0) = I_0(u_0) + I_1(u_1)$? Unfortunately, the answer is still no, because $I_1(u_1)$ depends upon x_1, which is an unknown quantity at $t = 0$.

To resolve this difficulty, we will begin by considering the *second* decision. From (7.62) and the remarks in the previous paragraph, the buyer's job at $t = 1$ is to maximize the function

$$J_1(u_1) = p(u_1 + x_1) - c_0 u_1 - \frac{(p + c_1)(u_1 + x_1 - a)^2}{2(b - a)} \tag{7.63}$$

over the interval $0 \le u_1 \le b - x_1$ (why not $0 \le u_1 \le b$?). This is a simple calculus problem. You can easily show that the maximum occurs at $u_1 = u_1^*$, where

$$u_1^*(x_1) = \begin{cases} \lambda_1 - x_1, & \text{if } x_1 \le \lambda_1; \\ 0, & \text{if } x_1 > \lambda_1; \end{cases} \tag{7.64}$$

and λ_1 is defined by

$$\lambda_1 = \left(\frac{p - c_0}{p + c_1}\right)(b - a). \tag{7.65}$$

Thus λ_1 is the quantity that the buyer should order for a single interval if the existing stock x_1 were zero. It is therefore the solution to the problem considered in Section 7.1, as you can easily verify by substituting (7.51) into (7.10). Now (7.64) yields a simple interpretation: if inventory at $t = 1$ already exceeds the quantity predicted as optimal by the simple-period model of Section 7.1, then order nothing; otherwise, raise the inventory to that level. Provided that he follows this policy at $t = 1$, the buyer should expect the maximum reward from the remaining interval. Denote this maximum reward by $\phi_1(x_1)$. Then on substituting (7.64) into (7.63), we obtain

$$\phi_1(x_1) = J_1(u_1^*(x_1)) = \begin{cases} c_0 x_1 + \dfrac{1}{2}(p - c_0)(a + \lambda_1), & \text{if } x_1 \le \lambda_1; \\ px_1 - \dfrac{1}{2}\dfrac{p + c_1}{b - a}(x_1 - a)^2, & \text{if } x_1 > \lambda_1. \end{cases} \quad (7.66)$$

This (continuously differentiable) function is sketched in Fig. 7.8. Note that the reward from $1 < t < 2$ is at least as great as the value $\phi_1(0)$ predicted by Section 7.1, because any c_0 costs associated with a surplus x_1 have already been paid at $t = 0$ (and hence, when $x_1 \le \lambda_1$, are simply added to the expected profit that the earlier analysis would predict for the second interval). Note also that *if* x_1 were at our disposal then we should certainly choose $x_1 = a + p(b - a)/(p + c_1)$, to maximize reward from $1 < t < 2$; but x_1 is *not* at our disposal, having been determined beyond our control by demand during $0 < t < 1$.

   Now, as far as the second decision is concerned, we are quite correct to say that $\phi_1(x_1)$ is the optimal reward from $1 < t < 2$, because the surplus at $t = 1$ is known to be x_1. As we move backward in time across $t = 1$, however, that surplus becomes the random variable X_1, causing optimal

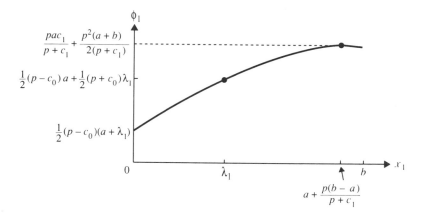

Fig. 7.8 Maximum reward ϕ_1 from the second interval, $1 < t < 2$, as a function of the initial surplus, x_1. The large dots denote the maximum of ϕ_1 and the point where the shape of the graph changes from linear to parabolic.

reward from $1 < t < 2$ to become the random variable $\phi_1(X_1)$, a utility. This cannot itself be maximized, but we can maximize its expected value, $E[\phi_1(X_1)]$. Thus reward from the second decision, *when viewed from the time of the first decision*, is $E[\phi_1(X_1)]$. We will call this *perceived optimal reward*. Thus perceived optimal reward is the expected value of a random variable, the realized value of which is the optimal reward from a period that begins when the value is realized and ends at the horizon. On using (7.54)–(7.57) in (A.38), a straightforward calculation (Exercise 7.14) reveals that if $u_0 \leq a + \lambda_1$ then

$$
\begin{aligned}
E[\phi_1(X_1)] &= \phi_1(0) \cdot \text{Prob}(X_1 = 0) + \int_0^{u_0 + x_0 - a} \phi_1(x) \frac{dx}{b-a} \\
&= \phi_1(0) \frac{b - u_0}{b - a} + \frac{1}{b-a} \int_0^{u_0 - a} \phi_1(x) dx \qquad\qquad (7.67a) \\
&= \frac{1}{2}(p - c_0)(a + \lambda_1) + \frac{c_0(u_0 - a)^2}{2(b - a)};
\end{aligned}
$$

whereas if $u_0 > a + \lambda_1$ then $E[\phi_1(X_1)]$

$$
\begin{aligned}
&= \frac{1}{2}(p - c_0)\left\{ \frac{(b - u_0)(a + \lambda_1) + a\lambda_1}{b - a} \right\} \\
&\quad + \frac{p(u_0 - a)^2}{2(b - a)} - \frac{p + c_1}{6(b - a)^2}\{(u_0 - 2a)^3 - (\lambda_1 - a)^3\}. \quad (7.67b)
\end{aligned}
$$

The quantity $E[\phi_1(X_1)]$ is independent of u_1 because, when looking ahead from $t = 0$, the buyer assumes that whatever decision is subsequently made at $t = 1$ will be the one that is then optimal, namely, (7.64). But (7.67) depends upon u_0, because X_1 depends upon u_0 (set $j = 1$ in (7.53)).

The total reward from $0 < t < 2$, when viewed from $t = 0$, is reward from $0 < t < 1$ plus perceived optimal reward from $1 < t < 2$; that is,

$$
J_2(u_0) = I_0(u_0) + E[\phi_1(X_1)].
$$

On using (7.54), (7.62) with $j = 0$ and (7.67), if $u_0 \leq a + \lambda_1$ then we obtain

$$
J_2(u_0) = (p - c_0)\left\{ u_0 + \frac{1}{2}(a + \lambda_1) \right\} - \frac{p - c_0 + c_1}{2(b - a)}(u_0 - a)^2; \qquad (7.68a)
$$

whereas if $u_0 > a + \lambda_1$ then we obtain

$$
\begin{aligned}
J_2(u_0) = (p - c_0)\left\{ u_0 + \frac{a\lambda_1 + (a + \lambda_1)(b - u_0)}{2(b - a)} \right\} - \frac{c_1(u_0 - a)^2}{2(b - a)} \\
- \frac{p + c_1}{6(b - a)^2}\{(u_0 - 2a)^3 - (\lambda_1 - a)^3\}.
\end{aligned}
$$
$$
(7.68b)
$$

Maximizing $J_2(u_0)$ is a straightforward calculus problem (Exercise 7.14). It is convenient to define the auxiliary parameters

$$\delta = \frac{(p - c_0)c_0}{(p + c_1)^2} - \frac{a(p - c_0 + c_1)}{(b - a)(p + c_1)}, \tag{7.69}$$

$$\Delta = \frac{(p - c_0)(p - c_0 + 2c_1) + c_1^2}{(p + c_1)^2} - \frac{2a(p - c_0 + c_1)}{(b - a)(p + c_1)}. \tag{7.70}$$

Then the maximum occurs at $u_0 = u_0^*$, where

$$u_0^* = \begin{cases} a + \dfrac{(p - c_0)(b - a)}{p - c_0 + c_1}, & \text{if } \delta \leq 0; \\[3mm] 2a + \left\{ \sqrt{\Delta} - \dfrac{c_1}{p + c_1} \right\} (b - a), & \text{if } \delta > 0. \end{cases} \tag{7.71}$$

You can easily check that u_0^* is greater than λ_1. What is the significance of this? See Exercise 7.15.

In summary, combining (7.64) with (7.71), we find that the optimal policy for horizon 2 is

$$\begin{array}{lll} t = 0 & \text{First decision} & \text{Order } u_0^*, \\ t = 1 & \text{Second decision} & \text{Order } u_1^*(x_1). \end{array} \tag{7.72}$$

Notice that the entire policy is *formulated* at $t = 0$, even though part of it is not *implemented* until $t = 1$. The reason for this, of course, is uncertainly in demand. A policy can be formulated as soon as the distribution of X_1 has been specified, but it cannot be fully implemented without certain knowledge of x_1.

In deriving this two-period policy, we have tacitly assumed that the periods are rather short. We have ignored the fact that the true worth of profits from the second period is not as great as that of profits from the first period, for the reasons described in Section 3.4. How would the effect of the discount rate be incorporated? See Exercise 7.16. For further discussion of the N-period buying problem, see Chapter 12.

7.6 Why Don't Fast-food Restaurants Guarantee Service Times Anymore?

Fast-food restaurants compete for customers in a bewildering variety of ways, ranging from claims of tastiness or nutritional value to gifts or simple price cutting. Many of those who eat in such places, however, particularly during a lunch break, are probably less concerned about the taste of their food than the speed with which it arrives. I don't mean by this that taste is unimportant. Rather, these people would argue that fast food tastes much the same wherever you go, so that the deciding factor in choosing a restaurant should be the speed of service. Wouldn't you think, therefore, that a fast-food restaurant could attract more customers by promising that food will be served within a given time limit, or else no charge will be made? Now, from time to time, various fast-food restaurants have tried such a policy, but the speed with which they have given it up suggests that it's never successful. Why? Let's try to answer this question by using a mathematical model.

Fig. 7.9 represents a hypothetical fast-food restaurant, which guarantees to serve your food within u minutes or else refund your money. We will refer to u as the guarantee. You enter this restaurant through the door on the left and approach the first available cashier. This cashier receives your order, transmits it to the kitchen, and takes your money. The time of day appears on your cash register receipt; this is the official beginning of your waiting time. You now walk across to the cashier on the right, who eventually receives your food from the kitchen and presents it to you. If it took longer than u minutes to arrive, then this cashier returns your money.

We can describe this restaurant by using the single-server queuing model developed in Sections 6.3 and 6.5. The single server will be the kitchen, where combined resources can be regarded as a kind of black box, which receives orders and exchanges them for food. We will assume that this service time is expcnentially distributed, with mean s. The queue the kitchen services consists of orders placed by the cashiers. We will assume that these arrive according to a Poisson process, with mean interarrival time c.[1] Then the mean number of orders per minute is $1/c$. To apply the model of Sections 6.3 and 6.5, we will also assume that meals leave the kitchen in the same sequence as that in which orders are placed by the cashiers.

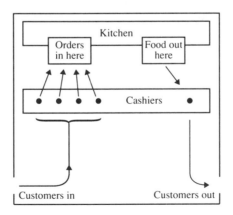

Fig. 7.9 A hypothetical fast food restaurant.

---

[1] See Supplementary Note 1.

Let a customer's waiting time be denoted by the random variable Y. Then, from (6.49),

$$\text{Prob}(Y > t) = e^{-(1/s - 1/c)t}. \tag{7.73}$$

Let the random variable $P(Y)$ denote the restaurant's profit from the customer whose waiting time is Y. Then, if r is the price of a meal and σ the cost of producing it, we have

$$P(Y) = \begin{cases} r - \sigma, & \text{if } Y \le u, \\ -\sigma, & \text{if } Y > u; \end{cases} \tag{7.74}$$

whence, on using (7.73), we have

$$E[P] = (r - \sigma)\,\text{Prob}(Y \le u) - \sigma\,\text{Prob}(Y > u)$$
$$= r - \sigma - re^{-(1/s - 1/c)u}. \tag{7.75}$$

Thus the expected profit per unit of time is

$$J(u) = \frac{1}{c}E[P] = \frac{1}{c}\left(r - \sigma - re^{-(1/s - 1/c)u}\right). \tag{7.76}$$

If Y exceeds u, then the restaurant must refund the customer's money. The sole reason for agreeing to this is that it should increase demand for the restaurant's product. An increase of demand will correspond to a decrease of interarrival time. Hence the effect of the guarantee can be incorporated into our model by allowing c to depend on u, i.e., by writing $c = c(u)$. Decreasing u will increase demand, or at least not decrease it; hence $c'(u) \ge 0$ (whenever c is differentiable). Let γ be the existing mean interarrival time, i.e., the one that would apply if the restaurant had offered no guarantee. Then $c(\infty) = \gamma$. It is convenient, moreover, to imagine that $c(0) = 0$, on the grounds that if $u = 0$ then the restaurant would have to give all its food away; and this would presumably lead to such large demand as to be virtually infinite. Of course, $u = 0$ now violates condition (6.10) for stationarity, namely,

$$s < c(u); \tag{7.77}$$

but this is without consequence, because the restaurant will choose the optimal u, i.e., the value u^* which maximizes (7.76), and it is clear that $J(u)$ cannot even be positive unless (7.77) is satisfied.

In practice, unless u is quite small, no new customers will be attracted to the restaurant, because the general public will respond to the guarantee only if u is less than the values of Y they have typically experienced elsewhere. It is therefore reasonable to assume the existence of a threshold, u_0, above which the guarantee has absolutely no effect:

$$c(u) = \gamma \qquad \text{if } u \ge u_0. \tag{7.78}$$

The value of γ will vary from restaurant to restaurant, but u_0 should be the same for all restaurants of a given type, say for all pizzerias or for all burger joints. Moreover, the speedier the service consumers expect in general, the lower in general the value of u_0. The simplest function satisfying (7.78) and the other conditions of the previous paragraph is the piecewise linear one sketched in Fig. 7.10. Giving $c(u)$ curvature for $u < u_0$ would not significantly alter the conclusions of this section, but would significantly complicate the mathematics. Henceforward, therefore, we adopt the form of Fig. 7.10.

In view of (7.77), we must seek the maximum of $J(u)$ not on $0 < u < \infty$ but on $su_0/\gamma < u < \infty$, where $J(u)$ is given by (7.76) and

$$c(u) = \begin{cases} \gamma u/u_0, & \text{if } su_0/\gamma < u < u_0; \\ \gamma, & \text{if } u_0 \leq u < \infty. \end{cases} \tag{7.79}$$

Now, if $su_0/\gamma < u < u_0$, then

$$J(u) = \frac{u_0(r - \sigma)}{\gamma u} \left\{ 1 - \alpha e^{-u/s} \right\}, \tag{7.80}$$

where

$$\alpha \equiv \frac{r}{r - \sigma} e^{u_0/\gamma}. \tag{7.81}$$

Note that $\alpha > 1$. Thus (7.80) has a unique maximum on $(su_0/\gamma, \infty)$ at $u = U^*$, where (Exercise 7.17) U^* satisfies

$$e^{U^*/s} = \alpha \left(1 + \frac{U^*}{s} \right) \tag{7.82}$$

and

$$J(U^*) = \frac{u_0(r - \sigma)}{\gamma(s + U^*)}. \tag{7.83}$$

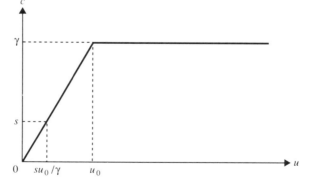

Fig. 7.10 Hypothesized response $c = c(u)$ of customer demand to the stimulus of a service guarantee.

It is also clear from (7.75) that $J(u)$ increases on $u_0 \leq u < \infty$, with maximum

$$J(\infty) = \frac{r - \sigma}{\gamma}. \tag{7.84}$$

Combining these two results, we find that the maximum of J on $su_0/\gamma < u < \infty$ occurs at $u = u^*$, where

$$u^* = \begin{cases} U^*, & \text{if } u_0 > U^* + s; \\ \infty, & \text{if } u_0 \leq U^* + s. \end{cases} \tag{7.85}$$

It follows immediately that the guarantee should not be advertised unless

$$\frac{u_0}{s} > \frac{U^*}{s} + 1. \tag{7.86}$$

Unless (7.86) is satisfied, the best option is to guarantee no more than that the food will be brought before the end of time, i.e., to offer no guarantee!

We can obtain the value of U^*/s, for given α, by solving (7.82) by the Newton–Raphson method. Don't forget that the equation has a negative root and that you want the positive one, so your initial guess must not be too close to zero. By obtaining U^*/s for varying α, we implicitly define a function. Let us denote it by $f = f(\alpha)$. Thus U^* is $sf(\alpha)$ mean kitchen service times. Part of the graph of $f = f(\alpha)$ is sketched in Fig. 7.11.

Now, (7.86) implies that the guarantee should be advertised only if $u_0 > s\{1 + f(\alpha)\}$. Because α depends on the three parameters r/σ, u_0, and γ, $u_0/\{1 + f(\alpha)\}$ is also a function of r/σ, u_0 and γ. Let us denote this function by s_c. Then the guarantee should be advertised only if

$$s < s_c(r/\sigma, u_0, \gamma). \tag{7.87}$$

For $r/\sigma = 1.25$ and for three values of γ, namely, $\gamma = 1$, $\gamma = 2$, and $\gamma = 5$, the graph of $s_c = s_c(r/\sigma, u_0, \gamma)$ is sketched in Figure 7.11(a) as a solid curve. Thus, for example, if the kitchen already receives orders at the rate of one per minute, and if a guarantee in excess of three minutes would have no effect on customer demand, then the kitchen's mean service time would have to be lower than 23.6 seconds for the guarantee to be profitable; in which case the optimal guarantee would be $f(\alpha) = f(100.4) \approx 6.64$ mean kitchen service times. We have placed Fig. 7.11(a) and (b) as shown to emphasize that (7.86) should be checked before obtaining U^*. Graphs corresponding to $r/\sigma = 1.25$ but other values of γ are readily added to Fig. 7.11(a). Moreover, by varying the value of r/σ, it is possible to obtain a whole series of similar diagrams, the graphs being higher or lower according to whether r/σ is higher or lower. As an indication of this process, Fig. 7.11(a) shows the (dashed) curve that corresponds to $r/\sigma = 2$ and $\gamma = 5$.

Taken as a whole, these diagrams suggest that if r/σ and γ are small, then the critical mean service time s_c, which the kitchen must better for a guarantee to be profitable, is unattainably low unless u_0 is large. But the

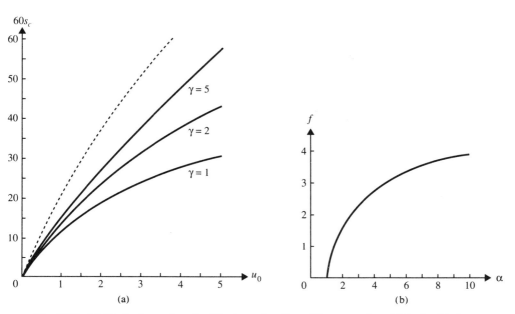

Fig. 7.11 (a) Typical graphs of critical service time $60s_c$ seconds, which the kitchen mean service time must not exceed for a guarantee to be profitable, as a function of u_0. The solid curves are drawn for $r/\sigma = 1.25$ and for various values of γ. The dashed curve is drawn for $r/\sigma = 2$ and $\gamma = 5$. (b) Graph of the function $f = f(\alpha)$. If (7.87) is satisfied, then the optimal guarantee is $f(\alpha)$ kitchen mean service times, where α is defined by (7.81).

fast-food industry is characterized by tight profit margins, heavy throughput, and high expectations of speedy service by customers—in other words, by low values of all three parameters r/σ, γ, and u_0. Thus our model predicts that the practice of guaranteeing service times is not, by itself, economically sound, and it is no surprise that it is soon discontinued by restaurants that adopt it in the first place.

The function $c = c(u)$ is an example of a *behavioral response function*, or simply response function. This represents an attempt to capture, in a single function, the characteristic response of people or other creatures to a *stimulus*, u, here the service guarantee. Another example of a behavioral response function appears in the following section, and the idea will be more fully discussed in Section 8.3. Meanwhile, try Exercise 7.18.

7.7 When Should One Barber Employ Another? Comparing Alternatives

Optimal decisions have been reached hitherto with the help of the calculus. This was appropriate because the range of feasible values for the decision variable u has been a rather large one. We grant that decision variables are usually restricted to integer values in practice. But if the number of feasible values is sufficiently large—say if an order quantity can lie anywhere between 0 and 100—then the calculus is still the most efficient mathematical technique for optimization. In other circumstances, however, when u is restricted to be an integer, the number of feasible values is so few that the most efficient optimization technique is to evaluate $J(u)$ for every feasible u and simply compare the alternatives. Strictly, the problem of the previous section probably belonged to this category, because the only guarantees a fast-food restaurant could realistically offer would form a very restricted set—say half a minute, one minute, 100 seconds, two minutes, and so on. But the simplicity of the behavioral response function made the calculus as efficient as comparing the alternatives.

We will illustrate the method of alternatives by taking up the first of the two questions raised at the end of Chapter 6. How would a store manager determine the optimal value for the length of a queue at the checkout? Remember that the queue length itself is a random variable, outside the manager's direct control. The only variable he can control directly is the number of servers. Let's call this u (instead of k, as in Section 6.4). Now, the manager could control u so that the expected queue length would always be three, irrespective of the prevailing value of the business intensity,

$$\rho = \frac{s}{c} = \frac{\text{MEAN SERVICE TIME}}{\text{MEAN INTERARRIVAL TIME}}. \qquad (7.88)$$

But three is a purely arbitrary choice. A more rational approach would be to select the value of u which minimizes the loss, $L(u)$, associated with queuing, when the number of servers is u. As observed in Chapter 3, this is equivalent to maximizing

$$J(u) = -L(u). \qquad (7.89)$$

Now, the true loss from queuing is borne by society as a whole, in terms of increased irritability and lost time for other activities, and is extremely difficult to quantify. We shall assume, however, that our businessman is concerned not so much about society's loss as about the loss of revenue that queuing causes, because customers may be enticed away by stores that promise speedier service. Unfortunately, even this more personal loss is difficult to quantify in general, and we will therefore concentrate on a particular example.

There are fewer "walk-in" barber shops in downtown areas than was once the case, but this particular small business simply refuses to die (Tallahassee has at least four). Let u be the number of barbers in such a business. It is usually the case that one of them owns the business; anyway, let's assume this. Let's also assume that the remaining $u - 1$ barbers are paid employees, earning w dollars per hour. Then the owner's loss per hour in terms of wages is simply $w(u - 1)$. If this were the only loss incurred, then the owner would minimize it by choosing $u = 1$, i.e., by employing no one.

But life isn't quite as simple as all that. If the owner is the only barber, then customers inside the shop will have longer to wait. Passers-by will look through the window and, noticing the length of the queue, perhaps decide that the queue is too long and continue to the next nearest barber shop. Potential customers are therefore lost.

Let's assume that, on average, a customer arrives at the barber shop every $60c$ minutes. In other words, the mean customer interarrival time is c hours, and the mean number of customers arriving per hour is $1/c$. If all agreed to enter, then the owner's average revenue per hour would be p/c dollars, where p is the price of a haircut. But all do not enter. A proportion of potential customers, say $r(u)$, decides that the queue in the shop is too long. These customers walk on by to do something else with their lunch break (perhaps buy a hamburger, see previous section). Note that r will be a decreasing function of u (why?). The corresponding loss of revenue per hour is $pr(u)/c$, provided that the u barbers can cut hair fast enough to cope with an average demand of $1/c$ haircuts per hour (they cannot claim to have lost revenue that they could not have earned in the first place). Hence the owner's total loss per hour, on average, will be

$$L(u) = \frac{p}{c} r(u) + w(u - 1). \tag{7.90}$$

If u is too small, then the owner will lose too much revenue from potential customers who turn away. If u is too large, then he will pay too much in wages. Somewhere in between lies the value, u^*, that minimizes loss.

We must still determine the form of $r(u)$. We will do this by assuming that the stationary u-server queuing model of Section 6.4 provides an adequate description of arrivals and service in the barber's shop. Let the random variable X denote the number of heads in the queue at any time; u of these are receiving attention, while $X - u$ are simply waiting. Let N denote the maximum number of heads there could ever be in the queue. In other words, assume that, if a potential customer looks through the window and sees $N - u$ people waiting, then he is bound to go away again. Then X has sample space $\{0, 1, \ldots, N\}$; and, on replacing k by u in (6.39) and using (6.25)–(6.26), we can evaluate the expression $\pi_j = \text{Prob}(X = j)$ for any value of j. It seems reasonable to suppose that the proportion $r(u)$ is equal to the proportion of time for which there are N heads in the queue, or in the language of Section 6.4, that $r(u) = \pi_N$. Hence (Exercise 7.19):

$$r(u) = \frac{\frac{u^u}{u!} \left(\frac{\rho}{u}\right)^N}{1 + \sum_{j=1}^{u} \frac{\rho^j}{j!} + \frac{u^u}{u!} \sum_{j=u+1}^{N} \left(\frac{\rho}{u}\right)^j}. \tag{7.91}$$

Observing now that the value of N will depend upon u—i.e., $N = N(u)$—and collecting together (7.90)–(7.92), we find that

$$J(u) = w(1 - u) - \frac{p}{c}r(u)$$

$$= \frac{p}{c}\left\{\frac{wc}{p}(1 - u) - \frac{\frac{u^u}{u!}\left(\frac{\rho}{u}\right)^{N(u)}}{1 + \sum_{j=1}^{u}\frac{\rho^j}{j!} + \frac{u^u}{u!}\sum_{j=u+1}^{N(u)}\left(\frac{\rho}{u}\right)^j}\right\}. \quad (7.92)$$

It still remains to specify the value of the integer $N(u)$ for integer values of u. In the terminology introduced at the end of the previous section, N is a behavioral response function and u is a stimulus. For now, let's simply suppose that, after thinking about it for several days, a barber produced the customer response function defined in Table 7.1. Does this seem unsatisfactory, like pulling a response function out of a hat? Perhaps you would like to ponder that question; we'll return to it in the following section.

For given values of s, c, w, and p, the maximum of $J(u)$ can now be determined by direct evaluation of $J(1)$, $J(2)$, $J(3)$, and $J(4)$. Suppose, for example, that the mean time for a haircut is 10 minutes and the mean time between potential customer arrivals is 12 minutes. Then $s = 1/6$, $c = 1/5$, and $\rho = 5/6$, from (7.88).[2] The corresponding values of $r(u)$ are easily calculated from (7.91) and are presented in Table 7.2; for example, $r(2) = \rho^5/(16 + 16\rho + 8\rho^2 + 4\rho^3 + 2\rho^4 + \rho^5)$. We see immediately that $J(1) < J(2)$ if $p > 1.14w$, and $J(2) < J(3)$ only if $p > 19.66w$.

It seems to me that the price of a haircut is quite likely to exceed an hour's wage by 14%, but could never exceed an hour's wage by 1866%—unless employees were being severely exploited! What this means is that a lone barber observing a potential business intensity of 5/6 should employ a second barber, and a barber who already has a helper should not employ

Table 7.1 Behavioral response function for a four-server barber shop. It is assumed that the shop isn't big enough to accommodate more than four barbers

u	$N(u)$
1	3
2	5
3	7
4	8

---

[2]Thus $\rho < 1$, implying in particular that $\rho < u$. Because the queue is finite, however, condition (6.38) is no longer necessary.

Table 7.2 Calculation of optimal number of barbers when $c = 1/5$, $\rho = 5/6$. The values for r are quoted to six decimal places so that you can check your own calculation.

u	$r(u)$	$J(u) - J(u + 1) = w - 5p\{r(u) - r(u + 1)\}$
1	0.186289	$w - 0.879p$
2	0.010419	$w - 0.051p$
3	0.000248	$w - 0.001p$
4	0.000016	

another. This result is strongly dependent, of course, on the values chosen for s and c and the form of the behavioral response function. But our model does at least indicate how such a decision can be made with the help of a mathematical model. You may wish to experiment further with choices of s, c and N that are more appropriate to a barber's shop you know.

7.8 On the Subjectiveness of Decision Making

We have now studied several decision models. In every case a single decision maker, whether a company or an individual, must

1. make assumptions about the future behavior of other people by supplying probability distributions or behavioral response functions,
2. assign utility to the possible consequences of his decisions, and
3. maximize reward from all feasible options.   (7.93)

In this chapter, utility has been either profit or negative loss, and reward has been its expected value. As we remarked at the beginning of the chapter, however, the reward could be some other statistic of a random variable (as in Exercise 7.13). Moreover, models that cater for several decision makers can also be constructed, though they are not discussed in this text.[3]

The models introduced in Chapter 7 are all prescriptive and therefore, as remarked in Section 4.15, essentially untestable. This applies in particular to (1). To appreciate this, let's suppose that each of two buyers would like to apply the model of Section 7.5. The first has been told by an economist that the economy is "buoyant;" he believes that demand will be uniformly distributed between 100 and 200. The second has been told by a psychic that a "crash" may be imminent; he believes that demand will be uniformly distributed between 0 and 100. Now, you know that economists are sometimes wrong, and you've heard that psychics are

---

[3]Models allowing more than one decision maker are known as games. For an introduction to game theory, see Dresher (1981), Vincent and Grantham (1981), Shubik (1982, 1984), and Mesterton-Gibbons (1992).

sometimes right, but a decision must be made *before* the future has had time to happen. Then which buyer has made the more realistic assumption about future behavior? In the final analysis, it's just impossible to be sure. We must simply accept that (1) involves a *subjective* assessment on the part of decision makers, in this case the buyers. Naturally, the more experienced a buyer, the more apt he is to guess correctly. But his assessment is nonetheless subjective, and other buyers may disagree. Note, moreover, the important point that if we intervened between the two buyers and said that economists know better than psychics, then we would still not eliminate the element of subjectiveness. We would simply transfer it to ourselves!

In these circumstances, it would appear desirable to design decision models with sufficient flexibility that two decision makers who agree on (2) but disagree on (1) can both use the same model to derive an optimal policy for their particular view. In this light, let us reconsider the model of the previous section. It is quite possible that one barber would agree with the behavioral response function $N(u)$ in Table 7.2, while another would hotly dispute it. Does it matter? Not at all. The prejudices of the second owner can easily be substituted for those of Table 7.2, because the model is flexible enough to allow an arbitrary choice of behavioral response function. Again, suppose that one of our buyers believed demand to have a distribution other than the uniform one. Does it matter? Not much. The model of Section 7.5 can easily be modified to deal with other distributions, though the mathematics may be more difficult.

To repeat what we said at the end of Chapter 4, whereas the ultimate test of a descriptive model's worth is its ability to predict observable data, the ultimate test of a prescriptive model's worth may well be its ability to accommodate alternative assumptions about future behavior. As we saw in the case of the barber shop, we obtain such flexibility by retaining as many free parameters as possible, as late as possible in the analysis. Occasionally, it is even unnecessary to estimate certain parameters, either because quantities of interest are somewhat insensitive to them or because they can be given their worst possible values (whatever that means in a particular context). Examples of the former will appear in Sections 10.6 and 12.3, and an example of the latter will appear in Section 8.1; while, in Sections 8.3 and 8.5–8.7, (1) and (2) of (7.93) will be discussed in greater depth.

Exercises

7.1 Verify (7.8).

7.2 (i) How can you be sure that a unique solution to (7.11) exists? Does your answer require any qualification if Z is normally distributed?
(ii) Verify (7.15).

*7.3 The expression for expected profit in (7.8) consists of three terms. The second two, namely $(p - c_0)u - k$, are the "sellout" profit, i.e., the retailer's profit if all items ordered are sold. Can you interpret the first term?

7.4 A bookseller is deciding whether to promote a new book, which he can buy from the publisher for $10 per copy (including shipment); holding costs are $1 per book. The bookseller reckons that, if he sells the books for $15 apiece, then demand (after advertising) is equally likely to fall anywhere between zero and 480. According to the model of Section 7.1, how many copies should he order if the costs of advertising are (i) $200 (ii) $375; and what is the expected profit?

7.5 How many copies should the bookseller in the previous exercise order according to the model of Section 7.3? What is the expected profit?

7.6 When is B sufficiently large for the normal approximation to the Poisson distribution to be incorporated into the model of Section 7.1? Compare (7.23) with $\Phi((k - B)/\sqrt{B})$ for relevant values of k and B.

7.7 By applying the shift technique of Chapter 2 to (7.10) or (7.11), show that u^* in Section 7.1 is an increasing function of the selling price but a decreasing function of c_0 and c_1. Interpret your answer in terms of Fig. 7.3. Repeat for u^* in Section 7.3.

7.8 Show that the maximum of the function defined by (7.41) is zero, at $k = 0$, whenever $\alpha \geq \alpha_0$. If $\alpha < \alpha_0$, show that the function has a local maximum at k^*, where k^* is defined by (7.43). Is this necessarily the global maximum of the function?

7.9 The book on marine biology in Section 7.4 costs the bookseller $11 per copy (including shipment), and holding costs are $1 per book. The bookseller, who will sell the book at the recommended price of $19, believes that there will never be more than 2500 potential buyers, no matter how much he publicizes it. The setup cost for the mailing and the author's expenses for the autograph session amount to $750; each brochure costs 50¢ to print and mail; and the mailing list of the local Marine Biology Society contains 1000 names and addresses. Show that the bookseller's optimal strategy, according to the model of Section 7.4, is to spend $1842 on advertising and order 665 copies of the marine biologist's book. Sketch the graph of $J(u^*(k))$ and use it to estimate the retailer's reward.

7.10 Suppose that the bookseller in the previous exercise believes demand to be distributed on $(0, b(k))$, not uniformly, but as shown in Fig. E7.1. Show that (7.34) is replaced by $u^*(k) = 0.31b(k)$; but that (7.35) is still true, provided that (7.36) is replaced by $\alpha_0 = 1.7775$. Hence show that the bookseller's

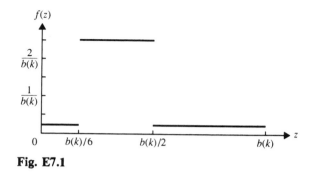

Fig. E7.1

optimal strategy is to spend $k^* \approx \$1914$ on advertising and order 528 copies of the book. What is his expected profit?

7.11 Why is the reward in Exercise 7.10 greater than in Exercise 7.9?

7.12 What is the optimal policy in Exercise 7.10 if the potential buyer function of Fig. 7.7 is used in place of the one in Fig. 7.5? Suppose that there are 2384 people on the mailing list, divided into six classes with $b_0 = 1000$, $b_1 = 600$, $b_2 = 360$, $b_3 = 216$, $b_4 = 130$, and $b_5 = 78$ (so that each class is approximately 40% smaller than the previous one); and that $\xi_i = 1 - 0.2i$, $0 \le i \le 5$.

*7.13 Suppose that the rose vendor in Section 7.2 would go home happy as long as his profit exceeded r. Would ordering the quantity that maximizes $E[P_u]$ still be his best strategy? If not, how many roses do you think he should order instead?

7.14 (i) Verify (7.67). *Hint:* Don't forget that

$$\int_0^{u_0-a} \phi_1(x)dx = \begin{cases} \int_0^{u_0-a} \left\{c_0 x + \frac{1}{2}(p - c_0)(a + \lambda_1)\right\} dx, & \text{if } u_0 - a \le \lambda_1; \\ \int_0^{\lambda_1} \phi_1(x)dx + \int_{\lambda_1}^{u_0-a} \phi_1(x)dx, & \text{if } u_0 - a > \lambda_1. \end{cases}$$

(ii) Verify (7.71).

*7.15 (i) Why is u_0^* in (7.71) always greater than the value λ_1 predicted by the single period model? First answer the question on the assumption that c_1 is small, then observe that u_0^*/λ_1 decreases with c_1.
(ii) Interpret the fact that u_0^* tends to a as $c_1 \to \infty$. (Note that $\delta < 0$ for large c_1 unless $a = 0$.)

*7.16 (i) Suppose that the unit of time in Section 7.5 is a year, so that the effect of the discount rate on future profits must be taken into account (see Section 3.4). Let the (real) annual interest rate be δ, and suppose that prices and costs are fixed at the beginning of each period and are constant in real terms. How must the analysis of Section 7.5 be altered to allow for the effect of the discount rate? At this stage, answer the question conceptually; do not work through the details.

(ii) To simplify the mathematics, set $a = 0$, so that demand in each interval is distributed uniformly between 0 and b. Hence determine the optimal u_0^* and u_1^* as a function of δ, b, p, c_0, and c_1.

(iii) Show that u_0^* is a decreasing function of the interest rate but is never less than λ_1. Interpret.

7.17 Show that (7.80) has a maximum on $(su_0/\gamma, \infty)$ at $U = U^*$, where U^* is the positive root of equation (7.82).

*7.18 A major airport has a single runway for international flights. At certain times of day, incoming planes must circle the airport for minutes, or even hours, before obtaining permission to land.

(i) Describe how you could model this circling as a stationary, single-server queue. State the assumptions involved.

(ii) An insurance company has observed that almost all the delay on incoming flights is caused by this circling before landing. It is debating whether to offer passengers an insurance against excessive delays. Specifically, for a premium of u dollars, the company would pay the sum of S dollars to any passenger whose plane arrived more than d hours after the appointed time. If the average cost of a transaction with customers is ϵS dollars, show that the insurance company can make money by adopting such a scheme only if d is sufficiently large. Try to obtain a critical value, d_c, as a function of ϵ and any other parameters you introduce. *Hint:* You will need a behavioral response function to describe the proportion of passengers who would pay for such insurance if the premium were u. You will also need to make an assumption about the time allowed in the flight schedule for circling and landing.

(iii) What is the optimal insurance premium?

(iv) Pick appropriate parameter values and decide whether the proposition is a viable one.

7.19 Obtain (7.91).

Supplementary Notes

7.1 This is not the same as the mean customer interarrival time, though it can be argued that the two should be roughly the same. To see this, let $\bar{c}$ denote the actual mean interarrival time for customers and let $\bar{s}$ be the mean service time for the cashiers (as opposed to the kitchen). Suppose that $j(\leq k)$ are currently busy, and let T_i be the service time for attendant i ($i \leq j$). Let the random variable U denote the interarrival time for placement of orders. Then $U = \min(T_1, \ldots, T_j)$; and

$$
\begin{aligned}
\mathrm{Prob}(U > t) &= \mathrm{Prob}(T_1 > t, T_2 > t, \ldots, T_j > t) \\
&= \mathrm{Prob}(T_1 > t) \times \mathrm{Prob}(T_2 > t) \times \cdots \times \mathrm{Prob}(T_j > t) \\
&= \left(e^{-t/\bar{s}}\right)^j = e^{-jt/\bar{s}}.
\end{aligned}
$$

Hence $c = \bar{s}/j$. In performing this calculation, we have regarded j as a fixed parameter whose value is unknown. But, from Exercise 6.7 (ii), the expected value of the number of busy servers is $\bar{s}/\bar{c}$. To the extent that this can be regarded as an approximation for j, then $\bar{c}$ can be regarded as an approximation for c.

IV THE ART OF APPLICATION

8 USING A MODEL: CHOICE AND ESTIMATION

You have already developed a number of mathematical models in the exercises, but you have always known that the chapter in which an exercise was set would offer significant clues to its solution. Then what can you do if an exercise belongs to no particular chapter? You must either use a "known" model—"known" meaning known to you—or build your own model. For economy of effort, you always use a known model if you know one; only if you don't do you build your own. The second of these two options is the subject of Chapter 9. In this chapter, we'll concentrate on the first.

Imagine, for example, that a bottle has floated to you from some distant desert island. The bottle contains a modelling exercise and a desperate message from a shipwrecked traveller, whose life will be spared by the natives only if she solves the problem contained in the bottle. Assuming that you would like to help this poor soul, how do you go about it? We'll assume that the natives have been sufficiently friendly as to pose a problem that a known model will solve. Your first concern is then to *choose* it. That choice may be tantamount to the solution. On the other hand, the model may involve one or more arbitrary parameters, in which case, your second concern is to *estimate* them.

Suppose, for example, that the castaway had sent you the island's census data for the last hundred years. You might stare at these data for several minutes and just get a feeling, deep down inside, that the natives obey a logistic growth law. So you choose model (1.62) from Section 1.9. It may be the wrong choice, but, at least for the moment, you've made it. The logistic growth law contains two arbitrary parameters R and K, so that your next concern is to estimate them. Figure 1.7 suggests how you might do this. Naturally, the goodness of fit of your line to the data points would determine whether your choice was the right one. Thus choice and estimation are really two sides of the same coin, though it's often helpful, while learning, to separate them.

But how do you actually choose? How do you produce that feeling inside, given only a verbal description of a problem? You can pare down your list of possible models by asking the following questions. Is the phenomenon you wish to describe *transient* or *permanent*? Should the information you are given be regarded as *certain* or *uncertain*? Your answers will help you decide whether to use a *dynamic* model or an *equilibrium* one, and whether the most appropriate description is *deterministic* or *probabilistic*. You should also ask yourself whether or not a *decision* is required. In the final analysis, however, there is no pat answer as to how a model is chosen—it's simply a matter of experience. This chapter should help you acquire it from your ongoing study of mathematical models.

Here, then, are some examples of choosing a model, interwoven with sections on estimating parameters.

8.1   Protecting the Cargo Boat. A Message in a Bottle

Your castaway has been captured by the natives of Nands, a nation of two islands with a simple agricultural lifestyle. The only crop that grows on Nor, the northern island, is X-grass. The only crop that grows on Sou, the southern island, is Z-fruit. Both foods are essential to the natives' simple diet. Consequently, the natives' cargo boat is always sailing back and forth between the islands, taking X-grass to Sou and Z-fruit to Nor.

Unfortunately, the sea between the islands is not entirely tranquil. It's roamed by pirates. The natives have cannons to defend their boat, and the pirates are so cowardly that they will always withdraw if they see one on board. But a cannon takes up so much space on board that not enough X-grass and Z-fruit would be shipped if a cannon went on every trip. The natives have solved this problem by taking a cannon with them only if their witch-doctors report that there are pirates in the area. They always unload the cannon at the end of a trip to make room for more cargo on the return. This tactic worked perfectly well for several months until, on the day that your castaway was captured, the witch-doctor on Sou reported pirates in the area, when all the cannons were lying on the beach on Nor.

The natives, who had four cannons, already appreciate that they must acquire some more. The question is, how many? Your castaway must help them decide; and we must choose a model to help her.

The boat sails back and forth for ever; therefore we're describing a *permanent* phenomenon. Because we don't know when a witch-doctor will report that there are pirates in the area, our information is *uncertain*. It therefore seems appropriate to choose a probabilistic model and seek its stationary state. Because nothing changes during a boat trip, we may measure time discretely, with each trip marking the passage of a unit of time.

All the above suggests that we use a Markov chain, with time being measured nonlinearly.

Let the random variable $X(n)$ denote the number of cannons, at time n, on the island where the boat is docked. Let N be the total number of cannons. Then we can model the boat's movements in relation to the cannons as a Markov chain with $N + 1$ states, $i = 0, 1, \ldots, N$. Suppose that $X = i$. If $i = 0$, then the only allowable transition is (from 0) to N, because all the cannons are on the other island. If $0 < i \leq N$, however, then there are two possibilities, according to whether a cannon is included in the cargo boat or not. If one is, then a transition is made from i to $N - i + 1$. If one is not, then a transition is made from i to $N - i$. Let U be the event that the witch-doctor on the island where the boat is docked reports pirates in the area, and let $q = \text{Prob}(U)$. We will assume that this probability is the same for both witch-doctors, perhaps because they are always right; but the value of q is completely unknown (to us). Then the transition matrix for the Markov chain, i.e., the matrix in which $p_{ij} = \text{Prob}(X(n + 1) = j \mid X(n) = i)$ occupies row i and column j, is

$$\mathbf{P} = \begin{bmatrix} 0 & 0 & 0 & \cdots & 0 & 0 & 1 \\ 0 & 0 & 0 & \cdots & 0 & 1-q & q \\ 0 & 0 & 0 & \cdots & 1-q & q & 0 \\ \vdots & \vdots & \vdots & \ddots & \vdots & \vdots & \vdots \\ 0 & 0 & 1-q & \cdots & 0 & 0 & 0 \\ 0 & 1-q & q & \cdots & 0 & 0 & 0 \\ 1-q & q & 0 & \cdots & 0 & 0 & 0 \end{bmatrix}. \tag{8.1}$$

You should satisfy yourself that the entries in this matrix are correct. Note, in particular, that the number of nonzero entries on the leading diagonal is precisely one, but still enough to ensure the existence of a stationary distribution. You can verify (Exercise 8.1) that it's given by

$$\pi^* = \frac{1}{N - q + 1}(1 - q, 1, 1, \ldots, 1, 1)^T. \tag{8.2}$$

Let V be the event that all the cannons are on the other island, and assume that the witch-doctors never know where the cannons are. Then U and V are independent events. The probability of being intimidated by pirates when all the cannons are on the other island is therefore

$$\text{Prob}(U \cap V) = \text{Prob}(U) \cdot \text{Prob}(V) = q\pi_0^* = \frac{q(1 - q)}{N + 1 - q}. \tag{8.3}$$

In particular, the event that occurred on the day your castaway was captured had a probability of $q(1 - q)/(5 - q)$. Clearly this was too high, and we must reduce (8.3) to an acceptable value by increasing the value of

N. Unfortunately, we have no information as to the value of q. We must therefore give q its *worst possible value*. It's a simple matter of the calculus (Exercise 8.2) to show that (8.3) has a maximum on $0 < q < 1$ at $q = N + 1 - \sqrt{N(N+1)}$, the value of the maximum being

$$m(N) = 2N + 1 - 2\sqrt{N(N+1)} = \frac{\sqrt{1+(1/N)} - 1}{\sqrt{1+(1/N)} + 1}. \tag{8.4}$$

Note that $m(0) = 1$, $m(\infty) = 0$. By choosing N such that m lies below some acceptable value, we guarantee that the probability of the boat being unprotected is also less than this value. Perhaps it is considerably lower, because q may not take its worst possible value.

The first 20 relevant values of $100m(N)$, the percentage of journeys for which the boat would be unprotected, are given in Table 8.1 correct to three significant figures. A Taylor expansion of (8.4) in terms of $1/N$ shows that the percentage is approximately $100/(4N + 1)$ for large values of N, and this will be correct to at least two significant figures for all $N \geq 24$.

Your castaway will be able to use Table 8.1 to convince the natives that their boat could not endure more than about 11 unprotected journeys in every 200 even if they acquired no new cannons; whereas their present strategy could never guarantee its safety on every journey, even if they acquired a hundred cannons. She might even suggest a figure to the natives. For example, she might suggest they acquire eight new cannons to reduce the long run percentage of unprotected journeys to 2%. But it would be up to the natives to balance the risk of a given number of unprotected journeys (or the cost, in terms of hunger pangs, of a given number of delays until the witch-doctor gave the all clear) against the cost of acquiring the cannons. If the natives had half a heart, then Table 8.1 might even be enough to secure your castaway's release. But if they did not relent, then she would have to send you a further message, with enough information to construct a reward function like those of the previous chapter, and hence determine a unique value for the optimal number of cannons.

Table 8.1 Worst possible percentage of journeys for which cargo boat would be unprotected, as a function of the number of cannons.

N	$100m(N)$	N	$100m(N)$	N	$100m(N)$	N	$100m(N)$
4	5.57	9	2.63	14	1.72	19	1.28
5	4.55	10	2.38	15	1.61	20	1.22
6	3.85	11	2.17	16	1.52	21	1.16
7	3.34	12	2.00	17	1.43	22	1.11
8	2.94	13	1.85	18	1.35	23	1.06

8.2 Oil Extraction. Choosing an Optimal Harvesting Model

A public outcry against nuclear power has led to the closure of several nuclear power stations and the suspension of plans to build several others. Oil is now relatively scarcer, and its (real) price is rising at an annual rate of 12%. A Texan farmer owns the land directly above a proven oil deposit but, because the price is still rising so rapidly, her oil is not for sale. The farmer believes that the (real) price of oil will ultimately level off at about three times the current price, as soon as alternative energy sources have been developed sufficiently to compensate for the loss of nuclear power. Before that time, however, our Texan farmer will have sold her oil, whenever she considers the price to be right. If she believes the cost of extracting each barrel of oil to be a fixed percentage of its selling price, and the (real) rate of interest that she can earn in a bank to be 9%, when should she consider that the price is right?

Notice, first of all, that it is by no means certain that the price of oil will level off as she thinks it will, or, for that matter, that the cost of extraction will be a fixed percentage of the amount extracted. Nevertheless, because our farmer *believes* that she is right, her information should be regarded as certain when making the optimal decision. In other words, a deterministic model is quite appropriate; and our first concern must be to choose between the three (deterministic) optimal harvesting models already known to us, namely, those encountered in Section 3.4, Section 3.7, and Section 4.8. None of these models was designed specifically for harvesting oil, but that does not mean that we won't be able to use one of them.

We arrive at our choice by elimination. The optimal fleet size model was designed for the harvesting of a renewable resource, from now until eternity. Oil is not renewable, at least on any respectable time scale, so we cannot use the model of Section 3.7. For the same reason, there is no such thing as an oil rotation time, so we cannot use the model of Section 4.8. The remaining candidate is that of Section 3.4. This would be an appropriate model if the oil could be extracted very rapidly. Let's assume that it can be, on the grounds that impatient, oilthirsty tycoons have been hovering over our Texan farmer day and night for several years. It is important to appreciate that the "improvement" on (3.40) obtained in Section 4.8 would not be an improvement here. In other words, a model can never be rejected absolutely; it can be rejected only *for a particular purpose*. Never discard a known model outright, because you never know when it might come in handy. One problem's reject may be another problem's solution.

Let $p(t)$ denote the (real) price at time t, N the number of barrels beneath the ground, and b the fraction of a barrel's price that it costs to extract it. Then the value of the oil is $V(t) = N(1 - b)p(t)$. According to (3.40), the farmer should have the oil extracted at time $t = s$, where

$$0.09 = \delta = \frac{V'(s)}{V(s)} = \frac{p'(s)}{p(s)}, \tag{8.5}$$

provided

$$\frac{V''(s)}{V(s)} - \delta^2 = \frac{p''(s)}{p(s)} - \delta^2 < 0. \tag{8.6}$$

The farmer's beliefs concerning the form of $p(t)$ can be summarized as $p(0) \leq p(t) \leq 3p(0)$ and

$$\frac{p'(0)}{p(0)} = 0.12, \qquad \lim_{t \to \infty} \frac{p(t)}{p(0)} = 3. \tag{8.7}$$

One way to satisfy (8.7) is to assume that price obeys the logistic growth law of Section 1.9, i.e., that

$$\frac{p(t)}{p(0)} = \frac{3}{1 + e^{-0.18t}}, \tag{8.8}$$

on setting $K = 3p(0)$ and $R = 0.18$ in Exercise 1.10. It is straightforward to show that (8.5) and (8.8) yield (Exercise 8.3):

$$s = \frac{\ln(2)}{0.18} \approx 3.85 \text{ years}, \tag{8.9}$$

with $p(s)/p(0) = 1.5$; (8.6) is satisfied because $-0.0081 < 0$. But the choice of (8.8) presupposes that the rate of price increase is still increasing at the time that the farmer makes her decision. If the rate of price increase is already falling, however, then we need to choose for $p(t)$ a form that is always curving downward. An appropriate choice might be

$$\frac{p(t)}{p(0)} = 3 - \frac{2}{(1 + Ct)}, \tag{8.10}$$

where C is a constant, the first of (8.7) determining the constant to be 0.06. It is straightforward (Exercise 8.3) to show that (8.10) would yield the earlier time

$$s = \frac{\sqrt{5} - 2}{0.18} \approx 1.31 \text{ years}, \tag{8.11}$$

with $p(s)/p(0) = 3(\sqrt{5} - 1)/(\sqrt{5} + 1)$, so that the price would have risen by less than 15%; (8.6) is satisfied because $-0.0181 < 0$. Thus, even if the farmer is absolutely right about the price of oil, the optimal extraction date could be as soon as just over a year away or as far into the future as almost four years away. Clearly, many other forms of $p(t)$ are consistent with (8.7). Therefore, before we could reach a firm conclusion, we would have to consult the farmer for further details of how she believes that $p(t)/p(0)$ will approach 3 asymptotically.

8.3 Models within Models. Choosing a Behavioral Response Function

A behavioral response function, which is used to embody everything assumed about the way some activity responds to a stimulus, may properly be regarded as a model in its own right—a functional model, as it is sometimes called.[1] A functional model can be used as an input to a known model, such as those we met in Sections 7.6 and 7.7. It becomes, in the process, a model within a model. Functional models, like the ones in which they are embedded, may be constructed by either a conceptual approach or an empirical one. The latter is the subject of the following section. In this section, we concentrate on the former.

The conceptual approach to constructing a functional model is greatly aided by what we might call the method of extremes. Let x be the stimulus and $r(x)$ the response. Then we wish to determine a curve $r = r(x)$ in the x-r plane. By the method of extremes, we simply attempt to enclose the feasible region in the x-r plane—i.e., the region in which $r = r(x)$ might possibly lie—by a boundary consisting of straight line segments. An acceptable form for the behavioral response function would then be $r = R(x, \theta)$, where θ is a parameter with the following property: by varying θ, it is possible to cover the entire feasible region. An appropriate value for θ, say $\theta = \theta^*$, yielding $r(x) = R(x, \theta^*)$, can then be estimated from data (either by eye or as described in the following section); or else θ can be retained as a parameter in the analysis. Two examples will suffice to indicate how the method is applied.

Consider first the airline insurance company in Exercise 7.18. It wishes to know what proportion r of travellers will pay x dollars for an insurance worth S dollars. By definition, $0 \leq r \leq 1$. If $x = 0$, then everyone will take out the insurance, implying $r(0) = 1$, but if $x \geq S$, then nobody will buy the insurance, implying $r = 0$ if $x \geq S$. Thus the behavioral response function is confined to the shaded region in Fig. 8.1(a), with the dots actually lying on the curve. We know more than this, however. The higher the value of the stimulus, x, the lower the proportion of travellers buying the insurance; hence $r'(x) \leq 0$. If a stimulus of one dollar induces, say, only 99% of travellers to buy the insurance, then the percentage induced by a stimulus of two dollars will surely not exceed 98%. By an extension of this argument, we have $r''(x) \leq 0$. It seems reasonable to assume, moreover, that there is some critical percentage of S, say $100\alpha\%$, above which the premium would be regarded as too high by everyone; $r(x) = 0$ if $x \geq \alpha S$. Thus, by a more refined application of the method of extremes, the feasible region is reduced from the shaded rectangle of

---

[1] You may wish to review your solutions to Exercise 4.12 and Exercise 7.18 before proceeding.

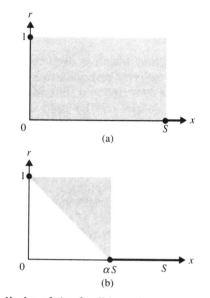

Fig. 8.1 Refining the limits of the feasible region.

Fig. 8.1(a) to the shaded triangle of Fig. 8.1(b), with the dots actually lying on the curve.

The region shaded in Fig. 8.1(b) is covered by the one-parameter family of curves

$$R(x, \theta) = 1 - \left(\frac{x}{\alpha S}\right)^{\theta}, \qquad 0 \leq x \leq \alpha S;$$
$$= 0, \qquad \alpha S < x < \infty. \tag{8.12}$$

The value $\theta \to \infty$ corresponds to the horizontal and vertical boundary of the triangular region, the remaining boundary of the triangle corresponding to $\theta = 1$. Note that (8.12) satisfies $r'(x) \leq 0$, $r''(x) \leq 0$. My guess is that the best value of θ almost certainly lies between 1 and 2 and is probably rather close to 1, but the value appropriate to any particular circumstance would have to be determined empirically (see Section 8.4). Naturally, (8.12) is rather crude, but a more complicated functional form would not retain the correct balance between the complexity of the mathematical model and our understanding of the intricate psychological processes that actually determine the public's response to a stimulus.

As a second example, we'll continue our study of island biogeography begun in Exercise 4.12. Both that exercise and the remainder of this section are very simplified treatments of the work of Gilpin and Diamond (1976) and Diamond and May (1976). In particular, I have simplified their results by all but ignoring the way in which extinction rates depend upon an island's area, and immigration rates upon distance from the colonizing source (the mainland territory or other islands). If interested, you should consult these authors for further details.

Let x denote the number of species of bird on an island and $E(x)$ the number of species that become extinct per unit of time; then E is a behavioral response function, with x as the stimulus. Because extinctions are impossible in the absence of species, $E(0) = 0$. Because an abundance of species will increase competition for available resources and hence encourage extinctions, $E'(x) \geq 0$. If extinctions occur at the rate of one per year when there are x species, then we would expect at least two per year when there are $2x$ species confined to the same finite area. By an extension of this argument, $E''(x) \geq 0$. When we combine these assumptions, one extreme of the feasible region will be the most downward sloping curve that satisfies $E''(x) \geq 0$, $E(0) = 0$, namely, a straight line through the origin, as shown in Fig. 8.2. Now, the main reason we expect $E'(x) \geq 0, E''(x) \geq 0$ is that the island has a finite amount of resources for the birds, enough, let's imagine, to support at most x_0 species. We might call x_0 the *species capacity* of the island. Although it's possible that there might be no extinctions if the number of species is less than or equal to x_0, if $x > x_0$ then extinctions are inevitable; the worst case occurs when the extinction rate suddenly becomes infinite at $x = x_0$. Thus the other extreme of the feasible region in Fig. 8.2 is given by an L-shaped curve. The curve $E = E(x)$ must pass through the dot and lie in the shaded region.

One set of curves that covers this region and satisfies $E(0) = 0, E'(x) \geq 0, E''(x) \geq 0$ is given by

$$E(x, \theta) = E_0 \left(\frac{x}{x_0}\right)^{\theta}, \tag{8.13}$$

where θ varies between 1 and ∞ and E_0 is the extinction rate at species capacity. You should check that $\theta = 1$ corresponds to the line through the origin in Fig. 8.2 and that $\theta \to \infty$ corresponds to the L-shaped curve.

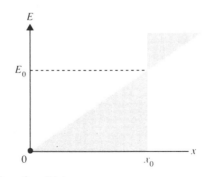

Fig. 8.2 Feasible region for $E(x)$.

Similarly, if $I(x)$ is the number of species immigrating per unit of time, then $I(0) = \nu$ and $I(x_m) = 0$, where ν is the maximum immigration rate and x_m is the maximum number of colonizing species available for immigration to the island. Because of competition for available resources, $I'(x) \leq 0$, and the fact that the best colonists will tend to arrive first suggests that $I''(x) \geq 0$. Hence the feasible region for $I(x)$ is the shaded one in Fig. 8.3, with dots denoting points through which the curve must pass.

One set of curves that covers this region is

$$I(x, \sigma) = \nu \left(1 - \frac{x}{x_m}\right)^{\sigma}, \qquad 0 \leq x \leq x_m;$$

$$= 0, \qquad x_m < x < \infty,$$

$$\text{(8.14)}$$

for σ lying between 1 and ∞. You should check that $\sigma = 1$ corresponds to the line between $(0, \nu)$ and $(x_m, 0)$ and that $\sigma \to \infty$ corresponds to the remaining L-shaped boundary of the triangle.

Gilpin and Diamond (1976) applied the response functions (8.13) and (8.14) to several islands, including, in particular, the Solomon island of Vatilau, for which the maximum number of (lowland) bird species available for immigration from the mainland is $x_m = 106$ (see p. 4132 of their paper). Using techniques similar to those we shall describe in the following section, Gilpin and Diamond determined best values of θ and σ to be $\theta = 2.37$ and $\sigma = 7.23$ (for details, see their paper). We will adopt these values here, denoting the corresponding response functions by $I(x)$ and $E(x)$. We will also adopt Gilpin and Diamond's estimate that $E_0/(\nu x_0^{\theta}) = 0.276 \times 10^{-5}$. With these parameter values, E/ν and I/ν are plotted against x in Fig. 8.4.

As discovered in Exercise 4.12, the dynamics of immigration and extinction are governed by the differential equation

$$\frac{dx}{dt} = I - E. \qquad \text{(8.15)}$$

Because $E(x)$ increases with x and $I(x)$ decreases, there is a unique equilibrium at $x = x^*$, defined implicitly by the equation $I(x^*) = E(x^*)$. Moreover, because the right-hand side of (8.15) is positive if $x < x^*$ and negative if

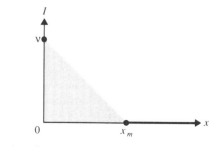

Fig. 8.3 Feasible region for $I(x)$.

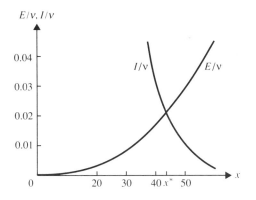

Fig. 8.4 Behavioral response functions for immigration, $I(x)$, extinction, $E(x)$, and graphical determination of equilibrium number of bird species for Solomon island of Vatilau. Based on the work of Gilpin and Diamond (1976).

$x > x^*$, our model implies that $x \to x^*$ as $t \to \infty$. It appears from the graphs in Fig. 8.4 that $x^* = 44$. According to Gilpin and Diamond (1976, p. 4133), the actual number of bird species on Vatilau is 41, suggesting that the island is close to its predicted equilibrium.

Now it's your turn to specify the form of a behavioral response function; see Exercise 8.4.

8.4   Estimating Parameters for Fitted Curves: An Error Control Problem

An error control problem arises whenever we attempt to compress the essential features of a set of data into a single functional relationship.[2] In Fig. 1.5, for example, we approximated a set of 37 economic data points

$$\{(\xi(t), \psi(t)); t = -9, \ldots, 27\}$$

by a linear relation of the form $\psi = u\xi$. Fitting the line by eye, we guessed that $u \approx 0.75$. But was this really the best value? In the present section, we will begin by considering the problem of fitting a straight line to data. We will then proceed to address nonlinear curve-fitting problems. Our treatment, however, is a simplified one. In particular, we shall ignore possible errors in the data itself, restricting our attention to errors that arise from the curve-fitting.

We would like the straight line in Fig. 1.5 to lie as close as possible to the data points. We must therefore construct a suitable measure of closeness

---

[2]You may wish to review Sections 1.6, 1.9, and your solution to Exercise 4.15 before proceeding.

and minimize it. This measure of closeness will depend upon u, the slope of the line, which is the only parameter we can control (because definitions (1.26)–(1.28) constrain the line to pass through the origin). Therefore, denote the measure by $L(u)$. How shall we define L? One possibility is as the sum of the perpendicular distances from the line to the data points; a second is as the sum of the vertical distances from the line to the data points. But each of these choices has the disadvantage that the resulting mathematical expression is difficult to minimize. It is therefore conventional to opt for a third measure of closeness, defining L to be the sum of the *squares* of the *vertical* distances from the line to the data points (this is a simpler expression than the sum of the squares of the perpendicular distances). Hence we have

$$L(u) = \sum_{t=-9}^{27} \{\psi(t) - u\xi(t)\}^2. \tag{8.16}$$

This is an infinitely differentiable function of u, with first derivative

$$L'(u) = -2 \sum_{t=-9}^{27} \xi(t) \{\psi(t) - u\xi(t)\} \tag{8.17}$$

and second derivative

$$L''(u) = 2 \sum_{t=-9}^{27} \xi(t)^2 > 0. \tag{8.18}$$

It follows readily from the calculus that L has an absolute minimum at $u = u^*$, where $L'(u^*) = 0$. From (8.17) we readily obtain

$$u^* = \frac{\sum_{t=-9}^{27} \psi(t)\xi(t)}{\sum_{t=-9}^{27} \xi(t)^2}. \tag{8.19}$$

Hence, from Table 1.1 and (1.28), we obtain

$$u^* \approx \frac{8.653}{11.79} \approx 0.734. \tag{8.20}$$

In fitting a straight line to the data in Fig. 1.5, only a single parameter had to be estimated. In fitting a straight line, say $y = u_1 - u_2 x$, to the U.S. population data in Fig. 1.7, however, two parameters must be estimated. Recall that the line in Fig. 1.7 was fitted (by eye) to the first 15 data points. We shall label these as $\{(x(t), y(t)): t = 1, \ldots, 15\}$ by defining $y(t) = D(t)/x(t)$, where D is defined by (1.55). Then the sum of squares of vertical distances between data points and the line $y = u_1 - u_2 x$ is

$$L(u_1, u_2) = \sum_{t=1}^{15} \{y(t) - u_1 + u_2 x(t)\}^2. \tag{8.21}$$

Let us define

$$\bar{x} = \frac{1}{15}\sum_{t=1}^{15} x(t), \qquad \bar{y} = \frac{1}{15}\sum_{t=1}^{15} y(t),$$

$$S_1^2 = \sum_{t=1}^{15}(x(t) - \bar{x})^2 = \sum_{t=1}^{15} x(t)^2 - 15\bar{x}^2,$$

$$S_{12} = \sum_{t=1}^{15}(x(t) - \bar{x})(y(t) - \bar{y}) = \sum_{t=1}^{15} x(t)y(t) - 15\bar{x}\bar{y}, \qquad (8.22)$$

$$S_2^2 = \sum_{t=1}^{15}(y(t) - \bar{y})^2 = \sum_{t=1}^{15} y(t)^2 - 15\bar{y}^2.$$

Then a straightforward calculation shows that

$$L(u_1, u_2) = 15(u_1 - \bar{x}u_2)(u_1 - \bar{x}u_2 - 2\bar{y}) + S_1^2 u_2^2 + 2S_{12}u_2 + S_2^2 + 15\bar{y}^2. \quad (8.23)$$

(More general expressions for a straight line fitted to k data points are obtained from (8.21)–(8.23) by replacing 15, wherever it appears, by k.) Now, let's imagine that the value of u_2 is fixed and regard (8.23) as a function of u_1 alone. Then it is easily shown that an absolute minimum occurs where

$$u_1 = u_1^*(u_2) = \bar{x}u_2 + \bar{y}. \qquad (8.24)$$

The corresponding value of the loss function is

$$L(u_1^*(u_2), u_2) = S_1^2 u_2^2 + 2S_{12}u_2 + S_2^2. \qquad (8.25)$$

Because the cluster of points in Fig. 1.7 stretches down and toward the right, it is clear that an x value in excess of the average will be associated with a y value that is below the average; thus S_{12} is negative. Hence (8.25) has an absolute minimum where

$$u_2 = u_2^* = \frac{-S_{12}}{S_1^2}. \qquad (8.26)$$

The corresponding value of u_1 is

$$u_1^*(u_2^*) = \bar{y} - \frac{S_{12}\bar{x}}{S_1^2}. \qquad (8.27)$$

Hence the optimal values of the parameters u_1 and u_2 are given by (8.27) and (8.26). From the data in Tables 1.3 and 1.4, it is straightforward to calculate that

$$\bar{x} = 52.4, \qquad \bar{y} = 0.23, \qquad S_1 = 163.4, \qquad S_{12} = -44.3. \qquad (8.28)$$

From (8.26)–(8.28), the optimal values of u_1 and u_2 are 0.317 and 0.00166, respectively, whence the capacity of the logistically growing population is $u_1/u_2 \approx 191$. Note that the parameter estimates 0.317 and 191 differ from the estimates 0.31 and 198 obtained by eye in Section 1.9. In fitting by eye, one is more likely to succeed in minimizing the sum of perpendicular distances, and one tends to overlook points that seem to lie outside the overall pattern (whereas minimizing L gives equal weight to all points); see Exercise 8.5. But because either of these visual "corrections" may actually be desirable, you should beware of concluding too readily that the estimates obtained in the present section are improvements upon the ones obtained in Chapter 1.

The fitted curve need not be a straight line. In the previous section, we discussed a method for suggesting a nonlinear curve that might reasonably be fitted to data, just as Fig. 1.5 and Fig. 1.7 suggested a straight line. Suppose that the response to a stimulus x_i has been measured as r_i, for $i = 1, \ldots, k$. Suppose, moreover, we have reason to believe that the essential features of the response's relationship to the stimulus can be captured by the (nonlinear) functional model

$$r(x) = R(x, u), \tag{8.29}$$

where u is a suitable parameter. Then a best value for u can be determined by minimizing the sum of squares of vertical differences

$$L(u) = \sum_{i=1}^{k} \{r_i - R(x_i, u)\}^2. \tag{8.30}$$

Because $L'(u)$ is a nonlinear function of u, however, it is no longer possible to determine a general expression for u^*, as in (8.19).

More generally, we might believe that the essential features of a relationship can be captured by a functional relationship of the form $y(x) = G(x, \mathbf{u})$, where $\mathbf{u} = (u_1, u_2, u_3, \ldots, u_m)^T$ is a vector of m control parameters. Suppose that we have k data points $\{(x_i, y_i): i = 1, 2, \ldots, k\}$. Then the sum of squares of vertical differences is

$$L(\mathbf{u}) = L(u_1, u_2, \ldots, u_m) = \sum_{i=1}^{k} \{y_i - G(x_i, \mathbf{u})\}^2. \tag{8.31}$$

Because $L(\mathbf{u})$ is now a nonquadratic function of several variables, it cannot be minimized by analytical methods; and we are obliged to turn to numerical ones. These numerical methods belong to a field that has come to be known as nonlinear programming. For the most part, nonlinear programming is beyond the scope of this text, but there do exist computer packages for computing the vector $\mathbf{u}^*$ that minimizes $L(\mathbf{u})$. Although they are not 100% reliable, you will find such routines adequate for the problems discussed here, for which I used the IMSL routine ZXSSQ. Your computing center will have a copy of this routine, or a similar one, and be able to show you how to use it (if necessary—the documentation that accompanies the routine makes it very simple to use). The routine will always require you to guess a value $\mathbf{u}_G$, near which you believe the minimum of L to lie. This is the point in m-dimensional space where the routine will start searching for the minimum. The better this guess, the sooner the minimum will be located.

As our first example of minimizing (8.31), let's take $m = 2$ and return to the problem of fitting the logistic growth model to the U.S. population data for the years 1790–1945. We have just attacked this problem by fitting a straight line to the data points in Fig. 1.7. But we could also have fitted the nonlinear logistic growth function

$$y(t) = G(t, \mathbf{u}) = G(t, u_1, u_2) = \frac{3.929u_1}{3.929u_2 + (u_1 - 3.929u_2)e^{-u_1 t}} \qquad (8.32)$$

directly to the data given in Tables 1.3 and 1.4. Here the control parameter u_1 corresponds to r in Section 1.9, and u_2 corresponds to r/K. If you supply the observed data and the function (8.32) to a routine such as ZXSSQ, then it will quickly inform you that $L(u_1, u_2)$ takes its minimum value where $u_1 = 0.316$ and $u_2 = 0.00168$, so that the optimal value of K is $u_1/u_2 = 188$. Again, these values differ from previous estimates (which provide a suitable $\mathbf{u}_G$). This time, however, the main reason for the discrepancy is that we are estimating the parameters by solving a different control problem!

For our second example, let's take $m = 3$ and return to Exercise 4.15(vii). There we conjectured that the mortality law

$$\mu(z) = ae^{cz} + b \qquad (8.33)$$

might yield a better fit to the life table data for U.S. females in 1966, for ages $z \geq 20$, than the Gompertz mortality law $\mu(z) = ae^{cz}$. The law defined by (8.33) is known by demographers as Makeham's law.

Recall that the probability $l(z)$ of surviving to age z is related to the mortality μ by

$$l(z) = \exp\left\{-\int_0^z \mu(\xi)d\xi\right\}. \qquad (8.34)$$

Let us define control parameters $\mathbf{u}$ by

$$u_1 = b, \qquad u_2 = \frac{a}{c}, \qquad u_3 = c. \qquad (8.35)$$

Then the nonlinear curve we shall fit to the life table data is given by

$$l(z) = G(z, \mathbf{u}) = G(z, u_1, u_2, u_3) = e^{-u_1 z}e^{u_2(1-e^{u_3 z})}. \qquad (8.36)$$

The observed pairs of values $\{(z_i, l_i): i = 1, 2, \ldots, 14\}$ for age and survival probability are reproduced, correct to 3 significant figures, as the second and third columns of Table 8.2. If you supply these values and the function (8.36) to a computer routine such as ZXSSQ, then it will quickly inform

Table 8.2 Fitting the mortality laws of Makeham and Gompertz to the life table data for U.S. females in 1966.

Index i	Age z_i	l_i	Survival probabilities Makeham	Gompertz
1	20	0.971	0.977	0.996
2	25	0.967	0.970	0.993
3	30	0.963	0.963	0.989
4	35	0.957	0.955	0.982
5	40	0.948	0.945	0.972
6	45	0.935	0.932	0.958
7	50	0.915	0.913	0.936
8	55	0.887	0.887	0.904
9	60	0.847	0.849	0.857
10	65	0.791	0.792	0.791
11	70	0.707	0.709	0.699
12	75	0.593	0.593	0.580
13	80	0.447	0.444	0.436
14	85	0.276	0.277	0.283

The observed values of l appear in column 3, those predicted by the best fitted Makeham law in column 4, and those predicted by the best fitted Gompertz law in column 5.

you that $L(u_1, u_2, u_3)$ takes its minimum at $\mathbf{u} = \mathbf{u}^*$, where

$$\mathbf{u}^* = 10^{-3} \times (1.087, 0.2534, 99.48)^T. \tag{8.37}$$

I actually took logarithms of (8.36) and fitted $\ln(l)$ to $u_2\{1 - \exp(u_3 z)\} - u_1 z$. My guess was $\mathbf{u}_G = (0, 0, 0.1)^T$. This was suggested by the attempt in Chapter 4 to fit Gompertz's law ($u_1 = b = 0$) to the same data. The corresponding minimum value of L is

$$L(\mathbf{u}^*) = 0.954 \times 10^{-4}. \tag{8.38}$$

The corresponding optimal values of the parameters a, b, and c are $a^* = u_2^* u_3^* = 0.252 \times 10^{-4}$, $b^* = u_1^* = 0.1087 \times 10^{-2}$, and $c^* = u_3^* = 0.9948 \times 10^{-1}$. Using the same computer routine but fixing b to be zero, we easily obtain the optimal values for fitting Gompertz's law; they are $a = 0.846 \times 10^{-4}$ and $c = 0.839 \times 10^{-1}$. The corresponding minimum value of L is 0.492×10^{-2}. Because this exceeds (8.38), we have established that Makeham's law gives a better fit to the data than Gompertz's law for ages $z \geq 20$. The predictions of Makeham's and Gompertz's laws are compared to the observations in the final columns of Table 8.2. It is clear from this comparison that Makeham's law gives an excellent fit to the data and represents a considerable improvement over the law of Gompertz.

8.5   Assigning Probabilities: A Brief Overview

There is broad agreement among mathematicians that probabilities can be assigned in at most three ways:

   i. conceptually, by theoretical arguments rooted in symmetry;

   ii. empirically, by relative frequencies of independent outcomes; (8.39)

   iii. intuitively, by inspired guesswork.

An example of (i) would be to say that the probability of obtaining a head when tossing a coin is 1/2, on the grounds that heads and tails are equally likely. An example of (ii) would be to say that the probability of a spruce budworm egg turning into a larva is 0.81, on the grounds that, on average, 38 of the 200 eggs an average female has been observed to lay have failed to hatch (because of predation, parasitism, and other phenomena); successive generations of spruce budworm can be assumed independent because adults do not survive from one breeding cycle to the next (Lack, 1980, p. 831). And an example of (iii) would be for a candidate to say, "I know in my bones I have a 60% chance of success," and then assign probability 0.6 to being successful. Who could argue? The first two ways can be described as objective: two rational people would be expected to agree on the values assigned. The third way can only be described as subjective: two rational people would be expected to disagree, however slightly.

What mathematicians and philosophers hotly dispute, however, is whether probability can be assigned in as many as three ways. When you toss a coin, for example, wouldn't theory suggest there is a very small probability, ϵ say, that the coin will land on its edge? (Have you seen the movie *Mr. Smith Goes to Washington?*) Then, by symmetry, the probability of obtaining a head should be $(1 - \epsilon)/2$, not 1/2. Therefore, when you say that Prob(Heads) = 1/2, mustn't the basis for your judgment be empirical rather than conceptual—aren't you saying that Prob(Edge) = 0 because you've never seen a coin land on its edge? This suggests that apparently conceptual assignments of probability are really empirical, after all. But are they even empirical? For example, 81% of spruce budworm eggs have been known to survive until hatching in the past. But probabilities are statements about the future. If a prophet told you that there was going to be a sudden surge in parasitism, predation, disease, or conflagration, would you still say that Prob(Egg becomes larva) = 0.81? If so, then your intuition is telling you that the prophet is wrong; and isn't the basis for your apparently empirical assignment of probability really intuitive after all? Similarly, you could argue that Prob(Heads) = 1/2 has an intuitive basis because you assume that the coin-tosser will not acquire sufficient skill as to influence the outcome of the toss, one way or the other.

Arguments such as these have led some mathematical philosophers to the view that the basis for assigning probabilities is always (iii), not (i) or (ii). Let's call these people the subjectivists. They would see nothing wrong if each of two candidates for the same position declared a 60% chance of success for herself and a 40% chance for her competitor. To subjectivists, it would be natural for two different observers to have two different sets of probabilities. Others, however—let's call them objectivists—would argue that at least one candidate would have to be wrong because $0.6 + 0.6 \neq 1$. In their view, one candidate's probability of success would have to be the other's probability of failure because there is a single "objective" probability for each event (which a supreme genius might be able to assign).

By good fortune, however, we need never decide whether subjectivists or objectivists have right on their side. Even if there did exist objective probabilities that a supreme genius would invariably assign to future events, no mortal decision maker would ever have either the theoretical insight or empirical evidence to know what they were. We will therefore adopt the view that, although (i) and (ii) may guide our thinking, the ultimate basis for assigning probabilities is (iii), simply by virtue of our ignorance. We will add, however, that there are some circumstances (e.g., tossing a coin) in which so many assignors would heartily agree with the values assigned by (i) or (ii) that the ultimate recourse to (iii) would seem rather petty. Our position, if we had to give it a name, might be labelled reluctant subjectivism.

Broadly speaking, a reluctant subjectivist would assign probabilities

1. solely according to (i) only
   (a) when there is so little information that mutually exclusive possibilities must be regarded as equally likely, or
   (b) when deducing the probability of a compound event from probabilities assigned to simpler ones;   (8.40)

2. solely according to (ii) in some circumstances;

3. by a combination of (i), (ii) and (iii) in most circumstances; or

4. solely according to (iii) only if all else fails.

As an example of (1a), suppose you knew that a woman was pregnant with one child but had absolutely no information about the child's sex. What probability would you assign to the child being a girl? This possibility excludes the possibility that the child is a boy, and together these are the only possibilities. Hence (1a) might lead you to say that Prob(Girl) = 1/2. Indeed if absolutely all that you knew was that the baby would be either a boy or girl, then what else could you assign? On the other hand, you might know that approximately 105 boys are born for every 100 girls[3] and prefer to assign Prob(Girl) = $100/205 \approx 0.4878$. But then you would have too much information to apply (1a)!

---

[3]*Encyclopaedia Britannica*, 15th edition, 1980, Volume 14, p. 817.

As an example of (1b), suppose that you are conducting a sequence of N independent trials, for each of which the probability of success is known to be p. Then the possibility of r successes in the N trials is

$$\frac{N!}{(N-r)!r!}p^r(1-p)^{N-r}. \tag{8.41}$$

Note, however, that the value of p itself must have been assigned by one of the other methods.

Examples of (2) and (3) are the subject of the following two sections. We conclude the present one with a remark about (4). Whenever possible, a purely subjective probability should remain a parameter in the analysis. This is what happened in Section 8.1, where the probability that pirates roamed in the waters around Nands was retained in the analysis as the parameter q, with the evident advantage that we were ultimately able to give q its worst possible value, thus minimizing the effects of our ignorance. Even if its worst possible value cannot be found, retaining a subjective probability as a parameter adds to a model's usefulness by keeping it flexible, as discussed in Sections 4.15 and 7.8. Further examples of retaining subjective probabilities as parameters will appear in Chapter 12.

8.6 Empirical Probability Assignment

We have so far encountered two mathematical models for which probabilities were assigned purely by empirical means, namely, the life table model in Exercises 4.15–4.17 and Exercise 5.11, and the workforce model in Section 6.6.[4] We omitted, however, to describe the data from which the relative frequencies were computed. The purpose of this section is to make amends. These two examples, together with an additional one, should suffice to indicate how probabilities are estimated for purely empirical distributions (parameter estimation for theoretical distributions is the subject of the following section). We begin with the life table.

You will recall from Exercise 4.15 that the probability $l(z)$ of survival to age z can be identified with the fraction of a birth cohort that survives to age z, and is therefore related to the specific death rate (or force of mortality) $\mu(z)$ by

$$-\frac{1}{l(z)}\frac{dl}{dz} = \mu(z). \tag{8.42}$$

You will also recall that life tables specify $l(z)$ only for certain discrete values of age. Let two consecutive such ages be denoted by $z = a$ and $z = a + h$. Then $l'(a + h/2)$ can be approximated by $\{l(a+h) - l(a)\}/h$, and $l(a + h/2)$ can be approximated by $\{l(a+h) + l(a)\}/2$. It is easy to show,

---

[4]You may wish to review Section 6.6 before proceeding.

by using a Taylor expansion of $l(z)$ about $z = a + h/2$, that the error involved in each approximation is $O(h^2)$. But $O(h^2)$ is not a small number if $h = 5$, as is so often the case in demographic work. Thus the approximation requires more careful justification. This is a book about mathematical modelling, not numerical analysis; therefore we will simply observe that each approximation would be exact if $l = l(z)$ were a linear function, and is therefore adequate if $l = l(z)$ does not depart too much from linearity. Thus, evaluating (8.42) at $z = a + h/2$, we have

$$\frac{2}{h} \frac{l(a) - l(a+h)}{l(a) + l(a+h)} \approx \mu\left(a + \frac{h}{2}\right), \tag{8.43}$$

whence

$$\frac{l(a+h)}{l(a)} \approx \frac{1 - \frac{h}{2}\mu\left(a + \frac{h}{2}\right)}{1 + \frac{h}{2}\mu\left(a + \frac{h}{2}\right)}. \tag{8.44}$$

Because $l(0) = 1$, approximations to the survival probabilities can be generated recursively from (8.44), provided that $\mu(a + h/2)$ can be estimated from census data.

Suppose that there were $C(z)$ females aged z in the 1966 U.S. population, in the sense that the number aged between a and $a + h$ would have been

$$\int_a^{a+h} C(z)dz. \tag{8.45}$$

Then the number of deaths between those ages would have been

$$\int_a^{a+h} \mu(z)C(z)dz. \tag{8.46}$$

Hence the fraction dying between ages a and $a + h$ would have been

$$\frac{\int_a^{a+h} \mu(z)C(z)dz}{\int_a^{a+h} C(z)dz}. \tag{8.47}$$

A rough approximation to the integral of $f(z)$ between $z = z_1$ and $z = z_2$ is given by the formula that earlier yielded (3.91):

$$\int_{z_1}^{z_2} f(z)dz \approx (z_2 - z_1)f\left(\frac{1}{2}(z_1 + z_2)\right). \tag{8.48}$$

This formula is exact if f varies linearly between $z = z_1$ and $z = z_2$; and the more linearly it varies, the better the approximation. As a special case, the formula is exact if f is constant; and the more nearly constant f is, the better the approximation. Now, suppose that population is more or less uniformly distributed between a and $a + h$, while force of mortality varies more or less linearly. Then $C(z)$ is approximately constant and $\mu(z)$ approximately linear, so that $\mu(z)C(z)$ is also approximately linear. Then both numerator and denominator of (8.47) are adequately approximated by (8.48); whence the fraction dying between ages a and $a + h$ is approximately

$$\frac{h\mu \left(a + \frac{h}{2}\right) C \left(a + \frac{h}{2}\right)}{hC \left(a + \frac{h}{2}\right)} = \mu \left(a + \frac{h}{2}\right). \tag{8.49}$$

Census data for U.S. females in 1966 are presented in Table 8.3. From the first line of the table, we see that 1,793,000 U.S. infant girls were alive in 1966 (at mid-year) and that 36,353 died. The fraction of infant girls dying was therefore

$$\frac{36,353}{1,793,000} \approx 0.02027.$$

Hence on using (8.49) with $a = 0$ and $h = 1$, we obtain

$$\mu(0.5) \approx 0.02027. \tag{8.50a}$$

Similarly, on using (8.49) with $a = 1$ and $h = 4$, we get

$$\mu(3) \approx \frac{6,706}{7,922,000} \approx 0.00085, \tag{8.50b}$$

and with $a = 5$ and $h = 5$ we get

$$\mu(7.5) \approx \frac{3,752}{10,226,000} \approx 0.00037. \tag{8.50c}$$

Table 8.3 Mortality data for U.S. females in 1966. Source: Keyfitz and Flieger, 1971, p. 354.

Age last birthday	Number of females (thousands)	Number of deaths
0	1793	36353
1–4	7922	6706
5–9	10226	3752
10–14	9542	2906
15–19	8806	5171
20–24	6981	4981
25–29	5840	5137
30–34	5527	7010
35–39	5987	10958
40–44	6371	17841
45–49	5978	25363
50–54	5498	34308
55–59	4839	44368
60–64	4174	56592
65–69	3476	77775
70–74	2929	102041
75–79	2124	118314
80–84	1230	115335

Continuing in this manner, we obtain the set of approximations to $\mu(z)$ contained in Table 8.4. From (8.44) with $a = 0$ and $h = 1$ we obtain

$$l(1) \approx \frac{1 - 0.5 \times 0.02027}{1 + 0.5 \times 0.02027} \approx 0.9800. \tag{8.51a}$$

From (8.44) with $a = 1$ and $h = 4$ we obtain

$$l(5) \approx \left(\frac{1 - 2 \times 0.00085}{1 + 2 \times 0.00085} \right) l(1) \approx 0.9767. \tag{8.51b}$$

From (8.44) with $a = 5$ and $h = 5$ we obtain

$$l(10) \approx \left(\frac{1 - 2.5 \times 0.00037}{1 + 2.5 \times 0.00037} \right) l(5) \approx 0.9749. \tag{8.51c}$$

Continuing in this way, we easily obtain the life table presented in Table 8.5, which you should verify for yourself by using a calculator (Exercise 8.6).

The final column of Table 8.5 agrees with that computed by Keyfitz and Flieger (1971, p. 354) to between two and four significant figures. The discrepancy is partly due to rounding error, but mainly due to the fact that demographers use more accurate approximations than (8.43) and (8.48). The values appearing in Table 8.5 are adequate for many purposes, however, and the accurate computation of life tables is as much as a matter of numerical analysis as of mathematical modelling. We will therefore not pursue it further; see Keyfitz (1968) and Maron (1982) for further details.[5]

Table 8.5 can be used to construct the transition matrix for Exercise 5.11. Let U be the event that an individual survives to age a, and V the event that she survives to age $a + h$. Then, clearly, $U \cap V = V$. Thus the conditional probability of surviving to age $a + h$, given that one has already survived to age a, is

$$\text{Prob}(V \mid U) = \frac{\text{Prob}(V \cap U)}{\text{Prob}(U)} = \frac{\text{Prob}(V)}{\text{Prob}(U)} = \frac{l(a + h)}{l(a)}. \tag{8.52}$$

Table 8.4 Mortality estimates.

z	$\mu(z)$	z	$\mu(z)$
0.5	0.02027	42.5	0.00280
3.0	0.00085	47.5	0.00424
7.5	0.00037	52.5	0.00624
12.5	0.00030	57.5	0.00917
17.5	0.00059	62.5	0.01356
22.5	0.00071	67.5	0.02237
27.5	0.00088	72.5	0.03484
32.5	0.00127	77.5	0.05570
37.5	0.00183	82.5	0.09377

[5]See Supplementary Note 8.1.

Table 8.5 Life table for
U.S. females, 1966.

z	$\dfrac{l(z)}{l(z-5)}$	$l(z)$
0		1.0000
1		0.9800
5	0.9767	0.9767
10	0.9982	0.9749
15	0.9985	0.9734
20	0.9971	0.9705
25	0.9965	0.9671
30	0.9956	0.9629
35	0.9937	0.9568
40	0.9909	0.9481
45	0.9861	0.9349
50	0.9790	0.9153
55	0.9693	0.8872
60	0.9552	0.8474
65	0.9344	0.7918
70	0.8941	0.7079
75	0.8398	0.5945
80	0.7555	0.4492
85	0.6202	0.2786

The values would differ slightly from those given in Exercise 5.11, however, for the reasons already stated. We shall need (8.52) in Section 9.7.

As our second example of purely empirical probability assignment, we will return to our Markov model for the structure of a workforce and estimate its transition matrix **P**. You will recall from Section 6.6 that there are seven states, denoting seven classes of occupation, and that $X(n)$ denotes the occupational class of an individual chosen randomly from the nth generation. Thus

$$p_{ij} = \text{Prob}(X(n+1) = j \mid X(n) = i) \tag{8.53}$$

is the probability that a son or daughter of a member of occupational class i will become a member of occupational class j. For convenience, the class definitions are recorded in Table 8.6.

We should first explain that the transition matrix in (6.63) was estimated, by David Glass and John Hall, about 1953; see Glass (1954, p. 183). At that time, the workforce was still predominantly male. Accordingly, Glass and Hall's data was based on a sample of 3497 pairs of fathers and sons and excluded the female workforce. Nowadays, you would replace father by parent(s), son by son or daughter (as reflected in Section 6.6), and you would have to exclude from your data any parent-son/daughter pairings where two parents belonged to separate occupational classes. The

Table 8.6 Numbers of fathers and sons in various occupational classes.

State	Class of occupation	Number of fathers	Number of sons
0	Professional and higher administrative	129	103
1	Managerial and executive	150	159
2	Higher grade supervisory and nonmanual	345	330
3	Lower grade supervisory and nonmanual	518	459
4	Skilled manual and routine nonmanual	1510	1429
5	Semi-skilled manual	458	593
6	Unskilled manual	387	424

numbers of fathers and sons in each class in the sample of 3497 pairs are shown in Table 8.6. Transitions between classes are recorded in Table 8.7. Here ν_i denotes the number of fathers in class i, and ν_{ij} denotes the number of fathers in class i with sons in class j. Thus, for example, of the $\nu_0 = 129$ fathers in class 0, $\nu_{04} = 18$ had sons in class 4; and of the $\nu_4 = 1510$ fathers in class 4, $\nu_{40} = 14$ had sons in class 0. The proportion of class 0 fathers with class 4 sons is therefore 18/129, and the proportion of class 4 fathers with class 0 sons is 14/1510. Proceeding in this manner, these and the 47 remaining proportions yield the matrix

$$
\begin{bmatrix}
50/129 & \underline{19/129} & 26/129 & 8/129 & 18/129 & 6/129 & \underline{2/129} \\
16/150 & 40/150 & 34/150 & 18/150 & \underline{31/150} & 8/150 & 3/150 \\
12/345 & 35/345 & 65/345 & 66/345 & 123/345 & 23/345 & 21/345 \\
11/518 & 20/518 & 58/518 & 110/518 & \underline{223/518} & 64/518 & 32/518 \\
14/1510 & 36/1510 & 114/1510 & 185/1510 & 714/1510 & 258/1510 & 189/1510 \\
0/458 & 6/458 & 19/458 & \underline{40/458} & 179/458 & 143/458 & 71/458 \\
0/387 & 3/387 & 14/387 & 32/387 & 141/387 & 91/387 & 106/387
\end{bmatrix}
$$

$$(8.54)$$

with ν_{ij}/ν_i in the jth column of row i. By evaluating the above fractions to three decimal places, you can check that (8.54) is equivalent to the transition matrix of (6.63). The apparent discrepancies between the underlined elements and the corresponding ones in (6.63) are simply due to the fact that the probabilities in each row of a transition matrix must sum to 1. When first rounded to three decimal places, the first row of (8.54) sums to 1.002, and rows 2, 4, and 6 each sum to 1.001. Accordingly, Glass and Hall made corrections to ensure that the estimate in (6.63) was indeed a transition matrix. (Why they chose to round 19/129 to 0.146, when it would appear more consistent to leave it as 0.147 and round down either the fifth or sixth element of the first row instead is a mystery to me but of little consequence; (6.64) obtains in either case.)

Table 8.7 Transitions between occupational classes over one generation.

Father's class i	Son's class ν_{i0}	ν_{i1}	ν_{i2}	ν_{i3}	ν_{i4}	ν_{i5}	ν_{i6}	Number of fathers ν_i
0	50	19	26	8	18	6	2	129
1	16	40	34	18	31	8	3	150
2	12	35	65	66	123	23	21	345
3	11	20	58	110	223	64	32	518
4	14	36	114	185	714	258	189	1510
5	0	6	19	40	179	143	71	458
6	0	3	14	32	141	91	106	387
Number of sons	103	159	330	459	1429	593	424	3497

Now that the origin of (6.63) has been revealed, let's understand what we have actually done. We have

1. assumed that intergenerational movement of the work force between occupational classes is a Markov process, i.e., that a son's occupational class in no way depends upon his grandfather's; and (8.55)
2. having made this assumption, we have used ν_{ij}/ν_i to estimate (8.53) from a sample of 3497 actual transitions.

The first step is a matter of choice, the second a matter of estimation; and a few remarks are in order.

If (1) is true, then p_{ij} exists quite independently of the sample data on 3497 transitions. All we have done is to *estimate* it. If our sample had consisted of a different 3497 pairs of fathers and sons, then our estimate would almost certainly have differed, too. Let's suppose that (1) is true and (with due respect to subjectivists) further suppose that there is some "true" probability of moving from class i to class j in one generation. Then this probability, which a supreme genius might be able to supply, is the objective one, and ν_{ij}/ν_i is merely our subjective view of it. The subjectiveness is introduced by the process of sampling. But we would like to think that the estimate $\nu_{ij}^{(2)}/\nu_i^{(2)}$ from a second sample of 3497 transitions would not differ greatly from ν_{ij}/ν_i and that $(\nu_{ij}^{(2)} + \nu_{ij})/(\nu_i^{(2)} + \nu_i)$ would be even better still. Indeed the estimation procedure we have used has the very desirable property of "maximum likelihood," as discussed in the following section.

But how can we be sure that (1) was an appropriate choice to begin with? How can we be sure that a son's occupational class depends only on his father's, not on whether his father rose or fell in rank? To justify our choice—or, which is exactly the same thing, to validate our model—we would have to propose tests for it in accordance with the guidelines devised in Chapter 4. One test could be based on the supposition that if such a Markov process had been operating for several generations, then π should now be approaching a stationary distribution. Two successive estimates of π, the jth entry of which records the probability that a worker chosen at random will be in occupational class j, are obtainable immediately from Table 8.6. We have

$$\pi(0)^T \approx \frac{1}{3497}(129, 150, 345, 518, 1510, 458, 387)$$
$$\approx (0.037, 0.043, 0.099, 0.148, 0.432, 0.131, 0.111), \quad (8.56)$$
$$\pi(1)^T \approx \frac{1}{3497}(103, 159, 330, 459, 1429, 593, 424)$$
$$\approx (0.029, 0.045, 0.094, 0.131, 0.409, 0.170, 0.121). \quad (8.57)$$

The relative closeness of (8.57) to (6.64) is some indication that our choice is at least along the right lines, though the appropriateness of the class definitions in Table 8.6 might be subject to further debate: see the remarks at the end of this section.

The success of the transition matrix estimation procedure we have just discussed is strongly dependent upon having a detailed data set, such as that contained in Table 8.7. If the only available information concerns state vectors, such as those in Table 8.6, then a different estimation procedure must be used. To illustrate this, we will consider a simple example.

Table 8.8 describes the movement through a hospital in New Haven, Connecticut, of 471 coronary patients. The data set, condensed from Kao (1974, p. 688), shows the number of coronary patients at three levels of care after a given number of moves between levels. The first level, labelled $i = 0$, contains the coronary care unit (CCU), the post-coronary care unit (PCCU), the intensive care unit (ICU), and the medical unit (MED). The second level, labelled $i = 1$, contains the surgical unit (SURG) and ambulatory unit (AMB). The third level, labelled $i = 2$, contains the extended care facility (ECF) and patients who have either gone home or died. In other words, level 2 contains all those patients who have left the hospital's heart-patient facility. Is it possible that the levels of care could be states of a Markov chain, with moves corresponding to transitions? If so, how do we estimate the transition matrix?

Table 8.8 Movement of patients through coronary facility.

Level of care, i	Number of moves, n				
	0	1	2	3	4
0	471	417	26	12	0
1	0	5	6	0	0
2	0	49	439	459	471

Notice how this problem differs from the one we have just considered. There, in the case of the workforce, we had information on how individuals moved between states. Here, we know how many patients occupy each state after each transition, but we do not know where they came from. Let us assume nevertheless that patient movements do indeed form a three-state Markov chain, with time being measured nonlinearly. Our problem is then to estimate the transition matrix $\mathbf{P}$ from knowledge of patient numbers at each level after n moves, for $n = 0, 1, 2, 3$, and 4. In fact, we will estimate $\mathbf{P}$ from the data for $n = 0, 1, 2$, and 3. Thus data for $n = 4$ could be used as a test of our model.

Notice that all patients enter the hospital at level 0. Hence

$$\pi(0) = (1, 0, 0)^T. \tag{8.58}$$

Moreover, patients do not re-enter the coronary facility after leaving it (at least not as part of this cohort of 471 patients). Hence, on using the fact that the rows of a transition matrix must sum to 1, the most general allowable form of the 3×3 transition matrix is

$$\mathbf{P} = \begin{bmatrix} u_1 & u_2 & 1 - u_1 - u_2 \\ u_3 & u_4 & 1 - u_3 - u_4 \\ 0 & 0 & 1 \end{bmatrix}, \tag{8.59}$$

where the entries of the vector $\mathbf{u} = (u_1, u_2, u_3, u_4)^T$, satisfying $0 \le u_1 \le 1$, $0 \le u_2 \le 1$, $0 \le u_3 \le 1$, and $0 \le u_4 \le 1$, are to be determined. On using (5.83) and (8.58), we find that

$$\pi(1)^T = \pi(0)^T \mathbf{P} = (u_1, u_2, 1 - u_1 - u_2)^T, \tag{8.60}$$

$$\pi(2) = \begin{bmatrix} u_1^2 + u_2 u_3 \\ u_2(u_1 + u_4) \\ (1 - u_1 - u_2)(u_1 + 1) + u_2(1 - u_3 - u_4) \end{bmatrix}, \tag{8.61}$$

$$\pi(3) = \begin{bmatrix} u_1(u_1^2 + u_2 u_3) + u_2 u_3(u_1 + u_4) \\ u_2 \{ u_1^2 + u_2 u_3 + u_4(1 + u_4) \} \\ (1 - u_1 - u_2)(u_1^2 + u_2 u_3 + u_1 + 1) \\ + u_2(1 - u_3 - u_4)(u_1 + u_4 + 1) \end{bmatrix}. \tag{8.62}$$

The model predicts that there will be $471\pi_k(n)$ patients at level k at time n. Let O_{kn} be the observed number of patients at level k at time n. If patient movements do indeed correspond to a Markov chain, then $471\pi_k(n)$ should be close to O_{kn}. Thus the differences $471\pi_k(n) - O_{kn}$ should be close to zero for all values of k and n. From (8.60)–(8.62), these differences are functions of u_1, u_2, u_3, and u_4; hence so is the sum of their squares. If we define

$$L(\mathbf{u}) = L(u_1, u_2, u_3, u_4) = \sum_{n=1}^{3} \sum_{k=1}^{3} \{ \pi_k(n) - O_{kn} \}^2, \tag{8.63}$$

then L is a quantity that should be close to zero if Table 8.8 is evidence of a Markov chain; and the closer L is to zero, the better the fit between theory and observation. Thus we calculate a best estimate of the matrix $\mathbf{P}$ by minimizing the function L.

As discussed in Section 8.4, minimizing L is a problem in nonlinear programming. If we blindly use a computer package like the IMSL routine ZXSSQ to minimize (8.63), then it will quickly inform us that the minimum occurs at $\mathbf{u} = \mathbf{u}^*$, where, to four significant figures,

$$\mathbf{u}^* = (0.8851, 0.01111, -65.01, -0.8446)^T \tag{8.64}$$

and that $f(\mathbf{u}^*) = 0.0002271$. This is clearly an absurd result because it fails to satisfy

$$0 \le u_j \le 1, \qquad 1 \le j \le 4, \tag{8.65}$$

and highlights the importance of constraints on the decision variables in nonlinear programming. Properly posed, the problem we face is the *constrained* nonlinear programming problem

$$\text{minimize } L(\mathbf{u}) = L(u_1, u_2, u_3, u_4) = \sum_{n=1}^{3} \sum_{k=1}^{3} \{\pi_k(n) - O_{kn}\}^2, \tag{8.66}$$

subject to

$$0 \le u_j \le 1, \qquad j = 1, 2, 3, 4, \tag{8.67}$$
$$1 - u_1 - u_2 \ge 0, \tag{8.68}$$
$$1 - u_3 - u_4 \ge 0. \tag{8.69}$$

This is just a special case of the more general nonlinear programming problem

$$\text{minimize } L(\mathbf{u}) = L(u_1, u_2, \ldots, u_n),$$

subject to

$$g_i(u_1, u_2, \ldots, u_n) \ge 0, \qquad i = 1, \ldots, m, \tag{8.70}$$
$$h_j(u_1, u_2, \ldots, u_n) = 0, \qquad j = 1, \ldots, q,$$

where g_i, $1 \le i \le m$, and h_j, $1 \le j \le q$, are (usually) differentiable functions of their arguments. You can easily verify that the linear programming problem (3.20) is just a special case of (8.70).

Constrained nonlinear programming remains an active area of ongoing research. The state of the art is nowhere near as complete as for linear programming, however, and computer routines for solving (8.70) are neither as common nor as reliable as for (3.20). Nevertheless, methods do exist for solving (8.70), among the most efficient being those known alternatively as *augmented Lagrangian* or *multiplier* methods. A discussion of these would be quite beyond our brief, however, and the interested reader is referred to Jeter (1986, Chapters 9 and 10) or Luenberger (1984, Chapters 10, 12, and 13).

Your computer library may not have a routine that copes with (8.70). Nevertheless, by using a multiplier method, it is possible to show that the minimum of (8.66) subject to (8.67)–(8.69) occurs at $\mathbf{u} = \mathbf{u}^*$, where, to four significant figures

$$\mathbf{u}^* = (0.5028, 0.1107, 0, 0)^T \qquad (8.71)$$

and $f(\mathbf{u}^*) = 0.3628$. Notice, however, that constraints (8.68) and (8.69) are inactive at the solution, so that we would obtain the same answer if we simply ignored them. The problem of minimizing (8.66) subject to (8.67) alone is much more likely to be covered by a library routine. In particular, if you have access to the IMSL package ZXMWD, then you can verify (8.71) for yourself.

On substituting (8.71) into (8.59), we find that our best estimate of $\mathbf{P}$ is

$$\mathbf{P} = \begin{bmatrix} 0.5028 & 0.1107 & 0.3865 \\ 0 & 0 & 1 \\ 0 & 0 & 1 \end{bmatrix}. \qquad (8.72)$$

On using (8.60)–(8.62), our model predicts that there will be $471\pi_k(n)$ patients at level k at time n, where $471\pi_k(n)$ is given in Table 8.9 for $1 \le k \le 3$, $1 \le n \le 4$. Comparing with Table 8.8, we see that the Markov model yields a very poor fit to the observations, and we have no hesitation in rejecting it.

Recall, however, that our three levels of health care were an amalgamation of nine: CCU, PCCU, ICU, MED, SURG, AMB, ECF, HOME, and DIED. If we knew patient numbers at each of these nine levels after a given number of moves, then we could attempt to model the coronary care unit as a nine-state Markov chain, with the 9×9 transition matrix $\mathbf{P}$ being estimated by the same procedure as we used to determine (8.72). It turns out (Exercise 8.7) that the fit is then a good one. Thus whether or not a dynamical system has the Markov property depends upon how we define the states; and it is wise to group data in as natural a way as possible.

Table 8.9 Predictions by best fit Markov model of movements of patients through coronary facility.

Level of care, k	Number of moves, n				
	0	1	2	3	4
0	417	237	119	60	30
1	0	52	26	13	7
2	0	182	326	398	434

8.7 Choosing Theoretical Distributions and Estimating Their Parameters

In Section 7.1 we constructed a buying model in which demand for a product was a random variable, there denoted by Z. Allowing the random variable to have an arbitrary distribution gave the model flexibility, but the model could not be used until the distribution had actually been specified, i.e., until probability had been assigned. Accordingly, we will now discuss the assignment of a distribution to a random variable.

We will begin with some definitions. Let X_1, X_2, X_3, ..., X_n denote n random variables, e.g., demands for a product in the last n buying periods. Assume that these random variables are independent and all have the same distribution as the random variable, X, whose distribution we wish to assign. Then a sample from that distribution consists of n realizations, one for each of the random variables X_1, X_2, X_3, ..., X_n, and is denoted by $\{x_1, x_2, ..., x_n\}$. Because X_1, X_2, X_3, ..., X_n are random variables, so are the quantities

$$\bar{X} \equiv \frac{1}{n} \sum_{i=1}^{n} X_i, \qquad (8.73)$$

$$S^2 \equiv \frac{1}{n-1} \sum_{i=1}^{n} (X_i - \bar{X})^2. \qquad (8.74)$$

The value the random variable (8.73) actually takes, or realizes, i.e.,

$$\bar{x} \equiv \frac{1}{n} \sum_{i=1}^{n} x_i, \qquad (8.75)$$

is known as the *sample mean*, and the value

$$s^2 \equiv \frac{1}{n-1} \sum_{i=1}^{n} (x_i - \bar{x})^2, \qquad (8.76)$$

that (8.74) realizes is known as the *sample variance*. The random variable S^2 has the desirable property that $E[S^2] = \sigma^2$. This simply means that if we take a sequence of samples from the distribution, and hence obtain a sequence of sample variances, then the average value of the latter sequence should be close to the value of σ^2. For large n, however (and it is always desirable that n be as large as possible), the divisor in (8.76) is approximately n, so that the sample variance is approximately the sample's average squared deviation from its mean.

As remarked in Section 8.5, assigning distributions may involve all of (8.39)'s three methods of probability assignment; namely, the conceptual, the empirical, and the intuitive. The principal determinant of the relative weight given each of the three methods is the size of any sample taken from the distribution. Broadly speaking, the larger the value of n, the more empirical and objective the assignment; the smaller the value of n, the more intuitive and subjective the assignment. Examples below will illustrate this. In all cases, however, the assignment will involve the following two steps, the first a matter of choice, the second a matter of estimation:

1. First choose a known theoretical probability distribution from which all probabilities can be derived immediately once values have been assigned to one or two parameters.   (8.77a)
2. Estimate the value(s) of the free parameter(s).   (8.77b)

In principle, one could use theoretical distributions with more than two free parameters; but their use is rarely justified in practice, and we will avoid them altogether.

A theoretical distribution embodying everything that is known about a certain random variable (e.g., demand) may properly be regarded as a model in its own right—a *statistical model*, as it is often called. This model can then be used as an input to another model, such as those encountered in Chapter 7. It becomes, in the process, a model within a model, just like the behavioral response functions of Section 8.3. Indeed the first of the statistical models derived in this section has effectively been used as an input already, in Section 5.4; and the second will be used as an input in Section 10.6.

When the sample size n is large, the frequency distribution of the sample—i.e., the frequencies with which various values of x occur in the sample—is used to perform step (8.77a); and (8.77b) is most easily achieved by setting the mean and/or variance of the distribution equal to the sample mean and/or variance. An example will clarify what is meant by this. Let X denote the time gap between cars in a stream of traffic, and let X_1, X_2, X_3, ..., X_{214} denote 214 particular time gaps that in a certain experiment realized values $\{x_1, x_2, x_3, \ldots, x_{214}\}$. The experiment in question was performed on the Arroyo Seco Freeway in October, 1950, between 2:00 and 2:30 in the afternoon. The 214 observed gaps totalled 1753 seconds. Thus

$$\sum_{k=1}^{214} x_k = 1753. \tag{8.78}$$

The sample frequency distribution is presented in Table 8.10. From this information we can easily deduce the *cumulative frequency distribution of the sample*, denoted by $F_s(x)$, i.e., the frequency with which gaps in the sample were less than or equal to x. For example, because 43 gaps were less than or equal to 2 seconds, we have $F_s(2) = 43$; and because 182 gaps were less than or equal to 14 seconds, we have $F_s(14) = 182$. Naturally, $F_s(0) = 0$, and because no gap exceeded 31 seconds, $F_s(31) = 214$. The sample cumulative frequency distribution is plotted in Fig. 8.5.

Table 8.10   Time gaps between cars in an experiment on the Arroyo Seco Freeway. The data are due to Gerlough and Barnes (1971, p. 39).

Gap x satisfying	Observed frequency	Gap x satisfying	Observed frequency
$0 < x \leq 1$	29	$16 < x \leq 17$	7
$1 < x \leq 2$	14	$17 < x \leq 18$	3
$2 < x \leq 3$	22	$18 < x \leq 19$	2
$3 < x \leq 4$	13	$19 < x \leq 20$	3
$4 < x \leq 5$	11	$20 < x \leq 21$	1
$5 < x \leq 6$	14	$21 < x \leq 22$	1
$6 < x \leq 7$	16	$22 < x \leq 23$	1
$7 < x \leq 8$	11	$23 < x \leq 24$	0
$8 < x \leq 9$	12	$24 < x \leq 25$	1
$9 < x \leq 10$	11	$25 < x \leq 26$	0
$10 < x \leq 11$	9	$26 < x \leq 27$	1
$11 < x \leq 12$	12	$27 < x \leq 28$	2
$12 < x \leq 13$	6	$28 < x \leq 29$	1
$13 < x \leq 14$	2	$29 < x \leq 30$	2
$14 < x \leq 15$	3	$30 < x \leq 31$	1
$15 < x \leq 16$	3		

Our sample is large, and so we would expect $F_s(x)$ to resemble $214 \cdot$ Prob$(X \leq x)$. Multiplication by a constant does not alter the shape of a curve; hence we would expect $F_s(x)$ to have the shape of Prob$(X \leq x)$. But $F_s(x)$ has a similar shape to the cumulative distribution function

$$\text{Prob}(X \leq x) = 1 - e^{-\lambda x} \qquad (8.79)$$

of the exponential distribution as sketched in Fig. 5.1, suggesting that the random variable X is distributed according to (8.79). This completes (8.77a) for the problem at hand.

Because the exponential has a single free parameter, namely, λ, we estimate it by equating the mean of the distribution to the mean of the sample. But the mean of the exponential distribution is $1/\lambda$, by (5.33), and the mean of the sample is

$$\frac{1}{214} \sum_{k=1}^{214} x_k. \qquad (8.80)$$

Thus, from (8.78) and (8.80), we estimate λ as $214/1753 = 0.122$. Having chosen a distribution and estimated its free parameter, we now propose

$$\text{Prob}(X \leq x) = 1 - e^{-214x/1753} \qquad (8.81)$$

as a statistical model of time gaps in traffic on the Arroyo Seco freeway. Note that if $F_s(x)$ had suggested a two-parameter theoretical distribution

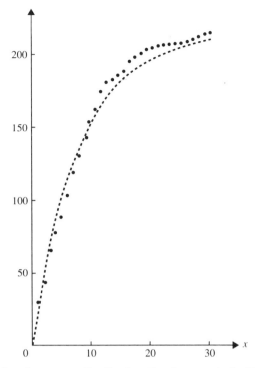

Fig. 8.5 Cumulative frequency distribution for the sample in Table 8.10, denoted by dots, together with the graph of $214 \cdot \text{Prob}(X \leq x)$, denoted by the dashed curve, where $\text{Prob}(X \leq x)$ is defined by (8.81).

(such as the normal distribution), then we would have estimated the parameters by setting both mean and variance of the distribution equal, respectively, to the sample mean and variance; see Exercise 8.8.

How good is the statistical model (8.81)? The function $214 \cdot \text{Prob}(X \leq x) = 214(1 - e^{-0.122x})$ is plotted in Fig. 8.5 alongside $F_s(x)$. The agreement appears to be quite acceptable. It is possible to be more precise than this, however; and we now provide a brief general description of a suitable "goodness of fit" test.

Let the sample space of a random variable X be subdivided into k intervals, and let p_j denote the theoretical probability that the value of X will lie in class j, $1 \leq j \leq k$. Suppose that a sample of size n is about to be taken (i.e., we are about to obtain n realizations of n random variables with the same distribution as X). Now, after the sampling has been performed, the frequency of realizations in class j, denoted by o_j, will be fully determinate. Before the sampling, however, the frequency of observations in class j is a random variable, which we will denote by O_j, $1 \leq j \leq k$. Note that the random variables $O_1, O_2, O_3, \ldots, O_k$ are not independent because

$$O_1 + O_2 + O_3 + \cdots + O_k = n. \tag{8.82}$$

Now, the quantity

$$\chi^2 \equiv \sum_{j=1}^{k} \frac{(O_j - np_j)^2}{np_j}, \tag{8.83}$$

being a function of random variables, is itself a random variable, with the positive real numbers as its sample space; but its distribution is extremely complicated. If n is large, however, then it can be shown that χ^2 has approximately the distribution with probability density function

$$f_\nu(\xi) \equiv \frac{z^{\nu/2-1}e^{-z/2}}{2^{\nu/2}\Gamma\left(\frac{\nu}{2}\right)}, \tag{8.84}$$

where

$$\nu \equiv k - m - 1, \tag{8.85}$$

where the *Gamma function* Γ is defined by[6]

$$\Gamma(s) \equiv \int_0^\infty x^{s-1}e^{-x}dx, \tag{8.86}$$

for $s > 0$, and where m is the number of parameters to be estimated in (8.77b); see, for example, Meyer (1972, p. 333).

    The distribution with probability density function f_ν is known as the *Chi-squared distribution* with ν *degrees of freedom*. A typical density function, namely f_{17}, is sketched in Fig. 8.6; f_ν has a similar shape whenever

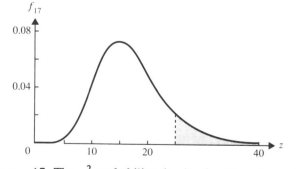

Fig. 8.6 f_ν for $\nu = 17$. The χ^2 probability density function always takes its maximum where $z = \nu - 2$, in this case 15.

---

[6]The Gamma function can be shown to satisfy $\Gamma(s) = (s - 1)\Gamma(s - 1)$. Thus $\Gamma(s) = (s-1)!$ if s is an integer, because $\Gamma(1)$ is clearly 1. It can also be shown that $\Gamma(1/2) = \sqrt{\pi}$, a result that is needed for (8.88).

$\nu > 2$. The Chi-squared distribution is widely tabulated; see for example Lindley and Miller (1953, p. 7). Hence we can compute the probability that if a large number of samples were taken from the distribution, then χ^2 would exceed the value it actually realized in our particular sample. If this probability is exceedingly small, then we can say that a value of χ^2 as high as or higher than the one that occurred could almost certainly not be due to chance. Then a value of χ^2 as large as the one we have obtained must be due to the fact that our statistical model does not provide a good fit to the data. We should not then accept it. But if this probability is more than, say, 0.1, then there is, say, a 1 in 10 chance that differences between theoretical frequencies and expected frequencies could be due to random variation; and we have no good reason to reject our statistical model.

Before we apply the Chi-squared test to (8.81) and Table 8.10, a few remarks are in order. First, the derivation of the above result would take us far beyond our brief; the interested reader is referred to Meyer (op. cit.). Second, because the supporting theory strictly applies only in the limit as $n \to \infty$, it isn't possible to say precisely how large n must be for the Chi-squared test to be applicable; but statisticians appear to agree that it is unwise to apply it if n is less than about 50 and that it is desirable for n to exceed 200. Third, and most important, applied statisticians usually insist that the sample space should be subdivided into intervals satisfying

$$np_j \geq 5, \qquad 1 \leq j \leq k. \tag{8.87}$$

We will follow this convention in our example, even though, as remarked by Kendall and Stuart (1967, p. 440), there seems to be no theoretical underpinning for (8.87); and circumstances are known when the validity of the Chi-squared test does not depend on it. In practice, if (8.87) is violated when intervals are first chosen, then a fresh subdivision is obtained by combining neighboring intervals, as in the following example.

We now return to our statistical model (8.81). Because $k = 19$, from Table 8.11, and because we have estimated a single parameter, namely λ, it follows from (8.85) that $\nu = 17$. Hence χ^2 has the distribution with density function

$$f_{17}(\xi) = \frac{7!}{15!} \sqrt{\frac{2^{13} \xi^{15}}{\pi e^\xi}}. \tag{8.88}$$

This is plotted in Fig. 8.6. The value actually realized by χ^2 is calculated in Table 8.11 as 22.06. From Lindley and Miller (1953), the probability that χ^2 exceeds 24.77 is 0.1. This area is shaded in Fig. 8.6. Because 22.06 lies in the unshaded region, there is a greater probability than 0.1 that differences between the observed frequencies o_j and expected frequencies np_j are due purely to chance. We should therefore not reject our statistical model. On the other hand, we would have been happier if the value of χ^2 lay somewhere near the peak of the distribution at $z = 15$, i.e., if the value

Table 8.11 Calculation of value realized by χ^2 for sample of values in Table 8.10.

j	Interval	o_j	np_j	$\dfrac{(o_j - np_j)^2}{np_j}$
1	$0 < x \leq 1$	29	24.59	0.7898
2	$1 < x \leq 2$	14	21.77	2.771
3	$2 < x \leq 3$	22	19.27	0.3882
4	$3 < x \leq 4$	13	17.05	0.9625
5	$4 < x \leq 5$	11	15.09	1.109
6	$5 < x \leq 6$	14	13.36	0.0309
7	$6 < x \leq 7$	16	11.82	1.476
8	$7 < x \leq 8$	11	10.46	0.0275
9	$8 < x \leq 9$	12	9.261	0.8099
10	$9 < x \leq 10$	11	8.197	0.9585
11	$10 < x \leq 11$	9	7.255	0.4197
12	$11 < x \leq 12$	12	6.421	4.847
13	$12 < x \leq 13$	6	5.683	0.0176
14	$13 < x \leq 14$	2	5.030	1.825
15	$14 < x \leq 16$	6	8.393	0.6821
16	$16 < x \leq 18$	10	6.574	1.785
17	$18 < x \leq 20$	5	5.150	0.0044
18	$20 < x \leq 23$	3	5.711	1.287
19	$x > 23$	8	12.91	1.869
Totals 19	$0 < x < \infty$	214	214	22.06

Entries in the fourth column are obtained from (8.81). For example, $np_3 = 214(e^{-0.244} - e^{-0.366}) = 19.3$. For gaps exceeding 14 seconds, intervals must be longer than a second to satisfy (8.87). Note that the table on pp. 38–39 of Gerlough and Barnes (op. cit.) contains several errors, which is why the calculation is presented in detail here.

actually realized by χ^2 were one of the most probable. But although we accept our statistical model with caution, we have at least demonstrated that it was not unreasonable, in the crossing control model of Section 5.4, to assume that cars passing a fixed point on an unrestricted highway may be regarded as arrivals in a Poisson process.

A few remarks are now in order. Achieving (8.77b) by equating means and/or variances involves a special case of what is often called the *method of moments*, the mean and variance of a distribution being its first two moments. But (8.77b) need not be achieved by the method of moments. Indeed statisticians usually recommend that whenever the χ^2 test is being applied parameters be estimated by a different method, namely, the method of maximum likelihood (see below). Moreover, these are not the only two methods by which (8.77b) can be achieved, and there is already a large literature on what has come to be known as *statistical inference*. The larger the value of n, however, the smaller the differences between results obtained by the various methods.[7] Thus, as almost always in statistics, difficulties arise mainly when the sample size n is small. Then the cumulative distribution of the sample need not suggest the shape of the underlying distribution, and the χ^2 test may not be used. In these circumstances, the assignment pendulum swings heavily across from the side of objectivity to that of subjectivity. When is n small? No precise answer can be given, though statisticians would appear to agree that n is small if $n < 30$. As if to leave the matter beyond all reasonable doubt, however, let's consider the second extreme by taking $n = 6$. The model that results is inevitably very subjective.

General (i.e., national) elections were held in the United Kingdom (of Great Britain and Northern Ireland) on the dates shown in Table 8.12. L denotes a victory by the Labour party, C a victory by the Conservative party. No other party has been in power since World War II. For ease of computation, the postwar years have been broken into quarters. The dates of transitions between quarters are year's end, winter/spring, mid-year, and Fall. Each general election has been assigned to its nearest transition date.

Table 8.12   Waiting times between postwar general elections in the United Kingdom. Data prior to 1970 are taken from Childs (1979), later data from personal memory.

Election date	l	Winner	Labour waiting time	Conservative waiting time
July, 1945	0	L	19	
March, 1950	19	L	6	
October, 1951	25	C		15
May, 1955	40	C		17
October, 1959	57	C		20
October, 1964	77	L	6	
March, 1966	83	L	17	
June, 1970	100	C		15
February, 1974	115	L	2	
October, 1974	117	L	19	
May, 1979	136	C		16
June, 1983	152	C		16
June, 1987	168	C		

[7]Indeed the method of moments and method of maximum likelihood yield *identical* results for some of the commoner distributions, in particular the normal and exponential, regardless of the value of n. For a concise introduction to statistical inference, see Chapter 10 of Ross (1980).

Thus the July, 1945, election is regarded as having taken place in mid-year, or at the end of the second quarter of 1945. Similarly, the February, 1974, election is regarded as having taken place in winter/spring, or at the end of the first quarter of 1974. The number of quarters that have elapsed at the time of each general election is denoted by l, with $l = 0$ corresponding to the July, 1945, election. These "quarter dates" appear in the second column of Table 8.12.

The government of the U.K. may stay in power for at most five years, or 20 quarters, but may call a general election at any time within that five-year period. Thus the waiting time between general elections is a random variable, taking integer values between 1 and 20. Let's denote this random variable by W_L, if Labour are in power, or by W_C, if Conservatives are in power. What kind of probability distribution is associated with W_L and W_C?

The last columns of Table 8.12 together reveal that since World War II an election has always been called either in the last year and a half or the first year and a half of the governing party's five-year term. A mid-term postwar general election has yet to take place. There is conceptual support for these empirical findings. In theory, a government is elected to do five years' work. If its majority is so slim that it cannot govern effectively, then it may return to the polls quickly, in the hope of obtaining a firmer mandate. Because the parliamentary majority is determined at the general election and can change only in the rare event of a by-election, it is most unlikely that a government will perceive its majority as adequate in the early years but not in the middle years. As a consequence, a mid-term election is virtually impossible. On the other hand, if a government waits a full five years before calling an election, then it virtually admits that it fears defeat at the polls. The confidence that gives to the opposition is likely to weaken the government's effectiveness in Parliament. Thus the five-year Conservative term from October, 1959, until October, 1964, should be regarded as an event that had low probability, even though it actually took place (note that the Conservatives were promptly defeated). But it is also clear from Table 8.12 that Labour are more prone to return to the polls early than Conservatives are. Thus W_C and W_L do not have identical distributions, even though they are not totally dissimilar.

Let $\text{Prob}(W_L = k)$ be denoted by $w_0(k)$ and $\text{Prob}(W_C = k)$ by $w_1(k)$. Because mid-term elections are so unlikely, I'm going to stick my neck out and say that

$$w_0(k) = 0 = w_1(k), \qquad 9 \leq k \leq 12. \tag{8.89}$$

Thus elections are impossible in the middle year of a government's potential life span. In practice, there is a tiny probability that such an election might occur, but (8.89) is not a bad approximation. In my opinion, the five-year Conservative term, from October, 1959, until October, 1964, was an aberration. Even though it actually happened, I don't consider it any more likely than a quick return to the polls by Conservatives. On the other hand, I do think that Conservatives are more likely to call an election after about four years than at any other time. Indeed, after election victory in 1987, the Conservative leader actually predicted seeking re-election in 1991! I will therefore take

$$w_1(k) = \begin{cases} v, & 1 \le k \le 8, k = 13, 14, 19, 20; \\ u, & 15 \le k \le 18. \end{cases} \tag{8.90}$$

I expect that u will be much larger than v (see below). Notice that I have made a four-and-a-half-year term as likely as a four-year term, even though there have been none of the former and two of the latter. This is largely to keep the mathematics simple while getting across the ideas. Nevertheless, we cannot be sure that $W_C = 18$ isn't just as likely as $W_C = 17$. Now, because

$$\sum_{k=1}^{20} w_1(k) = 1, \tag{8.91}$$

it follows immediately from (8.89) and (8.90) that

$$12v + 4u = 1. \tag{8.92}$$

I therefore assume the following one-parameter theoretical probability distribution for the random variable W_C, whose sample space is the integers between 1 and 20:

$$w_1(k) = \begin{cases} \dfrac{1}{12} - \dfrac{u}{3}, & \text{if } 1 \le k \le 8, k = 13, 14, 19, 20; \\ 0, & \text{if } 9 \le k \le 12; \\ u, & \text{if } 15 \le k \le 18, \end{cases} \tag{8.93}$$

where $0 \le u \le 1/4$ (why?). Thus (8.77a) has been accomplished for the Conservatives. How do we accomplish (8.77b)?

Perhaps the most popular method for estimating u is the method of "maximum likelihood." Suppose that (8.93) is indeed the probability distribution for W_C. Then what is the probability of the joint event

$$\left\{ W_C^{(1)} = 15, W_C^{(2)} = 17, W_C^{(3)} = 20, W_C^{(4)} = 15, W_C^{(5)} = 16, W_C^{(6)} = 16 \right\}, \tag{8.94}$$

if the random variable $W_C^{(i)}$ denotes the Conservatives' waiting time during their ith postwar term in office? If we assume that the random variables $W_C^{(i)}, 1 \le i \le 6$, are independently distributed according to (8.93), then the probability of the event (8.94) is

$$\phi(u) \equiv u \cdot u \cdot \left(\frac{1}{12} - \frac{u}{3} \right) \cdot u \cdot u \cdot u = u^5 \left(\frac{1}{12} - \frac{u}{3} \right). \tag{8.95}$$

The function ϕ, defined on $0 \leq u \leq 1/4$, is known as a *likelihood* and is sketched in Fig. 8.7. Bear in mind that the graph actually shows $10^5\phi$, so that the probability (8.95) is negligible, except perhaps in the neighborhood of the peak at $u = 5/24$, where ϕ assumes the scarcely enormous value of 0.000005451. Yet the joint event (8.94) actually happened! We might thus conclude that the most appropriate value for u in (8.93) is the one that maximizes (8.95), because that attaches *maximum likelihood* to the event that was realized. This value of u, namely, 5/24, is known as a *maximum likelihood estimate*. From (8.92), the corresponding value of v is 1/72, which is 15 times smaller. I thus propose the following statistical model for Conservative length of stay in office:

$$\text{Prob}(W_C = k) = \begin{cases} \dfrac{1}{72}, & \text{if } 1 \leq k \leq 8, k = 13, 14, 19, 20; \\ 0, & \text{if } 9 \leq k \leq 12; \\ \dfrac{5}{24}, & \text{if } 15 \leq k \leq 18. \end{cases} \tag{8.96}$$

The form (8.96) relies heavily upon my intuition that an early Conservative dissolution of Parliament is about as likely as a tardy one, even though there has never been an early dissolution. You might have a different idea. You might, for example, agree to take $w_1(k) = u$ for $15 \leq k \leq 18$, $w_1(k) = v$ for $1 \leq k \leq 8$, $k = 13, 14$, as before, but prefer to take $w_1(k) = z$ for $k = 19$ and 20, with z larger than v in mind. Because (8.91) then yields $10v + 4u + 2z = 1$, you would replace (8.95) by

$$\phi(u, z) = u^5 z \tag{8.97}$$

and maximize it subject to the constraints

$$0 \leq u \leq 1, \qquad 0 \leq z \leq 1, \qquad 1 - 2z - 4u \geq 0, \tag{8.98}$$

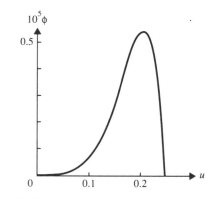

Fig. 8.7 Graph of the likelihood function defined by (8.95).

the last of which ensures that v is nonnegative. This is a nonlinear programming problem of the form (8.70). Its solution does not require a computer, however, because it is obvious from Fig. 8.8, in which the shaded region corresponds to (8.98). The curves above it are of the form $u^5z = $ constant, with the value of the constant increasing toward the top right. It is clear from this diagram that (8.97) has its maximum on the line $4u + 2z = 1$, at the point where the line is tangent to one of the curves. This point is easily found to be given by $u = 5/24$ and $z = 1/12$ (Exercise 8.9). The corresponding value of v is zero, yielding the alternative statistical model

$$\text{Prob}(W_\text{C} = k) = \begin{cases} 0, & \text{if } 1 \le k \le 14; \\ \dfrac{5}{24}, & \text{if } 15 \le k \le 18; \\ \dfrac{1}{12}, & \text{if } 19 \le k \le 20. \end{cases} \tag{8.99}$$

Now z is certainly larger than v; but it is also impossible for Conservatives to call an early election, and the probability that they might call one after about four years of government remains unchanged. For this reason, I prefer the first of the two statistical models because a single early election call by Conservatives would cause (8.99) to be rejected in favor of (8.96) or some other statistical model.

Although Labour's early dissolutions all occurred within the first year and a half of a parliamentary lifespan, I see no special reason they might not have occurred at any time during the first two years. Likewise, although the late dissolutions all occurred within the last year of a parliamentary lifes-

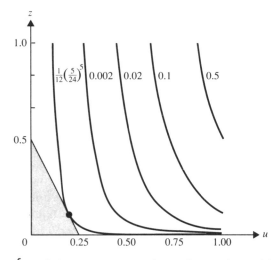

Fig. 8.8 Curves $u^5z = k$ in the u-z plane, for various values of k, as indicated on the sketch. The shaded region corresponds to (8.98).

pan, I see no special reason they might not have occurred sooner. Moreover, I see no special reason to distinguish one early quarter from another, nor one late quarter from another. I will therefore assume that $w_0(k) = u$ for $1 \leq k \leq 8$ and $w_0(k) = v$ for $13 \leq k \leq 20$. From (8.89) and (8.91), however, it then follows that $8u + 8v = 1$. Accordingly, I assume the following form of distribution for Labour stays in power:

$$w_0(k) = \begin{cases} u, & \text{if } 1 \leq k \leq 8; \\ 0, & \text{if } 9 \leq k \leq 12; \\ \dfrac{1}{8} - u, & \text{if } 13 \leq k \leq 20. \end{cases} \qquad (8.100)$$

This accomplishes (8.77a) for Labour. It is now straightforward (Exercise 8.10) to show that the maximum likelihood estimate of u is $1/16$. Hence I propose

$$\text{Prob}(W_\text{L} = k) = \begin{cases} \dfrac{1}{16}, & \text{if } 1 \leq k \leq 8; \\ 0, & \text{if } 9 \leq k \leq 12; \\ \dfrac{1}{16}, & \text{if } 13 \leq k \leq 20, \end{cases} \qquad (8.101)$$

as a statistical model of length of Labour terms in office.

Though we shall find them useful in Chapter 10, the models (8.96) and (8.101) are extremely subjective, and you may not agree with them. They assume in particular that the distributions of W_L and W_C do not change with time, which may not be true; yet there is no way to test this assumption. No matter how much you know or might learn about British politics, you could hardly construct more objective statistical models from so little data. What you can do, however, is to transfer the subjectivity from me to you, replacing (8.96) and (8.101) by models of your own. You are most welcome to try. But an easier task would be Exercise 8.15.

8.8   Choosing a Utility Function. Cautious Attitudes to Risk

A buyer who seeks to maximize the expected value of her company's profit from an item, over any given period, may be acting rationally only if the number of periods is large.[8] To see this, let the random variable P_u denote profit from a single period when the quantity ordered is u. Let $P_u^{\min}$ and $P_u^{\max}$ denote, respectively, the minimum and maximum values that the random variable P_u can take. Let u^* denote the quantity that maximizes $E[P_u]$. Then $P_{u^*}^{\min}$ is almost certainly negative. Suppose, for example, that demand, Z, is uniformly distributed between a and b. Let p, c_0, and c_1

---

[8]You may wish to review Section 7.1 before proceeding.

denote, respectively, unit selling price, unit cost price, and unit holding cost. Suppose that delivery cost is either zero or absorbed into the cost price, i.e., that $k = 0$ in the terminology of Section 7.1. Then, from (7.6), we have

$$P_u(Z) = \begin{cases} pZ - c_0 u - c_1(u - Z), & \text{if } Z \le u; \\ (p - c_0)u, & \text{if } Z > u; \end{cases} \tag{8.102}$$

and from either (7.10) or (7.63) we have

$$u^* = a + \left(\frac{p - c_0}{p + c_1}\right)(b - a). \tag{8.103}$$

It follows readily that $P_u^{max} = (p - c_0)u$, $P_u^{min} = (p + c_1)a - (c_0 + c_1)u$ and $P_{u^*}^{min} < 0$ if $(p + c_0 + 2c_1)a < (c_0 + c_1)b$, i.e., if a is sufficiently small. Thus, because of the uncertainty in P_u, choosing the optimal value for u cannot rule out the possibility that a negative profit be realized. If the number of periods is large, then periods when $P_{u^*} > E[P_{u^*}]$ are likely to compensate for periods when $P_{u^*} < E[P_{u^*}]$, including those for which $P_{u^*} < 0$, so that maximization of expected profit is a rational *long-term* policy. If the number of periods is small, however, then the company stands to suffer if $P_{u^*}^{min} < 0$, because later periods may not be able to compensate. In such circumstances, shouldn't the buyer order $u < u^*$ to reduce the loss that the company would suffer if the worst possible outcome were actually realized? In other words, shouldn't she adopt a more cautious attitude to risk?

In an attempt to indicate how the buyer might do so, we will confine our attention to cases when she orders for a single period only. Moreover, to keep the mathematics as simple as possible, we will assume henceforward that $a = 0$; then certainly $P_{u^*}^{min} = -(c_0+c_1)u^* < 0$. Let's define a *normalized* profit by

$$R_u \equiv \frac{P_u - P_b^{min}}{P_b^{max} - P_b^{min}}, \tag{8.104}$$

where $P_b^{min} = -(c_0 + c_1)b$ and $P_b^{max} = (p - c_0)b$ are, respectively, the least and the greatest profit that could possibly be realized from the period; $0 \le R_u \le 1$. Thus an actual profit of zero corresponds to a normalized profit of θ, where, on setting $P_u = 0$ in (8.104), we have

$$\theta = \frac{c_0 + c_1}{p + c_1}. \tag{8.105}$$

Because P_b^{min} and P_b^{max} are independent of u, the value of u that maximizes $E[P_u]$ is the same as that which maximizes $E[R_u]$. Thus the long-term buyer maximizes $E[R_u]$. The short-term buyer, however, must have an alternative goal. Let's simply suppose that she maximizes $E[M(R_u)]$, where M is a *merit function* taking values between 0 and 1. For the long-term buyer, we have $M(R) = R$. Our problem is to determine a form of M for the short-term buyer.

The long-term buyer's merit function is shown in Fig. 8.9(a) as the solid straight line. The value θ, corresponding to the break-even profit $P_u = 0$, is arbitrarily marked on the diagram. Now, high values of R are associated with high risk, because only by ordering a quantity close to b can the buyer hope to obtain them; but these are precisely the circumstances in which the company will suffer a heavy loss if demand is unexpectedly low. Hence, whereas the long-term buyer can justifiably be 90% satisfied with $R = 0.9$, we would expect the short-term buyer to be 90% satisfied with rather less than this; thus $M(\theta_1) = 0.9$ for some $\theta_1 < 0.9$. We have marked a possible point $(\theta_1, M(\theta_1))$ as a dot on the diagram in Fig. 8.9(a). By deciding how much profit will make her 90% satisfied, the buyer can attach a merit value to the (normalized) profit θ_1. Continuing in this way, the buyer can attach a merit value to all values of R for which $\theta < R < 1$. The point $(\theta_2, M(\theta_2))$, for example, is indicated (arbitrarily) on the graph. The shape of the associated merit function is also indicated. Note that it lies completely above the line

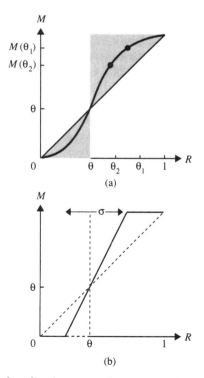

Fig. 8.9 (a) Feasible region for short-term buyer merit function. (b) Possible form of merit function, used in subsequent analysis.

$M(R) = R$ in this range because the short-term buyer is obliged to be more satisfied with less.

In the range $0 < R < \theta$, however, we would expect the short-term buyer's merit curve to lie below that for the long-term buyer, because the actual profit here is negative. A very negative profit is a worse outcome for the short-term buyer, who must therefore assign it a lower merit value. Again the probable shape of the merit curve is indicated. Summarizing, we have $M''(R) > 0$ for $R < \theta$ and $M''(R) < 0$ for $R > \theta$, assuming that M is twice differentiable. The shading in the diagram indicates the feasible region for short-term merit functions.

As usual when applying the method of extremes (Section 8.3), we must now find a one-parameter family of curves that covers the shaded region. To keep the mathematics tractable, it will be convenient to relax the assumption that M be twice differentiable and use the piecewise linear form depicted in Fig. 8.9(b), namely,

$$M(R) = \begin{cases} 0, & \text{if } 0 \le R < \theta(1 - \sigma); \\ \theta + \dfrac{R - \theta}{\sigma}, & \text{if } \theta(1 - \sigma) \le R < \theta(1 - \sigma) + \sigma; \\ 1, & \text{if } \theta(1 - \sigma) + \sigma \le R \le 1. \end{cases} \tag{8.106}$$

Note that we still have $M''(R) \ge 0$ for $R < \theta$ and $M''(R) \le 0$ for $R > \theta$; but now, as it were, all the "twice-differentiability" is compressed into the two corners at $\theta(1 - \sigma)$ and $\theta(1 - \sigma) + \sigma$ (where there is infinitely much of it). The risk parameter σ satisfies $0 < \sigma \le 1$. The greater its value within that range, the less cautious the buyer, the value $\sigma = 1$ being the one our long-term buyer would adopt. The buyer's attitude to risk is a subjective matter; therefore we retain σ as a parameter in the analysis.

From (8.102), (8.104), and (8.105):

$$R_u(X) = \begin{cases} \dfrac{X}{b} + \theta\left(1 - \dfrac{u}{b}\right), & \text{if } 0 \le X < u; \\ \dfrac{u}{b} + \theta\left(1 - \dfrac{u}{b}\right), & \text{if } u \le X \le b. \end{cases} \tag{8.107}$$

From (8.106) and (8.107), and from the fact that θ, σ, and u/b all lie between 0 and 1, we find that $M = 0$ only if $0 < X < (u - b\sigma)\theta$; $M = 1$ if either $X > u > b\sigma$ or $u > X > (u - b\sigma)\theta + b\sigma$ (implying $u > b\sigma$); and $0 < M < 1$ if either $X < u$ and $(u - b\sigma)\theta < X < (u - b\sigma)\theta + b\sigma$ or $X > u$ and $u < b\sigma$. Thus, if $u \le b\sigma$, then we have (Exercise 8.11)

$$M(R_u(X)) = \begin{cases} \dfrac{X - \theta u}{b\sigma} + \theta, & \text{if } 0 \le X < u; \\ \dfrac{u(1 - \theta)}{b\sigma} + \theta, & \text{if } u \le X \le b; \end{cases} \tag{8.108}$$

but if $u > b\sigma$, then we have

$$M(R_u(X)) = \begin{cases} 0, & \text{if } 0 \le X < (u - b\sigma)\theta; \\ \dfrac{X - \theta u}{b\sigma} + \theta, & \text{if } (u - b\sigma)\theta \le X < (u - b\sigma)\theta + b\sigma; \quad (8.109) \\ 1, & \text{if } (u - b\sigma)\theta + b\sigma \le X \le b. \end{cases}$$

A straightforward integration (Exercise 8.11) now yields

$$
\begin{aligned}
J(u) &\equiv E[M(R_u(Z))] = \int_0^b M(R_u(z)) \frac{dz}{b} \\
&= \begin{cases} \theta + \dfrac{u(1 - \theta)}{b\sigma} - \dfrac{u^2}{2b^2\sigma}, & \text{if } u \le b\sigma; \\ 1 - \dfrac{\sigma}{2} + \theta\left(\sigma - \dfrac{u}{b}\right), & \text{if } u > b\sigma. \end{cases}
\end{aligned} \quad (8.110)
$$

Maximizing J is a simple calculus problem (Exercise 8.11). We find that the maximum occurs at $u = v^*(\sigma)$, where

$$v^*(\sigma) = \begin{cases} b\sigma, & \text{if } 0 < \sigma \le 1 - \theta; \\ (1 - \theta)b, & \text{if } 1 - \theta < \sigma \le 1. \end{cases} \quad (8.111)$$

Note, on setting $a = 0$ in (8.103) and using (8.105) and (8.111), that $u^* = v^*(1)$, as required. The function v^* is sketched in Fig. 8.10(a); the corresponding maximum expected utility,

$$J(v^*(\sigma)) = \begin{cases} 1 - \dfrac{\sigma}{2}, & \text{if } 0 < \sigma \le 1 - \theta; \\ \theta + \dfrac{(1 - \theta)^2}{2\sigma}, & \text{if } 1 - \theta < \sigma \le 1, \end{cases} \quad (8.112)$$

is sketched in Fig. 8.10(b) and the maximum expected normalized profit,

$$E[R_{v^*(\sigma)}(X)] = \begin{cases} \theta + \sigma(1 - \theta) - \dfrac{1}{2}\theta^2, & \text{if } 0 < \sigma \le 1 - \theta; \\ \theta + \dfrac{1}{2}(1 - \theta)^2, & \text{if } 1 - \theta < \sigma \le 1, \end{cases} \quad (8.113)$$

appears in Fig. 8.10(c). Thus, as $\sigma \to 0$, the buyer orders nothing and makes no profit (normalized profit θ) but is perfectly happy because she has assigned no merit to any negative profit and full merit to any positive profit, however small, as is clear from Fig. 8.9(b) on taking the limit $\sigma \to 0$. The moral in this is Thoreau's: happiest are those who desire least! As σ increases from zero, a buyer becomes more daring. The optimal order quantity increases and so does the expected profit, but the expected merit decreases. The more narrowly the buyer defines maximum satisfaction (by increasing σ) the less attainable it becomes.

Figure 8.10 yields a qualitatively accurate portrayal of the effect of decreasing aversion to risk (increasing σ) on the optimal buying policy,

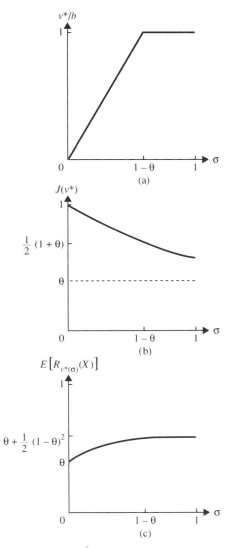

Fig. 8.10 (a) Optimal order quantity (8.111) for a single period as a function of increasing risk, when demand is uniformly distributed between 0 and b and $1 - \theta = (p - c_0)/(p + c_1)$. (b) Graph of $J(v^*(\sigma))$, defined by (8.112). (c) Graph of $E[R_{v^*(\sigma)}(X)]$, defined by (8.113).

though the corner in Fig. 8.10(a) would disappear if we chose a smoother merit function, like that drawn in Fig 8.9(a). Although our analysis of using the utility function to embody attitudes to risk has been deliberately simplified, it includes most of the important ideas. If you wish to pursue the matter further, however, then you should consult the literature on decision theory, which will indicate in particular how a buyer might be able to determine her personal (or her company's) value of σ. See, for example, the books by Holloway (1979) and Lindley (1985).

Exercises

8.1 Verify (8.2).

8.2 Verify (8.4).

8.3 (i) Verify (8.9). *Hint*: Use the fact that $p'(t)/p(t) = 0.18(1 - p(t)/3p(0))$.
(ii) Verify (8.11).

**8.4 Use the method of Section 8.3 to devise a set of behavioral response functions for the fast-food restaurant model of Section 7.6.

8.5 (i) Verify (8.23)–(8.27).
(ii) Suppose that the datum for 1850 ($t = 6$) is removed from the data set that was used to obtain (8.28). Show that the line that best fits the reduced data set, in the sense of Section 8.4, is closer than the line obtained in that section to the one that is drawn in Fig. 1.7.
(iii) Suppose that the data for the years 1840–1860 ($5 \leq t \leq 7$) are removed from the data set that was used to obtain (8.28). Show that the line that best fits the data, in the sense of Section 8.4, is closer even than (ii) to the line drawn in Fig. 1.7.
(iv) Comment.

*8.6 (i) Verify Table 8.5.
(ii) Pine trees in a Scottish forest are grouped by girth into six classes, as described in Section 5.8. The following data for a single census period are taken from Usher (1966, p. 361):

Girth class	Number of trees remaining at start of period	Number of trees remaining in class during period
0	4461	3214
1	2926	2029
2	1086	813
3	222	171
4	27	17

Use these data to estimate transition matrix S for the model of Section 5.8; i.e., estimate (5.93).

*8.7 After up to four moves through a coronary unit, patient numbers at nine levels of health care are given below. The data set, which was used to obtain Table 8.8, is taken from Kao (1974, p. 688).

Level of care	Number of moves, n				
	0	1	2	3	4
CCU	471	0	7	0	0
PCCU	0	349	2	12	0
ICU	0	4	8	0	0
MED	0	64	9	0	0
SURG	0	2	3	0	0
AMB	0	3	3	0	0
ECF	0	0	25	26	27
HOME	0	3	341	360	369
DIED	0	46	73	73	75

(i) Using data for $1 \leq n \leq 3$ only, find the best fit Markov model by using the procedure of Section 8.6 to estimate **P**.

(ii) Compare your model's predictions for $n = 4$ with the observations. *Note*: For this exercise, you will need access to a nonlinear programming routine. Even if there is no suitable package in the computer library, there is almost certainly at least one faculty member somewhere on campus who can lend you such a package if you ask her nicely (the physics department is often a better bet than mathematics). Alternatively, you may wish to write your own.

*8.8 There are numerous fields from which I could have drawn an example to illustrate the fitting of a two-parameter statistical model by the method of moments. Because I was thinking about elections while writing Section 8.7, however, the following sprang most readily to mind.

As we saw in Section 8.7, only two political parties, Labour and the Conservatives, have governed the United Kingdom since World War II. As indicated by Table 8.12, Labour were elected to govern in 1945. The election was decided on the basis of 640 parliamentary constituencies (there are now only 635). From each constituency, a candidate was elected by simple majority to be a member of the House of Commons; and Labour enjoyed victory in the greater number of constituencies. There were 576 constituencies in which Labour and the Conservatives each fielded a candidate for Parliament. Of votes cast for either Labour or Conservative in these 576 constituencies, the proportions accruing to Labour were as follows, the data being taken from Kendall and Stuart (1950, p. 187):

Proportion, x	Number of constituencies	Proportion, x	Number of constituencies
$0.00 \leq x < 0.05$	1	$0.50 \leq x < 0.55$	64
$0.05 \leq x < 0.10$	0	$0.55 \leq x < 0.60$	89
$0.10 \leq x < 0.15$	0	$0.60 \leq x < 0.65$	90
$0.15 \leq x < 0.20$	0	$0.65 \leq x < 0.70$	60
$0.20 \leq x < 0.25$	3	$0.70 \leq x < 0.75$	37
$0.25 \leq x < 0.30$	8	$0.75 \leq x < 0.80$	19
$0.30 \leq x < 0.35$	29	$0.80 \leq x < 0.85$	20
$0.35 \leq x < 0.40$	28	$0.85 \leq x < 0.90$	5
$0.40 \leq x < 0.45$	59	$0.90 \leq x < 0.95$	2
$0.45 \leq x < 0.50$	62	$0.95 \leq x < 1.00$	0

Let the random variable X denote Labour's proportion of the vote, assumed identically distributed in each constituency. Use the method described in Section 8.7 to propose a statistical model for X's distribution. *Hint*: At some stage, you will have to make an appropriate assumption or approximation because the above sample frequency distribution does not yield as much information as perhaps you would like.

8.9 Verify that the solution of the constrained nonlinear programming problem (8.97)–(8.98) occurs where $\mathbf{u} = (u, z)^T = (5/24, 1/12)$.

8.10 For the data in Table 8.10, find the maximum likelihood estimate of the parameter u in (8.100).

8.11 Verify the calculations in Section 8.8.

**8.12 For what proportion of the 21st century will the president of the United States be a Democrat, for what proportion a Republican? Attempt to answer this question by choosing a known model and finding data to estimate its parameters. State any assumptions you make about voting behavior in the United States.

*8.13 A chemical in solution occupies the region on either side of a porous membrane of thickness h. Assume that the solution on each side of the membrane is perfectly mixed, with ϕ_o denoting the concentration of chemical outside the membrane and ϕ_i the concentration inside.
(i) In which direction will there be a flux of chemical if $\phi_i > \phi_o$?
(ii) In which direction will there be a flux of chemical if $\phi_o > \phi_i$?
(iii) Use a *known* model to suggest a general expression for the outward flux per unit area. Can you propose a suitable name for any new parameter(s) you introduce in this expression? What are the dimensions of the parameter(s)?

(iv) The concentration of glucose inside a spherical cell with permeable wall of inner radius a is initially 10% of the concentration of glucose outside the cell. Write down a differential equation describing the subsequent uptake of glucose by the cell. To which known mathematical model does this equation correspond? Assuming that this glucose uptake does not affect the ambient concentration, deduce the time that elapses before the glucose concentration inside the cell reaches 90% of that outside (as a function of a, h, and any parameter(s) you have introduced).

*8.14 An irreversible chain reaction between n reactants $A_0, A_1, \ldots, A_{N-1}$ and product A_N is represented symbolically by

$$A_0 \longrightarrow A_1 \longrightarrow A_2 \longrightarrow \cdots \longrightarrow A_{N-1} \longrightarrow A_N.$$

Let $x_j(t) = [A_j]$ be the concentration of A_j at time t in moles per unit volume (of solvent), and let k_j denote the forward rate of reaction from A_j to A_{j+1} (in moles per unit of time). Answer the following with the help of what you have learned about chemical dynamics in Sections 1.10 and 1.11.
(i) Write down a set of differential equations that describes this chemical reaction.
(ii) Which known set of equations from Part III is mathematically equivalent to this (deterministic) model?
(iii) Why, however, will the previously calculated solution to those equations be incorrect for present circumstances?

*8.15 In a United Kingdom general election on April 9, 1992 the Conservative party was returned to power. What, if any, effect does this result have on the statistical models in Section 8.7 for length of stay-in-office?

Supplementary Notes

8.1 For values of z other than those appearing in Table 8.5, $l(z)$ can be estimated by using *numerical interpolation*. This is the process by which the graph of a function $l = l(z)$ is approximated by polynomials that are constrained to pass through known data points, such as those appearing in Table 8.5. If $l = l(z)$ is sufficiently smooth and if sufficient data are known, then arbitrary accuracy can be obtained by using polynomials of successively higher and higher order. Rarely, however, are both conditions satisfied; and in practice there is usually little justification for using polynomials of higher order than the third.

Linear interpolation, i.e., use of a first order polynomial, is often adequate for practical purposes. Suppose, for example, that we wished to estimate $l(4)$. The first order polynomial constrained to satisfy the data for $z = 1$ and $z = 5$ is clearly $l(1) + 0.25 \times \{l(5) - l(1)\}(z - 1)$. Because $l(1) = 0.98003$ and $l(5) = 0.97672$, from Keyfitz and Flieger's data, we obtain $l(4) \approx 0.9775$.

Care is needed in selecting known data points. For example, the second order polynomial constrained to satisfy the data for $z = 0$, $z = 1$, and $z = 5$ is

$$\phi(z) = 0.2(z - 1)(z - 5) + 0.25z\{0.2l(5)(z - 1) - l(1)(z - 5)\}.$$

This yields $l(4) \approx 0.9661$, which is absurd, because it is lower than $l(5)$, and clearly a worse approximation than previously obtained by linear interpolation. The absurdity arises because ϕ has positive curvature (everywhere); whereas $l = l(z)$ has negative curvature at $z = 4$. Thus the quadratic polynomial does not adequately model the survival probability. We have made a poor choice of three data points through which to constrain the quadratic to pass. We should have begun by selecting the nearest datum to $z = 4$ (i.e., $z = 5$), then the nearest on the other side of $z = 4$ (i.e., $z = 1$), then the next nearest on the other side again ($z = 10$). The quadratic polynomial that satisfies the data for $z = 1$, $z = 5$, and $z = 10$ is clearly

$$\frac{1}{36}l(1)(z - 5)(z - 10) - \frac{1}{20}l(5)(z - 1)(z - 10) + \frac{1}{45}l(10)(z - 1)(z - 5).$$

Because $l(10) = 0.97493$, this yields the approximation $l(4) \approx 0.9774$; and you would probably agree that any additional accuracy is scarcely worth the extra labor of computation. For an extended description of the leapfrogging scheme suggested above and the efficient Newtonian scheme for computing interpolants, see Maron (1982). An unfortunate (though understandable) omission from Maron's text is that of Aitken interpolation, for which see Fröberg (1969).

9 BUILDING A MODEL: ADAPTING, EXTENDING, AND COMBINING

The models encountered in Parts I to III, though useful and important, are but a small fraction of known mathematical models and an even smaller fraction of all the models there could be. Then what do you do if you meet a problem to which none of these models applies? By and large, there are only three things you can do: adapt, extend, or combine.

First, you may recognize that the new problem resembles an old problem in some abstract way, as though the old one had merely donned a fresh disguise. Then the old model can be *adapted* to your present purpose. The plant model of Section 1.3 and the population model of Section 1.9 are just two adaptations of the same logistic growth model. The momentum flux law of Section 2.6 is an adaptation of the temperature flux law of Section 2.5. And in Section 5.6 we adapted the idea of a metered model to a probabilistic view. You've seen several other examples of adaptations. How many can you recall?

Second, you may recognize that today's problem, although having new features, also has features in common with yesterday's. Then yesterday's model can be *extended* to account for the new ones. The model of bird flight in Section 3.1 is an extension of the equilibrium rowing model of Section 2.2 (to account for induced power losses). The model was further extended in Chapter 4 to account for a tailwind. And a Markov chain can be viewed as an extension of a birth and death chain. You've seen several other examples of extensions. See how many you can recall (don't forget the exercises).

Third, you may recognize that different aspects of today's problem are separately covered by models you know. Then the known models can be combined. The traffic model of Section 4.7 is a combination of two quite separate assumptions about the way that drivers brake and accelerate. The growth/harvest model in Exercise 1.12 is a combination of a pure growth model and the pure harvesting model of the previous exercise. And, of course, the classic example of a combination is that of a metered model. Again, can you think of others?

If you had paused in class to recall such examples then you would have been bound to discover that one person's combination is another person's extension or adaptation. Indeed experienced modellers adapt, extend, and combine simultaneously. They neither are aware of which label most aptly describes what they are doing, nor consider it especially useful to know. But while you are learning, the labels, though subjective, are useful. Moreover, although all are part of improving a model, I reserve that phrase for critical mode, when you review your model with the eyes of a harsh parent. We are switching to creative mode, so put validation out of your mind, for now is the time to play and experiment. You can always validate your models later; then, but only then, it's compulsory.

Remember, too, that modelling is fraught with trial and error. Don't give up easily, and be disposed to learning from your mistakes. But what if, after so much effort, all three options fail? Well, I suppose you could ask your professor. But you may not need to if you study the following examples well!

9.1   How Many Papers Should a News Vendor Buy? An Adaptation

In the simplest circumstances, adapting a model hardly differs from using a known one. Nevertheless, understanding the simplest adaptations is an important step toward understanding more complex ones.

Suppose that you're selling newspapers on the corner of a street in a large city. Every day you buy quantity u; the cost is c_0 cents per paper. You sell as many as you can for price p; this demand, Z, is a random variable. At the end of the day, you are able to return the unsold papers to the distributor; the unit salvage price is s cents. How many papers should you order each day?

This ought to remind you of the rose vendor in Section 7.2. In one sense, the problem is just as simple: there are no holding costs. A day is too short to worry about unearned interest, the stock is hardly worth insuring; and if you need an extra bit of sidewalk for stacking the newspapers early in the morning, then it's absolutely free of rent. But, in another sense, the problem is more difficult: no mention was made in Chapter 7 of a salvage price. Does this mean that you need a new model?

The answer is no, for a moment's thought reveals that a salvage price at the end of the day is mathematically equivalent to a negative holding cost, at the end of the time unit in the model of Section 7.1. Hence, on setting $c_1 = -s$, the optimal daily order u^* can be immediately deduced from (7.12) as the solution of the equation

$$\int_{u^*}^{\infty} f(z)dz = \frac{c_0 - s}{p - s}. \tag{9.1}$$

Note that the right-hand side must lie between 0 and 1 because $s < c_0 < p$ (why?). Then, assuming that the delivery cost is $k = 0$, the optimal reward is

$$J(u^*) = (p - s) \int_0^{u^*} zf(z)dz, \tag{9.2}$$

on using (7.12).

For an application of this result, see Exercise 9.1.

9.2   Which Trees in a Forest Should be Felled? A Combination

In Sections 3.4 and 4.8, we calculated an optimal rotation time for a forest, denoted by s^*. We assumed that a forest had N trees at any given time and that their ages were identical, because they had been planted at the same time. At the end of each s^*-year cycle, or rotation, the forest was completely harvested, then replanted. The value of the parameter s^* was found to be of the order of 50 years or more. Thus the forest yielded income only once or twice a century. What would a forester do if he needed income from the forest more frequently; say, every σ years, where σ is much smaller than s^*?

A possible option would be to harvest H trees every σ years and plant H replacements at the same time, so that the number of trees in the forest would remain constant. In this way, the forester would derive an income every σ years. By keeping the size of the forest constant, he could hope to sustain this income indefinitely. Let's assume that the parameter σ is already known. Then the problem the forester faces is, in some respects, the mirror image of that encountered in Sections 3.2 and 4.8. There, the forester knew how many trees to fell, namely, the entire forest; the question was when. Here, the forester knows when to cut, namely, every σ years; the question is which and how many. Don't forget that, by assumption, if the forester knows how many trees to fell, then he also knows how many to plant.

Now, if the forester genuinely hopes to derive a sustained income, then it is not enough merely to keep the size of the whole forest constant. The foresters in Chapters 3 and 4 did that and rarely saw an income. The problem there was that, because the trees were equal in age, trees hardly ever reached the age for felling (but then all together). Thus our new forester must keep trees of several sizes (and hence ages) in his forest, so that some of them will always be ripe for cutting. For the sake of definiteness, and because age and size are not quite the same thing, we will assume henceforward that our forester classifies trees according to girth. Thus a balanced girth distribution must be maintained, so that after each harvest, enough trees remain of sufficient girth as to reach harvestable girth σ years later. The easiest way to maintain this balance is to keep a fixed number of trees in each class.

To be quite specific, let's suppose that there are $N + 1$ girth classes, labelled $i = 0, 1, 2, \ldots, N$, where i increases with girth; $i = 0$ is the "seedling" class, consisting of trees that have just been planted, and $i = N$ corresponds to maximum girth. This means that we cannot use N to denote the total number of trees. Let's denote it by M instead. Let each σ years correspond to a unit of time, and let $u_i(l)$ denote the proportion of trees in class i at the beginning of cycle $l + 1$ (so that $M u_i(l)$ is the actual number in class i). Then, by definition,

$$\sum_{i=0}^{N} u_i(l) = 1. \tag{9.3}$$

At the end of each cycle, to maintain the girth balance the forester must restore each class to the size it had at the beginning of that cycle. In other words, the forest must satisfy

$$u_i(l) = u_i(l+1), \qquad i = 0, 1, 2, \ldots, N, \quad l = 0, 1, 2, \ldots. \tag{9.4}$$

Let's take a look at what happens during cycle $l + 1$, i.e., during the interval $l < t < l + 1$. For most of this time—i.e., during the interval $l < t < l + 1 - \epsilon$, where ϵ is very small—the trees are simply growing. It is possible that during this period, some trees grow enough to move from one girth class into the one above. Hence the proportion of trees in class i at time t, denoted by $u_i(t)$, will not remain constant throughout $l \leq t < l + 1 - \epsilon$. In particular, $u_i(l + 1 - \epsilon)$ will not be the same as $u_i(l)$. Therefore, let's denote $u_i(l + 1 - \epsilon)$ by γ_i, $0 \leq i \leq N$.

Just before the beginning of cycle $l+2$, the forester harvests H trees. Let v_i denote the proportion of trees taken from class i, $1 \leq i \leq N$, so that $H v_i$ is the actual number taken. We will assume that no trees are harvested from the seedling class (or, if you prefer, $v_0 = 0$). Note that $v_1 + v_2 + \cdots + v_N = 1$. At the same time as these trees are harvested, the forester adds H trees to the seedling class $i = 0$. The combined effect of this harvesting and replanting is to keep both the girth distribution and the total number of trees in the forest constant. In other words:

$$M\gamma_0 + H = M u_0; \tag{9.5a}$$

$$M\gamma_i - H v_i = M u_i, \quad 1 \leq i \leq N. \tag{9.5b}$$

Let us define $(N + 1)$-dimensional column vectors γ, $\mathbf{v}$, $\mathbf{r}$, and $\mathbf{u}$ by

$$\begin{aligned} \gamma &= (\gamma_0, \gamma_1, \ldots, \gamma_N)^T, & \mathbf{v} &= (0, v_1, \ldots, v_N)^T, \\ \mathbf{r} &= H(1, 0, 0, \ldots, 0)^T, & \mathbf{u} &= (u_0, u_1, \ldots, u_N)^T. \end{aligned} \tag{9.6}$$

Then (9.5) may be written in vector form as

$$M\gamma - Hv + r = Mu. \tag{9.7}$$

The first term in (9.7) represents the girth distribution after growth (during the interval $l \leq t < l + 1 - \epsilon$), the second term harvesting, the third term replanting. The term on the right-hand side of the equality represents the restored girth distribution.

Our new harvesting model is essentially a metered model like the ones we met in Sections 1.7 and 1.8. The long-term dynamics are described by (9.4)—if dynamics is the right word because, in the long term, the forest is kept in static equilibrium! To describe the short-term dynamics, let us agree to dub a vector like Mu a tree-count vector, because each entry records a number of trees in each class (whereas entries of u record proportions). Then, if Y is used to denote the tree-count vector *within* each cycle, the initial tree-count vector is $Y = Mu$. As a result of growth, the tree-count vector changes to $Y_a = M\gamma$. As a result of harvesting, the tree-count vector changes to $Y_b = Y_a - Hv$. As a result of replanting, the tree-count vector changes to $Y_c = Y_b + r$. This is the final tree-count vector within each cycle; hence $Y_c = Mu$. Eliminating Y_a, Y_b, and Y_c from these relations, we obtain once more (9.7).

To continue our study, we need an expression for γ in terms of u (or, which is the same thing, for Y_a in terms of Y). In other words, we need an expression for the girth distribution after one cycle of pure growth (no harvesting) in terms of the girth distribution at the beginning of that cycle. Now, we have already seen a model for pure tree growth, namely, the one introduced in Section 5.8. We may therefore wonder whether the latter model can be *combined* with (9.7) to yield an expression for u.

In Section 5.8, the cycle was six years long ($\sigma = 6$) and there were six girth classes ($N = 5$); which therefore, henceforward, we assume. We predicted that if $\pi(l)$ denoted the girth distribution after l cycles, then the girth distribution after $l + 1$ cycles would be $\pi(l + 1)$, where $\pi(l + 1)^T = \pi(l)^T S$ and the matrix S is defined by

$$S = \begin{bmatrix} a_0 & 1 - a_0 & 0 & 0 & 0 & 0 \\ 0 & a_1 & 1 - a_1 & 0 & 0 & 0 \\ 0 & 0 & a_2 & 1 - a_2 & 0 & 0 \\ 0 & 0 & 0 & a_3 & 1 - a_3 & 0 \\ 0 & 0 & 0 & 0 & a_4 & 1 - a_4 \\ 0 & 0 & 0 & 0 & 0 & 1 \end{bmatrix} \tag{9.8}$$

Strictly, $\pi(l)$ is a probability distribution vector. If M is large, however, then it is reasonable to interpret $\pi_i(l)$ as the proportion of forest actually observed in class i at time l. With this interpretation, we may identify u with $\pi(l)$ and γ with $\pi(l + 1)$. In other words, we may suppose that one cycle of pure growth is described by the equation $\gamma^T = \mathbf{u}^T \mathbf{S}$. Thus, on using (9.8), we obtain

$$\gamma_0 = a_0 u_0,$$
$$\gamma_i = (1 - a_{i-1})u_{i-1} + a_i u_i, \qquad 1 \le i \le 4, \qquad (9.9)$$
$$\gamma_5 = (1 - a_4)u_4 + u_5.$$

From (9.5b) with $N = 5$, $Hv_i = M(\gamma_i - u_i)$ trees must be harvested from class i, $1 \le i \le 5$, to keep the girth distribution stationary. Hence, on using (9.9), we get

$$Hv_i = M\left\{(1 - a_{i-1})u_{i-1} - (1 - a_i)u_i\right\}, \qquad 1 \le i \le 4,$$
$$Hv_5 = M(1 - a_4)u_4. \qquad (9.10)$$

Clearly, we cannot take a negative harvest from classes 1 to 5 because trees can be added only to class 0. Thus $v_i \ge 0$, $1 \le i \le 5$, implying in particular that

$$(1 - a_{i-1})u_{i-1} - (1 - a_i)u_i \ge 0, \qquad 1 \le i \le 4. \qquad (9.11)$$

Subject to this constraint, the total harvest is (verify)

$$H = \sum_{i=1}^{5} Hv_i = M(1 - a_0)u_0. \qquad (9.12)$$

(You can also obtain this result from (9.5a) and (9.9) as $H = M(u_0 - \gamma_0) = M(1 - a_0)u_0$.)

Let's now assume that the matrix $\mathbf{S}$ corresponds to $a_0 = 0.72$, $a_1 = 0.69$, $a_2 = 0.75$, $a_3 = 0.77$, and $a_4 = 0.63$, i.e., the values obtained in Exercise 8.6 and used in Section 5.8. Then, on using (9.3), (9.11), and (9.12), the sustainable harvest every six years is

$$0.28 M u_0 \qquad (9.13)$$

subject to

$$-0.28u_0 + 0.31u_1 \le 0,$$
$$-0.31u_1 + 0.25u_2 \le 0,$$
$$-0.25u_2 + 0.23u_3 \le 0, \qquad (9.14)$$
$$-0.23u_3 + 0.37u_4 \le 0,$$
$$u_0 + u_1 + u_2 + u_3 + u_4 + u_5 = 1, \qquad (9.15)$$

and, of course,

$$u_0 \ge 0, \qquad u_1 \ge 0, \qquad u_2 \ge 0, \qquad u_3 \ge 0, \qquad u_4 \ge 0, \qquad u_5 \ge 0. \quad (9.16)$$

We conclude immediately that the sustainable harvest is not unique. The girth balance of the forest could be maintained, for example, either by the distribution

$$\mathbf{u} = (0.25, 0.2, 0.2, 0.2, 0.1, 0.05)^T \tag{9.17}$$

or by the distribution

$$\mathbf{u} = (0.3, 0.2, 0.15, 0.15, 0.05, 0.15)^T. \tag{9.18}$$

Distribution (9.17) would yield a total harvest of $0.07M$ trees every six years, with harvest vector

$$H\mathbf{v} = M(0, 0.008, 0.012, 0.004, 0.009, 0.037)^T, \tag{9.19}$$

and (9.18) would sustain a total harvest of $0.084M$ trees, with harvest vector

$$H\mathbf{v} = M(0, 0.022, 0.0245, 0.003, 0.016, 0.0185)^T. \tag{9.20}$$

You should verify that (9.17)–(9.20) satisfy (9.10) and (9.12)–(9.16).

Now, although (9.18) sustains a greater yield than (9.17), it may not be a more profitable distribution because trees in girth class 5 have the most commercial value, and (9.19) has twice as many trees in this class as (9.20). Moreover, there may be distributions that are more profitable than either. How, then, shall we determine the most profitable of all possible distributions satisfying (9.14)–(9.16)? Let p_i denote the selling price of a tree from class i, and c_i the cost of felling it and planting a replacement. Let $\rho_i = p_i - c_i$ denote profit per tree. Then the profit from each harvest is

$$
\begin{aligned}
\sum_{i=1}^{5} \rho_i H v_i &= \sum_{i=1}^{4} \rho_i H v_i + \rho_5 H v_5 \\
&= \sum_{i=1}^{4} \rho_i M \left\{ (1 - a_{i-1}) u_{i-1} - (1 - a_i) u_i \right\} + \rho_5 M (1 - a_4) u_4 \\
&= M \left\{ \rho_1 (1 - a_0) u_0 + \sum_{i=1}^{4} (\rho_{i+1} - \rho_i)(1 - a_i) u_i \right\},
\end{aligned}
\tag{9.21}
$$

on using (9.10) and rearranging. Suppose, for example, that only trees in class 5 have commercial value. Then $\rho_1 = \rho_2 = \rho_3 = \rho_4 = 0$, so that (9.21) reduces to $M\rho_5(1 - \alpha_4)u_4 = 0.37M\rho_5 u_4$. Even though we do not know its value, $0.37M\rho_5$ is just a constant. Multiplication by a constant does not alter the values of the variables for which a utility function obtains its maximum. Hence the problem facing our forester is equivalent to the following optimization problem: find $\mathbf{u} = (u_0, u_1, u_2, u_3, u_4, u_5)^T$ which maximizes

$$J(\mathbf{u}) = J(u_0, u_1, u_2, u_3, u_4, u_5) = u_4 \tag{9.22}$$

subject to constraints (9.14)–(9.16). By writing down the relevant matrix of coefficients $\mathbf{A}$, and column vectors $\mathbf{b}$ and $\mathbf{c}$, you can easily verify that this problem belongs to the class of linear programming problems discussed in Section 3.2. The optimal value of $\mathbf{u}$, i.e., $\mathbf{u}^*$, can thus be obtained by using a computer package like IMSL's ZX3LP. On using this program, or a similar one, you will easily find that

$$\mathbf{u}^* = (0.2001, 0.1807, 0.2241, 0.2436, 0.1514, 0)^T, \qquad (9.23)$$

with $J(\mathbf{u}^*) = 0.1514$. Notice that we avoided calling $\mathbf{u}$ a control because the forester cannot control the number of trees in each class directly. Rather, he controls these quantities indirectly, through harvesting and planting. Of greater interest, therefore, is the optimal harvest vector $H\mathbf{v}^*$. From (9.10) and (9.23) we obtain

$$H\mathbf{v}^* = (0, 0, 0, 0, 0, 0.056M)^T. \qquad (9.24)$$

Thus the forester's optimal policy is to have no trees in girth class 5 at the beginning of a cycle and harvest all trees that have entered class 5 by the end of the cycle. The associated harvest $0.056M$ is the one that maximizes the forester's profit.

    If trees in class 5 are the only ones with commercial value, then it isn't surprising that the optimal policy is one in which only trees in class 5 are harvested. It is surprising, however, that, if trees in class 4 yield more than a certain fraction of the profit from those in class 5, then the optimal policy is to harvest all trees when they enter class 4; and thus, by preventing trees from ever entering class 5, to ignore completely the commercial value of the latter. Or is it? See Exercise 9.2.

9.3 Cleaning Lake Ontario. An Adaptation

In Section 1.1, we constructed a model that predicted how pollution would decay in a lake if all pollution input had suddenly ceased. Let $x(t)$ denote the concentration of contaminant at time t, V the lake volume (assumed constant), and r the flux of water through the lake, i.e., the volume of water that flows in or out per unit of time. Then the model of Section 1.1 predicts that

$$x(t) = x(0)e^{-t/T}, \qquad (9.25)$$

where $T = V/r$ is the drainage time—i.e., the time it would take the water to flow away at rate r if water inflow suddenly ceased. Where does all this pollutant go? It flows out of the lake at the rate

$$-\frac{d}{dt}\{Vx(t)\} = rx(t), \qquad (9.26)$$

on using (9.25), or on using (9.28) below with zero pollution inflow. We might, for example, apply this model to Lake Erie. Then, if we use a subscript E to denote Lake Erie, our model predicts that its pollution flows away at the rate $r_E x_E(t)$, where

$$x_E(t) = x_E(0)e^{-t/T_E} \qquad (9.27)$$

and $T_E = V_E/r_E$ is the drainage time for Lake Erie. From Section 1.1, T_E is approximately 2.6 years.

We have already seen in Section 4.2, however, that (9.25) cannot be applied to Lake Ontario, even if, as we shall assume, all pollution inflow to the Great Lakes as a whole has ceased. This is because the pollution that was already in Lake Erie will continue to flow into Lake Ontario. Because the existing model cannot be used, let's therefore adapt the model

$$
\begin{array}{ccccc}
\text{Rate of change} & = & \text{Pollution} & - & \text{Pollution} \\
\text{of pollution} & & \text{inflow} & & \text{outflow}
\end{array}
$$

$$
\frac{d}{dt}\{Vx(t)\} \quad = \quad ? \quad - \quad rx(t) \tag{9.28}
$$

to our present purpose. In this equation, $x(t)$ now denotes concentration of contaminant in Lake Ontario, V its volume, and r its water outflow. We will assume that the water inflow to Lake Ontario (including rainfall) balances the water outflow (including evaporation), so that the water inflow is also r. We know from Section 4.2 that 5/6 of this inflow is the outflow from Lake Erie; i.e., $r_E = 5r/6$. The pollution outflow from Erie is $r_E x_E(t)$, and so the pollution inflow to Ontario must be $r_E x_E(t) = 5rx_E(t)/6$. We can therefore replace the question mark in (9.28) by $5rx_E(t)/6$. Dividing by r, we thus obtain the equation

$$
T\frac{dx}{dt} + x = \frac{5}{6}x_E(t), \tag{9.29}
$$

where $T \equiv V/r$ is the drainage time for Lake Ontario. From Section 4.2, T is about 7.8 years. On using (9.27), the solution of this differential equation is (Exercise 9.3)

$$
x(t) = e^{-t/T}\left\{x(0) - \frac{5T_E x_E(0)}{6(T - T_E)}\left(e^{(1/T - 1/T_E)t} - 1\right)\right\}. \tag{9.30}
$$

Let's assume that $T = 3T_E$, which is an excellent approximation, and define

$$
\lambda = \frac{x_E(0)}{x(0)}. \tag{9.31}
$$

We thus obtain

$$
x(t) = x(0)e^{-t/T}\left\{1 + \frac{5\lambda}{12}\left(1 - e^{-2t/T}\right)\right\}. \tag{9.32}
$$

We do not know the value of λ, but Erie is more polluted than Ontario, so that $\lambda > 1$. Let $t_{0.05}$ denote the 5% cleaning time for Lake Ontario—i.e., the time it takes to reduce $x(t)$ to $0.05x(0)$—and define

$$
z = e^{-t_{0.05}/T}. \tag{9.33}
$$

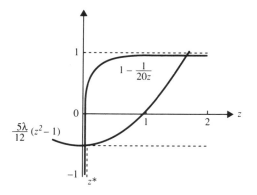

Fig. 9.1 Graphical determination of $z^*(\lambda)$.

Then $0 < z < 1$, and from (9.32),

$$\frac{5\lambda}{12}(z^2 - 1) = 1 - \frac{1}{20z}.$$ (9.34)

It is clear from Fig. 9.1 that this equation has a single solution, z^*, satisfying $0 < z^* < 1$. For any given λ, it can be found numerically by the Newton–Raphson method. In this way, we can determine

$$t_{0.05} = -7.8\ln(z^*(\lambda))$$ (9.35)

as a function of λ. The results are shown in Table 9.1 for values of λ between 1 and 2. Notice, however, from Fig. 9.1 that z^* is small and gets smaller and smaller as λ increases. For larger λ, therefore, it is quite permissible to ignore the term z^2 in (9.34) on the grounds that it is negligibly small.

Table 9.1 Five percent cleaning times for Lake Ontario, rounded to three significant figures.

λ	$t_{0.05}$
1.0	26.1
1.2	26.5
1.4	26.9
1.6	27.3
1.8	27.7
2.0	28.1

Then (9.34) is easy to solve and yields

$$t_{0.05} = -7.8\ln(z^*(\lambda)) \approx 7.8\ln\left\{\frac{5(5\lambda + 12)}{3}\right\}. \tag{9.36}$$

You can easily check that this approximation is acceptable even for the low values of λ in Table 9.1. (Note that the table appears to suggest a linear relationship $\partial t_{0.05}/\partial\lambda = 2$, but this is an illusion caused by rounding. From (9.36) we obtain directly that $\partial t_{0.05}/\partial\lambda \approx 7.8/(\lambda + 2.4)$. To one significant figure, this equals 2 for any λ satisfying $1 \leq \lambda \leq 2$.)

9.4 Cleaning Lake Ontario. An Extension

In the previous section we tacitly assumed that pollution was uniform throughout our lake at any given time. But might not pollution vary in intensity from one end of the lake to the other? We can extend our model to allow for this possibility. Before doing so, we remark once again that the creative process of extending a model may be hindered in the early stages by undue attention to validation. Therefore we shall assume freely, suspending judgment upon those assumptions until the end of the section.

Consider, therefore, a lake with constant vertical cross section A, occupying the region $a \leq z \leq b$. Thus the length of the lake is $b - a$, and any horizontal section is rectangular (Fig. 9.2). The primary inflow of water is at $z = a$, and the primary outflow is at $z = b$. The concentration of contaminant (in grams per cubic meter, say) at time t now also depends upon z, and so we will denote it by $\phi(z, t)$. Let $F(z, t)$ denote the flow (or "flux") of pollution at station z at time t, i.e., the amount of pollution (in grams) that crosses the section of area A in a unit of time. Let $Q(z, t)$ be the horizontal water flux—i.e., the volume of water (in cubic meters) that crosses the section of area A per unit of time—in the direction of increasing z. Because the volume crossing per unit of time is simply the velocity of water times the cross-sectional area of the lake, the flow velocity is given by

$$v(z, t) = \frac{Q(z, t)}{A}. \tag{9.37}$$

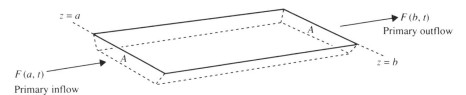

Fig. 9.2 Ideal lake occupying $a \leq z \leq b$. Vertical cross section A is constant. Any horizontal cross section is rectangular.

Moreover, a moment's thought reveals that

$$
\begin{array}{ccccc}
\text{Flux of pollution} & & \text{Amount of} & & \text{Volume of water} \\
\text{per unit of time} & = & \text{pollution} & \times & \text{flowing per unit} \\
& & \text{per unit} & & \text{of time} \\
& & \text{volume of water} & &
\end{array}
\tag{9.38}
$$

$$
F(z,t) \quad = \quad \phi(z,t) \quad \times \quad Q(z,t)
$$

for all $z(a \le z \le b)$ and all $t(\ge 0)$.

The volume of water contained in the infinitesimal interval from $z - \frac{1}{2}\delta z$ to $z + \frac{1}{2}\delta z$ is $A\delta z$. Hence the amount of pollution contained therein is $A\delta z \{\phi(z,t) + O(\delta z)\} = A\phi(z,t)\delta z + o(\delta z)$, on using the notation introduced at the beginning of Chapter 1. Thus the amount of pollution in the entire lake is

$$
x(t) = \int_a^b A\phi(z,t)dz,
\tag{9.39}
$$

the error terms vanishing during the limit process of integration. Instead of (9.28) we obtain

$$
\begin{array}{ccc}
\text{Rate of change} & = & \text{Pollution} \quad - \quad \text{Pollution} \\
\text{of pollution} & & \text{inflow} \qquad \text{outflow}
\end{array}
\tag{9.40}
$$

$$
\frac{\partial}{\partial t}\int_a^b A\phi(z,t)dz = F(a,t) - F(b,t).
$$

We use the symbol $\partial/\partial t$ in place of d/dt for differentiation with respect to time, because the quantities involved now also depend upon z. Similarly, we shall use $\partial/\partial z$ in place of d/dz to denote differentiation with respect to z. On using Leibnitz's rule, we deduce from (9.40) that

$$
\int_a^b A\frac{\partial \phi}{\partial t}dz = - \int_a^b \frac{\partial F}{\partial z}dz,
\tag{9.41}
$$

whence

$$
\int_a^b \left(A\frac{\partial \phi}{\partial t} + \frac{\partial F}{\partial z} \right) dz = 0.
\tag{9.42}
$$

Now imagine the lake to consist of many sublakes, each with the same cross-sectional area A; and let a typical sublake stretch from $z = z_0 - \epsilon$ to $z = z_0 + \epsilon$. The entire argument that led to (9.42) is also valid for the sublake; hence

$$
\int_{z_0-\epsilon}^{z_0+\epsilon} \left(A\frac{\partial \phi}{\partial t} + \frac{\partial F}{\partial z} \right) dz = 0.
\tag{9.43}
$$

If $\phi(z,t)$ has continuous derivatives at $z = z_0$, then the only way in which (9.43) can remain true in the limit as $\epsilon \to 0$ is for the integrand to vanish at $z = z_0$. For let $\omega(z,t) \equiv A\partial\phi(z,t)/\partial t + \partial F(z,t)/\partial z$. Then we wish to show that $w(z_0, t) = 0$. Suppose $w(z_0, t) \neq 0$. Then either $w(z_0, t) > 0$ or $w(z_0, t) < 0$. If $w(z_0, t) > 0$ then, because w is continuous, there exists $\eta > 0$ such that $z_0 - \eta < z < z_0 + \eta$ implies $w(z, t) > 0$. Choose $0 < \epsilon < \eta$, permissible in the limit as $\epsilon \to 0$. Then the left-hand side of (9.43) is positive, a contradiction. For $w(z_0, t) < 0$, the argument above can be repeated with $-w$ in place of w. We thus establish that $w(z_0, t) = 0$. But z_0 is arbitrary. Hence

$$A\frac{\partial \phi}{\partial t} + \frac{\partial F}{\partial z} = 0 \qquad \text{for all } a \leq z \leq b, \, 0 \leq t < \infty, \tag{9.44}$$

or on using (9.38),

$$A\frac{\partial \phi}{\partial t} + \frac{\partial}{\partial z}(Q\phi) = 0, \qquad a \leq z \leq b, \, 0 \leq t < \infty. \tag{9.45}$$

In practice, $\phi(z, t)$ may be discontinuous across isolated curves in the z-t plane, but (9.44) is still valid everywhere between those curves. (An example is provided by the solid curves in Fig. 9.4, for which $\phi(z, t)$ is discontinuous across the curve $z = vt$ in the z-t plane.)

Equation (9.45) is a partial differential equation for $\phi(z, t)$, whose solution can be found once $Q(z, t)$ has been specified. For partial differential equations, as for ordinary differential equations, initial or boundary conditions are needed to make the solution unique. In this particular case, it is sufficient to prescribe the boundary values $\phi(a, t)$ and the initial values $\phi(z, 0)$.

To make matters simple, let's assume that

$$Q(z, t) = R = \text{constant}, \tag{9.46}$$

for all $a \leq z \leq b$ and $0 \leq t < \infty$. Now, the primary inflow to Lake Ontario is from Lake Erie, through the Niagara river. Let this inflow correspond to $z = a$, and let the lake's discharge into the St. Lawrence River correspond to $z = b$. Then, from (9.38), (9.46), and Section 9.3, the pollution inflow to Ontario is $r_E x_E(t) = F(a, t) = R\phi(a, t)$; whence

$$\phi(a, t) = \frac{r_E x_E(t)}{R}. \tag{9.47}$$

All we know about existing pollution levels is that Erie is more polluted than Ontario. Therefore, we simply define

$$\phi(z, 0) = P(z) \tag{9.48}$$

and assume that

$$P(z) < x_E(0), \qquad a \leq z \leq b. \tag{9.49}$$

The solution of (9.45) subject to (9.46)–(9.48) is (Exercise 9.5)

$$\phi(z, t) = \begin{cases} \dfrac{r_E}{R} x_E(t - A(z - a)/R), & \text{if } t > A(z - a)/R; \tag{9.50a} \\ P(z - Rt/A), & \text{if } t < A(z - a)/R. \tag{9.50b} \end{cases}$$

In practice, we would expect $\phi(z, t)$ to be continuous everywhere, including $z = a$; thus $P(a) = r_E x_E(0)/R$, and either of the two strict inequalities in (9.50) can be interpreted as a weak one. In theory, however, the wave phenomena apparent in Fig. 9.4 do not require the continuity of ϕ at $z = a$. (In Fig. 11.5 and Fig. 11.3, for example, $\phi(z, t)$ and $v(z, t)$ are discontinuous across the curve $z = v_{max}t$.)

Now the volume of the lake $a \leq z \leq b$ is $V = A(b - a)$, so that the drainage time is $V/R = A(b - a)/R$. The cleaning time is bound to exceed this, and hence exceeds $A(z - a)/R$ for any value of z. Thus the 5% cleaning time will be obtained from (9.50a), or on using (9.27), from

$$\phi(z, t) = \frac{r_E}{R} x_E(0) \exp\left\{ -\frac{(t - A(z - a)/R)}{T_E} \right\}, \qquad (9.51)$$

where $t > A(z - a)/R$. Notice that (9.51) is an increasing function of z. Thus the 5% cleaning time, which varies with z, will be longest at the St. Lawrence end, $z = b$. The longest time is perhaps the one that is most of interest to environmentalists. Therefore, let's denote by $t_{0.05}$ the time it takes to reduce $\phi(b, t)$ to 5% of its initial value $\phi(b, 0) = P(b)$. Then, from (9.51), we have

$$0.05P(b) = \frac{r_E}{R} x_E(0) \exp\left\{ -\frac{(t_{0.05} - V/R)}{T_E} \right\}, \qquad (9.52)$$

which on defining

$$\mu \equiv \frac{x_E(0)}{P(b)} \qquad (9.53)$$

yields

$$t_{0.05} = \frac{V}{R} + T_E \ln\left(\frac{20\mu r_E}{R} \right). \qquad (9.54)$$

Note that $\mu > 1$ from (9.49).

Now we have been assuming all along that total (horizontal and vertical) water inflow balances total water outflow; both are equal to the constant r. But we have assumed in this section that the horizontal water flow is also a constant, namely, R. We are therefore obliged to assume that rainfall and evaporation are both equal to $r - R$. Moreover, if we assume that the only river inflow to Lake Ontario is from Lake Erie, then we are also obliged to assume that $r_E = R$; otherwise (9.46) would be false at $z = a$. Thus $R = r_E = 5r/6$, and (9.54) becomes

$$t_{0.05} = \frac{6}{5} T + T_E \ln(20\mu), \qquad (9.55)$$

where $T = V/r$ is the Ontario draining time. Because $T = 3T_E$ from Section 9.3, (9.55) is approximately

$$t_{0.05} = \left(2.2 + \frac{1}{3} \ln \mu \right) T. \qquad (9.56)$$

Comparison with the model of the previous section is facilitated by assuming that $P(z)$ = constant. Then the parameter μ defined by (9.53) is identical to the parameter λ defined by (9.31). When we divide (9.56) by (9.36), the ratio of the later model's 5% cleaning time to that of the previous section is

$$\Delta(\lambda) = \frac{2.2 + \frac{1}{3}\ln\lambda}{\ln\left\{\frac{5}{3}(5\lambda + 12)\right\}}, \tag{9.57}$$

which is tabulated in Table 9.2 for values of λ between 1 and 8. Notice how slowly this function varies: its value never differs much from 2/3 for moderate values of λ. Hence, even though we have no information about λ (other than $\lambda > 1$), we can say with confidence that cleaning Lake Ontario according to the model of the present section is always about a third faster than cleaning it according to the model of Section 9.3.

But why should this be so? Why, moreover, should the cleaning time predicted by (9.56) be smaller than even the cleaning time predicted by the model we rejected in Section 4.2, for moderate values of $\lambda(1 < \lambda < 11)$? The answer is that the slower cleaning of Section 9.3 is due to the assumption of perfect mixing, which makes ϕ independent of z. To understand this, let us consider the ideal lake of length $4v$ meters depicted in Fig. 9.3, where $v = R/A$ is the velocity of water. In a unit of time, water travels v meters to the right. We will assume that pollution input has ceased but that, initially, each v meters of lake contains 16 grams of pollutant. Pollution levels are then recorded at the end of each unit of time. In the left-hand column of Fig. 9.3 it is assumed that immediately before a reading is taken—i.e., immediately after the pollution in the last v meters of lake has been lost—a magic stick swoops down from the sky and stirs the lake perfectly. In the right-hand column of the figure, no such mixing occurs. You can readily see that in the first case the lake will never be completely clean and will

Table 9.2

λ	$\Delta(\lambda)$
1.0	0.658
1.5	0.671
2.0	0.675
2.5	0.675
3.0	0.674
3.5	0.672
4.0	0.669
5.0	0.664
6.0	0.658
7.0	0.653
8.0	0.648

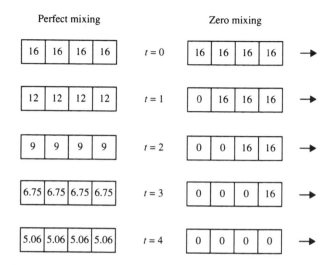

Perfect mixing   Zero mixing

| 16 | 16 | 16 | 16 | $t = 0$ | 16 | 16 | 16 | 16 | →

| 12 | 12 | 12 | 12 | $t = 1$ | 0 | 16 | 16 | 16 | →

| 9 | 9 | 9 | 9 | $t = 2$ | 0 | 0 | 16 | 16 | →

| 6.75 | 6.75 | 6.75 | 6.75 | $t = 3$ | 0 | 0 | 0 | 16 | →

| 5.06 | 5.06 | 5.06 | 5.06 | $t = 4$ | 0 | 0 | 0 | 0 | →

Fig. 9.3 The difference in cleaning time between an unmixed lake (right-hand column) and one that is perfectly mixed. The lake is assumed to be $4v$ meters long; water flows to the right; and the numbers in the boxes represent the amounts of pollution in lengths of v meters.

take a long time (11 time units, see Exercise 9.6) to reach a lower pollution level than 5% of the initial one. In the second case, however, the lake is completely clean after only 4 time units. The diagram further suggests that in the absence of fresh pollution, a lake of length $b - a$ meters should be absolutely clean after a time $t_{0.0} = (b - a)/v = A(b - a)/R$ if no mixing takes place. We can verify this by supposing that the pollution input from Lake Erie to Ontario can magically be prevented ($x_E = 0$). Then, according to (9.50),

$$\phi(z, t) = \begin{cases} P(z - vt), & \text{if } t < \dfrac{z - a}{v}; \\ 0, & \text{if } t > \dfrac{z - a}{v}, \end{cases} \qquad (9.58)$$

so that there are no positive values of ϕ for $t > t_{0.0}$.

We have made our first encounter with an important phenomenon, namely, that of a propagating wave. In (9.58) we have the simplest example of a wave that propagates with velocity v in the direction of increasing z. For the initial pollution level at $z = z_0$ is $P(z_0)$. According to (9.58), the same pollution level will subsequently be found wherever $P(z - vt) = P(z_0)$, i.e., wherever $z - vt = z_0$ or $z = vt + z_0$. But this is just the equation of motion that the cross section $z = z_0$ would have if it moved to the right with velocity v. The effect is illustrated in Fig. 9.4 by the solid curves for an

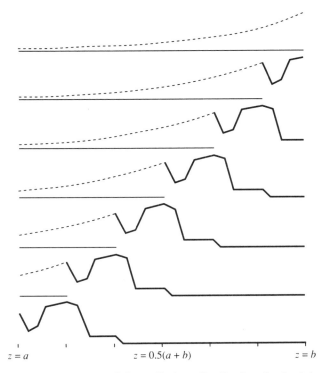

Fig. 9.4 Seven "photographs" of the pollution distribution in the lake $a \leq z \leq b$, taken at times $j(b-a)/6v$ for $j = 0, 1, \ldots, 6$; z increases to the right; time increases upward. The solid curves alone are a pictorial representation of (9.58). The dashed curves and solid curves together are a pictorial representation of (9.50).

arbitrary humpbacked initial pollution distribution $P(z)$. Equation (9.50) is also that of a wave propagating to the right with velocity v because (9.50) is also a function of the single argument $z - vt$ (you should satisfy yourself that this is so). It is represented pictorially in Fig. 9.4 by a series of "photographs," the first taken at $t = 0$ and the later ones every $(b-a)/6v$ units of time. See how Ontario's old pollution (solid curve) is convected out at velocity v, dragging behind it the fresh pollution (dashed curve) that is decaying in Lake Erie.

The real Lake Ontario is crudely sketched in Fig. 9.5. Its horizontal section is not rectangular, and its vertical section is not constant. Moreover, according to the article on the Great Lakes in *Encyclopaedia Britannica* by Professor Beeton, only 26 of Ontario's 34 inches of annual rainfall evaporate, so that rainfall and evaporation cannot both be equal to $r/6$. Does this mean that our model is useless?

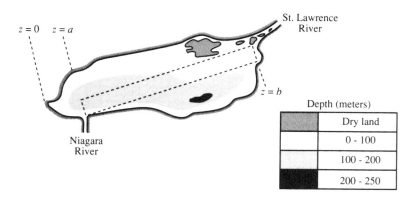

Fig. 9.5 Cartoon map of Lake Ontario (sketched freehand after the map of Ontario in the Great Lakes article in *Encyclopaedia Britannica*).

Of course not. It simply means that predictions from the model must be applied with caution and that further refinements are desirable; and this statement will always be true, no matter how much the model is refined and tested. Some refinements are quite straightforward and are broached in Exercise 9.7. Other refinements are more involved but are broached in Chapter 11. Still further requirements are possible but lie far beyond the scope of this text.

What about our expectation in Section 1.1 that poor mixing would prolong the cleaning? Has this been scotched by the present model with zero mixing? By no means. A lake that is fed by, and discharges into, a major river will have strong central currents sandwiched between more sluggish regions where water recirculates. A speculative rectangle has been superimposed on Fig. 9.5 to denote the region of strong central flow. Cleaning would be rapid here and might well be adequately modelled by (9.54) if we took V in that formula to denote the volume of the sublake corresponding to the rectangle. Because of poor mixing between this region and the more sluggish outer ones, however, the pollution in the outer regions can escape only by "diffusing" (a slow process, Section 9.5) into the central core and being swept out—so that these regions take longer to clean than if the lake were thoroughly mixed.

We remark finally on a further distinction between the present model and that of Section 9.3. There, horizontal fluxes of water could not be distinguished from vertical ones, so that pollution could escape by evaporation. Here, the fluxes are separated, so that pollution can escape only by river. In this respect, the present model is definitely an improvement, because most pollutants are known not to leave with water vapor, the major exception being DDT; see, for example, Harte (1985, p. 37).

9.5   Pure Diffusion of Pollutants. A Combination

Toward the end of the previous section we remarked that lakes may have strong central currents sandwiched between more sluggish flows, with minimal mixing between the two regions. Pollution in the central core will be swept out rapidly; but how does pollution leave the sluggish regions?

To make matters simple, suppose that the sluggish regions are as sluggish as possible: the water is at rest. Assume, moreover, that the central core is thoroughly clean at all times, on the grounds that no new pollution is being added to the lake; that which has diffused across from the sluggish regions has negligible concentration, in view of the rapidity of the central current. Let us further assume that, *initially*, the sluggish regions have constant concentration of contaminant, ϕ_0. (These assumptions greatly simplify the mathematics with little loss of generality, in the sense that the important features of the solution would be the same for any other initial distribution.) The water flow is therefore as shown in Fig. 9.6(a), and the initial pollution distribution is as shown in Fig. 9.6(b). The rectangle represents the central core of the lake; the shape of the outer regions is arbitrary.

Now consider the rectangular sublake, with constant vertical cross-sectional area B, which occupies the region $y \geq 0$, $\xi_1 \leq z \leq \xi_2$, shaded in Fig. 9.6(c). This region is finite, of course, but if we are mainly interested in how the pollution in the vicinity of $y = 0$ interacts with the adjacent core, then it is quite permissible to pretend that the lake extends forever in the direction of increasing y. In this subregion, therefore, the remarks of the previous paragraph ensure that

$$\begin{aligned}
\phi(0, t) &= 0, & 0 &< t < \infty, \\
\phi(y, 0) &= \phi_0, & 0 &< y < \infty,
\end{aligned}$$

(9.59)

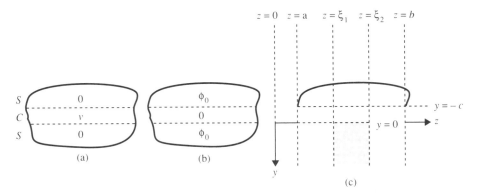

Fig. 9.6 Idealized lake with rapid core between regions at rest. C denotes core, S denotes sluggish.

where $\phi(y, t)$ denotes the concentration of pollution at station y at time t. As we will discover shortly, the initial discontinuity at $y = 0$ poses no mathematical problems and is therefore a convenient idealization of what would, in practice, be a very sharp pollution gradient.

Now the sublake $\xi_1 \leq z \leq \xi_2$, $0 \leq y < \infty$ with constant vertical cross section B should remind you of the sublake $-c \leq y \leq 0$, $a \leq z \leq b$, which we considered in the previous section. The only difference is that, whereas in Section 9.4 pollution moved in the z direction, now it will move in the y direction instead. Therefore, if $G(y, t)$ denotes the flux of pollution in the direction of increasing y—i.e., the amount of pollution per unit time crossing the section with area B—then the entire development from (9.39) to (9.44) remains valid, provided only that we replace A by B, F by G, and z by y. Thus

$$B\frac{\partial \phi}{\partial t} + \frac{\partial G}{\partial y} = 0, \qquad 0 < y < \infty, \ 0 < t < \infty. \qquad (9.60)$$

But why should pollution flow (have a flux) to begin with?

To understand this, we have to combine the ideas we borrowed from Section 9.3 with those we developed in Sections 2.5 and 2.6. Temperature is just the concentration of heat. In Section 2.5, we observed that heat flows from regions of high heat concentration to regions of low heat concentration, and that the flux per unit area is proportional to minus the gradient of heat concentration. Velocity is just the concentration of momentum (per unit mass). In Section 2.6, we observed that momentum flows from regions of high momentum concentration to regions of low momentum concentration, and that the flux per unit area is proportional to minus the gradient of momentum concentration. Is it therefore not reasonable to assume that pollution will flow from regions of high pollution concentration to regions of low pollution concentration, and that the flux per unit area is proportional to minus the gradient of the concentration? Let γ be the constant of proportionality. Then the actual flux is given by

$$G(y, t) = -B\gamma\frac{\partial \phi}{\partial y}. \qquad (9.61)$$

It is important to understand the physical difference between this pollution flux and the pollution flux

$$F(z, t) = Av\phi(z, t) \qquad (9.62)$$

of Section 9.4. In the previous section, pollution was swept out of the lake with the flow of water, as though the pollutant particles took piggyback rides on adjacent particles of water. We call this process *convection*. But here there is simply no flow of water; piggyback rides are not available. How, then, is pollution transported? The answer is that in the absence, as it were, of public transport, the pollution moves privately, by virtue of its molecular motion. The water may be motionless at the visible level, but it moves continually at the invisible, microscopic level. The microscopic motion of the water molecules does not concern us because it plays no part in cleaning the lake, but the microscopic motion of the pollutant molecules can be imagined to occur in discrete time steps, in each of which a pollutant molecule leaps forward or backward with equal probability. Imagine that there are infinitely many stations along the y-axis where a molecule can be; label them $m = 0, 1, 2, \ldots$. Interpret the initial pollutant profile

$$\phi(y, 0) = 0 \qquad \text{if } y = 0,$$
$$\phi(y, 0) = \phi_0 \qquad \text{if } y > 0 \tag{9.63}$$

as meaning that there are no molecules at station 0 but 100σ molecules at station m, for each $m \geq 1$. This corresponds to the first row of Table 9.3, where the factor σ is omitted and molecules are shown as a percentage of the original number. This table should be interpreted as merely the first few rows (time) and columns (stations) of an infinite matrix. Because a molecule at station m leaps to either $m - 1$ or $m + 1$ with equal probability, during the first time step roughly 50 of every 100 molecules that were at m will have moved to $m - 1$ and the remainder will have moved to $m + 1$. Imagine that the 50 molecules leaping from 1 to 0 are swept away instantly by the rapid core, all others remaining in the system until at least the next time step. After the first time step, the molecule distribution is as shown in the first row of Table 9.3. Be sure you appreciate that none of the molecules at (say) $m = 2$ at time $l = 1$ were there at $l = 0$; 50% came from $m = 1$, and the other fifty percent from $m = 3$. The evolution of the molecule distribution over the next six time steps is given by the remaining rows of Table 9.3. You may extend this table for higher values of m and l yourself if you wish, but the pattern is already clear. There is a slow migration of molecules to the left—slow because, having taken a step to the left, an individual molecule is just as likely to step back to the right as move on to the left at the subsequent time step. This slow migration of pollutant molecules is known as molecular diffusion. At the macroscopic

Table 9.3 First seven time steps of schematic representation of molecular diffusion.

l \ m	0	1	2	3	4	5	6	7	8
0	0	100.0	100.0	100.0	100.0	100.0	100.0	100.0	100.0
1	0	50.0	100.0	100.0	100.0	100.0	100.0	100.0	100.0
2	0	50.0	75.0	100.0	100.0	100.0	100.0	100.0	100.0
3	0	37.5	75.0	87.5	100.0	100.0	100.0	100.0	100.0
4	0	37.5	62.5	87.5	93.75	100.0	100.0	100.0	100.0
5	0	31.25	62.5	78.13	93.75	96.88	100.0	100.0	100.0
6	0	31.25	54.69	78.13	87.5	96.88	98.44	100.0	100.0
7	0	27.34	54.69	71.09	87.5	92.97	98.44	99.22	100.0

or visible level, where the pollution appears as a continuous substance because the individual molecules are invisible, this molecular diffusion manifests itself as a flow in the direction of decreasing concentration of contaminant. This diffusive flux is represented microscopically by Table 9.3 and macroscopically by (9.61). The constant of proportionality, γ, is known as the (molecular) *diffusivity* and has the dimensions of LENGTH^2/TIME, as you should verify for yourself. The diffusivity is typically a small number because it represents the magnitude of the *net* molecular motion, and most of the molecules' individual leaps just cancel each other out.

The size of the diffusivity is the basis for ignoring molecular diffusion in the longitudinal direction. It will be clear from the arguments above that there is a diffusive flux in the z direction also, namely, $-\gamma \partial\phi/\partial z$ per unit area. Thus, on using (9.62), the total flux of pollutant in the z direction is

$$A\left(v\phi - \gamma\frac{\partial\phi}{\partial z}\right). \tag{9.64}$$

The first of the two terms in (9.64) so dominates the second when γ is small,[1] however, that we may ignore the second one entirely. In the molecular world, public transport is always better than private!

Substituting (9.61) into (9.60) we get

$$\frac{\partial\phi}{\partial t} = \gamma\frac{\partial^2\phi}{\partial y^2}. \tag{9.65}$$

Both this equation and initial and boundary conditions (9.59) are satisfied (Exercise 9.8) by

$$\phi(y,t) = \frac{2\phi_0}{\sqrt{\pi}}\int_0^{\frac{y}{2\sqrt{\gamma t}}} e^{-\xi^2}d\xi. \tag{9.66}$$

This solution is plotted in Fig. 9.7 as a function of y for various fixed values of t. The 5% cleaning time at station y, $t_{0.05}$, is given by $\phi(y, t_{0.05}) = 0.05\phi_0$. Thus on using (9.66), we obtain

$$t_{0.05} = \frac{y^2}{4\gamma\beta^2}, \tag{9.67}$$

where β is (uniquely) defined by

$$\int_0^{\beta} e^{-\xi^2}d\xi = 0.025\sqrt{\pi}. \tag{9.68}$$

To four significant figures (Exercise 9.9), we have $\beta = 0.04434$; whence

$$t_{0.05} \approx \frac{127y^2}{\gamma}. \tag{9.69}$$

---

[1] More accurately, when the dimensionless parameter $v(b-a)/\gamma$ is large. This parameter is a first cousin of the Reynolds number we met in Section 4.10.

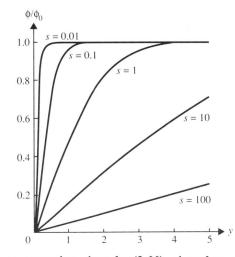

Fig. 9.7 Pollution concentration given by (9.66), plotted as a function of y when $t = s/\gamma$, for various values of s.

Thus cleaning time increases with the square of distance from the rapidly flowing core. The precise value of γ depends, of course, on the chemical properties of the pollutant but is probably of the order of 10^{-8} square miles per year.[2] At this rate, it would take about 12.7 billion years to clean the water even one mile from the core, almost certainly longer than most environmentally conscious people would be prepared to wait! Fortunately, no lake is ever completely stagnant, especially one with a rapid core; and the effect of the mixing will be to accelerate the cleaning time. But our model shows clearly how little we can rely on pure diffusion as a purifying process. It is much too slow. This is also apparent in Fig. 9.7, where the pattern suggested by Table 9.3 is clearly discernible.

   In our studies of lake purification, we have encountered three different cleaning times. These are listed in Table 9.4 as rough, order of magnitude estimates, together with verbal descriptions of the mechanisms that produced them. Because γ is so small, we find

$$T_D < T_1 \ll T_2, \tag{9.70}$$

where $\ll$ denotes "very much less than." In practice, all three mechanisms would occur together, but it is important to begin modelling them by isolating their separate effects. Thus the simple mathematical models of Sec-

---

[2] Or $10^{-5}\,\text{cm}^2/\text{sec}$. The diffusivity of sugar in water at room temperature, for example, is only $0.3 \times 10^{-5}\,\text{cm}^2/\text{sec}$ (explaining why coffee has to be stirred), and that of common salt is $1.1 \times 10^{-5}\,\text{cm}^2/\text{sec}$. See Landau et al. (1967, p. 332).

Table 9.4 Order of magnitude estimates of 5% cleaning time, by various mechanisms, for ideal lake into which all pollution inflow has ceased.

Cleaning mechanism	Order of magnitude	Symbol	Reference in text
Convection, no mixing	Drainage time	T_D	(9.54)
Convection, perfect mixing	$3T_D$	T_1	(1.11)
Pure diffusion	Area/γ	T_2	(9.69)

tions 1.1, 9.3–9.5 may be regarded as paradigms, which describe the effects of convection, diffusion and mixing in their purest form. Once understood in isolation, these basic mechanisms provide the foundations for more advanced, more realistic models, such as the ones derived in Section 11.6 and Exercise 11.15, in which all three mechanisms can be combined.

We conclude this section by noting that, strictly, we have not described the mixing mechanism—we have merely made the assumption of perfect mixing. The invisible spoon that stirs the water is a phenomenon known as *turbulence*. Certain features of it can broadly be described by assuming the concentration flux to have the form (9.61), with a very much larger value of γ (the turbulent, or "eddy," diffusivity) than would be appropriate for molecular diffusion; but, beyond that, modelling turbulence is extremely difficult and remains an active topic of ongoing research.

9.6 Modelling a Population's Age Structure. A First Attempt

In planning for its future, a nation needs to know, among other things, how many schools to build, how many hospitals to build, how many teachers and doctors to train, and how much tax to collect for social security and medical care.[3] All these things require a knowledge of the population's age structure—how many old people there will be, how many children there will be, the size of the potential labor force. Whether the logistic model of population growth is true or false, the best it can do is to predict a *total* population at time t, denoted by $x(t)$. It cannot predict the number of people aged z within that population. We must therefore build a new model, one that is capable of providing such information.

The model we develop here is an *adaptation* of the lake core model of Section 9.4. Let us think of life as a fast flowing river, along which distance is measured in years. The usual term for distance along the river of life is age; we'll denote it by z. Let us take the extremely morbid but nevertheless convenient view that an individual is just a particle of contaminant in life's

[3]Before proceeding, you may wish to review your solutions to Exercises 4.15–4.17 and the first part of Section 8.6.

river. Accordingly, let $\phi(z, t)$ denote the amount of age-z "contaminant" in unit length of life's river at time t (also measured in years), in the sense that the number of people between a and b years of age at time t is

$$\int_a^b \phi(z, t)dz. \tag{9.71}$$

In particular, if ω is the greatest age to which anyone can live (say, 115 years), then the nation's population magnitude at time t is given by

$$x(t) = \int_0^\omega \phi(z, t)dz, \tag{9.72}$$

and life's river extends from $z = 0$ to $z = \omega$. For many purposes, a more useful measure of the population's age structure than (9.71) is the proportion of the total population aged z at time t. We therefore define

$$\pi(z, t) \equiv \frac{\phi(z, t)}{x(t)}. \tag{9.73}$$

Then the proportion of the population between ages a and b is

$$\int_a^b \pi(z, t)dz, \tag{9.74}$$

and

$$\int_0^\omega \pi(z, t)dz = 1. \tag{9.75}$$

Our lake core, or river, model cannot be adapted in its present form because it assumes that all "contaminant" enters the river at $z = 0$ and leaves again at $z = \omega$. If, as we shall assume in this section, our nation is closed to immigration, then it is certainly true that everyone enters life's river at $z = 0$. Indeed we shall define the *instantaneous annual birth rate* to be

$$B(t) \equiv \phi(0, t). \tag{9.76}$$

This is the annual rate at which babies are added to the population at time t and may be regarded as 365 times the number of babies born during the interval $t-1/730 \le \xi < t+1/730$. It is allowed to vary within a calendar year and is therefore not the same thing as the "crude" birth rate, or number of babies per year; the latter would be

$$\int_{t_0}^{t_0+1} B(t)dt \tag{9.77}$$

for the year beginning at $t = t_0$.

But only a very fortunate few leave the river of life at the ripe old age of $z = \omega$. Most cross its banks, as it were, somewhere in the interval $0 < z < \omega$. We must therefore *extend* our river model, allowing "contaminant" to be removed from the river anywhere between $z = 0$ and $z = \omega$. Let the removal rate per unit river length be $d(z, t)$ at station z. Then the number of deaths per unit time at time t, between ages z_1 and $z_2 (> z_1)$, is

$$\int_{z_1}^{z_2} d(z, t)dz. \tag{9.78}$$

In particular, by analogy with (9.76), the *instantaneous annual death rate* is

$$D(t) \equiv \int_0^\omega d(z, t)dz. \tag{9.79}$$

Again, D should not be confused with the "crude" death rate, which is its average value over the year in question.

Now, the river model (9.40) was developed in terms of concentration of contaminant, i.e., amount of contaminant *per unit volume*; whereas ϕ measures "contaminant" per unit length. If the river has unit cross section, however, then the two quantities coincide. Accordingly, setting $A = 1$ in (9.40) and incorporating extension (9.78), the river of life satisfies

$$\begin{array}{ccccc} \text{Rate of change} & = & \text{Population} & - & \text{Population} & - & \text{Death} \\ \text{of population} & & \text{inflow} & & \text{outflow} & & \text{rate} \end{array}$$

$$\frac{\partial}{\partial t} \int_{z_1}^{z_2} \phi(z, t)dz = F(z_1, t) - F(z_2, t) - \int_{z_1}^{z_2} d(z, t)dz, \tag{9.80}$$

$$z_1 \leq z \leq z_2,$$

where $F(z, t)$ is the flux of people of age z in the direction of increasing age. It now follows, in the same way as (9.44) follows from (9.40), that

$$\frac{\partial \phi}{\partial t} + \frac{\partial F}{\partial z} + d = 0, \qquad 0 \leq z \leq \omega, \ 0 \leq t < \infty. \tag{9.81}$$

But the number of people crossing age z per unit of time is simply the number of people per unit of age (or river length) at station z, because both time and distance are measured in years. Hence

$$F(z, t) = \phi(z, t). \tag{9.82}$$

Let us define

$$\mu \equiv \frac{d(z, t)}{\phi(z, t)} \tag{9.83}$$

to be the fraction dying of people aged z (usually called force of mortality by demographers). We will assume that μ is a known function and depends only on z. We thus assume that d is proportional to ϕ. Hence, from (9.81)–(9.83), we obtain the following partial differential equation for $\phi(z, t)$:

$$\frac{\partial \phi}{\partial t} + \frac{\partial \phi}{\partial z} + \mu(z)\phi = 0, \qquad 0 \leq z \leq \omega, \ 0 \leq t < \infty. \tag{9.84}$$

Exercise 9.10 shows how to obtain the unique solution of (9.84) when $B(t) = \phi(0, t)$ and

$$C(z) \equiv \phi(z, 0) \tag{9.85}$$

are given.

By making μ independent of ϕ, we exclude the possibility that a highly populated nation may suffer increased mortality at all ages, due to crowding and competition for resources conducive to health. But there is little evidence that this effect is important for human populations at the present, though it may become so in the near future if population increase does not slow down. By making μ independent of t, we ignore improvements in health care and safety. For certain nations at certain times, this may not be such a bad assumption, especially if we restrict our attention to a span of two or three decades. Recall Exercise 4.15(i), where mortality was found to depend more strongly upon age than upon time. There is, however, another factor on which mortality depends and which has so far been ignored. What is it? We'll return to it shortly.

Unfortunately, the solution given in Exercise 9.10 is of limited use because it requires B to be known in advance as an explicit function of time. This is most unlikely, however, because B depends in quite a complicated way on the behavior of women of childbearing age. Now, we could define a *fertility function*, $m(z, t)$, which gave the number of babies born to people aged z, per unit of time, as a fraction of the total within that age group. Then the number of babies born to people aged between a and b would be

$$\int_a^b m(z, t)\phi(z, t)dz. \tag{9.86}$$

But now a difficulty arises. Because children are almost invariably born to parents of different ages, which parent's age should be counted? The demographer's traditional resolution of this difficulty, which we will adopt, is to count the age of the mother. But we have to go further than this. If m is given as a fraction, then it must be as a fraction of women aged z, because we have agreed that a baby is born to his mother's age group. Thus $\phi(z, t)$ in (9.86) must denote the *female* population aged z. But if ϕ denotes the female population, then we cannot add sons to it. Henceforward, therefore, we shall use m to denote *daughters* born to women aged z, as a fraction of their total number, and use $\phi(z, t)$, $\pi(z, t)$, $B(t)$, $D(t)$, $C(z)$, and $x(t)$ to describe the female population only. Then it is clear from (9.86) that

$$B(t) = \int_0^\omega m(z, t)\phi(z, t)dz. \tag{9.87}$$

What about the guys? Let $S(z, t)$ denote the sex ratio, i.e., the ratio of males aged z to females aged z at time t. Then, provided S is known or can be estimated, the number of males aged between a and b can be deduced from $\phi(z, t)$ as

$$\int_a^b S(z, t)\phi(z, t)dz, \tag{9.88}$$

and the total population is simply

$$x(t) + \int_0^\omega S(z, t)\phi(z, t)dz. \tag{9.89}$$

In these circumstances, knowing the structure of the female population would imply knowledge of the male structure, too. It is often assumed that S is independent of time, i.e., that $S = S(z)$. In the remainder of this section, however, we will concentrate on the female population.

It still remains to determine ϕ, by solving (9.84) subject to (9.76) and (9.85), i.e., in view of (9.87) subject to

$$\phi(0, t) = \int_0^\omega m(z, t)\phi(z, t)dz \tag{9.90}$$

and (9.85). Now, fertility certainly does change with time for sociological reasons. But we have effectively restricted ourselves to short-term prediction by taking $\mu = \mu(z)$, and variation of m with t may be less important over the span of a decade or two than variation with z. Let us therefore assume that m is independent of t. Moreover, a woman can bear children only within a fairly narrow age band, say $\alpha \leq z \leq \beta$. Accordingly, let us take

$$m(z) = 0 \quad \text{if} \quad z < \alpha \quad \text{or} \quad z > \beta. \tag{9.91}$$

Then, on using (9.84), (9.85), (9.90), and (9.91), the problem we face is to solve

$$\frac{\partial \phi}{\partial t} + \frac{\partial \phi}{\partial z} + \mu(z)\phi = 0, \qquad 0 \leq z \leq \omega, \ 0 \leq t < \infty, \tag{9.92a}$$

$$\phi(0, t) = \int_\alpha^\beta m(z)\phi(z, t)dz, \tag{9.92b}$$

$$\phi(z, 0) = C(z). \tag{9.92c}$$

Any attempt to find the general solution of this problem would take us far beyond the scope of this text. Instead, we look for a particular solution of the form

$$\phi(z, t) = C(z)e^{rt}, \tag{9.93}$$

where r is a constant. It now follows immediately from (9.72) that

$$x(t) = x(0)e^{rt}, \tag{9.94}$$

i.e., that the total female population is growing exponentially at specific rate r. In Section 1.9 we rejected (9.94) as a long-term model of U.S. population growth. But we also found that it was quite acceptable in the short term, certainly for two or three decades. We have effectively restricted our model to making short-term predictions by assuming that μ is independent of t; therefore there is nothing in (9.94) to suggest that (9.93) is not a useful assumption. Proceeding, we observe from (9.73), (9.93), and (9.94) that

$$\pi(z, t) = \frac{C(z)}{x(0)} \qquad (9.95)$$

is independent of z, i.e., that the proportion of the population in each age group is independent of time. When π is independent of time, demographers say that the population has a *stable age structure*, or is *quasi-stable*. But the population is not stable in the sense of Chapter 2 because according to (9.93), each age group and hence the entire population is growing at the constant specific rate r.

On substituting (9.93) into (9.92a) and (9.92b) and on cancelling the factor e^{rt}, we thus obtain

$$\frac{dC}{dz} + \{\mu(z) + r\} C = 0, \qquad (9.96)$$

$$\int_\alpha^\beta m(z)C(z)dz = C(0); \qquad (9.97)$$

(9.92c) is satisfied automatically. The ordinary differential equation (9.96) is easily integrated, yielding (verify)

$$C(z) = C(0)e^{-rz}e^{-\int_0^z \mu(\xi)d\xi}$$
$$= C(0)e^{-rz}l(z), \qquad (9.98)$$

where l denotes the probability of surviving to age z, as defined by (8.34). Note that C decreases with z. Thus a quasi-stable population always has a greater proportion of young people than old, a greater proportion of people entering middle age than leaving it, and so on.

On substituting (9.98) into (9.97) and on cancelling the factor $C(0)$, we obtain

$$\int_\alpha^\beta e^{-rz}m(z)l(z)dz = 1. \qquad (9.99)$$

Once α, β, l, and m have been specified, the left-hand side of (9.99) becomes a function of r alone, usually denoted $\Psi(r)$ by demographers. The growth rate r can then be deduced as the solution of the equation $\Psi(r) = 1$.

As an example, let us consider the U.S. female population. The numbers of U.S. female births in 1966 are recorded in Table 9.5 according to the age of the mother. The first three columns of the table are taken from Keyfitz and Flieger (1971, p. 354). The fourth column is computed by using the fact that the male-to-female sex ratio at birth was 1.0485 in 1966 (Keyfitz and Flieger, 1971, p. 355). Thus, for example, the 6.981

Table 9.5 Births of daughters to U.S. females in 1966.

Age group	Thousands of females	Births	Daughters born	Representative age, z	Estimate of $m(z)$	Estimate of $l(z)$
10–14	9542	8128	3968	12.5	0.416×10^{-3}	0.974
15–19	8806	621426	303357	17.5	0.344×10^{-1}	0.972
20–24	6981	1297990	633629	22.5	0.908×10^{-1}	0.969
25–29	5840	872786	426061	27.5	0.730×10^{-1}	0.965
30–34	5527	474542	231653	32.5	0.419×10^{-1}	0.960
35–39	5987	252526	123274	37.5	0.206×10^{-1}	0.952
40–44	6371	74440	36339	42.5	0.570×10^{-2}	0.941
45–49	5978	4436	2165	47.5	0.362×10^{-3}	0.925

Data taken from Keyfitz and Flieger (1971, p. 354). I did wonder how many daughters are actually born to girls aged 10 to 12—if, as I suspect, very few, then 14 might be a more representative age for age group 10–14 than 12.5. But, in the absence of hard information, I have left things as they are. Similarly for women aged close to 50.

million women in the U.S. population who in 1966 had reached their twentieth, but not their twenty-fifth, birthday gave birth to 1,297,990 babies, of whom $1,297,990/(1 + 1.0485) = 633,629$ were daughters. From (9.71), (9.86), and (9.93), this gives us

$$\int_{20}^{25} C(z)e^{rt}dz = 6981 \times 10^3, \qquad \int_{20}^{25} m(z)C(z)e^{rt}dz = 633,629. \quad (9.100)$$

Hence, dividing and cancelling the factor e^{rt}, we get

$$\frac{\int_{20}^{25} m(z)C(z)dz}{\int_{20}^{25} C(z)dz} = 0.0908. \quad (9.101)$$

Now, as remarked already in Section 8.6, if $f(z)$ varies sufficiently linearly across the interval $z_1 \le z \le z_2$, then a rough approximation to the integral

$$\int_{z_1}^{z_2} f(z)dz$$

is

$$(z_2 - z_1)f\left(\frac{1}{2}(z_1 + z_2)\right). \quad (9.102)$$

Depending upon the values of z for which data are available, it is sometimes more convenient to use the alternative approximation

$$\frac{1}{2}(z_2 - z_1)\{f(z_1) + f(z_2)\}. \quad (9.103)$$

Because this is a text on modelling, not numerical analysis, we shall not delve into the errors involved in using either of these formulae. We shall simply observe that both are exact when f varies linearly with z; and hence they should yield acceptable approximations whenever z_1 is close enough to z_2 for f to be almost linear.[4]

Let us assume that $C(z)$ is sufficiently constant and $m(z)$, and hence $m(z)C(z)$, sufficiently linear for (9.102) to render an acceptable approximation to the integrals on the left-hand side of (9.101). Then, setting $z_1 = 20$ and $z_2 = 25$, we have $m(22.5) \approx 0.0908$. Applying similar arguments to the interval $40 \leq z \leq 45$, we obtain $m(42.5) \approx 36,339/6,371,000 \approx 0.570 \times 10^{-2}$. Continuing in this manner, we obtain the penultimate column of Table 9.5. Thus m is specified.

An excerpt from the 1966 life table appeared in Exercise 4.15(v) (and is repeated as Table 9.8). It gave estimates of $l(10)$, $l(15)$, $l(20)$, ..., from which rough estimates of $l(12.5)$, $l(17.5)$, $l(22.5)$, ..., can be obtained by taking averages, i.e., by "linear interpolation." For example, $l(12.5) \approx 0.5 \times \{l(10) + l(15)\} = 0.5 \times \{0.97493 + 0.97345\} \approx 0.974$. Proceeding in this manner, we obtain the final column of Table 9.5. Thus l is specified.

We can now obtain a rough approximation to the left-hand side of (9.99) by setting $\alpha = 12.5$, $\beta = 47.5$, and using (9.103). Because m varies considerably across the interval $12.5 < z \leq 47.5$, we should not set $z_1 = 12.5$ and $z_2 = 47.5$ in (9.103). Rather, we break that interval into seven equal subintervals, each of length 5. Then, setting $z_1 = 12.5 + 5j$, $z_2 = 17.5 + 5j$, and

$$f(z) = e^{-rz}m(z)l(z) \tag{9.104}$$

in (9.103), we obtain the approximation

$$\Psi(r) = \int_\alpha^\beta e^{-rz}m(z)l(z)dz = \int_{12.5}^{47.5} f(z)dz = \sum_{j=0}^{6} \left\{ \int_{12.5+5j}^{17.5+5j} f(z)dz \right\}$$

$$\approx \sum_{j=0}^{6} \frac{5}{2} \{f(12.5 + 5j) + f(17.5 + 5j)\}$$

$$= \frac{5}{2}f(12.5) + 5\sum_{j=1}^{6} f(12.5 + 5j) + \frac{5}{2}f(47.5) \tag{9.105}$$

$$= \zeta(r),$$

say. Using (9.104) and Table 9.5, we can readily evaluate ζ for any value of r. The resulting function ζ is plotted in Fig. 9.8 and appears to take the value 1 near $r = 0.01$. By the Newton–Raphson algorithm or otherwise,[5]

---

[4]Approximation (9.102) is usually dubbed the mid-point rule, (9.103) the trapezoidal rule. The errors associated with these approximations are studied, for example, by Fröberg (1969) and Maron (1982).

[5]In this particular case it is more efficient to use a different numerical method, but the details fall outside our brief. See Keyfitz (1968, p. 111).

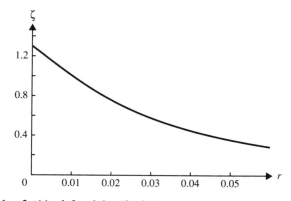

Fig. 9.8 Graph of $\zeta(r)$, defined by (9.105).

the solution of $\zeta(r) = 1$ is found more accurately to be 0.964×10^{-2}. Because ζ is an approximation to Ψ, the solution of $\zeta(r) = 1$ should approximate the solution of $\Psi(r) = 1$. A more accurate value than 0.964×10^{-2}, obtained from a more sophisticated approximation than (9.105) to the integral in (9.99), is $r = 0.970 \times 10^{-2}$ (Keyfitz and Flieger, 1971, p. 355). According to our model, therefore, the U.S. population in 1966 was growing at an instantaneous annual rate of 0.97%.

How good is this model—does it fit the facts? According to our model, $C(z)$ is a decreasing function. Thus, in particular, our model predicts fewer women between the ages of 40 and 45 than between the ages of 35 and 40. Table 9.5 shows clearly that this was not the case in 1966. So something is wrong. At least one of our assumptions must be false.

Now, in solving (9.92), we have assumed that the population is quasi-stable. It can be shown, however, that all solutions of (9.92) will tend to the form of a stable age structure as $t \to \infty$; see for example, Impagliazzo (1985, p. 51). Thus (9.93) may not itself be unduly restrictive.[6] But we have also assumed that $m = m(z)$ and $\mu = \mu(z)$; and the solution of (9.92) need not converge to a quasi-stable one if m or μ depends on t. These assumptions could account for the discrepancy between theory and observation. The dominant cause of the discrepancy, however, is probably that we have ignored immigration—i.e., entering a nation's river of life across its banks—and this is certainly invalid for the United States. It is therefore doubtful whether our model can describe even the asymptotic structure of the U.S. population. Does this make it useless?

---

[6]There is something of an analogy with Section 3.7. There we assumed a particular form of control so that we could solve our problem using calculus. More sophisticated mathematics (Clark, 1976) showed, however, that our solution was indeed the optimal one.

Of course not. For one thing, the United States is just one nation. The model can easily be applied to other nations by using appropriate fertility and mortality data. For example, there is evidence to suggest that toward the end of the nineteenth century in Denmark, fertility and mortality were fixed, immigration was negligible, and growth was indeed quasi-stable; see Impagliazzo (1985, p. 168). More recently, it appears that Venezuela's population may have been quasi-stable in the middle of the sixties; see Keyfitz and Flieger (1971, p. 23).

It is as a paradigm (see Section 4.14), however, that the quasi-stable population model is most valuable. It enables us to answer the following question: if all immigration suddenly ceased, and if fertility and mortality suddenly became fixed, then what would be the ultimate structure of a growing U.S. population, and how fast would it grow? If we like the answer the model gives us, then we may not wish to do anything about fertility, mortality, and immigration; if we don't like the answer, then we may! The answer to this question, as posed in 1966, appears as the uppermost of the solid curves in Fig. 9.9. Since Keyfitz and Flieger's 1966 data do not cover the age range $85 \leq z \leq 115$, the dashed extension of the curve is purely speculative. The same diagram also shows $C(z)/C(0)$ for $r = 0.02$ and $r = 0.04$, with $l(z)$ fixed according to the same 1966 life table; and it is clear that (9.98) can be used to sketch the equivalent curves for any other value of r. We are therefore able to determine how the structure of the quasi-stable population would change if mortality remained fixed but

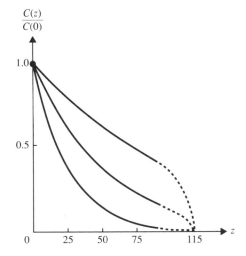

Fig. 9.9 Structure of the quasi-stable U.S. female population for various values of r, when mortality is fixed according to the 1966 life table. (A multiplicative constant cannot alter proportions, and (9.95) implies $\pi(z) = x(0)C(z)$.) The dashed curves are purely speculative but remind us that nobody lives beyond ω!

changing fertility altered the value of r. We see, in particular, that a high value of r would mean so many young people that the labor force might be hard pressed to provide for their education and welfare; whereas a low value of r would result in a more uniform distribution.

Now try Exercises 9.11, 9.12, and 9.13.

9.7 Modelling a Population's Age Structure. A Second Attempt

Given the mathematical difficulties inherent in solving (9.92), there would appear to be little hope of solving the equations you derived in Exercise 9.11. Yet immigration is clearly important in modelling U.S. age structure; so we must find a way to include it. Now, there is rarely only one way to model any phenomenon, and whenever our first attempt leads to disappointment we should always ask ourselves: have we adopted the best approach? Could we make further progress by adapting (or extending) a different model? In modelling a population's age structure, there is indeed an alternative to the river of life, and we will now describe it.

If, as is often said, workers entrenched in a huge bureaucracy can turn into just so much "dead wood," then why not think of people just as trees that move? (Might this description of ourselves be marginally more flattering than being regarded as river contaminants?) We have already developed a model for the structure of a forest in Section 9.2. There, we divided the trees into $N + 1$ classes, with Mu_i trees in class i at the beginning of each time unit, $0 \leq i \leq N$. At the end of the time unit, there were $M\gamma_i$ trees in class i. Then Hv_i trees were harvested from it and r_i trees added to it (with $r_i = 0$ for $1 \leq i \leq N$). Thus, at the end of the time unit, the number of trees in class i was $M\gamma_i - Hv_i + r_i$, $0 \leq i \leq N$. The forester kept the number of trees in each class constant; therefore we had

$$M\gamma_i - Hv_i + r_i = Mu_i, \qquad 0 \leq i \leq N. \tag{9.106}$$

This is just the ith component of the vector equation (9.7). Moreover, the vector $\gamma = (\gamma_0, \gamma_1, \gamma_2, \ldots, \gamma_N)^T$ was related to the vector $\mathbf{u} = (u_0, u_1, u_2, \ldots, u_N)^T$ by an equation of the form $\gamma^T = \mathbf{u}^T S$, or

$$\mathbf{G}^T = \mathbf{M}^T \mathbf{S}, \tag{9.107}$$

where $\mathbf{G} = M\gamma$, $\mathbf{M} = M\mathbf{u}$ and S is a 6×6 matrix, in which entry (i, j) is the probability of moving from class i to class j in a unit of time.

This tree model must be modified in several ways before it can be applied to a human population, in this case the U.S. female one. The first modification is rather trivial. We are interested in the population's age structure, not its girth structure, and so we must divide the population into $N + 1$ classes of increasing age, with $i = 0$ corresponding to the youngest girls and $i = N$ to the oldest women. Second, "replanting" is no longer the prerogative of a single manager but depends upon the behavior of women of childbearing age, so that r_0 must be modified accordingly. Third, because it is unrealistic to assume that all moving trees will survive to class N, we must allow for deaths at an earlier age; but this assumption may not even have been realistic for the motionless trees of Section 9.2, and you may already have proposed an appropriate modification of the matrix S in Exercise 9.2(vi). Fourth, how shall we interpret "harvesting?" Because deaths are to be incorporated into the vector γ, let us think of "harvesting" as synonymous with emigration. But the population is also subject to *im*migration, the net "harvest" in each age class being simply the difference between the two. When describing population growth it is more natural to think of immigration as positive and emigration as negative, and so we shall define $y_i = -Hv_i$ to be the net number of females, in millions, immigrating into class i in a unit of time; y_i can still be negative, in principle, though it is positive for the United States at the present time. Fifth, and most important, the U.S. female population is not constant but growing. We must alter (9.106) accordingly.

Let k denote time, measured linearly; we use k, rather than l, because we will use l to denote a survival probability. We will refer to the period of time between $k = j - 1$ and $k = j$ as period j. Because we have been using 1966 data, let us suppose that $k = 0$ corresponds to mid-1966. If necessary, the origin of time is easily altered. Where appropriate, we will use t to denote continuous time *within* a period; then the unit of continuous time will be a year. Thus, if τ years is the length of a period, then period j corresponds to the interval $(j - 1)\tau < t < j\tau$.

Let us make the following definitions:

$u_i(k)$: proportion of population in class i at beginning of period $k + 1$;

$M_i(k)$: number of millions in class i at beginning of period $k + 1$;

$M(k)$: number of millions in population at beginning of period $k + 1$;

$G_i(k)$: number of millions in class i at end of period $k+1$ after deaths have been counted, but not births or (net) immigration; (9.108)

$G(k)$: number of millions in population at end of period $k+1$, not counting births or (net) immigration;

$y_i(k)$: number of millions immigrating into class i during period $k + 1$;

$r_i(k)$: number of millions of females added to the population at the end of period $k + 1$, as a result of births during the year.

It follows immediately from these definitions that, for all $k = 0, 1, 2, \ldots$, $r_i(k) = 0$ if $1 \leq i \leq N$,

$$M_i(k) = M(k)u_i(k) \qquad (9.109)$$

if $0 \leq i \leq N$, and

$$M(k) = \sum_{i=0}^{N} M_i(k), \qquad G(k) = \sum_{i=0}^{N} G_i(k). \qquad (9.110)$$

Moreover, the number of deaths in period $k + 1$ is

$$M(k) - G(k), \qquad (9.111)$$

and the number of millions in class i at the beginning of period $k + 2$ is $G_i(k) + y_i(k) + r_i(k)$. But the latter, by definition, is $M(k + 1)u_i(k + 1) = M_i(k + 1)$. Hence, for all $k = 0, 1, 2,\ldots$, and for $0 \leq i \leq N$, we have $M_i(k + 1) = G_i(k) + y_i(k) + r_i(k)$. We can write this more succinctly in vector form as

$$\mathbf{M}(k + 1) = \mathbf{G}(k) + \mathbf{y}(k) + \mathbf{r}(k), \qquad k = 0, 1, 2, \ldots, \qquad (9.112)$$

where $\mathbf{M}$ is the $(N + 1)$-dimensional column vector with M_i in row i, $0 \leq i \leq N$, and $\mathbf{G}$, $\mathbf{y}$, $\mathbf{r}$ are similarly defined.

If $\mathbf{y}(k)$ is known, perhaps because it has been set by the U.S. government, then (9.112) enables us to deduce one period's age structure from that of a previous period as soon as $\mathbf{G}$ and $\mathbf{r}$ are known. We therefore proceed to specify them. To make matters simple, however, we shall suppose henceforward that $N = 5$, i.e., that there are precisely six age classes. No new concepts are needed to specify $\mathbf{G}$ and $\mathbf{r}$ for a larger number of age classes, although the arithmetic may become significantly messier.

What should be the length of a period? In the model of Section 9.2 it was six years because that was the census period. This suggests that our period should be ten years because that is the census period for the U.S. population. But immigration is controlled on a yearly basis, suggesting that the period should be one year. Which should we choose? You should pause to answer this question before continuing.

It turns out that one of these choices is vastly superior to the other. If we knew which it was, then we would certainly make it. In the process of modelling, however, you don't always know in advance which is the better of two alternatives. Then you pick one and hope for the best. If it turns out to be a bad one, then something will tell you, and you can still try the other (which may also be a bad one, in which case you must think again). In this spirit, let us begin by making the wrong choice. Let the length of a period be one year.

Let us assume for the time being that the six age classes are defined as in Table 9.6. Our choice of class boundaries is arbitrary, but that hardly matters because our immediate purpose is to discover a wrong choice. Purely as a temporary measure, let us append a seventh class, $i = 6$, defined to consist of females who have died. Let us suppose, again temporarily, that these seven classes correspond to the states of a Markov chain for which the ith entry in the distribution vector is the probability that a female

Table 9.6 Definitions of six age classes.

Class	Ages z included	Class	Ages z included
$i = 0$	$0 \le z < 1$	$i = 3$	$30 \le z < 50$
$i = 1$	$1 \le z < 10$	$i = 4$	$50 \le z < 70$
$i = 2$	$10 \le z < 30$	$i = 5$	$70 \le z < 115$

chosen randomly from the population will belong to class i. Let a_i denote the probability that a female now in class i will still be in class i after one year, $0 \le i \le 5$; $a_0 = 0$, by definition of class 0. Let b_i denote the probability that a female now in class i will be in class $i + 1$ after one year, $0 \le i \le 4$. Then the transmission matrix for the Markov chain is

$$
\bar{\mathbf{S}} \equiv \begin{bmatrix}
0 & b_0 & 0 & 0 & 0 & 0 & 1 - b_0 \\
0 & a_1 & b_1 & 0 & 0 & 0 & 1 - a_1 - b_1 \\
0 & 0 & a_2 & b_2 & 0 & 0 & 1 - a_2 - b_2 \\
0 & 0 & 0 & a_3 & b_3 & 0 & 1 - a_3 - b_3 \\
0 & 0 & 0 & 0 & a_4 & b_4 & 1 - a_4 - b_4 \\
0 & 0 & 0 & 0 & 0 & a_5 & 1 - a_5 \\
0 & 0 & 0 & 0 & 0 & 0 & 1
\end{bmatrix} . \tag{9.113}
$$

Note that this matrix can be partitioned into a 6×6 matrix $\mathbf{S}$, a six-dimensional column vector $\mathbf{d}$, a row vector $\mathbf{0}^T$ of zeroes and a 1×1 matrix containing the number 1 by writing

$$
\bar{\mathbf{S}} = \begin{bmatrix} \mathbf{S} & \mathbf{d} \\ \mathbf{0}^T & 1 \end{bmatrix} , \tag{9.114}
$$

where

$$
\mathbf{S} \equiv \begin{bmatrix}
0 & b_0 & 0 & 0 & 0 & 0 \\
0 & a_1 & b_1 & 0 & 0 & 0 \\
0 & 0 & a_2 & b_2 & 0 & 0 \\
0 & 0 & 0 & a_3 & b_3 & 0 \\
0 & 0 & 0 & 0 & a_4 & b_4 \\
0 & 0 & 0 & 0 & 0 & a_5
\end{bmatrix} , \quad
\mathbf{d} \equiv \begin{bmatrix}
1 - b_0 \\
1 - a_1 - b_1 \\
1 - a_2 - b_2 \\
1 - a_3 - b_3 \\
1 - a_4 - b_4 \\
1 - a_5
\end{bmatrix} , \quad
\mathbf{0} \equiv \begin{bmatrix}
0 \\
0 \\
0 \\
0 \\
0 \\
0
\end{bmatrix} .
$$

$$\tag{9.115}$$

If we include dead females, then our extended population has fixed size $M(k)$ during almost all of year $k + 1$, because births and immigration are not counted until the very end of the year. Accordingly, let us confine our attention to year $k + 1$, and let the seven-dimensional vector

$$
\bar{\mathbf{u}} = (u_0(k), u_1(k), \ldots, u_5(k), 0)^T \tag{9.116}
$$

record the respective proportions of our extended population in each of the seven classes at time k. Note from (9.109) that (9.116) is related to the six-dimensional vector $\mathbf{M}$ by

$$M(k)\bar{\mathbf{u}} = \begin{bmatrix} \mathbf{M}(k) \\ 0 \end{bmatrix}. \tag{9.117}$$

By analogy with Section 9.2, identify (9.116) with the state distribution vector at time k of the Markov chain with transition matrix (9.113). Then the transpose of the state distribution vector at time $k + 1$ is

$$\bar{\mathbf{u}}^T \bar{\mathbf{S}}, \tag{9.118}$$

by (5.72). Identify this vector with the actual proportions in the seven classes of our extended population just before time $k + 1$. Then the actual numbers in the various classes of the extended population are given by the seven-dimensional vector

$$M(k)(\bar{\mathbf{u}}^T \bar{\mathbf{S}})^T = M(k)\bar{\mathbf{S}}^T \bar{\mathbf{u}}. \tag{9.119}$$

Hence, from (9.108), (9.111), (9.115)–(9.117), and (9.119):

$$\begin{bmatrix} \mathbf{G}(k) \\ M(k) - G(k) \end{bmatrix} = M(k)\bar{\mathbf{S}}^T \bar{\mathbf{u}} = \begin{bmatrix} \mathbf{S}^T \mathbf{M}(k) \\ \mathbf{d}^T \mathbf{M}(k) \end{bmatrix}. \tag{9.120}$$

The seventh component of the vector equation (9.120) records the total number of deaths, but that is only of passing interest to us. The first six components of (9.120), however, give us our desired expression for $\mathbf{G}$:

$$\mathbf{G}(k) = \mathbf{S}^T \mathbf{M}(k). \tag{9.121}$$

Although (9.121) is equivalent to (9.107) mathematically, by including deaths it represents a considerable improvement. As remarked in Exercise 9.2, the matrix $\mathbf{S}$ is no longer stochastic. But although a component has been adapted from the probabilistic model of Section 5.8, both this model and that of Section 9.2 are essentially deterministic. Hence the fact that $\mathbf{S}$ is stochastic in Section 9.2 is coincidental; and the fact that it is not in (9.121) is without consequence. Having thus reduced the problem of determining $\mathbf{G}$ to that of determining $\mathbf{S}$, we have no further need of the seventh class. We therefore dispense with it.

Let us turn our attention, now, to $\mathbf{r}$. Note that $r_0(k)$ was defined in (9.108) to be the number of millions of females added to the population *at the end* of period $k + 1$. In other words, if we define $B_0(k)$ to be the number of births during period $k + 1$, in millions, then $r_0(k)$ and $B_0(k)$ are not the same thing. We will first find an expression for $B_0(k)$, then relate it to $r_0(k)$.

The number of births will depend upon the number of women of childbearing age and hence, in view of Table 9.6, on M_2 and M_3. We might therefore suppose that $B_0 = f(M_2, M_3)$, where f is a function to be determined. But what do we actually mean by this? Do we mean $B_0(k) = f(M_2(k), M_3(k))$, $B_0(k) = f(M_2(k + 1), M_3(k + 1))$, or something else entirely? Upon reflection, it appears most reasonable to assume that

$B_0 = f(m_2, m_3)$, where $m_i(k)$ is defined to be the *average* number in class i in period $k+1$. It also seems reasonable to assume that f increases with m_2 and m_3, i.e., that $\partial f/\partial m_2 > 0$, $\partial f/\partial m_3 > 0$. The simplest hypothesis embodying these assumptions is that $f = \theta_2 m_2 + \theta_3 m_3$, where θ_2, θ_3 are positive and independent of m_2 and m_3. But (9.103) yields the approximation

$$m_i(k) \equiv \frac{1}{\tau} \int_{k\tau}^{(k+1)\tau} M_i(t)dt \approx \frac{1}{2}\left(M_i(k) + M_i(k+1)\right), \qquad (9.122)$$

where τ is the number of years in a period (in this case, one). We therefore assume that

$$B_0(k) = \frac{1}{2}\theta_2 \left(M_2(k) + M_2(k+1)\right) + \frac{1}{2}\theta_3 \left(M_3(k) + M_3(k+1)\right). \quad (9.123)$$

What fraction of this number actually survives until the end of year $k+1$? Suppose that an infant is born at time $k+t$ in year $k+1$, where $0 \le t \le 1$. Then, from Section 8.6, the probability that she will survive until time $k+1$ is $l(1-t)$, where $l(z)$ is the probability of survival to age z. This probability will vary among individuals; but, from (9.103), its average value

$$\int_0^1 l(1-t)dt = \int_0^1 l(x)dx \approx \frac{1}{2}\{l(0) + l(1)\}. \qquad (9.124)$$

Hence the number of survivors is approximately $B_0(k) \cdot (l(0) + l(1))/2$. We therefore assume that

$$r_0(k) = c_2 \left(M_2(k) + M_2(k+1)\right) + c_3 \left(M_3(k) + M_3(k+1)\right), \qquad (9.125)$$

where

$$c_i \equiv \frac{1}{4}(l(0) + l(1))\theta_i, \qquad i = 2, 3. \qquad (9.126)$$

Then, from (9.108), we deduce that

$$\mathbf{r}(k) = \mathbf{C}(\mathbf{M}(k) + \mathbf{M}(k+1)), \qquad (9.127)$$

where the 6×6 matrix $\mathbf{C}$ is defined by

$$\mathbf{C} \equiv \begin{bmatrix} 0 & 0 & c_2 & c_3 & 0 & 0 \\ & & _5\mathbf{0}_6 & & & \end{bmatrix}, \qquad (9.128)$$

$_5\mathbf{0}_6$ being the 5×6 zero matrix. Inserting (9.121), (9.127) into (9.112) and rearranging, we obtain

$$(\mathbf{I} - \mathbf{C})\mathbf{M}(k+1) = (\mathbf{S}^T + \mathbf{C})\mathbf{M}(k) + \mathbf{y}(k), \qquad (9.129)$$

where $\mathbf{I}$ is the 6×6 identity matrix. You can easily verify that the square of $\mathbf{C}$ is the 6×6 zero matrix, whence $\mathbf{I} - \mathbf{C}$ has inverse $\mathbf{I} + \mathbf{C}$. Hence, premultiplying (9.129) by $\mathbf{I} + \mathbf{C}$ gives

$$\mathbf{M}(k+1) = \mathbf{A}\mathbf{M}(k) + \mathbf{Q}\mathbf{y}(k), \qquad (9.130)$$

where

$$A \equiv S^T + C + CS^T,$$
$$Q \equiv I + C.$$

(9.131)

If $M(0)$ is known, then the matrix equation (9.130) enables us to determine $M(k)$ by recursion, for any $k = 1, 2, \ldots$, and for any S, C, and $y(k)$.

It still remains to estimate C and S. Let's begin with C. The numbers of 1966 births to U.S. females in classes 2 and 3 are recorded in Table 9.7, being deduced from Table 8.3 and the fact that the male/female sex ratio was 1.0485. We see that 1,367,015 daughters were born to the $31, 169 \times 10^3$ females aged 10 to 30. We therefore estimate that

$$\theta_2 = \frac{1,367,015}{31,169 \times 10^3} = 0.43858 \times 10^{-1}.$$

(9.132)

Similarly, we estimate that

$$\theta_3 = \frac{393,431}{23,863 \times 10^3} = 0.16487 \times 10^{-1}.$$

(9.133)

But $l(0) = 1$, and from Table 9.8 we have $l(1) = 0.98003$. We therefore estimate that

$$c_2 = 0.217 \times 10^{-1}, \qquad c_3 = 0.816 \times 10^{-2},$$

(9.134)

on substituting (9.132)–(9.133) into (9.126). Having determined C, let us turn our attention to S.

What is the probability that a female in class 0 at time k will survive to be counted (in class 1) at time $k + 1$, i.e., survive for a further year? If she is aged z at time k, where $0 \le z < 1$, then her probability of surviving one more year is simply

$$\frac{l(z + 1)}{l(z)},$$

(9.135)

by (8.52). But our model assigns the same probability of being in class 1 at time $k + 1$—namely, b_0—to all females in class 0. We must therefore

Table 9.7 Births and deaths in 1966 U.S. female population, classified according to Table 9.6.

Age	Number of females (thousands)	Number of daughters	Number of deaths
0–1	1793		36353
1–10	18148		10458
10–30	31169	1367015	18195
30–50	23863	393431	61172
50–70	17987		213043
70+	6977		471101

Table 9.8 Survival probabilities to age z for U.S. females, based on mortality prevailing in 1966.

Age z	$l(z)$	Age z	$l(z)$
0	1.00000	45	0.93486
1	0.98003	50	0.91519
5	0.97672	55	0.88698
10	0.97493	60	0.84706
15	0.97345	65	0.79115
20	0.97058	70	0.70671
25	0.96711	75	0.59260
30	0.96285	80	0.44656
35	0.95677	85	0.27641
40	0.94805		

This table first appeared in Exercise 4.15, was further discussed in Section 8.6, and is repeated here for convenience.

replace (9.135) by a number that is independent of z yet representative of the whole of class 0. One way to achieve this is to average (9.135) over all values of z between 0 and 1. We obtain

$$b_0 = \int_0^1 \frac{l(z+1)}{l(z)}dz \approx \frac{l(1.5)}{l(0.5)}, \tag{9.136}$$

on using (9.102). Approximating the graph of $l = l(z)$ in $0 < z < 1$ by a straight line, we see from Table 9.8 that $l(0.5) \approx 0.9900$; whereas a straight line approximation in $1 < z < 5$ yields $l(1.5) \approx 0.9796$. We therefore estimate that $b_0 = 0.9895$. Note, however, that the integral in (9.136) is not the only expression we can use for estimating b_0. An alternative will be mentioned toward the end of the section.

Similarly, our model assigns the same probability of being in class $i+1$ at time $k+1$, namely, b_i, to *all* females in class i, $1 \leq i \leq 4$. But surviving for a year no longer means the same thing as being counted in a higher class. Anyone less than 29 years old in class 2, for example, would survive only to remain in class 2. In other words, the probability that a female in class 2 at time k will still be alive at time $k+1$ is not b_2 but $a_2 + b_2$. Setting this equal to the average class-2 survival probability, we obtain

$$a_2 + b_2 = \frac{1}{20}\int_{10}^{30} \frac{l(z+1)}{l(z)}dz \approx \frac{l(21)}{l(20)}, \tag{9.137}$$

on using (9.102). From Table 9.8, $l(20) = 0.9706$; whereas from a straight line approximation to $l = l(z)$ in $20 < z < 25$, $l(21) \approx 0.9699$. We thus estimate that $a_2 + b_2 = 0.9993$. We can obtain similar estimates for $a_1 + b_1$, $a_3 + b_3$, and $a_4 + b_4$. But four equations between the eight parameters a_1,

..., a_4, b_1, ..., b_4 are not enough to determine them. Let us therefore attempt to estimate b_1, ..., b_4 by some other means.

To simplify matters, let us imagine that all females aged between j and $j + 1$ are concentrated at age $j + 1/2$, precisely. Let a female be chosen randomly from class 1. Let V be the event that she survives for one year, and let W be the event that she is aged between 9 and 10. Then

$$b_1 = \text{Prob}(V \cap W) = \text{Prob}(V \mid W) \cdot \text{Prob}(W), \qquad (9.138)$$

on using (A.11) and because the female can enter class 2 within a year only if she is at least 9 years old. Given our assumption about the concentration of ages, we deduce from (8.52) that

$$\text{Prob}(V \mid W) = \frac{l(10.5)}{l(9.5)} \approx 0.9997, \qquad (9.139)$$

using linear interpolation from Table 9.8 to estimate $l(9.5)$ and $l(10.5)$. We therefore estimate that $b_1 = 0.9997 \times \text{Prob}(W)$. But what is the value of $\text{Prob}(W)$? If we knew that there were ν_1 females aged 1 to 9 and ν_2 females aged 9 to 10, then we would estimate that $\text{Prob}(W) = \nu_2/(\nu_1 + \nu_2)$. But we do not have such information about the distribution of ages within class 1 and are thus unable to make a sensible estimate of b_1. Nor, for similar reasons, can we estimate b_2, ..., b_4. Yet we were able to estimate b_0, because class 0 has the length of a period, making it impossible for an infant to remain in her class for more than a period and ensuring that $a_0 = 0$. (Incidentally, it was precisely in order to make this point that I began by defining classes in the bizarre manner of Table 9.6.)

In general, if classes 0, 1, ..., $N-1$ have the length of a period, then a_0, a_1, ..., a_{N-2} and a_{N-1} are all identically zero, enabling b_i to be determined as the average probability that a member of class i will survive a period, $0 \leq i \leq N - 1$. On the other hand, if class N contains all ages beyond a certain age, as in Table 9.6, then its length is irrelevant; b_N is identically zero, so that the average probability of surviving a period can be set equal to a_N. Thus if we wish to estimate all parameters but insist on keeping a year as the length of a period, then we will need at least 51 classes to avoid confusing old age with childbearing ages, which have qualitatively different effects on the growth of the population. I think you'll agree that 51 is a prohibitively large number. If, on the other hand, we adopt ten years as the length of a period, then six classes will be enough to achieve the same goals, although seven or eight would be more informative. Thus, with ease of accountancy in mind, $\tau = 10$ is vastly superior to $\tau = 1$. But there is no special reason why the period should equal a census period, and five years is generally preferred by demographers.

To simplify calculations, however, we will indeed adopt ten years as our period. Moreover, we will continue to use only six age classes. Therefore, let us redefine classes as in Table 9.9 and regroup the data accordingly.

Table 9.9 Alternative classification of U.S. female population, with births and deaths in 1966.

Class	Ages included	Number of females (thousands)	Number of daughters	Number of deaths
$i = 0$	$0 \leq z < 10$	19941		46811
$i = 1$	$10 \leq z < 20$	18348	307324	8076
$i = 2$	$20 \leq z < 30$	12821	1059691	10118
$i = 3$	$30 \leq z < 40$	11514	354927	17968
$i = 4$	$40 \leq z < 50$	12349	38504	43204
$i = 5$	$50 \leq z < 115$	24964		684144

There are now four childbearing classes, $i = 1, \ldots, 4$, instead of the two we had previously; so we replace (9.123) by

$$B_0(k) = \frac{1}{2} \sum_{i=1}^{4} \theta_i \{M(k) + M_i(k+1)\}, \qquad (9.140)$$

where $\theta_1, \ldots, \theta_4$ are independent of $\mathbf{M}$. The period is now ten years instead of one. So we replace (9.124) by the average value of the probability that an individual born at time $10k + t$ in period $k + 1$, where $0 \leq t \leq 10$, will survive $10 - t$ years to be counted in class 0, i.e., by

$$\frac{1}{10} \int_0^{10} l(10 - t)dt = \frac{1}{10} \int_0^{10} l(x)dx \approx l(5), \qquad (9.141)$$

on using (9.102). From (9.140) and (9.141), (9.127) and (9.129)–(9.131) will remain intact if we redefine the 6×6 matrix $\mathbf{C}$ to be

$$\mathbf{C} \equiv \begin{bmatrix} 0 & c_1 & c_2 & c_3 & c_4 & 0 \\ & & _5\mathbf{0}_6 & & & \end{bmatrix}, \qquad (9.142)$$

where

$$c_i = \frac{1}{2}l(5)\theta_i, \qquad 1 \leq i \leq 4. \qquad (9.143)$$

Let us now proceed to determine θ_i, and hence c_i, for $1 \leq i \leq 4$. By assumption, $\theta_1 m_1$ daughters are born to class 1 in a period of ten years. Hence, crudely, daughters are born to class 0 at the rate of $\theta_1 m_1/10$ per year. We can therefore interpret $\theta_1/10$ as the fixed probability that a female chosen randomly from class 1 will give birth to a daughter within a year. But we know from Table 9.8 that 0.307324 million daughters were born to the 18.348 million females in class 1 in 1966. We can therefore estimate $\theta_1/10$ to be $0.307324/18.348 = 0.16750 \times 10^{-1}$. Repeating this argument for each of the remaining classes, we obtain $\theta_1 = 0.16750$, $\theta_2 = 0.82653$, $\theta_3 = 0.30826$, and $\theta_4 = 0.31180 \times 10^{-1}$. From Table 9.8, $l(5) = 0.97672$.

Hence, combining (9.142) and (9.143) with definition (9.131), we obtain

$$\mathbf{Q} = \mathbf{I} + \mathbf{C} = \begin{bmatrix} 1 & 0.08180 & 0.40364 & 0.15054 & 0.01523 & 0 \\ 0 & 1 & 0 & 0 & 0 & 0 \\ 0 & 0 & 1 & 0 & 0 & 0 \\ 0 & 0 & 0 & 1 & 0 & 0 \\ 0 & 0 & 0 & 0 & 1 & 0 \\ 0 & 0 & 0 & 0 & 0 & 1 \end{bmatrix}. \qquad (9.144)$$

We identify b_0 with the average period-survival probability for class 0. Thus

$$b_0 = \frac{1}{10} \int_0^{10} \frac{l(z+10)}{l(z)} dz \approx \frac{l(15)}{l(5)} = 0.99665, \qquad (9.145)$$

on using (9.102) and Table 9.8. Identifying b_1 with the average period-survival probability for class 1, we obtain

$$b_1 = \frac{1}{10} \int_{10}^{20} \frac{l(z+10)}{l(z)} dz \approx \frac{l(25)}{l(15)} = 0.99349; \qquad (9.146)$$

and b_2, b_3, b_4 are similarly estimated. A little more care is required to estimate a_5 because so many ages are included in class 5. Let ω denote maximum life potential. Then the average period-survival probability for class 6 is

$$\frac{1}{\omega - 50} \int_{50}^{\omega} \frac{l(z+10)}{l(z)} dz = \frac{1}{\omega - 50} \int_{50}^{\omega - 10} \frac{l(z+10)}{l(z)} dz, \qquad (9.147)$$

because $l(z) = 0$ for $z \geq \omega$. But associating this quantity with a_5 would produce a number that, although admittedly independent of z, would hardly be representative of class 5. The reason is simply that (9.147) attributes as much weight to very old age as to late middle age; whereas we would expect the majority of people in class 5 to belong to the lower end of the age class. In 1966, for example, only 1.3% of the population was over 80 years old. We can obtain a number more representative of class 5 by averaging the survival probability $l(z+10)/l(z)$ over only part of the age class. I have chosen to average it over ages $50 \leq z < 80$. Whatever arbitrariness I have thereby introduced can later be eliminated, at least in principle, simply by adding more and more age classes. Thus

$$a_5 = \frac{1}{30} \int_{50}^{80} \frac{l(z+10)}{l(z)} dz. \qquad (9.148)$$

The range of integration is three times as big as for $b_0, \ldots, b_4$, so divide it into three before approximating. Then

$$a_5 = \frac{1}{30} \left\{ \int_{50}^{60} \frac{l(z+10)}{l(z)} dz + \int_{60}^{70} \frac{l(z+10)}{l(z)} dz + \int_{70}^{80} \frac{l(z+10)}{l(z)} dz \right\}$$

$$\approx \frac{1}{30}\left\{\frac{l(65)}{l(55)}+\frac{l(75)}{l(65)}+\frac{l(85)}{l(75)}\right\} \tag{9.149}$$
$$= 0.70248,$$

on using (9.102) and Table 9.8. This completes estimation of the matrix $\mathbf{A}$:

$$\mathbf{A} = \begin{bmatrix} c_1 b_0 & c_1 + c_2 b_1 & c_2 + c_3 b_2 & c_3 + c_4 b_3 & c_4 & 0 \\ b_0 & 0 & 0 & 0 & 0 & 0 \\ 0 & b_1 & 0 & 0 & 0 & 0 \\ 0 & 0 & b_2 & 0 & 0 & 0 \\ 0 & 0 & 0 & b_3 & 0 & 0 \\ 0 & 0 & 0 & 0 & b_4 & a_5 \end{bmatrix} \tag{9.150}$$

$$= \begin{bmatrix} 0.08153 & 0.48281 & 0.55257 & 0.16542 & 0.01523 & 0.0 \\ 0.99665 & 0.0 & 0.0 & 0.0 & 0.0 & 0.0 \\ 0.0 & 0.99349 & 0.0 & 0.0 & 0.0 & 0.0 \\ 0.0 & 0.0 & 0.98931 & 0.0 & 0.0 & 0.0 \\ 0.0 & 0.0 & 0.0 & 0.97710 & 0.0 & 0.0 \\ 0.0 & 0.0 & 0.0 & 0.0 & 0.94878 & 0.70248 \end{bmatrix}. \tag{9.151}$$

Because

$$\mathbf{M}(0) = (19.941, 18.348, 12.821, 11.514, 12.349, 24.964)^T \tag{9.152}$$

from Table 9.9, we are now in a position to perform the recursion (9.130).

Let's begin by assuming that there is no immigration, i.e., that $\mathbf{y} = \mathbf{0}$. Then (9.130) yields

$$\mathbf{M}(k+1) = \mathbf{A}\mathbf{M}(k). \tag{9.153}$$

You can easily verify that the solution of this matrix recurrence relation is

$$\mathbf{M}(k) = \mathbf{A}^k \mathbf{M}(0), \qquad k = 1, 2, \ldots. \tag{9.154}$$

Values of $\mathbf{M}(k)$ are given in Table 9.10 for $k = 0, 1, \ldots, 10$. Rather than quote the components of $\mathbf{M}$—namely, $M_i(k)$, $0 \le i \le 5$—we quote the proportions in each class—namely, $u_i(k)$, $0 \le i \le 5$—together with the grand total $M(k)$; the components can then be deduced from (9.108). A population that grew at specific rate r would be $\exp(10r)$ times bigger after a decade and so the final column of Table 9.10 is an estimate of specific growth rate in units of YEAR^{-1}, the same units as were used in Section 9.6. Note that the proportions $\mathbf{u}$ and the specific growth rate appear to converge to a quasi-stable state, in which a fixed proportion of the population resides in each age class (for example, 16% between the ages of 10 and 20), and all classes grow at the constant rate $r = 0.9631 \times 10^{-2}$. You can verify this by continuing recursion (9.153) to convergence, as implied by the last row of Table 9.10. The truth of this convergence can be established rigorously by applying the theory of linear algebra; see, for example, Chapter 5 of Impagliazzo (1985), in particular his Theorem 5.19 (p. 111) and Theorem 5.36 (p. 124).

Table 9.10   Solution of recursion (9.153).

k	$u_0(k)$	$u_1(k)$	$u_2(k)$	$u_3(k)$	$u_4(k)$	$u_5(k)$	$M(k)$	$\frac{1}{10}\ln\left(\frac{M(k)}{M(k-1)}\right)$
0	0.1995	0.1836	0.1283	0.1152	0.1236	0.2498	99.94	
1	0.1772	0.1791	0.1643	0.1143	0.1014	0.2637	111.0	0.1046×10^{-1}
2	0.1890	0.1574	0.1586	0.1448	0.0995	0.2507	124.5	0.1155×10^{-1}
3	0.1829	0.1685	0.1398	0.1403	0.1265	0.2420	139.2	0.1116×10^{-1}
4	0.1784	0.1636	0.1503	0.1242	0.1231	0.2604	155.1	0.1077×10^{-1}
5	0.1794	0.1603	0.1466	0.1341	0.1094	0.2702	172.0	0.1036×10^{-1}
6	0.1782	0.1619	0.1442	0.1313	0.1186	0.2658	190.0	0.9945×10^{-2}
7	0.1773	0.1608	0.1456	0.1291	0.1161	0.2709	209.8	0.9941×10^{-2}
8	0.1774	0.1602	0.1448	0.1306	0.1144	0.2725	231.5	0.9802×10^{-2}
9	0.1771	0.1604	0.1445	0.1300	0.1158	0.2721	255.1	0.9724×10^{-2}
10	0.1770	0.1602	0.1447	0.1297	0.1153	0.2732	281.1	0.9707×10^{-2}
...	...	...	...	...	...	...	...	...
∞	0.1768	0.1600	0.1444	0.1297	0.1151	0.2740	∞	0.9631×10^{-2}

Because of rounding to four significant figures, the components of u may not sum precisely to 1 (for example, they sum to 0.9998 when $k = 7$). The apparent discrepancy between the final two columns is also due to rounding error.

You may already have observed that the value of r implied by Table 9.10 is close to that obtained in the previous section. You can make it more precise by increasing the number of age classes and by using cubic rather than linear polynomials to approximate $l = l(z)$ when estimating integrals. Perhaps you would like to pursue this. But the picture that emerges won't differ drastically from the one implied by Table 9.10.

In a refined form, i.e., with five-year age classes and cubic interpolation, the model we have just developed is essentially the one demographers use for population projections. It is known as the Leslie model, in honor of the man who first proposed it. There is, however, one crucial difference. We have chosen throughout to approximate b_j by

$$b_j = \frac{1}{\tau}\int_{j\tau}^{(j+1)\tau}\frac{l(z+\tau)}{l(z)}dz, \tag{9.155}$$

as being both independent of z and representative of class j. As remarked above, however, there is nothing unique about (9.155), and demographers generally prefer to use

$$b_j = \frac{\int_{j\tau}^{(j+1)\tau} l(z+\tau)dz}{\int_{j\tau}^{(j+1)\tau} l(z)dz} \tag{9.156}$$

instead. The rationale behind (9.156) is essentially twofold—that if the population were quasi-stable, then

$$\frac{\int_{j\tau}^{(j+1)\tau} C(z+\tau)dz}{\int_{j\tau}^{(j+1)\tau} C(z)dz} \tag{9.157}$$

would be the ratio of the number of females in class $j+1$ to the number in class j; and that $C(z)$ may in turn be approximated by $l(z)$, the approximation becoming exact as $r \to 0$. Now, the latter approximation can be improved; see Chapter 11 of Keyfitz (1968). But the population need not be quasi-stable, and I remain to be convinced that (9.155) is not indeed preferable to (9.156). On the other hand, the two formulae become equivalent if we use (9.102) to approximate integrals.

If we were interested only in the quasi-stable age structure of a closed population, of course, then we would have been quite content with the model of Section 9.6. One advantage of the Leslie model is that it enables us to study immigration without inducing complicated mathematics; see Exercise 9.14. It also enables us to study changes with time of fertility and mortality. Allowing fertility to change with time would cause c_i, and hence **Q**, to depend upon k. Allowing mortality to change with time would cause b_i, and hence **A**, to depend upon k. Hence recursion (9.130) would have to be generalized to

$$\mathbf{M}(k+1) = \mathbf{A}(k)\mathbf{M}(k) + \mathbf{Q}(k)\mathbf{y}(k), \qquad k = 0, 1, 2, \ldots. \tag{9.158}$$

The computer wouldn't mind a bit. It would be just as happy to crank out solutions to (9.158) as to (9.130). Unfortunately, you would have to supply it with updates for the matrices **A** and **Q**. Where would you obtain them? Certainly not from the U.S. government, which is hard pressed to say what $\mathbf{y}(k)$ is, even though it's officially supposed to know. Rather, you would have to obtain these updates from other models, i.e., by combining (9.158) with assumptions about future fertility and mortality; and perhaps this is a matter you would like to pursue.

Exercises

9.1 A vendor buys newspapers for 25¢ apiece, sells them for 50¢ apiece, and recoups 10¢ for each unsold newspaper at the end of a 10-hour day. If customers arrive according to a Poisson process at an average rate of one every two minutes, how many newspapers should the vendor order each day, according to the model of Section 9.1? A second vendor has a less favorable arrangement whereby his profit per item just equals the loss he incurs from each unsold newspaper at the end of a β-hour day. If customers arrive according to a Poisson process at an average rate of one every α minutes, how many newspapers should he order each day? Show that the answer given by the model of Section 9.1 conforms to intuition. *Hint*: Use Chapter 7's normal approximation to the Poisson distribution.

*9.2 (i) Suppose that in the forest of Section 9.2 only trees in classes 4 and 5 have commercial value. Let the profit from a tree in class 4 be a fraction λ of the profit from a tree in class 5. What is then the forester's optimization problem?

(ii) Using the IMSL routine ZX3LP, or a similar linear programming routine, show that there is a critical value $\lambda_c \approx 0.849$ such that if $\lambda > \lambda_c$, then the forester's optimal policy is to have no trees in class 5, have no trees in class 4 at the beginning of a cycle and harvest all trees that enter class 4 by the end of the cycle. What is the optimal policy if $\lambda < \lambda_c$?

(iii) Does it surprise you that when $\lambda > \lambda_c$, it is optimal to prevent trees from reaching the most valuable girth class? Comment.

(iv) Observe that whether $\lambda > \lambda_c$ or $\lambda < \lambda_c$, the optimal policy is such that all trees of the widest existing girth class are harvested at the end of a cycle. By assuming the optimal policy to be of this form, can you derive your results in (ii) analytically? (This may help clarify your answer to (iii)).

(v) Why are profits not discounted, as in Section 3.4?

(vi) The harvesting model of Section 9.2 does not allow for the possibility that trees might die before reaching maturity (class 5). Can you alter the matrix S to allow for this? *Hint*: Although S originated in Section 5.8 as a stochastic matrix (one with rows summing to 1), why need S no longer be stochastic in the present model?

9.3 Verify (9.30).

9.4 According to the model of Section 9.3, find an approximation to the $\alpha\%$ cleaning time for Lake Ontario, where α is small.

9.5 Verify that (9.50) satisfies (9.45), (9.47), and (9.48) when r is given by (9.46). In the z-t plane in Fig. E9.1, $\phi(z,t)$ is given along $z = a$ and along $t = 0$; the dot denotes an arbitrary point (z,t) satisfying $z > a$, $t > 0$. Show that $\phi(z,t)$ can be determined by drawing an appropriate line from (z,t) to the boundary and using the given information. Show that this method produces the complete solution (9.50). The lines are known as *characteristics*. We shall discuss them further in Section 11.3. *Hint*: According to (9.50), on which lines will $\phi(z,t)$ be constant?

9.6 How many units of time must elapse before we can be sure that pollution in the imaginary lake of Fig. 9.3 has been reduced to no more than $\alpha\%$ of its initial level, if there is perfect mixing across the lake?

*9.7 (i) We remarked in Section 9.4 that if total water inflow to Lake Ontario balances total water outflow (which assume), then (9.46) is inconsistent with

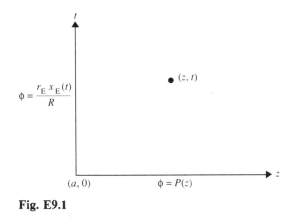

Fig. E9.1

the fact that only 26 of every 34 inches of rain evaporate. Can you propose
a new form for $Q(z, t)$ that would be consistent with this observation? *Hint*:
Assume that the excess of rainfall over evaporation increases uniformly
from west to east.

(ii) Solve (9.45), with this new form of Q, subject to (9.47) and (9.48). Show
that the 5% purification time is slightly reduced from (9.54) to

$$0.977\frac{V}{r_E} + T_E \ln(19.1\mu).$$

Note: You may wish not to attempt this part until after you have covered
Section 11.3. The characteristics are slightly nonlinear; x is no longer constant along them, as in Exercise 9.5, but satisfies a certain ordinary differential equation.

(iii) We also remarked in Section 9.4 that Lake Ontario's cross section varies
with z. If $A = A(z)$, is (9.45) still valid?

9.8 The *error function* is defined by

$$\text{Erf}(\theta) \equiv \frac{2}{\sqrt{\pi}} \int_0^\theta e^{-\xi^2} d\xi.$$

(i) Show that (9.66) can be written $\phi(y, t) = x_0 \cdot \text{Erf}(y/2\sqrt{\gamma t})$ and verify that
it satisfies (9.59). *Hint*: Use the facts that

$$\int_{-\infty}^\infty e^{-x^2} dx = \sqrt{\int_{-\infty}^\infty \int_{-\infty}^\infty e^{-x^2-y^2} dx\,dy} = \sqrt{\int_0^{2\pi} \int_0^\infty e^{-r^2} r\,dr\,d\theta} = \sqrt{\pi}$$

and that e^{-x^2} is an even function.

(ii) Show that (9.66) satisfies (9.65).

9.9 The right-hand side of (9.68) is a small number. Therefore β, and hence all values of ξ that the integrand takes, are small. Determine an approximate value for β by using the first two terms of the Taylor series for $e^{-\xi^2}$.

9.10 (i) Verify by direct substitution that the solution of the partial differential equation

$$\frac{\partial \phi}{\partial t} + \frac{\partial \phi}{\partial z} + \mu(z)\phi = 0, \qquad 0 \leq z \leq \omega, \ 0 \leq t < \infty, \tag{a}$$

subject to the boundary conditions

$$\begin{aligned} \phi(0,t) &= B(t), & 0 \leq t < \infty, \\ \phi(z,0) &= C(z), & 0 \leq z \leq \omega, \end{aligned} \tag{b}$$

is

$$\phi(z,t) = \begin{cases} B(t-z)\exp\left\{-\int_0^z \mu(\xi)d\xi\right\}, & \text{if } t \geq z; \\[2ex] C(z-t)\exp\left\{-\int_{z-t}^z \mu(\xi)d\xi\right\}, & \text{if } t < z. \end{cases} \tag{c}$$

Assume that $B(0) = C(0)$, so that $\phi(z,t)$ is continuous along $z = t$.
(ii) It is clear from (c) that along any straight line of the form $z - t = $ constant, ϕ is a function of z alone. Deduce from (a) that ϕ satisfies the ordinary differential equation

$$\frac{d\phi}{dz} + \mu(z)\phi = 0. \tag{d}$$

Hint: Use the fact that $dt/dz = 1$ along these straight lines.
(iii) Give the statement $dt/dz = 1$ a verbal interpretation.
Note: From a mathematical point of view, the straight lines $z - t = $ constant are just characteristics, as in Exercise 9.5; and it is precisely because of (d) that (a) is so easy to solve. From a demographic viewpoint, however, these are the *life-lines* of individuals in the population, and a selection of them is shown, dashed, in Fig. E9.2. Life-lines belong to one of two categories, labelled I and II in the diagram. In $z > t$ lies the (Category I) life-line of a person who was already alive at time $t = 0$. It has the form $z = t + a$, where a is the person's age at $t = 0$. In $z < t$ lies the (Category II) life-line of an individual yet to be born at $t = 0$. It has the form $z = t - t_0$, where $t = t_0$ is the time of that individual's birth. Thus a life-line simply records the age z of an individual at any given time t, assuming that he has been born and has survived; but survival, of course, is not guaranteed. From (d), a large group of individuals born at the same instant (or, more realistically, neighboring instants) would find their number declines with age at the specific rate

$$-\frac{1}{\phi}\frac{d\phi}{dz} = \mu(z) \tag{e}$$

as they followed their life-lines (or, more realistically, their bundle of neighboring life-lines). Such a group is called a *birth cohort* and was discussed in Exercises 4.15–4.17, where ϕ was denoted instead by l.

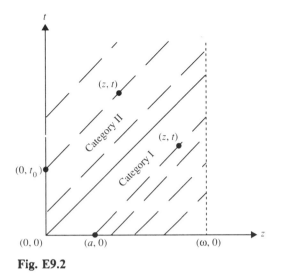

Fig. E9.2

*9.11 Can you extend (9.80) to allow for immigration? What is the new form of (9.92)? *Note*: You are not expected to solve your equations!

**9.12 Because this is 1988 (or later), and not 1966, the information contained in Fig. 9.9 is more than 20 years out of date. Can you update it? In other words, if immigration to the United States ceased today, and if fertility and mortality were suddenly fixed, what would be the ultimate structure of the U.S. female population, and how fast would it grow? Your library will have the most recent U.S. census data.

**9.13 Suppose that fertility, $m(z)$ in Section 9.6, is fixed according to the most recent U.S. census data and that immigration has ceased. How might the population's ultimate structure depend upon changes in mortality? Apply the same analysis to other countries, in particular those where AIDS is rampant. For the latter, use Keyfitz and Flieger (1971) if you cannot find more recent data for $m(z)$. *Note*: You cannot assume that r is fixed, because it depends upon l through (9.99). Choose a model for $\mu(z)$, with two or three free parameters, which can be used to capture the changes in mortality. (Your model is thus a combination.)

*9.14 (i) State a behavioral assumption you make about immigrants if you use the recursion

$$\mathbf{M}(k+1) = \mathbf{A}\mathbf{M}(k) + \mathbf{Q}\mathbf{y}(k), \qquad k = 0, 1, \ldots, \tag{a}$$

as in Section 9.7, to describe the evolution of a nation's age structure.

(ii) Suppose that this nation is subject to constant immigration, i.e., that

$$\mathbf{y}(k) = \mathbf{z}, \qquad k = 0, 1, \ldots,$$

where $\mathbf{z}$ is independent of k. Show that

$$\mathbf{M}(k) = \mathbf{A}^k \mathbf{M}(0) + (\mathbf{I} - \mathbf{A})^{-1}(\mathbf{I} - \mathbf{A}^k)\mathbf{Q}\mathbf{z}, \qquad k = 0, 1, \ldots. \qquad \text{(b)}$$

(iii) Which is the more useful, (a) with $\mathbf{y}(k) = \mathbf{z}$ or (b)?

(iv) Imagine that fertility and mortality in the U.S. female population have been fixed at 1966 levels. What would have been the effect on age structure of a constant influx of one million females in each age class, from 1966 onwards?

(v) Can you adapt the Leslie model to describe the effect on age structure of a sharp temporary increase in a nation's fertility (a "baby boom")?

*9.15 Suppose that the owner or manager of a game preserve wishes to generate a sustained income from the annual harvest of a herd of deer. The manager will achieve this goal by maintaining the herd at a constant age structure, controlling the harvest by issuing licenses to kill so many deer in each age class. Describe how you could combine the models of Sections 9.2 and 9.7 to derive an optimal policy. Restrict your attention to females for the reasons given in Section 9.6.

Table 9.11   These data for Scottish red deer are taken from Beddington and Taylor (1973, p. 807).

Age class	Probability of survival to next birthday	Proportion giving birth to females	Mean body weight in kilograms
$0 \leq z < 1$	0.907	0.000	23
$1 \leq z < 2$	0.987	0.000	36
$2 \leq z < 3$	0.992	0.266	46
$3 \leq z < 4$	0.990	0.282	49
$4 \leq z < 5$	0.987	0.338	51
$5 \leq z < 6$	0.992	0.380	51
$6 \leq z < 7$	0.953	0.383	53
$7 \leq z < 8$	0.942	0.391	54
$8 \leq z < 9$	0.880	0.339	54
$9 \leq z < 10$	0.719	0.237	51
$10 \leq z < 11$	0.730	0.164	50
$11 \leq z < 12$	0.668	0.170	49
$12 \leq z < 13$	0.668	0.169	46
$13 \leq z < 14$	0.668	0.169	45
$14 \leq z < 15$	0.668	0.169	44
$15 \leq z < 16$	0.668	0.169	43
$16 \leq z < 17$	0.668	0.169	42
$z \geq 17$	0.668	0.169	41

**9.16 (Continuation of Exercise 9.15.) Using the IMSL routine ZX3LP, or a similar linear programming routine, maximize yield in terms of body weight for a game preserve in which the data in Table 9.11 describe the herd.

*9.17 A certain species of furred mammal, which is hunted by the human species, is confined to a fixed habitat. Its life cycle is characterized by three seasons: the hunting season (which is strictly regulated by law), a breeding season, and a season during which the mammal is subject to natural mortality. By adapting and extending an appropriate section of the textbook, develop a mathematical model that is capable of describing the population's long-term dynamics.

*9.18 In Section 3.7 we constructed a model that predicted an optimal size for a fishing fleet. We assumed that the boats harvested a single species of fish. Can you extend that model to apply to two species of fish, which are being harvested together by the same boats? Attempt to *formulate* optimal control problems that describe different biological relationships between the two species (solving these problems may be extremely difficult).

V TOWARD MORE ADVANCED MODELS

10 FURTHER DYNAMICAL SYSTEMS

Our first nine chapters have established a blueprint for constructing and evaluating mathematical models. By applying it repeatedly, we can make our models as accurate or as flexible as we please, and the models of Part V are merely a further step in an ever beckoning direction. Each model is developed in accordance with the philosophy expounded in Chapter 9, i.e., by adapting, extending, and combining known models.

We begin by discussing dynamical systems. Sections 10.1–10.3 adopt the deterministic approach, Sections 10.4–10.6 the probabilistic one.

10.1 How Does a Fetus Get Glucose from Its Mother?

Glucose reaches a fetus by being transported across the placenta from the mother.[1] Table 10.1 presents values of glucose concentration in a ewe and her fetus during a three-hour period. The experiment was performed by W. F. Widdas.

The placenta is a membrane; glucose is a chemical. Thus Exercise 8.13 suggests that the flux of glucose should be proportional to the concentration difference $g_m - g_f$ between mother and fetus. If so, then the flux of glucose should decrease with time, as suggested by the final column of Table 10.1. Widdas found, however, that the flux of glucose was relatively constant from 30 to 150 minutes. Thus our steady state diffusion model, first encountered in Section 2.5, is unable to explain the outcome of the experiment. An alternative model must be developed.

Let's begin by borrowing from biochemistry, where it is observed that almost all (biochemical) reactions are *catalyzed*. This simply means that

---

[1] Before proceeding, you may find it useful to review Section 1.10, Section 1.11, and your solution to Exercise 8.13.

Table 10.1 Maternal and fetal glucose concentration in an experiment on sheep.

Time (minutes)	Maternal glucose concentration, g_m	Fetal glucose concentration, g_f	Difference $g_m - g_f$
30	528	185	343
60	344	169	175
90	280	150	130
120	228	132	96
150	183	113	70
180	157	91	66

All quantities are expressed in units of milligram per 100 milliliters. The data are due to W. F. Widdas and are taken from Bray and White (1966, p. 212).

the conversion of a substance, A, into a second substance, B, can proceed only in the presence of a third substance, E—which, however, is not itself a part of the reaction. Moreover, the quantity of substance E is usually a tiny fraction of that of substances A and B; this fact will be important later. The substance E is called the catalyst, and in biochemistry the role of catalyst is played by *enzymes*. For the present section, it is necessary only to know what an enzyme *does* (it catalyzes). What an enzyme *is* amounts to a much longer story, which is told in biochemistry texts. Strictly, an enzyme's role is to accelerate a reaction that would otherwise be very slow. As a first approximation, however, it is convenient to imagine that it facilitates a reaction that would otherwise be impossible.

Widdas proposed that transport of glucose across the placenta can happen only in the presence of an enzyme. In other words, glucose molecules need piggyback rides on enzyme molecules. According to this hypothesis, the substance actually transported across the placenta is an enzyme–glucose complex consisting of enzyme molecules and glucose molecules bound together. Let y_m denote its concentration (mass per unit volume) on the maternal side of the placenta and y_f its concentration on the fetal side. Then we can modify the model of Exercise 8.13 by proposing that the flux of glucose from mother to fetus is

$$F = k_{eg} \left(\frac{y_m - y_f}{h} \right) \tag{10.1}$$

per unit area of placenta, where k_{eg} is the chemical conductivity of the enzyme–substrate complex and h is the width of the placenta. It remains to find an expression for y in terms of glucose concentration g.

With this in mind, we will suppose that the enzyme–glucose complex EG is formed as part of a reaction that converts the glucose G into some other substance P, the product of the reaction. This enzyme-catalyzed reaction is written symbolically as either

$$E + G \longleftrightarrow EG \longrightarrow E + P \tag{10.2}$$

or

$$G \longrightarrow P, \tag{10.3}$$

according to whether we are interested in the reaction mechanism or just the overall effect. The arrows in (10.2) indicate that the first step of the reaction is reversible, whereas the second is not. (To be more precise, the rate of conversion from P to EG is negligible compared to the rate of conversion from EG to P.)

At time t, let ϵ denote the molar concentration of enzyme, $x(t)$ the molar concentration of glucose (whether in G form or EG form), $y(t)$ the molar concentration of the EG complex, and $z(t)$ the molar concentration of product P; ϵ is constant because E is not itself part of the overall reaction. Then the molar concentration of *free* enzyme—enzyme that is *not* a part of the EG complex—is $\epsilon - y$; the molar concentration of *free* glucose is $x - y$. Using the ideas developed in Sections 1.10 and 1.11, we find that the forward rate of the first step of (10.2) is $k_1(\epsilon - y)(x - y)$, where k_1 is a constant, and the backward rate of the same step is $k_2 y$, where k_2 is constant; the forward rate of the second step of (10.2) is $k_3 y$, for some constant k_3, and the backward rate is zero. Hence the rates of change of G, EG, and P are given, respectively, by:

$$-\frac{d(x-y)}{dt} = k_1(\epsilon - y)(x - y) - k_2 y, \tag{10.4}$$

$$\frac{dy}{dt} = k_1(\epsilon - y)(x - y) - k_2 y - k_3 y, \tag{10.5}$$

$$\frac{dz}{dt} = k_3 y. \tag{10.6}$$

Note that $(10.6) + (10.5) - (10.4)$ yields

$$-\frac{dx}{dt} = \frac{dz}{dt}, \tag{10.7}$$

the mass conservation equation for the overall reaction (10.3); and (10.5) yields the rate of change of free enzyme to be

$$\frac{d(\epsilon - y)}{dt} = -\{k_1(\epsilon - y)(x - y) - k_2 y\} + k_3 y \tag{10.8}$$

$$= \quad \text{net loss from step 1} \quad + \text{gain from step 2.}$$

It is convenient to define the constant

$$K_M \equiv \frac{k_3 + k_2}{k_1}. \tag{10.9}$$

Then (10.4) and (10.5) together yield the pair of nonlinear ordinary differential equations

$$\frac{dx}{dt} = -k_3 y \tag{10.10a}$$

$$\frac{dy}{dt} = k_1 \{(\epsilon - y)(x - y) - K_M y\}. \tag{10.10b}$$

This dynamical system is of the form (1.24) with $m = 4$ and $n = 2$.

Explicit analytic solution of (10.10) does not seem possible. Instead, therefore, we trace the path of the point with coordinates $(x(t), y(t))$ in the phase-plane $0 < x < \infty$, $0 \leq y < \epsilon$. The dashed curve depicted in Fig. 10.1 is a branch of the hyperbola

$$x = \frac{K_M y}{\epsilon - y} + y. \tag{10.11}$$

On this curve, dy/dt equals zero, so that any solution of (10.10) will cross (10.11) horizontally. Above the hyperbola $dy/dt < 0$; below it, $dy/dt > 0$.

Initially all the enzyme is free. Hence $y(0) = 0$. Initially, therefore, dx/dt is zero and $dy/dt = k_1 \epsilon x(0)$ is positive. Let's assume that glucose is sufficiently plentiful as to make this positive number large in some sense (we will shortly be more precise). Then $(x(t), y(t))$ reaches the vicinity of the hyperbola (10.11) rapidly, as indicated in the diagram. As long as $(x(t), y(t))$ lies below the hyperbola, the negative sign of dx/dt and positive sign of dy/dt will ensure continued movement toward it, and eventually the phase trajectory will cross the hyperbola horizontally. Thereafter $(x(t), y(t))$ remains close to the hyperbola, because any tendency for negative dx/dt to move $(x(t), y(t))$ above the curve is immediately corrected by negative dy/dt. In other words, once the vicinity of the hyperbola has been reached, the subsequent movement of $(x(t), y(t))$ toward the origin (as $t \to \infty$) is so close to the hyperbola as to be virtually on it. We may therefore replace

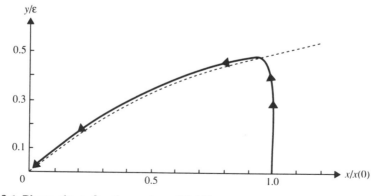

Fig. 10.1 Phase-plane for the system (10.10).

the differential equation (10.10b) by the nondifferential equation (10.11). In doing so, we recognize that x and y are changing with time but in such a way that x is (virtually) exactly balanced by $K_M y/(\epsilon - y) + y$. We might say that these two quantities are in "dynamic equilibrium." Such a dynamic equilibrium, which is frequently dubbed a *quasi-equilibrium*, is implied by any nondifferential relation. Recall, for example, that labor L, capital K, and output Q all varied with time in the economic growth model of Section 1.6, but in such a way that there was a dynamic equilibrium between $Q(t)$ and the quantity $aL(t)^\gamma K(t)^{1-\gamma}$, where a is a constant; see (1.32). The quasi-equilibrium (10.11) has been observed in the laboratory and is actually achieved in a fraction of a second, suggesting that it is indeed valid to assume $x(0)$ large (Bray and White, 1966, p. 209). We will therefore assume henceforward that (10.11) applies. Then the concentration of EG complex as a function of x is given by

$$y(t) = \frac{2\epsilon x(t)}{K_M + x(t) + \epsilon + [(K_M + x(t) + \epsilon)^2 - 4\epsilon x(t)]^{1/2}}. \qquad (10.12)$$

In practice, ϵ is so small compared to $K_m + x$ that the terms involving ϵ in the denominator can be ignored. We then obtain the excellent approximation

$$y(t) = \frac{\epsilon x(t)}{K_M + x(t)}, \qquad (10.13)$$

for the concentration of the EG complex.

But what do we mean by $x(0)$ large—large compared to what? The approximations above are so important that it is instructive to provide an alternative derivation, using dimensional analysis. First observe from (10.10) that x, y, ϵ, and K_M all have the dimensions of a concentration, and $k_1\epsilon$, k_2, and k_3 have the dimensions of 1/[TIME]. We can therefore define the following dimensionless quantities:

$$\bar{x} = \frac{x}{x(0)}, \qquad \bar{y} = \frac{y}{\epsilon}, \qquad \bar{t} = k_1 \epsilon t,$$

$$a = \frac{k_3}{k_1 x(0)}, \qquad b = \frac{k_2}{k_1 x(0)}, \qquad q = \frac{\epsilon}{x(0)}. \qquad (10.14)$$

Note from (10.9) that $K_M/x(0) = a+b$. You can readily show (Exercise 10.1) that the dimensionless form of (10.10) becomes

$$\frac{d\bar{x}}{d\bar{t}} = -a\bar{y},$$

$$q\frac{d\bar{y}}{d\bar{t}} = (1 - \bar{y})(\bar{x} - q\bar{y}) - (a + b)\bar{y}. \qquad (10.15)$$

If q is so small that the terms involving q in the second of these equations can be neglected, then we obtain

$$(1 - \bar{y})\bar{x} = (a + b)\bar{y}. \qquad (10.16)$$

You can easily verify that this is the dimensionless form of (10.13). Thus the answer to the question at the beginning of this paragraph is that $x(0)$ must be large compared to ϵ for a quasi-equilibrium approximation to be valid. Fig. 10.1 is drawn for $a = b = 0.5$ and $q = 0.1$. The heads of the five arrows correspond to the times $\bar{t} = 0.05$, $\bar{t} = 0.1$, $\bar{t} = 1$, $\bar{t} = 5$, and $\bar{t} = 10$, respectively.

We now return to the experiment of W. F. Widdas. Because x is the molar concentration of glucose and g is the mass concentration, we have $g = m_G x$, where m_G is the mass of a mole of glucose. Hence

$$y = \frac{\epsilon g}{g + \alpha}, \tag{10.17}$$

where $\alpha \equiv m_G K_M$ is a constant. Then, according to our improved model (10.1), the flux of glucose across the placenta from mother to fetus is

$$F = \frac{k_{eg}\epsilon}{h} \left\{ \frac{g_m}{g_m + \alpha} - \frac{g_f}{g_f + \alpha} \right\}. \tag{10.18}$$

At low concentrations of glucose, such that $|g| \ll \alpha$, we obtain approximately $F \approx k_{eg}\epsilon(g_m - g_f)/\alpha h$, which is proportional to $g_m - g_f$. From Table 10.1, we already know that this is inadequate. At high concentrations, however, for which $|g| \gg \alpha$, a Taylor expansion of (10.18) in powers of α/g_m and α/g_f yields (Exercise 10.2) the approximation

$$F \approx \frac{k_{eg}\epsilon\alpha}{h} \left(\frac{1}{g_f} - \frac{1}{g_m} \right). \tag{10.19}$$

This is proportional to $(1/g_f - 1/g_m)$. From Table 10.2, we see that the hypothesis of Widdas is quite compatible with the observation that F was constant between 30 and 150 minutes.

In addition to explaining the data of Widdas, our model suggests that the rate of conversion of glucose (or any other substance, say substance A)

Table 10.2 Comparison of (10.19) with data of Widdas.

Time (minutes)	$100 \left(\dfrac{1}{g_f} - \dfrac{1}{g_m} \right)$
30	0.35
60	0.30
90	0.31
120	0.32
150	0.35
180	0.46

Adapted from Bray and White (1966, p. 212).

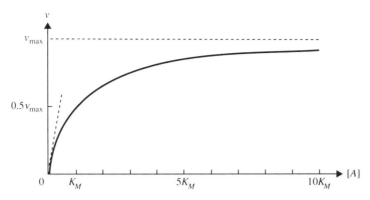

Fig. 10.2 The Michaelis–Menten law.

in an enzyme-catalyzed reaction is given by (10.6) and (10.13) to be

$$\frac{dz}{dt} = \frac{k_3 \epsilon x}{x + K_M}. \tag{10.20}$$

The rate of conversion increases linearly with x for small values of x but soon levels off to the asymptotic value $k_3 \epsilon$; see Fig. 10.2.

It is common practice to call the reaction rate a velocity (even though it does not have the dimensions of velocity) and to write (10.20) in the form

$$v = \frac{v_{max}[A]}{[A] + K_M}, \tag{10.21}$$

where $v_{max} \equiv k_3 \epsilon$ and $[A]$ is the molar concentration of substance A. In this form, it is known as the *Michaelis–Menten* law, and K_M is known as the Michaelis constant. The ability of many plants and other organisms to take up substrate is adequately described by the Michaelis–Menten law, which can be regarded as a sort of behavioral response to the stimulus of substrate. For this reason, it is widely used in biochemistry; see, for example, Thornley (1976, p. 14) and Bray and White (1966, pp. 212–213).

10.2 A Limit-Cycle Ecosystem Model

A mathematical model of a predator–prey ecosystem was first attempted in Section 1.4, where it produced metastable population cycles.[2] We validated this model in Section 4.12 and discovered that, although predator–prey oscillations are observed in nature, they can be explained satisfactorily only as forming a stable equilibrium cycle, or limit cycle, as defined in Chapter 2.

---

[2] You may wish to review Sections 1.4, 4.12, and your solution to Exercise 4.14 before proceeding.

In Section 4.12, we attempted to improve the model of Section 1.4 by modifying our assumptions about the way in which the prey population interacted with itself. The resulting model,

$$\frac{dx}{dt} = x\{a_1(1 - \epsilon x) - b_1 y\}, \tag{10.22}$$

$$\frac{dy}{dt} = y(-a_2 + b_2 x), \tag{10.23}$$

where $x(t)$ and $y(t)$ denote the prey and predator population levels at time t, was a failure from the point of view of explaining predator–prey oscillations because it shrank the metastable cycles of Section 1.4 to a static equilibrium. Clearly, to obtain a limit cycle, our assumptions about the way in which prey and predators interact must be modified once more.

Now, in Section 4.12, we assumed that predators would die off exponentially in the absence of prey. As we found in Exercise 4.14, however, this presupposed that the predators would not vary their diet. In the absence (or simply scarcity) of their preferred food source, might not the predators eat something else? To be sure, the predators will thrive only if there is an abundance of their favorite prey, but a scarcity of it need not imply that their demise is unavoidable. One way of incorporating these ideas into our mathematical model would be to assume that the predator population grows logistically, *but with a capacity proportional to the prey population*. Then (10.23) would be replaced by

$$\frac{dy}{dt} = a_2 y \left(1 - \frac{y}{b_2 x}\right), \tag{10.24}$$

where a_2, b_2 are positive constants and $b_2 x$ is the capacity. You may wish to go back and compare this equation with (1.62).

We also assumed in Section 4.12, and it is clear from (10.22), that predators remove prey at the rate $b_1 xy$ per unit of time. Thus, on average, each predator removes $b_1 x$ prey per unit of time. This assumption is suspect for large x because even if predators are extremely greedy they still have only a limited ability to capture and devour their prey. A much more reasonable assumption is that, on average, each predator removes $\phi(x)$ prey per unit of time, where $\phi(x) \approx b_1 x$ for small x but $\phi(x) \to c_1$ as $x \to \infty$, c_1 being the maximum rate of removal. A wide variety of functions ϕ will satisfy these two conditions. If we imagine that the predator is taking up substrate (food) when removing prey, however, then the wide applicability of the Michaelis–Menten law (10.21) suggests that an appropriate form might be

$$\phi(x) = \frac{c_1 x}{x + (c_1/b_1)}. \tag{10.25}$$

Adopting (10.25), we replace (10.22) by

$$\frac{dx}{dt} = a_1 x \left(1 - \frac{x}{K}\right) - y\phi(x) = x\left\{a_1\left(1 - \frac{x}{K}\right) - \frac{b_1 c_1 y}{b_1 x + c_1}\right\}, \qquad (10.26)$$

where we now use $K = 1/\epsilon$ to denote the prey capacity.

Let us propose that (10.24) and (10.26) are an improved mathematical model. If this new model gives rise to limit cycles, then according to the philosophy set forth in Chapter 4, we have reason to believe that (10.24) may be appropriate and (10.25) may be a suitable behavioral response function. At least until a more stringent test could be proposed, we would have no good reason to reject the model. If (10.24) and (10.26) do not yield limit cycles, however, then we must reject our model for predator–prey oscillations as surely as we rejected the previous model (consisting of (10.22) and (10.23)) in Section 4.12.

Equations (10.24) and (10.26) can be scaled according to the procedure described at the end of Chapter 1. Dimensionless variables ξ, ψ, and τ are defined by

$$\xi = \frac{x}{K}, \qquad \psi = \frac{y}{b_2 K}, \qquad \tau = a_2 t. \qquad (10.27)$$

You can easily check (Exercise 10.3) that the dimensionless forms of (10.24) and (10.26) are

$$\frac{d\xi}{d\tau} = \xi\left\{\lambda(1 - \xi) - \frac{\theta\psi}{\xi + \mu}\right\}, \qquad (10.28)$$

$$\frac{d\psi}{d\tau} = \psi\left\{1 - \frac{\psi}{\xi}\right\}, \qquad (10.29)$$

where we have defined three dimensionless parameters:

$$\lambda \equiv \frac{a_1}{a_2}, \qquad \theta \equiv \frac{c_1 b_2}{a_2}, \qquad \mu \equiv \frac{c_1}{b_1 K}. \qquad (10.30)$$

Because a_1 is the maximum specific growth rate of prey when there are no predators and a_2 is the specific growth rate of predators when there are infinitely many prey, we can interpret λ as a ratio of specific growth rates under ideal conditions. From (10.25), c_1/b_1 is the prey population for which the removal rate has fallen to half the maximum removal rate (in essence, the Michaelis constant for predation), and K is the prey capacity in the absence of predation. Hence μ, the ratio of these two quantities, is a dimensionless measure of predator effectiveness in removing prey. If μ is small, then the predator's ability to capture and devour will start to saturate at a fraction of the prey capacity. The parameter θ is best regarded as measuring the predator's maximum prey removal rate c_1 relative to its ideal specific growth rate a_2, the parameter b_2 being dimensionless to begin with.

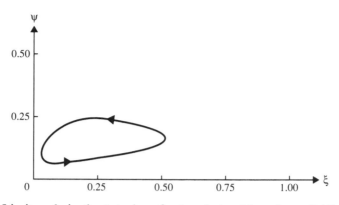

Fig. 10.3 Limit cycle in the ξ-ψ plane for $\lambda = 5$, $\theta = 10$, and $\mu = 0.15$.

Equations (10.28) and (10.29) can be solved numerically for any values of λ, θ and $\mu(> 0)$. For the particular values $\lambda = 5$, $\theta = 10$, and $\mu = 0.15$, but for any pair $(\xi(0), \psi(0))$ of initial population levels, the point $(\xi(\tau), \psi(\tau))$ in the ξ-ψ plane approaches the limit cycle depicted in Fig. 10.3. If $(\xi(0), \psi(0))$ lies on this closed curve then so does $(\xi(\tau), \psi(\tau))$, for all $\tau > 0$. The corresponding fluctuations with time of $\xi(\tau)$ and $\psi(\tau)$ are shown in Fig. 10.4. You might like to verify this result for yourself by using a computer package to integrate equations (10.28) and (10.29) numerically for various values of $(\xi(0), \psi(0))$.

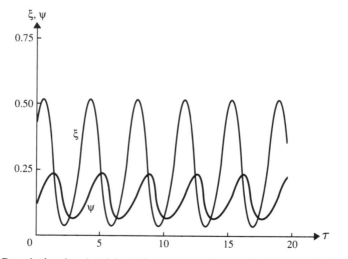

Fig. 10.4 Population levels $\xi(\tau)$, $\psi(\tau)$ corresponding to the limit cycle in Fig. 10.3. Note that the period of oscillation is approximately four units of dimensionless time. For a discussion of this, see May (1976, p. 8).

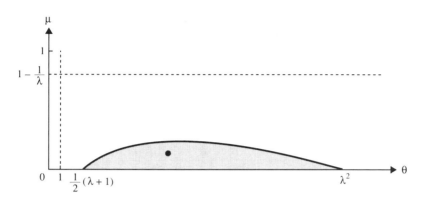

Fig. 10.5 Region in the θ-μ plane giving rise to limit cycles. The figure is drawn for $\lambda = 5$. The dot corresponds to the limit cycle depicted in Fig. 10.3.

We have thus demonstrated that a limit cycle exists for at least one set of values of the parameters λ, θ, and μ. More generally, however, it can be shown that a limit cycle exists whenever the parameters θ and μ lie within the region of the θ-μ plane that is shaded in Fig. 10.5. This region is bounded above by an arc of the ellipse

$$\mu\lambda(2\lambda + 1)^2 + (\lambda + 1)(\lambda - \mu\lambda - \theta)^2 = (\theta - 1)\{(\lambda - 1)\theta - \mu\lambda\} \qquad (10.31)$$

and below by the $\mu = 0$ axis. The region lies to the right of the line $\theta = 1$ and below the line $\mu = 1 - 1/\lambda$. Thus limit cycles arise if μ is sufficiently small, λ exceeds 1, and θ lies between $\theta = (\lambda + 1)/2$ and $\theta = \lambda^2$. If, on the other hand, the point (θ, μ) lies outside the shaded region in Fig. 10.5, then the ecosystem reaches a stationary equilibrium at $(\xi, \psi) = \gamma(1, 1)$, where the constant γ is defined by

$$\gamma \equiv \frac{1}{2}\left\{1 - \mu - \frac{\theta}{\lambda} + \sqrt{(1 - \mu - \theta/\lambda)^2 + 4\mu}\right\}. \qquad (10.32)$$

Thus, with the above interpretation of the parameter μ, our model suggests that predator–prey oscillations arise in the lynx–hare ecosystem because the lynx is relatively ineffective at capturing and devouring the hare.[3]

10.3   Does Increasing the Money Supply Raise or Lower Interest Rates?

In Section 10.1, we found that the differential equation

$$\frac{dy}{dt} = k_1\{(\epsilon - y)(x - y) - K_M y\}, \qquad (10.33)$$

---

[3] See Supplementary Note 10.1.

which was satisfied by the concentrations $x(t)$ and $y(t)$ of glucose and an enzyme–glucose complex in an enzyme-catalyzed reaction, evolved so quickly toward the quasi-equilibrium

$$y(t) = \frac{\epsilon x(t)}{K_M + x(t)} \tag{10.34}$$

that there would be scant justification for preferring (10.33) to (10.34) as a model component.[4] We present this as evidence that some dynamic phenomena are as adequately described by nondifferential as by differential equations, with the added advantage that the mathematics is usually somewhat simpler; and in this spirit we now discuss a wholly nondifferential model of the effect of money circulation on interest rates.

There is an ongoing debate between "Keynesian" and "Monetarist" economists, as to who has the right answer to all kinds of questions. Consider, for example, the question of money supply, or the amount of money in circulation. By and large, Keynesians believe that increasing the money supply will lower interest rates, and Monetarists believe that increasing the money supply will raise interest rates. How can they both be right? We will now attempt to provide an answer to this question as an illustration of the usefulness of nondifferential models. This is a book on the process of modelling, however, and not a text on economics; therefore we will deliberately avoid many subtle twists to Keynesian and Monetarist arguments, as well as the vexed question of which kinds of bank deposit should be regarded as "in circulation" (in addition to hard currency). For a lively elementary discussion of these matters, see Ritter and Silber (1977), in particular Chapters 2 and 4. The following model is freely adapted from Brems (1980).

Let $S(t)$ denote an economy's *real* money supply at time t, i.e., the number of real dollars then in circulation. $S(t)$ is the number of paper dollars in circulation at time t divided by the price index $P(t)$; and $S(t)$ must be carefully distinguished from $M(t)$, the value at time t of money deposited in a bank, as defined in Section 3.4.[5] Let $D(t)$ be the economy's real demand for money at time t, i.e., the number of real dollars the economy requires to carry out its business. Let $Q(t)$ denote the economy's output at time t, $I(t)$ its rate of investment, and $\nu(t)$ the nominal interest rate.

---

[4]You may wish to review Sections 1.6 and 3.4 before proceeding.

[5]If only hard currency were regarded as being in circulation, then dollars included in S and dollars included in M would be mutually exclusive. If all money were regarded as being in circulation, and if all deposits were regarded as being made by a single hypothetical depositor in the same bank (hence same interest rate), then M should be S minus hard currency. In general, however, definitions of S lie between these two extremes and there is no simple relationship between M and S.

The more it produces, the more money an economy requires to purchase its goods. Hence $D(t)$ increases with $Q(t)$, or

$$\frac{\partial D(t)}{\partial Q(t)} > 0. \tag{10.35}$$

On the other hand, consumers and manufacturers may not always have the money they would like to spend, in which case they must consider borrowing it. Whether they do actually borrow or not depends, of course, on the rate of interest. If it is low, then their desire to borrow is apt to become an actual borrowing requirement; if high, then it is apt to remain a desire. In other words, the (real) demand for money decreases with the interest rate or

$$\frac{\partial D(t)}{\partial \nu(t)} < 0. \tag{10.36}$$

The simplest hypothesis consistent with (10.35) and (10.36) is that

$$D(t) = D_0 + a_2 Q(t) - a_3 \nu(t), \tag{10.37}$$

where D_0, a_2, and a_3 are positive constants. Both Keynesian and Monetarist economists appear happy with (10.37). Moreover, it seems reasonable to assume that the demand for money is always exactly balanced by the supply; for any imbalance between the two would be quickly corrected by a change of interest rate. Hence $D(t) = S(t)$; whence, from (10.37):

$$D_0 + a_2 Q(t) - a_3 \nu(t) = S(t). \tag{10.38}$$

In Section 1.6 we assumed that rate of investment was proportional to output or

$$I(t) = \sigma Q(t). \tag{10.39}$$

We will retain this assumption here. But we also observe that the money a manufacturer would like to spend to make additions to her capital stock may not be available at the time she desires it, because the goods she is producing may not have been sold and paid for. In the short term, therefore, investment may require borrowing. The attractiveness of borrowing as a means of investment will depend, of course, on the interest rate: I will decrease with ν, or

$$\frac{\partial I(t)}{\partial \nu(t)} < 0. \tag{10.40}$$

The simplest hypothesis consistent with (10.40) is that

$$I(t) = I_0 - a_1 \nu(t), \tag{10.41}$$

where I_0 and a_1 are positive constants. Let us therefore assume this.

In (10.38), (10.39), and (10.41) we have three equations for the four functions $Q(t)$, $I(t)$, $\nu(t)$, and $S(t)$. As explained in Section 1.6, we may in principle regard any of these as an exogenous function and the remaining three as endogenous functions. We are exploring the effect of the money supply on the interest rate, however, and so the natural choice for exogenous function is S. On solving these equations for $D(t)$, $I(t)$, and $\nu(t)$ in terms of S, we obtain in particular (verify) that

$$\{\sigma a_3 + a_1 a_2\}\, \nu(t) = a_2 I_0 + \sigma\,\{D_0 - S(t)\}\,; \qquad (10.42)$$

whence it immediately follows that

$$\frac{\partial \nu(t)}{\partial S(t)} = -\frac{\sigma}{a_1 a_2 + a_3 \sigma} < 0. \qquad (10.43)$$

Thus increasing the money supply decreases the interest rate. It appears at first sight that the Keynesians have won.

If things were really quite that simple, however, would the Monetarists persist in the opposite view? When two schools of thought adhere to opposite viewpoints for many years, isn't it usually the case that both are at least partly in the right? Shouldn't we therefore suspect that the case for the Keynesians is not as clear cut as it seems at first sight? If so, then at least one of the assumptions that led to (10.43) must be false. Which one?

According to Brems (1980, p. 19), the false assumption is (10.41). Recall from Section 3.4 that the real interest rate δ, the nominal interest rate ν, and the inflation rate r are related by:

$$\nu(t) = \delta(t) + r(t). \qquad (10.44)$$

Thus, according to (10.41), a high inflation rate discourages investment. A few moments thought, however, suggests a fallacy in this thinking. A manufacturer is able to borrow the money she needs for investment purchases because she will be able to pay it back when she sells her product. But isn't the price of that product (to the consumer) rising at the rate of inflation? A part of the nominal interest rate, (10.44), is due to inflation. Then doesn't that part correspond to a free loan, in the sense that the cost of repaying money to the bank is borne not by the producer but by the general rise in prices to the consumer? If this is the case, then shouldn't producers be somewhat indifferent to inflation when assessing the cost of investment? Accordingly, Brems (1980) proposes that investment is discouraged by an increase in the *real* rate of interest, rather than by an increase in the nominal rate; i.e., he replaces (10.41) by

$$I(t) = I_0 - a_1 \delta(t). \qquad (10.45)$$

In (10.38), (10.39), (10.44), and (10.45), we now have four equations for the six functions $Q(t)$, $\nu(t)$, $S(t)$, $I(t)$, $\delta(t)$, and $r(t)$. If we wish to keep $S(t)$ as the sole exogenous function, then we are obliged to obtain a fifth equation.

Let Q_{max} denote the maximum output the economy is capable of producing (so that $Q(t) \leq Q_{max}$), and suppose that firms in the economy are always trying to raise their prices. How successful are they likely to be? We will argue that the answer depends on the quantity $Q_{max} - Q(t)$, which we shall dub the excess capacity. Suppose that demand for the economy's product exceeds $Q(t)$. This appears to be a golden opportunity for firms to raise their prices because the existing output can then be rationed out to those buyers who are prepared to pay more. But a moment's thought reveals that this ruse will work only if $Q(t)$ is so close to Q_{max} that production cannot be increased. If $Q(t)$ is very much less than Q_{max}, then any attempt by one firm to raise its prices is likely to be thwarted by other firms producing more, for the same price, to satisfy all demand. Thus any firm will be reluctant to raise prices when excess capacity is high.[6] Accordingly, Brems (1980) assumes that the rate of inflation is a decreasing function of excess capacity, i.e., that

$$r(t) = r_0 - a_4(Q_{max} - Q(t)), \tag{10.46}$$

where r_0 and a_4 are positive constants. This is our fifth equation.

At first sight it seems rather curious, because it appears to say that increasing production increases inflation. Haven't we just said the opposite? The answer is both yes and no. Increasing production will temporarily restrain inflation, but at the same time this production increase will decrease the economy's excess capacity. The nearer the economy to its production limit, the more successful firms will be in attempting to raise prices, and this is the sense in which inflation increases with output.

From (10.38), (10.39), and (10.44)–(10.46), we easily obtain (verify):

$$\{\sigma a_3 + a_1 a_2 - a_1 a_3 a_4\}\, \nu(t)$$
$$= a_2\{I_0 + a_1(r_0 - a_4 Q_m)\} + (\sigma - a_1 a_4)\{D_0 - S(t)\}. \tag{10.47}$$

It follows immediately that

$$\frac{\partial \nu(t)}{\partial S(t)} = -\frac{\sigma - a_1 a_4}{\sigma a_3 + a_1 a_2 - a_1 a_3 a_4}. \tag{10.48}$$

Thus increasing the money supply decreases the interest rate only if

$$\sigma < a_1\left(a_4 - \frac{a_2}{a_3}\right) \qquad \text{or} \qquad \sigma > a_1 a_4, \tag{10.49}$$

---

[6]This argument would break down if only one firm—a monopoly—produced the entire output of the economy. For certain products (e.g., jumbo jets) this might well be so, but we are considering all products together as the single product $Q(t)$, with $P(t)$ being the average price index, so we can safely assume that there are many firms.

and increasing the money supply actually increases the interest rate if

$$a_1 \left(a_4 - \frac{a_2}{a_3} \right) < \sigma < a_1 a_4. \tag{10.50}$$

Now, a_4 is a measure of the sensitivity of an economy to excess capacity. If the economy is depressed—if manufacturers are having difficulty selling their output even at existing prices—then they will be reluctant to raise prices even if there is evidence that excess capacity has diminished; only a significant decrease of excess capacity would encourage them to do so. Hence a_4 is small in a depressed economy. If the economy is booming, however—if demand is high—then even a slight decrease of excess capacity might trigger a price rise. Hence a_4 is large in a booming economy. It appears that increasing the money supply will always lower the interest rate in a sufficiently depressed economy. But in a booming economy, where (10.50) may be satisfied, increasing the money supply may raise the interest rate.

Our simple paradigm has thus explained why increasing the money supply can raise or lower interest rates. It is much too simple a model, however, to predict which will happen for any given economy; and the fact that Monetarists and Keynesians still argue is some indication that a satisfactory predictive model does not yet exist. Perhaps you would like to build one! If not, a less ambitious project would be to tackle Exercise 10.4.

10.4   Linearizing Time: The Semi-Markov Process. An Extension

In Section 5.7 we used a Markov chain to describe the behavior of four birds in a cage.[7] The birds spent almost all of their time at the ends of the cage and were allowed to change ends only at discrete instants $n = 1, 2, 3, \ldots$. We found that if there were i birds at one end of the cage at one such instant, then the probability that there would be j birds at the same end at the next instant was p_{ij}, where, for $i, j = 0, \ldots, 4$, p_{ij} denotes entry (i, j) in the transition matrix (5.89), namely,

$$\mathbf{P} = \begin{bmatrix} \frac{1}{2} & \frac{1}{2} & 0 & 0 & 0 \\ \frac{1}{8} & \frac{1}{2} & \frac{3}{8} & 0 & 0 \\ 0 & \frac{1}{4} & \frac{1}{2} & \frac{1}{4} & 0 \\ 0 & 0 & \frac{3}{8} & \frac{1}{2} & \frac{1}{8} \\ 0 & 0 & 0 & \frac{1}{2} & \frac{1}{2} \end{bmatrix}. \tag{10.51}$$

We counted time purely in terms of instants of transition. Thus, if the initial instant $n = 0$ corresponded to 12 o'clock and the first three transitions

---

[7] You may wish to review Section 5.7 before proceeding.

were made at times 12:05, 12:07, and 12:10, then these three times would be denoted by $n = 1$, $n = 2$, and $n = 3$; even though they corresponded to real time increments of 5 minutes, 2 minutes, and 3 minutes, respectively. Thus time was being counted nonlinearly; whereas real time is linear. This is perfectly adequate if we wish only to know the distribution of birds after, say, 20 transitions—but how can we determine the distribution of birds after 20 minutes? Our existing Markov model gives no information about this. We must therefore extend it to allow a conversion from nonlinear time to linear time.

We will assume henceforward that one minute is the unit of linear time. As in Chapter 5, we will denote linear time by the integer l and the number of birds in the North end of the cage at time l by the random variable $X(l)$. We will assume that the birds remain in state i for some precise number of minutes then make an instant transition to state j. You could imagine that the intervening trapeze act took zero time. The fact that it takes finite time is of little consequence because trapeze time can simply be regarded as the last part of the time that the birds spend in state i. Moreover, the choice of one minute as the unit of linear time is purely for illustration. If real birds were apt to make transitions after a fraction of a minute, then a second could be used as the unit of linear time instead.

Of great consequence, however, is the assumption we now make, namely, that the origins of linear and nonlinear time coincide. Figure 10.6 depicts how the transitions described above would appear in linear and nonlinear time. Notice that $l = 0$ corresponds to $n = 0$; i.e., a transition has taken place at the initial instant. Now, it's just possible that a purely deterministic relationship between l and n might exist (for example, l equals the integer part of $140n^3/(43n^2 + 3n - 18)$ would account for Fig. 10.6), but even if such a relationship existed, it's most unlikely that we would know what it was. We shall therefore assume that the time for which birds wait in state i after each transition is a discrete random variable W_i, taking values $l = 1, 2, \ldots$. Note that W_i cannot take the value 0. At least one unit of linear time elapses between each transition, and *precisely* one unit of nonlinear time elapses. A possible set of distributions for the five random variables W_0, W_1, W_2, W_3, and W_4 would be the one that makes linear

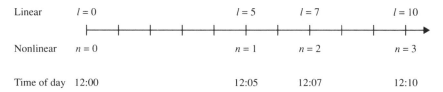

Fig. 10.6 Possible realization of transitions in bird cage in linear and nonlinear time.

and nonlinear time identical, namely,

$$\text{Prob}(W_i = l) = \delta_{l1}, \qquad l = 1, 2, \ldots, i = 0, 1, \ldots, 4, \qquad (10.52)$$

where δ_{l1} is the Kronecker delta defined by (5.36). More generally, however, we will denote the waiting time distribution by

$$\text{Prob}(W_i = l) = w_i(l), \qquad l = 1, 2, \ldots, i = 0, 1, \ldots, 4. \qquad (10.53)$$

We will assume throughout that the waiting time in state i is independent of when the state was entered.

Before proceeding, it will be convenient to define random variables $T(l)$ and D_i. Let $T(l)$ denote the time since the last transition, when l units of linear time have elapsed in all. Then our assumption that the origins of linear and nonlinear time coincide may be written simply as

$$T(0) = 0. \qquad (10.54)$$

Let D_i denote the state next visited, given that the current state is i. Then

$$\text{Prob}(D_i = j) = \text{Prob}(X(n+1) = j \mid X(n) = i) = p_{ij}, \qquad (10.55)$$

on using the Markov property (5.81). We will assume throughout that the next state visited is independent both of states previously visited and of time of entry to the current state. We will assume in this section that the choice of next state is independent of waiting time in the current one, i.e., that

$$\text{Prob}(D_i = j \mid W_i = l) = \text{Prob}(D_i = j) = p_{ij}. \qquad (10.56)$$

In Section 10.5, however, we shall relax this assumption.

Now, the quantity we would like to derive in this section is the probability of being in state j precisely l linear time units after entering state i, regardless of the number of transitions (at most l) that have occurred in all. We will assume throughout that this quantity does not depend on when state i is entered. In other words,

$$\text{Prob}(X(l+s) = j \mid i \text{ entered at } s) = \text{Prob}(X(l+s) = j \mid X(s) = i, T(s) = 0)$$
$$(10.57)$$

is independent of s. Note that this bears some resemblance to the Markov property and would be identical to it if we eliminated the restriction $T(s) = 0$. We will therefore refer to it as the *semi-Markov* property. Because (10.57) is independent of s, we might just as well set s equal to zero. Let us therefore define

$$r_{ij}(l) \equiv \text{Prob}(X(l) = j \mid i \text{ entered at } 0)$$
$$= \text{Prob}(X(l) = j \mid X(0) = i, T(0) = 0), \qquad (10.58)$$

for values of i and j between 0 and 4. The fact that (10.57) and (10.58) are equal will be important below.

Let us also define

$$\pi_i(l) \equiv \text{Prob}(X(l) = i \mid T(0) = 0). \tag{10.59}$$

Because we have assumed (10.54) to begin with, it may be most useful to think of $\pi_i(l)$ simply as the (unconditional) probability that the birds are in state i at time l. We will denote the column vector with (10.59) in row i by $\pi(l)$.

We will denote by $\mathbf{R}(l)$ the square matrix with (10.58) in row i and column j. Thus determining $\mathbf{R}$ is the major task of this section. Central to this task is the expansion (A.15), which we reiterate here in a form convenient for our present purpose. Let U, V, U_0, U_1, U_2, ..., U_M be events in an arbitrary sample space S, such that U_0, U_1, U_2, ..., U_M are mutually disjoint and together exhaust S (i.e., one of these $M + 1$ events must happen). In the notation of set theory, $U_i \cap U_j = \varnothing$, where $\varnothing$ is the empty set, for all $i \neq j$ satisfying $0 \leq i, j \leq M$; and $U_0 \cup U_1 \cup U_2 \cup \cdots \cup U_M = S$. Then

$$\text{Prob}(U \mid V) = \sum_{k=0}^{M} \text{Prob}(U \mid U_k \cap V) \cdot \text{Prob}(U_k \mid V). \tag{10.60}$$

As our first application of this formula, let V be the event that $T(0) = 0$, U the event that $X(l) = i$, U_k the event that $X(0) = k$, and take $M = 4$. Then $U_k \cap V$ is the event that $X(0) = k$, $T(0) = 0$ (i.e., $X(0) = k$ *and* $T(0) = 0$). It follows immediately from (10.58)–(10.60) that

$$\pi_i(l) = \sum_{k=0}^{4} r_{ki}(l) \cdot \pi_k(0) \tag{10.61}$$

or, in matrix form,

$$\pi(l)^T = \pi(0)^T \mathbf{R}(l). \tag{10.62}$$

Now let V be the event that $X(0) = i$, $T(0) = 0$ (i.e., the event that state i is entered at time zero), U the event that $X(l) = j$, U_0 the event that $W_i > l$, U_k the event that $W_i = k$, for $1 \leq k \leq l$, and take $M = l$. Then it follows from (10.58) and (10.60) that

$$r_{ij}(l) = \text{Prob}(X(l) = j \mid W_i > l, X(0) = i, T(0) = 0)$$
$$\times \text{Prob}(W_i > l \mid X(0) = i, T(0) = 0)$$
$$+ \sum_{k=1}^{l} \text{Prob}(X(l) = j \mid W_i = k, X(0) = i, T(0) = 0)$$
$$\times \text{Prob}(W_i = k \mid X(0) = i, T(0) = 0). \tag{10.63}$$

Now, if the birds are in state i at time zero and $W_i > l$, then they must still be in state i at time l. Thus $\text{Prob}(X(l) = j \mid W_i > l, X(0) = i, T(0) = 0)$ equals 1 if $i = j$ and zero if $i \neq j$. Moreover, because waiting time does not depend on time of entry, we have $\text{Prob}(W_i = k \mid X(0) = i, T(0) = 0) = w_i(k)$, from (10.53). Thus (10.63) yields

$$r_{ij}(l) = \delta_{ij} \sum_{k=l+1}^{\infty} w_i(k) + \sum_{k=1}^{l} w_i(k)$$
$$\times \text{Prob}(X(l) = j \mid W_i = k, X(0) = i, T(0) = 0), \qquad (10.64)$$

where δ_{ij} is the Kronecker delta defined by (5.36).

To obtain an expression for $\text{Prob}(X(l) = j \mid W_i = k, X(0) = i, T(0) = 0)$, we now let U be the event that $X(l) = j$; V the event that $W_i = k$, $X(0) = i$, $T(0) = 0$; U_m the event that $D_i = m$ (i.e., that m is the next state entered after i), and take $M = 4$. Then $U_m \cap V$ is the event that $D_i = m$, $W_i = k$, $X(0) = i$, $T(0) = 0$. But if the birds enter state i at time zero, if they wait for time k in it, and if the next state visited is m, then it follows that state m is entered at time k. In other words, $U_m \cap V$ is also the event that $X(k) = m$, $T(k) = 0$. Hence

$$\text{Prob}(U \mid U_k \cap V) = \text{Prob}(X(l) = j \mid X(k) = m, T(k) = 0)$$
$$= \text{Prob}(X(l - k) = j \mid X(0) = m, T(0) = 0)$$
$$= r_{mj}(l - k), \qquad (10.65)$$

on using (10.58) and the fact that it equals (10.57) by the semi-Markov property. Moreover, the identity of the next state is independent of previous states and time of entry to the current one, and so we have

$$\text{Prob}(U_m \mid V) = \text{Prob}(D_i = m \mid W_i = k, X(0) = i, T(0) = 0)$$
$$= \text{Prob}(D_i = m \mid W_i = k)$$
$$= p_{im}, \qquad (10.66)$$

on using (10.56). From (10.60), (10.65), and (10.66), it now follows immediately that

$$\text{Prob}(X(l) = j \mid W_i = k, X(0) = i, T(0) = 0) = \sum_{m=0}^{4} r_{mj}(l - k) \cdot p_{im}. \quad (10.67)$$

Substituting into (10.64), we have

$$r_{ij}(l) = \delta_{ij} \sum_{k=l+1}^{\infty} w_i(k) + \sum_{k=1}^{l} w_i(k) \sum_{m=0}^{4} p_{im} \cdot r_{mj}(l - k), \qquad (10.68)$$

for $0 \leq i, j \leq 4$. Moreover, because every transition takes at least one unit of time, we must have

$$r_{ij}(0) = \delta_{ij}, \qquad 0 \leq i, j \leq 4. \qquad (10.69)$$

Equations (10.68) and (10.69) are sufficient to determine the matrix **R**, by recursion, for any value of l. In the special case where w_i is defined by (10.52), i.e., where linear and nonlinear time coincide, (10.68) yields

$$r_{ij}(l) = \sum_{m=0}^{4} p_{im} r_{mj}(l-1),$$ (10.70)

the matrix form of which is

$$\mathbf{R}(l) = \mathbf{P} \cdot \mathbf{R}(l-1).$$ (10.71)

The matrix form of (10.69) is just $\mathbf{R}(0) = \mathbf{I}$, the identity matrix; it then follows from (10.71) that $\mathbf{R}(l) = \mathbf{P}^l$, in agreement with the results of Chapter 5.

Before obtaining the bird distribution after 20 minutes, it remains to specify distributions for the waiting times $W_0, \ldots, W_4$. Let's suppose that the budgerigars are philosophically inclined and like to engage in dialogues. If there is an even number of birds at each end of the cage, then the birds can be expected to remain in that state for quite a while, each having a partner in dialogue; whereas if there is an odd number of birds at each end of the cage, then one of the four birds is likely to get restless. It therefore seems reasonable to assume that

$$w_1(l) = w_3(l)$$
$$w_0(l) = w_2(l) = w_4(l),$$ (10.72)

with $E[W_1] \ll E[W_0]$. Note that this does not contradict our earlier assumption, in Section 5.7, that each bird is equally likely to be next on the trapeze. For the sake of illustration, let's take

$$w_0(l) = \frac{1}{10}, \qquad l = 1, 2, \ldots, 10,$$
$$= 0, \qquad l \geq 11$$ (10.73)
$$w_1(l) = \frac{1}{2}, \qquad l = 1, 2,$$
$$= 0, \qquad l \geq 3.$$ (10.74)

We are now in a position to calculate $\pi(20)$ in linear time.

In Section 5.7, we supposed that the birds were all at the South end of the cage initially, so that the initial state of the system (number of birds at North end) was $i = 0$. We shall do the same here; hence

$$\pi(0)^T = (1, 0, 0, 0, 0).$$ (10.75)

Note that because the zero of linear time must correspond to a transition, (10.75) implies not only that $X(0) = 0$ but also that $T(0) = 0$. Setting $l = 1$ in (10.68) and using (10.69) yields

$$r_{ij}(1) = \delta_{ij} \sum_{k=2}^{10} w_i(k) + w_i(1) p_{ij}, \qquad i, j = 0, \ldots, 4.$$ (10.76)

In matrix form, on using (10.73) and (10.74), we obtain

$$
\mathbf{R}(1) =
\begin{bmatrix}
\frac{19}{20} & \frac{1}{20} & 0 & 0 & 0 \\
\frac{1}{16} & \frac{3}{4} & \frac{3}{16} & 0 & 0 \\
0 & \frac{1}{40} & \frac{19}{20} & \frac{1}{40} & 0 \\
0 & 0 & \frac{3}{16} & \frac{3}{4} & \frac{1}{16} \\
0 & 0 & 0 & \frac{1}{20} & \frac{19}{20}
\end{bmatrix}.
\tag{10.77}
$$

From (10.62) and (10.75), we have

$$
\pi(1)^T = \pi(0)^T \mathbf{R}(1) = \left(\frac{19}{20}, \frac{1}{20}, 0, 0, 0 \right).
\tag{10.78}
$$

Again, putting $l = 2$ in (10.68), we obtain

$$
r_{ij}(2) = \delta_{ij} \sum_{k=3}^{10} w_i(k) + w_i(2)p_{ij} + w_i(1)\{\mathbf{PR}(1)\}_{ij}, \qquad i,j = 0,\dots,4, \tag{10.79}
$$

where $\{\mathbf{PR}(1)\}_{ij}$ denotes entry (i,j) in the product of matrices $\mathbf{P}$ and $\mathbf{R}(1)$. On using (10.73) and (10.74), we thus obtain

$$
\mathbf{R}(2) = \frac{1}{3200}
\begin{bmatrix}
2882 & 288 & 30 & 0 & 0 \\
440 & 1425 & 1320 & 15 & 0 \\
5 & 144 & 2902 & 144 & 5 \\
0 & 15 & 1320 & 1425 & 440 \\
0 & 0 & 30 & 288 & 2882
\end{bmatrix},
\tag{10.80}
$$

$$
\pi(2)^T = \pi(0)^T \mathbf{R}(2) = \frac{1}{1600}(1441, 144, 15, 0, 0).
\tag{10.81}
$$

Continuing in this manner and performing 18 further recursions of (10.68), we find

$$
\mathbf{R}(20) =
\begin{bmatrix}
0.261134 & 0.154495 & 0.495506 & 0.062465 & 0.026401 \\
0.135504 & 0.121364 & 0.589276 & 0.092934 & 0.060921 \\
0.082584 & 0.108480 & 0.617872 & 0.108480 & 0.082584 \\
0.060921 & 0.092934 & 0.589276 & 0.121364 & 0.135504 \\
0.026401 & 0.062465 & 0.495506 & 0.154495 & 0.261134
\end{bmatrix}.
\tag{10.82}
$$

You should verify this matrix by using a computer. The first row is the quantity we sought, the distribution of birds after 20 minutes.

Of greatest significance, however, is the distribution of birds after a great deal of time has elapsed. A large number of recursions of (10.68) will indicate that

$$\mathbf{R}(\infty) = \lim_{l \to \infty} \mathbf{R}(l) = \frac{1}{112} \begin{bmatrix} 11 & 12 & 66 & 12 & 11 \\ 11 & 12 & 66 & 12 & 11 \\ 11 & 12 & 66 & 12 & 11 \\ 11 & 12 & 66 & 12 & 11 \\ 11 & 12 & 66 & 12 & 11 \\ 11 & 12 & 66 & 12 & 11 \end{bmatrix} \tag{10.83}$$

and

$$\boldsymbol{\pi}(\infty)^T = \lim_{l \to \infty} \boldsymbol{\pi}(l)^T = \boldsymbol{\pi}(0)^T \mathbf{R}(\infty) = \frac{1}{112}(11, 12, 66, 12, 11)$$
$$\approx (0.098, 0.107, 0.589, 0.107, 0.098), \tag{10.84}$$

which we may call the *limiting distribution* of our extended Markov model. It is not a stationary distribution, however, because $\boldsymbol{\pi}(0) = \boldsymbol{\pi}(\infty)$ does not imply that $\boldsymbol{\pi}(l) = \boldsymbol{\pi}(\infty)$ for all l; for you can verify directly that if $\boldsymbol{\pi}(0) = \boldsymbol{\pi}(\infty)$, then $\boldsymbol{\pi}(1) = (0.1, 0.1, 0.6, 0.1, 0.1)^T$ and $\boldsymbol{\pi}(2) = (0.104, 0.084, 0.625, 0.084, 0.104)^T$. The reason for this is simply that the dynamic equations (10.68) for $r_{ij}(l)$ have coefficients depending explicitly on l; this is analogous to the fact that the solution $x(t) = te^{-t} + x(0)e^{-t}$ of the differential equation

$$\frac{dx}{dt} + x = e^{-t} \tag{10.85}$$

has the limiting value $x(\infty) = 0$ for any value of $x(0)$, even though $x(t) = 0$ is not itself a solution of the equation.

Although the linear time limiting distribution $\boldsymbol{\pi}(\infty)$ is not a stationary distribution, it does bear an important relationship to the stationary distribution

$$\boldsymbol{\pi}^* = \left(\frac{1}{16}, \frac{1}{4}, \frac{3}{8}, \frac{1}{4}, \frac{1}{16} \right)^T \tag{10.86}$$

of the original Markov model in Section 5.7, which is simply the limiting distribution in *nonlinear time*. To see this, we first observe from (10.72)–(10.74) that the expected waiting times in the various states satisfy $E[W_0] = E[W_2] = E[W_4]$, $E[W_1] = E[W_3]$ with

$$E[W_0] = \sum_{l=1}^{\infty} lw_0(l) = \frac{1}{10} \sum_{l=1}^{10} l = 5.5, \tag{10.87}$$

$$E[W_1] = \sum_{l=1}^{\infty} lw_1(l) = 1 \cdot \frac{1}{2} + 2 \cdot \frac{1}{2} = 1.5. \tag{10.88}$$

Now, (10.86) merely records the proportions of nonlinear time, *in the long run*, spent in each of the various states; these proportions are in the ratio

$1 : 4 : 6 : 4 : 1.$[8] But each unit of nonlinear time spent in state i corresponds, in the long run, to $E[W_i]$ units of linear time. Hence the proportions of *linear* time spent in each of the various states are in the ratio $E[W_0]$: $4E[W_1] : 6E[W_2] : 4E[W_3] : E[W_4]$ or $11 : 12 : 66 : 12 : 11$. To convert these ratios to proportions, we merely divide by their sum, 112. Hence $\pi(\infty)$ simply records the long run proportions of linear time spent in each of the various states.

Our extended model of birds in a cage is a special case of what has come to be known as a semi-Markov process. In linear time, it is no longer true that the future depends only on the present; it depends on the past through the waiting time. In nonlinear time, on the other hand, the model is identical to the one we developed in Section 5.7. It is therefore customary to say that the (nonlinear time) Markov chain is *embedded* in the (linear time) semi-Markov process; see, for example, Ross (1983, p. 130).

It is easy to see how this model can be generalized to systems with $N + 1$ states: essentially, one just replaces 4 by N, wherever it appears. But the assumption we made in (10.56) restricts its applicability. Therefore, in the following section, we proceed to relax it, thereby obtaining a more general and more useful version of our semi-Markov model.

10.5   A More General Semi-Markov Process. Further Extension

In the previous section, the probability of transition from state i to state j did not depend on the time spent waiting in state i. In particular, the matrix of transition probabilities, (10.51), was derived (in Section 5.7) by assuming that, having performed on the trapeze, a bird is equally likely to fly back down to either end of the cage, irrespective of how long the birds have spent in their current state (recall that they do not arrive in a new state until the trapezing bird has actually reached an end of the cage). In other words, if V denotes the event that a bird flies to the opposite end of the cage after trapezing, then we obtained **P** by assuming that

$$\text{Prob}(V) = \frac{1}{2}. \tag{10.89}$$

In practice, however, a bird may be more likely to fly back down to the opposite end after a long wait than after a short wait. Then $\text{Prob}(V)$ would be an increasing function of the waiting time. For the sake of illustration, let's just assume that

$$\text{Prob}(V) = \frac{l}{l+1}, \tag{10.90}$$

---

[8]These are the coefficients of z in the expansion of $(1 + z)^4$. More generally, if there were N birds in the cage, then the corresponding ratios would be the coefficients of z in the expansion of $(1 + z)^N$.

where l is the number of units of linear time that the birds have spent in the state they are about to leave; Prob(V) is the same for all birds. In the special case (10.52) where linear and nonlinear time correspond, (10.90) agrees with our earlier assumption, (10.89).

The probability that if the birds are now in state i, then their next state will be j, is still p_{ij} and has nothing to do with linear time l. But the probability that the birds are about to enter state j, *given that they have spent time l in state i*, will depend on l through (10.90). Suppose, for example, that $i = 1$ and that one of the birds is about to go on the trapeze. Thus the birds are about to leave state 1, having spent time l in it (minus the virtually zero time that it takes a bird to trapeze). Let U be the event that the bird that trapezes is the loner in the North end of the cage. Because all birds are equally likely to trapeze, Prob(U) = 1/4. Then, because V is independent of U, the probability that the birds now move to state 0 is Prob($V \cap U$) = Prob(V) $\cdot$ Prob(U). Let's denote this quantity by $q_{10}(l)$, i.e., define $q_{10}(l) = \text{Prob}(D_1 = 0 \mid W_i = l)$. Then, clearly,

$$q_{10}(l) = \frac{l}{4(l+1)}.$$ (10.91)

On the other hand, the probability that the birds now move to state 1 is simply the probability that the bird that trapezes flies back to where it came from. Thus, if we define $q_{11}(l) = \text{Prob}(D_1 = 1 \mid W_i = l)$, then

$$q_{11}(l) = 1 - \text{Prob}(V) = \frac{1}{l+1}.$$ (10.92)

More generally, let us define $\mathbf{Q}(l)$ to be the square matrix in which row i and column j contain the conditional probability

$$q_{ij}(l) \equiv \text{Prob}(D_i = j \mid W_i = l),$$ (10.93)

where $0 \leq i, j \leq 4$. Then, proceeding as above and using arguments like those that produced $\mathbf{P}$ in Section 5.7, we obtain (Exercise 10.5):

$$\mathbf{Q}(l) = \begin{bmatrix} \frac{1}{l+1} & \frac{l}{l+1} & 0 & 0 & 0 \\ \frac{l}{4(l+1)} & \frac{1}{l+1} & \frac{3l}{4(l+1)} & 0 & 0 \\ 0 & \frac{l}{2(l+1)} & \frac{1}{l+1} & \frac{l}{2(l+1)} & 0 \\ 0 & 0 & \frac{3l}{4(l+1)} & \frac{1}{l+1} & \frac{l}{4(l+1)} \\ 0 & 0 & 0 & \frac{l}{l+1} & \frac{1}{l+1} \end{bmatrix}, \qquad l = 1, 2, \ldots. \quad (10.94)$$

Note that the rows of this matrix all sum to 1, as of course they must.

The absolute probability of going from i to j (in a unit of nonlinear time), or the probability that j will be the next state after i when nothing is known about the length of stay in i (other than its distribution), is now readily obtained from (10.60). Let U be the event that $D_i = j$, V the sample space S, U_l the event that $W_i = l$, and take $M \to \infty$ in that formula. Then

$$p_{ij} = \text{Prob}(D_i = j)$$

$$= \sum_{l=0}^{\infty} \text{Prob}(D_i = j \mid W_i = l) \cdot \text{Prob}(W_i = l)$$

$$= \sum_{l=1}^{\infty} q_{ij}(l) \cdot w_i(l), \tag{10.95}$$

because $\text{Prob}(W_i = 0) = 0$. Note that if q_{ij} is independent of l, then q_{ij} is equal to p_{ij} because the sum of the $w_i(l)$ over all l is 1, for all i. This corresponds to the special case of the semi-Markov model considered in the previous section, for which (10.56) is satisfied.

On using (10.73), (10.74), (10.94), and (10.95) you can verify (Exercise 10.5) by straightforward calculation that the absolute transition matrix for our (further extended) birdcage model is given by

$$\mathbf{P} = \begin{bmatrix} \gamma & 1-\gamma & 0 & 0 & 0 \\ \frac{7}{48} & \frac{5}{12} & \frac{7}{16} & 0 & 0 \\ 0 & \frac{1}{2}(1-\gamma) & \gamma & \frac{1}{2}(1-\gamma) & 0 \\ 0 & 0 & \frac{7}{16} & \frac{5}{12} & \frac{7}{48} \\ 0 & 0 & 0 & 1-\gamma & \gamma \end{bmatrix}, \tag{10.96}$$

where we have defined

$$\gamma = \frac{1}{10} \sum_{l=0}^{10} \frac{1}{l+1} \approx 0.202. \tag{10.97}$$

The matrix $\mathbf{P}$, together with the initial condition $\pi(0)^T = (1,0,0,0,0)$ and the recursion $\pi(n+1)^T = \pi(n)^T \mathbf{P}$, suffices for a complete probabilistic description of the birds' state distribution vector $\pi(n)$ in nonlinear time n; or, which is exactly the same thing, gives a complete description of the embedded Markov chain.

To obtain the equivalent description in linear time, we must modify the recursion (10.68) obtained in the previous section. Notationally, the modification is a slight one; we merely replace p_{ij} in (10.66) and (10.67) by (10.93). Thus (10.68) is replaced by

$$r_{ij}(l) = \delta_{ij} \sum_{k=l+1}^{\infty} w_i(k) + \sum_{k=1}^{l} w_i(k) \sum_{m=0}^{4} q_{im}(k) r_{mj}(l-k), \qquad i,j = 0,\dots,4. \tag{10.98}$$

The notational simplicity of this modification, however, belies its practical importance; for the extension of the semi-Markov process to cope with transitions whose outcomes depend upon the waiting time is an extremely useful one.

With $w_i(l)$ given by (10.72)–(10.74) and $\mathbf{Q}$ by (10.94), the first two recursions of (10.98) yield the same $\mathbf{R}(1)$ as in (10.77), but[9]

$$\mathbf{R}(2) = \frac{1}{9600} \begin{bmatrix} 8486 & 1024 & 90 & 0 & 0 \\ 1520 & 3475 & 4560 & 45 & 0 \\ 15 & 512 & 8546 & 512 & 15 \\ 0 & 45 & 4560 & 3475 & 1520 \\ 0 & 0 & 90 & 1024 & 8486 \end{bmatrix}. \qquad (10.99)$$

The recursion can be continued indefinitely; as in the previous section, you will find that $\mathbf{R}(l)$ tends to a limit as $l \to \infty$, each row of which is the limiting distribution

$$\pi(\infty)^T = \frac{1}{8(113 - 36\gamma)} \, (77, 144(1 - \gamma), 462, 144(1 - \gamma), 77)$$
$$\approx (0.09104, 0.13586, 0.54621, 0.13586, 0.09104), \quad (10.100)$$

on using (10.97). It is left to you as Exercise 10.7 to show that (10.100) bears the same relationship to the stationary distribution of the embedded Markov chain as in Section 10.4.

Finally, the semi-Markov process is readily extended to cope with an arbitrary number of states, say, $N + 1$; we merely replace 4 by N in (10.98) to obtain

$$r_{ij}(l) = \delta_{ij} \sum_{k=l+1}^{\infty} w_i(k) + \sum_{k=1}^{l} w_i(k) \sum_{m=0}^{N} q_{im}(k) r_{mj}(l - k), \qquad i, j = 0, \ldots, N. \qquad (10.101)$$

This describes the most general semi-Markov process we shall meet in this book. The birds have served their purpose by introducing it. Now let us bid them farewell.

10.6   Who Will Govern Britain in the Twenty-First Century? A Combination

In Exercise 8.12, you were asked to predict the proportions of the twenty-first century for which the United States would be governed by Democrat and Republican presidents.[10] Let's suppose that you chose a 2-state Markov chain as your model and used election data from 1860 onward, obtaining figures of about 5/8 for Republicans and 3/8 for Democrats. Leaving aside the question of whether the future is independent of the past in U.S. voting behavior, you would have been able to estimate these proportions from a

---

[9]See Supplementary Note 10.2.
[10]You may wish to review the second part of Section 8.7 before proceeding.

simple Markov chain because, if presidential elections correspond to transitions, then linear and nonlinear time coincide, an election being held every four years. In the United Kingdom, however, the timing of an election is a random variable, with a maximum value of 5 years, as we saw in Section 8.7. Thus linear and nonlinear time do not coincide. Because there are still only two major political parties in Britain, Labour and the Conservatives, we can still use a 2-state Markov chain to model the governing process; but to predict the proportion of the twenty-first century for which, say, Conservatives will be in power, that chain must be embedded in a semi-Markov process.

Let $X(l)$ be a random variable taking values 0 and 1. If $X = 0$ then Labour govern, if $X = 1$ then Conservatives. Let each general election correspond to a transition, let the random variable W_0 denote the number of quarters (at most 20) for which a Labour government waits in power, and let W_1 be the corresponding waiting time for a Conservative government. Statistical models for the distributions of W_0 and W_1 were obtained in Section 8.7 (where W_0 was denoted by W_L and W_1 by W_C). Thus, from (8.101) and (8.96), we obtain

$$\text{Prob}(W_0 = l) = w_0(l) = \begin{cases} \dfrac{1}{16}, & 1 \le l \le 8, 13 \le l \le 20; \\ 0, & 9 \le l \le 12, 21 \le l < \infty. \end{cases} \quad (10.102)$$

$$\text{Prob}(W_1 = l) = w_1(l) = \begin{cases} \dfrac{1}{72}, & 1 \le l \le 8, l = 13, 14, 19, 20; \\ \dfrac{5}{24}, & 15 \le l \le 18; \\ 0, & 9 \le l \le 12, 21 \le l < \infty. \end{cases} \quad (10.103)$$

We shall incorporate (10.102) and (10.103) without change into the semi-Markov process described by (10.101), with $N = 1$. Our model will therefore be a combination. Our principal task is to specify the 2×2 matrix $Q(l)$. As in Section 8.7, my choice will be very subjective because so few data are available.

There have been 14 British general elections and hence 13 transitions of power since World War II. The results of the first 13 elections and the lengths of the subsequent terms in office are recorded in Table 8.12. (See Exercise 10.15 for the exception.) The corresponding transitions are recorded in Table 10.3. Despite the paucity of data, patterns emerge. Labour terms in office are divided equally between long ones and short ones. After short terms, Labour have had twice as many victories (two) as defeats (one). After long terms, Labour have had twice as many defeats (two) as victories (one). This suggests that we take $q_{00}(l) = 2/3$ and $q_{01}(l) = 1/3$ if l is short, and that we take $q_{00}(l) = 1/3$ and $q_{01}(l) = 2/3$ if l is long. There are far too few data to justify this choice on empirical grounds. But we may regard it as a mathematical statement of

Table 10.3   Changes of government in postwar Britain. State 0 denotes Labour, state 1, Conservatives.

Labour in power		Conservatives in power	
Waiting time	Transition	Waiting time	Transition
2	$0 \rightarrow 0$	15	$1 \rightarrow 1$
6	$0 \rightarrow 1$	15	$1 \rightarrow 0$
6	$0 \rightarrow 0$	16	$1 \rightarrow 1$
17	$0 \rightarrow 1$	16	$1 \rightarrow 1$
19	$0 \rightarrow 0$	17	$1 \rightarrow 1$
19	$0 \rightarrow 1$	20	$1 \rightarrow 0$

some politician's belief that Labour are twice as likely to lose an election after a long term as after a short one, and in this sense I find it acceptable. Again, Conservative terms in office have always been long ones, followed by twice as many victories as defeats. This suggests that we take $q_{11}(l) = 2/3$ and $q_{10}(l) = 1/3$, at least when l is long; but I am going to assume this for all values of l.

For the sake of simplicity, I will regard a term as short if $1 \le l \le l_c$ and long if $l_c < l \le 20$, where l_c is a parameter. Then my choice of $\mathbf{Q}(l)$ is given by

$$\mathbf{Q}(l) = \begin{bmatrix} \frac{2}{3} & \frac{1}{3} \\ \frac{1}{3} & \frac{2}{3} \end{bmatrix}, \qquad \text{if } 1 \le l \le l_c;$$

$$\mathbf{Q}(l) = \begin{bmatrix} \frac{1}{3} & \frac{2}{3} \\ \frac{1}{3} & \frac{2}{3} \end{bmatrix}, \qquad \text{if } l_c < l \le 20. \tag{10.104}$$

What is the value of l_c? Maybe all we can say for sure about it is that we would expect it to be fairly close to 10; but, perhaps surprisingly, this rather vague estimate turns out to be adequate, as we shall see below!

The (absolute) transition matrix $\mathbf{P}$, which gives the transitional probabilities when nothing is known about the length of term in office, now follows readily from (10.95) and (10.102)–(10.104). We obtain (Exercise 10.11):

$$\mathbf{P} = \begin{bmatrix} \frac{1}{3} + \frac{l_c}{48} & \frac{2}{3} - \frac{l_c}{48} \\ \frac{1}{3} & \frac{2}{3} \end{bmatrix}, \qquad \text{if } 1 \le l_c \le 8; \tag{10.105}$$

$$\mathbf{P} = \begin{bmatrix} \frac{1}{2} & \frac{1}{2} \\ \frac{1}{3} & \frac{2}{3} \end{bmatrix}, \qquad \text{if } 9 \le l_c \le 12; \tag{10.106}$$

$$\mathbf{P} = \begin{bmatrix} \frac{1}{4} + \frac{l_c}{48} & \frac{3}{4} - \frac{l_c}{48} \\ \frac{1}{3} & \frac{2}{3} \end{bmatrix}, \qquad \text{if } 13 \le l_c \le 20. \tag{10.107}$$

It follows readily that the stationary distribution of the embedded Markov chain is given (verify) by

$$(\pi^*)^T = \left(\frac{16}{48 - l_c}, \frac{32 - l_c}{48 - l_c}\right), \qquad \text{if } 1 \le l_c \le 8;$$

$$(\pi^*)^T = \left(\frac{2}{5}, \frac{3}{5}\right), \qquad \text{if } 9 \le l_c \le 12; \qquad (10.108)$$

$$(\pi^*)^T = \left(\frac{16}{52 - l_c}, \frac{36 - l_c}{52 - l_c}\right), \qquad \text{if } 13 \le l_c \le 20.$$

Because (verify) $E[W_0] = 21\frac{1}{2}$ and $E[W_1] = 9\frac{1}{6}$, it follows (verify) from the analytical method of Section 10.4 that the limiting distribution of the semi-Markov process is given by

$$\pi(\infty)^T = \left(\frac{1008}{3920 - 91l_c}, \frac{91(32 - l_c)}{3920 - 91l_c}\right), \qquad \text{if } 1 \le l_c \le 8;$$

$$\pi(\infty)^T = \left(\frac{42}{133}, \frac{91}{133}\right), \qquad \text{if } 9 \le l_c \le 12; (10.109)$$

$$\pi(\infty)^T = \left(\frac{1008}{4284 - 91l_c}, \frac{91(36 - l_c)}{4284 - 91l_c}\right), \qquad \text{if } 13 \le l_c \le 20.$$

According to our model, the long-run proportion of time for which Labour are in power increases from 0.263 to 0.409 as l_c increases from 1 to 20, and that for which Conservatives are in power decreases from 0.737 to 0.591. Thus, even if we were not prepared to hazard a guess at l_c, the quantities of interest would have been restricted to fairly narrow bands. It seems quite reasonable to suppose, however, that l_c lies either in or very close to the middle of its range, $9 \le l_c \le 12$. Then our model predicts that in the long run, say in the twenty-first century, Conservatives will govern for about 68% of the time and Labour for 32%.

Only time can tell whether this prediction is correct. Indeed it may never be possible to test it, because the political fabric of the United Kingdom could change so drastically as to render it immediately invalid. It should be emphasized once more, however, that even if the prediction turns out to be correct, it does not necessarily mean that power sharing in Britain evolves according to a semi-Markov process and that my choices of w_0, w_1, and Q were good ones. It simply means that you have no good reason to reject my model, at least until you can apply a more stringent test.

The principal difficulty in applying the semi-Markov model to United Kingdom power sharing is that transitions occur so infrequently as to make data sparse. For applications involving more frequent transitions, see Exercises 10.12 and 10.14.

Exercises

10.1 Verify (10.15).

10.2 Verify (10.19).

10.3 Verify (10.28) and (10.29).

*__10.4__ According to the model of Section 10.3, what are the effects of an increase in money supply upon inflation, investment, output, and real interest rate? State one or two assumptions that are implied by this model. Is it realistic? Discuss some of its limitations.

10.5 Verify (10.94) and (10.96).

10.6 Verify (10.98).

10.7 (i) Find the stationary distribution of the Markov chain embedded in the semi-Markov process of Section 10.5.
(ii) Use the analytical method of Section 10.4 to verify that (10.100) is the limiting distribution of the semi-Markov process.

*__10.8__ As we remarked at the time, the semi-Markov model of Section 10.4 does not possess the Markov property (in linear time), because the future depends on the past through the waiting time. The new model required an extra assumption, namely, that a transition took place at the initial instant. But if waiting times were such that it made no difference whether a system had entered some state at $t = 0$ or had been there for some time, then because any moment could be regarded as $t = 0$, the future would be independent of the past again; and the Markov property, that the future depends only on the present, would be restored. Unfortunately, the only waiting time distribution with the required property is a continuous one, the exponential distribution

$$\text{Prob}(W_i > t) = e^{-t/\mu_i},$$

where $\mu_i = E[W_i]$ is the expected length of a stay in state i; whereas the model of Section 10.4 assumes that waiting times are discrete. Restoring the Markov property, therefore, automatically implies an *adaptation* of that model to allow for continuous time. The resulting model is known as a *continuous-time Markov jump process*. Derive it.

10.9 Having attempted Exercise 10.8, show that, for the elevator model of Section 5.5, (5.65) is a special case of the formula

$$\mathbf{R}(t) = \mathbf{I} + \sum_{m=1}^{\infty} \frac{t^m}{m!} \{\mathbf{D}(\mathbf{P} - \mathbf{I})\}^m,$$

where $\mathbf{D}$ is defined in the solution to Exercise 10.8.

**10.10 Derive the continuous-time Markov jump process as an extension of the birth and death process of Section 5.5. (This derivation is an alternative to adapting the discrete-time semi-Markov process of Section 10.4, as in Exercise 10.8.)

10.11 Verify (10.105)–(10.107).

**10.12 A few years ago, General Motors became very worried that its share of the automobile market in the United States was shrinking. What should GM do in such circumstances? How much effort should it concentrate on persuading existing GM car owners to stay with GM when buying a new car, and how much on persuading owners of other cars to switch to GM?

You may be able to answer these questions with the help of a semi-Markov model in which transitions correspond to auto purchases. Begin with a two-state model in which $X = 0$ implies that a randomly chosen car owner is driving a GM car and $X = 1$ implies that she is not. Choose a form of **Q** containing one or two free parameters, which can be adjusted according to the direction of GM's marketing effort. Later, you can increase the number of states to distinguish between, say, domestic and foreign competitors, or between specific companies in the automobile industry.

**10.13 By first combining the models of Sections 1.6, 3.3, and 10.3, develop a more realistic model of economic growth than we have yet discussed in this textbook.

**10.14 Construct a semi-Markov model of an elevator. Extend your model to deal with two elevators, side by side.

*10.15 In a United Kingdom general election on April 9, 1992 the Conservative party was returned to power. What, if any, effect does this result have on the model in Section 10.6?

Supplementary Notes

10.1 A rigorous derivation of these more general results is beyond the scope of the present text. Nevertheless, we can motivate them by using only the results of Chapter 2. The system (10.28)–(10.29) is of the form (2.12) with $u(\xi, \psi) = \xi(\lambda(1 - \xi) - \theta\psi/(\xi + \mu)), v(\xi, \psi) = \psi(1 - \psi/\xi)$. You can easily verify that $u(\gamma, \gamma) = 0 = v(\gamma, \gamma)$ and, furthermore, that $\xi^* = \gamma$, $\psi^* = \gamma$ is the only static equilibrium satisfying $\xi > 0$ and $\psi > 0$; it always exists, but may be stable or unstable. Observing that $\partial u/\partial \xi = \lambda(1 - 2\xi) - \mu\theta\psi/(\xi + \mu)^2$ and setting $\xi = \gamma = \psi$, we find that the quantity g_{11} defined by (2.14) is given by $g_{11} = \lambda(1 - 2\gamma) - \mu\theta\gamma/(\mu + \gamma)^2$. Similarly, $g_{12} = -\theta\gamma/(\gamma + \mu)$, $g_{21} = 1$, and $g_{22} = -1$. Thus (2.19) becomes $f(\omega) = 0$, where

$$f(\omega) \equiv \omega^2 + \omega(1 - \lambda + 2\lambda\gamma + \mu\lambda(1 - \gamma)/(\gamma + \mu)) + \mu\lambda(1 - \gamma)/(\gamma + \mu) + \lambda\gamma.$$

Because $f(0) > 0$, ω has a positive real part if and only if $f'(0) < 0$. Thus the static equilibrium is unstable when $2\lambda\gamma^2 + \gamma(\lambda\gamma - \lambda + 1) + \mu < 0$; or, on using (10.32), when $\mu\lambda(2\lambda + 1)^2 + (\lambda + 1)(\lambda - \mu\lambda - \theta) < \lambda(\theta - 1)^2 + (\theta - 1)(\lambda - \mu\lambda - \theta)$. Comparing with (10.31), we see that the limit cycle exists if the static equilibrium is unstable. It is easy to understand why the limit cycle cannot exist if $\xi = \gamma = \psi$ is stable: there is then no "source" from which trajectories can converge toward the inside of the limit cycle trajectory. It is more difficult to establish that the limit cycle must exist if $\xi = \gamma = \psi$ is unstable, but your numerical calculations will at least demonstrate that this result is plausible.

10.2 Because $\mathbf{Q}(1)$ is identical to (10.51), the difference between $r_{ij}(2)$ in (10.98) and $r_{ij}(2)$ in (10.79) is simply $w_i(2) \cdot \{q_{ij}(2) - q_{ij}(1)\}$. Hence $\mathbf{R}(2)$ in (10.99) is obtained from $\mathbf{R}(2)$ in (10.80) simply by adding the matrix

$$\frac{1}{240} \begin{bmatrix} -4 & 4 & 0 & 0 & 0 \\ 5 & -20 & 15 & 0 & 0 \\ 0 & 2 & -4 & 2 & 0 \\ 0 & 0 & 15 & -20 & 5 \\ 0 & 0 & 0 & 4 & -4 \end{bmatrix}$$

11 FURTHER FLOW AND DIFFUSION

In this our penultimate chapter, we will construct more comprehensive models for the transport of "quantities" through a *medium*, any expanse of matter in which things (the quantities) move. The medium can either move or be at rest. The only moving media we shall describe are air and water, which are examples of fluids; but our models can be adapted to describe other substances in motion (paint being stirred or lava erupting from a volcano). A medium at rest might still be a fluid, but it could also be a solid, like the wall of a building. Quantities might, for example, be chemical substances dissolved in a fluid, but they might also be properties of a fluid which is itself the medium, e.g., mass or momentum (both per unit volume).

The first five sections of Chapter 11 describe a medium at rest. Section 11.1 is an extension of Section 2.5's model of heat flow; the medium is a wall, and heat the transported quantity. In Sections 11.2–11.4 the medium is a highway and traffic the transported quantity, and in Section 11.5 the medium is a (stagnant) canal and pollution the transported quantity. These four sections describe extensions and adaptations of models we built in Chapter 9.

In Section 11.6, we incorporate these developments into a generic model of flow and diffusion, which is spatially one-dimensional. In the next two sections, we apply this model to moving media. In Section 11.7, the medium is (river) water and the transported quantities are concentrations of chemicals. In Section 11.8, the medium is air (in an organ pipe) and the transported quantities are velocity and density. Exercise 11.15 invites you to extend our one-dimensional generic model to three dimensions (in space).

From a mathematical viewpoint, models of flow and diffusion most often result in partial differential equations. Except in the simplest circumstances, however, the equations that arise are not amenable to analytical solution. In practice, therefore, if you are constructing mathematical models of physical phenomena, then it is necessary to know how partial differential equations are solved numerically. Nevertheless, numerical techniques for solving partial differential equations are more properly regarded as related mathematics than as part of the modelling process itself and are therefore not discussed in this textbook. For an introduction to this important topic, see, for example, Ames (1977).

11.1 Unsteady Heat Conduction. An Adaptation

One of the major themes of Chapter 2 was that stable equilibria could be identified in either of two ways: directly by solving equations that are independent of time, or indirectly by solving dynamic equations, then taking the limit as time tends to infinity.[1] A direct method was used in Section 2.5 to derive the equilibrium heat flux through a window. Thus temperature was assumed to be independent of time. In this section, by contrast, we will use an indirect method to obtain the equilibrium heat flux through a window. Accordingly, we require a model for unsteady heat flow. Let us therefore begin by borrowing a little from the science of physics.

Let H be the amount of heat per unit mass contained in a certain substance (glass, perhaps), and let ϕ be its temperature. Heat is a form of energy, and so H has dimensions L^2/T^2, from Table 1.6; ϕ has dimensions C. Thus heat and temperature are not the same thing, if only because the equation *temperature = heat* would not be dimensionally consistent. Then how are they related? The hotter a substance is—the higher its temperature—the more heat it contains. The simplest hypothesis consistent with this observation is that heat is proportional to temperature or

$$H = c\phi, \tag{11.1}$$

where c is a constant of proportionality.

This equation has been found acceptable and is used by physicists. Because $[H] = L^2/T^2$ and $[\phi] = C$, it follows immediately from (11.1) that c, which is known as the *specific heat*, has dimensions $L^2/(T^2 \cdot C)$. Specific heat is a measure of a substance's (or medium's) ability to derive heat energy from an increase in temperature. An analogy you may find useful is that if you give the same number of chocolate eclairs to each of two people, one of whom is characteristically thin and the other of whom is prone to obesity, then the latter will derive a greater quantity of fatness from the same input of eclairs and can be viewed as having the higher specific fatness. Physicists actually distinguish between c_P, the specific heat at constant pressure, and c_V, the specific heat at constant volume, but we will

---

[1] You may wish to review Sections 2.5 and 9.5 before proceeding.

simply use c to denote specific heat, supposing throughout that the pressure is atmospheric. For further elaboration of the important distinction between heat and energy, see, for example, Dugdale (1966).

Let us now adapt the model of Section 9.5 to describe the flow of heat through a single-glazed window. Let x measure distance perpendicular to the window, and let D be the width of the pane of glass, with $x = 0$ inside the building and $x = D$ outside it. Let $\phi(x, t)$ be the temperature of the glass at station x at time t. Then, in view of (2.45), the outward flux of heat through the window, per unit area, at station x at time t is

$$F(x, t) = -k_G \frac{\partial \phi}{\partial x}, \tag{11.2}$$

where k_G is the thermal conductivity of the glass.

Consider a piece of glass with cross-sectional area A occupying the region $x_1 \leq x \leq x_2$ (Fig. 11.1), and let ρ be its mass per unit volume. If H were constant, then the amount of heat contained in the infinitesimal volume $A\delta x$ in Fig. 11.1 would be HEAT PER UNIT MASS·MASS PER UNIT VOLUME· VOLUME $= H \cdot \rho \cdot A\delta x$; but because H depends upon x and t, it is more correct to write the heat content as $H\rho A\delta x + o(\delta x)$. Then, by analogy with (9.40), and because the error term vanishes during the limit process of integration, we can express the heat energy balance for the region $x_1 \leq x \leq x_2$ as

Rate of change of heat = Heat inflow − Heat outflow

$$\frac{\partial}{\partial t} \int_{x_1}^{x_2} H(x, t)\rho A dx = AF(x_1, t) - AF(x_2, t). \tag{11.3}$$

It now follows, in the same way as (9.44) followed from (9.40), that

$$\rho \frac{\partial H}{\partial t} + \frac{\partial F}{\partial x} = 0 \tag{11.4}$$

throughout the glass and for all values of time. Substituting from (11.1) and (11.2), we thus obtain the partial differential equation

$$\frac{\partial \phi}{\partial t} = \kappa \frac{\partial^2 \phi}{\partial x^2}, \tag{11.5}$$

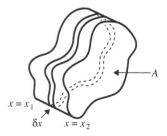

Fig. 11.1 A piece of glass.

where

$$\kappa \equiv \frac{k_G}{\rho c}. \qquad (11.6)$$

Equation (11.6) is known as the (unsteady) heat conduction equation. Its mathematical structure is identical to that of (9.65), which differs from (11.5) only insofar as γ, x, and y are replaced by κ, ϕ, and x. Thus κ is a diffusivity, having the dimensions of area per time; and the mechanism of conduction, by which heat is transported across a medium, is equivalent to the mechanism of diffusion, by which pollution flows down a concentration gradient.

The mechanisms are equivalent but not identical. The difference is that, whereas in Section 9.5 the molecules actually drifted in the direction of the concentration gradient, now they just oscillate about where they are but transfer their kinetic energy by means of collisions to neighboring particles with lower energy. Energy, not mass, is the quantity that drifts; therefore the process is much more rapid than the process of Section 9.5. It is largely for this reason that we preserve the linguistic distinction between conduction and diffusion. Conduction is rapid diffusion, or, if you prefer, diffusion is slow conduction; but each is equivalent to a flux of the concentrated quantity down the concentration gradient. For illustration, the thermal diffusivity of pure water at room temperature is about $1.4 \times 10^{-3} \text{cm}^2/\text{sec}$, more than two orders of magnitude greater than the typical diffusivity we encountered in Section 9.5. But conduction is still slow compared to convection.

Returning to the window, let T_1 be the temperature inside the building and T_2 ($< T_1$) the temperature outside, so that

$$\begin{aligned} \phi &= T_1 \qquad \text{if } x = 0 \\ \phi &= T_2 \qquad \text{if } x = D; \end{aligned} \qquad (11.7)$$

and suppose that the temperature throughout the glass has reached a steady state. Then, on setting $\partial/\partial t = 0$ in (11.5), we obtain

$$\kappa \frac{d^2 T}{dx^2} = 0, \qquad (11.8)$$

which agrees with (2.52). The solution of (11.8) subject to (11.7) is the steady temperature distribution

$$\phi = T_s(x) \equiv T_1 + (T_2 - T_1)\frac{x}{D}. \qquad (11.9)$$

You should verify that the heat flux given by (11.9) agrees with (2.50) on setting $D = 2d$.

Suppose that the temperature outside the building now suddenly drops to T_3 ($< T_2$). Then the temperature within the glass is no longer governed by (11.9). Rather, this is merely the initial temperature distribution for an unsteady heat flow governed by (11.5). We must therefore solve (11.5) subject to the boundary conditions

$$\phi(0, t) = T_1, \tag{11.10}$$

$$\phi(D, t) = T_3, \tag{11.11}$$

and the initial condition

$$\phi(x, 0) = T_s(x). \tag{11.12}$$

The solution of (11.5) subject to (11.10)–(11.12) is

$$\phi(x, t) = T_1 + (T_3 - T_1)\frac{x}{D} + \frac{2(T_3 - T_2)}{\pi} \sum_{n=1}^{\infty} \frac{(-1)^n}{n} e^{-\kappa n^2 \pi^2 t / D^2} \sin\left(\frac{n\pi x}{D}\right).$$

$$\tag{11.13}$$

You can easily see that (11.13) satisfies (11.10) and (11.11), and you can verify by differentiation that (11.13) satisfies (11.5) because every term in the infinite series is itself a solution. The fact that (11.13) also satisfies (11.12) depends upon a result from the theory of Fourier series, namely, that for $0 < x < D$,

$$\frac{\pi x}{2D} = \sum_{n=1}^{\infty} \frac{(-1)^{n+1}}{n} \sin\left(\frac{n\pi x}{D}\right). \tag{11.14}$$

Given this result, you can verify for yourself that the initial condition is satisfied. A proof that the series (11.13) converges will be found in any substantial text on Fourier analysis and does not belong in a text on mathematical modelling; see, for example, Chapter 7 of Carslaw (1930).

Taking the limit of (11.13) as $t \to \infty$ we see that the temperature approaches a new steady state:

$$T_s^{\text{new}}(x) = T_1 + (T_3 - T_1)\frac{x}{D}, \tag{11.15}$$

in perfect agreement with (2.50). Moreover, because the terms in the infinite series are dominated by the first term and decrease so rapidly, we see that the glass has virtually evolved toward its new equilibrium after time t_1, where t_1 is a low multiple of $D^2/\kappa\pi^2$.

The terms in (11.13) that fade away as $t \to \infty$ are known as *transients*. In this section, the steady state that then emerges is a static one, but if the external temperature varied cyclically, then we might expect the steady state to be a limit cycle. Clearly, when seeking such cycles directly, it is no longer permissible to set $\partial/\partial t = 0$ and solve (11.8); rather, we must seek an appropriate solution of the unsteady equation (11.5). An example of this emerges in Exercise 11.1, which you should now attempt.

11.2 How Does Traffic Move after the Train Has Gone By? A First Look

We have already seen two models of traffic flow in a single lane. One was an equilibrium model (Section 2.3) in which traffic density was independent of both time and distance along the road.[2] The other was a nonequilibrium model (Section 1.13) in which a driver responded only to the car in front of him. Now, the latter is probably perfectly adequate in a tunnel, because it is difficult to see the brake lights of the car in front of the car in front of you (or at least that was true in the old days, before manufacturers began to affix additional brake lights below the rear window or on the roof). On the open road, however, we would expect a driver to look ahead and react to changes in traffic conditions, such as variations of traffic density with distance along the road. But if traffic density depends upon distance, and in such a way that drivers react to conditions far ahead of them, then neither of our existing models will suffice. We are obliged to invent a new one.

An adaptation suggests itself immediately if we think of cars as pollutants and roads as rivers of traffic. As we said in Chapter 9, the lake core model of Section 9.4 is essentially a river model, and we have already adapted it once to describe population "flow." To adapt the same model for our present purpose, all we need do is to regard traffic density as a concentration of contaminant—but per unit length, rather than per unit volume. As shown in Section 9.6, this is accomplished simply by setting A, the cross-sectional area of the "river," equal to 1.

Accordingly, let $\phi(z, t)$ denote the density of cars, or the number of cars per unit roadlength, on a single lane of highway at station z at time t. Suppose that the road between $z = z_1$ and $z = z_2$ $(> z_1)$ has no entrances or exits and that passing is not allowed. Then the only way cars can enter the region $z_1 \leq z \leq z_2$ is from "upstream," i.e., by crossing $z = z_1$; and the only way they can leave is by going "downstream," i.e., by crossing $z = z_2$. Accordingly, we can replace (9.40) by

Rate of change of traffic = Traffic inflow − Traffic outflow

$$\frac{\partial}{\partial t} \int_{z_1}^{z_2} \phi(z, t)\, dz = F(z_1, t) - F(z_2, t), \tag{11.16}$$

where $F(z, t)$ is the flux of traffic, or the number of cars per unit time passing station z at time t. This leads, in the same way as (9.40) yielded (9.44), to

$$\frac{\partial \phi}{\partial t} + \frac{\partial F}{\partial z} = 0. \tag{11.17}$$

We must now choose F to mimic the way in which drivers behave. Clearly, this will depend upon the kind of road and the traffic conditions.

---

[2]You may wish to review Sections 2.3, 3.5, and 9.4–9.6 before proceeding.

Suppose that the road has a single lane in each direction, one going north and one going south, and is traversed from west to east, at $z = 0$, by a railroad. We'll consider only the northbound traffic, in which direction z increases. Now imagine that a long, slow train has just passed by. There is therefore a long line of cars (bumper to bumper) in $z < 0$, and the northbound lane in $z > 0$ is completely clear. Let the gates finally open at time $t = 0$. If we are interested solely in the vicinity of the crossing, then it is legitimate to pretend that the road extends to infinity on either side of $z = 0$. Hence the initial conditions are

$$\phi(z,0) = \begin{cases} 0, & \text{if } z > 0; \\ \phi_{\max}, & \text{if } z < 0, \end{cases} \tag{11.18}$$

where $\phi_{\max} = 1/\{\text{CAR LENGTH}\}$. Moreover, cars cannot pass after the gates have opened because of the queue in the opposite direction; and we will simply assume that there are no entrances or exits near the crossing. The assumptions embodied in (11.16) are thus satisfied.

The initial density profile (11.18) is sketched in Fig. 11.2 as the dashed curve, together with some solid curves that suggest the shape that the profile should take at later times, as the traffic begins to flow. Doesn't this diagram remind you of Fig. 9.7? It is the mirror image of the diagram that we would have obtained in Section 9.5 if the lake pollution had been allowed to diffuse into $y < 0$, and it suggests the hypothesis

$$F = -\Gamma \frac{\partial \phi}{\partial z}, \tag{11.19}$$

where $\Gamma \ (> 0)$ is a constant. From (11.17) and (11.19) we then obtain the partial differential equation

$$\frac{\partial \phi}{\partial t} = \Gamma \frac{\partial^2 \phi}{\partial z^2}. \tag{11.20}$$

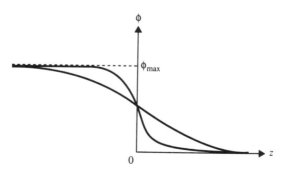

Fig. 11.2 Evolution of (11.18) according to (11.20). The dashed profile is (11.18). The solid curves are for two later times.

You can readily show (Exercise 11.2) that (11.18) and (11.20) are both satisfied by

$$\phi(z,t) = \frac{\phi_{max}}{\sqrt{\pi}} \int_{z/2\sqrt{\Gamma t}}^{\infty} e^{-\xi^2} d\xi, \qquad -\infty < z < \infty, 0 < t < \infty. \qquad (11.21)$$

With the help of Exercise 11.2, notice that $\phi(z,t) \to \phi_{max}/2$ as $t \to \infty$. Moreover, by repeating the argument that led to (9.69), you can show that the road at station z would be within 5% of this limiting density $\phi_{max}/2$ after time $127z^2/\Gamma$—*if* the model were correct.

Is the model correct? For a short time after the gates have opened, in the immediate vicinity of the crossing (11.21) may well lead to density profiles with roughly the right shape (see Fig. 11.4). Indeed, were this not so, we would never have adopted (11.19) in the first place. The model also predicts, however, that

$$F(z,t) = -\Gamma\frac{\partial\phi}{\partial z} = \frac{\Gamma\phi_{max}}{2\sqrt{\pi\Gamma t}} e^{-z^2/4\Gamma t} \qquad (11.22)$$

tends to zero as $t \to \infty$, and on replacing x by ϕ in Exercise 2.5, we have in general that

Number of cars passing per unit of time	=	Number of cars per unit length of road	×	Length of road covered per unit of time	
$F(z,t)$	=	$\phi(z,t)$	×	$v(z,t)$,	(11.23)

where v is the velocity of traffic flow. Note, from (9.37) with $A = 1$, that this is just the traffic flow analogue of (9.38). Thus $v(z,\infty) = F(z,\infty)/\phi(z,\infty) = 2F(z,\infty)/\phi_{max} = 0$; i.e., the model predicts that traffic slows to a halt for $\phi < \phi_{max}$. We know that this is not correct. Our model must therefore be rejected.

Knowing this in advance, of course, I could have omitted this section; but I wanted to illustrate yet again the importance of trial and error in constructing mathematical models. Yet the moral of this section is not, as you might think, that models must sometimes be rejected. Rather, it's that successful models are often the results of failures, as Section 11.4 will demonstrate. Meanwhile, try Exercise 11.3.

11.3   How Does Traffic Move after the Train Has Gone By? A Second Look

In this section, we will take a different approach to modelling the effect of a railroad crossing on traffic flow. Recall that in our road tunnel model of Chapter 2 we found an equilibrium velocity v for any traffic density $\phi(z,t)$ in the range $0 \leq \phi \leq \phi_{max}$, where t denotes time and z distance. Because the flow was in equilibrium, $\phi(z,t) = \phi$ was a constant. But the value of that constant was arbitrary (within the range $0 \le \phi \le \phi_{max}$). For each such value, the velocity was given by a relationship of the form

$$v = v(\phi) \tag{11.24}$$

and the traffic flow by

$$F = F(\phi) = \phi v(\phi). \tag{11.25}$$

Now, let's *extend* these ideas by assuming that relationships (11.24) and (11.25) are valid even when the traffic flow is not in equilibrium, i.e., by assuming that

$$v(z, t) = v(\phi(z, t)), \qquad F(z, t) = F(\phi(z, t)). \tag{11.26}$$

The rationale behind this assumption is that if density changes are not too violent, then traffic conditions will adjust continually, as though passing quickly through a series of stable equilibria, in such a way as to maintain "dynamic equilibrium" between F and $\phi v(\phi)$. Be sure you appreciate that (11.26) is an assumption and represents a considerable extension of (11.24) and (11.25). As usual, the quality of this assumption will be determined by how well the model's predictions accord with reality.

Let us now combine the nondifferential dynamic model (11.26) with the model (11.17) for traffic flow in a single lane, which we developed in the previous section. We obtain

$$\frac{\partial \phi}{\partial t} + \frac{\partial F}{\partial z} = \frac{\partial \phi}{\partial t} + F'(\phi)\frac{\partial \phi}{\partial z} = 0, \tag{11.27}$$

where, as usual, a prime denotes differentiation with respect to argument. Hence, if (11.26) is a good assumption, then the solution of (11.27), subject to the initial conditions (11.18), will provide an adequate description of how traffic flows in the immediate aftermath of a long, slow train. In principle, $v(\phi)$ can have any form that gives $F(\phi)$ the overall shape of the fundamental diagram of Exercise 2.5, but we shall use the form that gave such good agreement with data in Section 4.6. On rearranging (4.26), therefore, we have

$$v(\phi) = \begin{cases} v_{max}, & \text{if } 0 \le \dfrac{\phi}{\phi_{max}} \le e^{-v_{max}/\lambda}; \\[3mm] \lambda \ln\left(\dfrac{\phi_{max}}{\phi}\right), & \text{if } e^{-v_{max}/\lambda} \le \dfrac{\phi}{\phi_{max}} \le 1; \end{cases} \tag{11.28}$$

where λ is the velocity that maximizes the traffic flux. From (11.25) and (11.28) we easily deduce that

$$F'(\phi) = \begin{cases} v_{max}, & \text{if } 0 \le \dfrac{\phi}{\phi_{max}} \le e^{-v_{max}/\lambda}; & (11.29a) \\[3mm] \lambda\left\{\ln\left(\dfrac{\phi_{max}}{\phi}\right) - 1\right\}, & \text{if } e^{-v_{max}/\lambda} \le \dfrac{\phi}{\phi_{max}} \le 1. & (11.29b) \end{cases}$$

The solution of (11.27) subject to (11.18) and (11.29) is (Exercise 11.4)

$$
\begin{aligned}
\frac{\phi(z,t)}{\phi_{max}} &= 1, & \text{if } -\infty < z \le -\lambda t; \\
&= e^{-(z/\lambda t+1)}, & \text{if } -\lambda t \le z \le (v_{max}-\lambda)t; \qquad (11.30)\\
&= e^{-v_{max}/\lambda}, & \text{if } (v_{max}-\lambda)t \le z \le v_{max}t; \\
&= 0, & \text{if } v_{max}t < z < \infty.
\end{aligned}
$$

The velocities corresponding to each of the four regions in the z-t plane defined by (11.30) are given in Fig. 11.3. Note that both $\phi(z,t)$ and $v(z,t)$ are continuous except along the line $z = v_{max}t$, which corresponds to the motion of the leading vehicle into the clear road ahead of the crossing. Notice also that cars cross the rails with velocity λ, corresponding to maximum flux. Thus one easy way to measure the road's capacity would be to stand at the crossing for a minute or two after a train goes by and count the number of vehicles.

Figure 11.3 can be interpreted in either of two ways: by fixing z and letting t vary, or by fixing t and letting z vary. First, therefore, consider an observer standing at $z = z_0$. What he sees is represented in the diagram by a vertical column of dashes. The first car passes the observer at time $t = z_0/v_{max}$. From then until time $t = z_0/(v_{max} - \lambda)$, he sees a short platoon of cars, density $\phi_{max}\exp(-v_{max}/\lambda)$, moving at maximum speed v_{max}. Thereafter, the density of cars is always increasing but never exceeds

$$
\phi^* \equiv \frac{\phi_{max}}{e}, \qquad (11.31)
$$

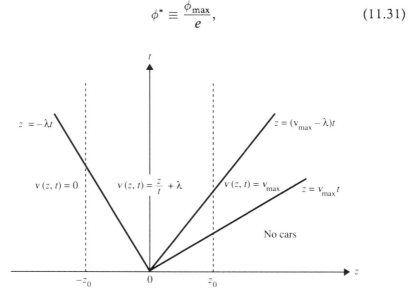

Fig. 11.3 Representation in the z-t plane of the traffic flow immediately after the crossing gates have opened.

where $e = \exp(1)$, and the speed of cars is always decreasing, but never falls below the optimal value λ. Therefore, in terms of the fundamental diagram introduced in Exercise 2.5, traffic is light ahead of the crossing. Now consider a second observer at $z = -z_0$. The car at this station is unable to move for time z_0/λ. Thus cars take a finite time to begin moving again after the train has gone by. This is in perfect agreement with reality. Thereafter, the density decreases but never falls below ϕ^*, and the velocity increases but never exceeds λ. In terms of the fundamental diagram, traffic is heavy behind the railroad crossing.

Second, we can examine the density profile at a series of fixed times as though we were photographing the entire road in the vicinity of the crossing. Three such profiles are sketched in Fig. 11.4. You can see that they are not totally unlike the profiles of the previous section, though the focal point has shifted down from 0.5 to $1/e$. But there is a crucial difference: there are now corners, where ϕ/ϕ_{max} takes the values $\exp(-v_{max}/\lambda)$ and 1. Thus our new model predicts that information—specifically the fact that the train has passed—cannot propagate more than a finite distance, either ahead of or behind the crossing, in any finite time (imagine that the observer at $z = -z_0$ does not realize that the gates are open until the car

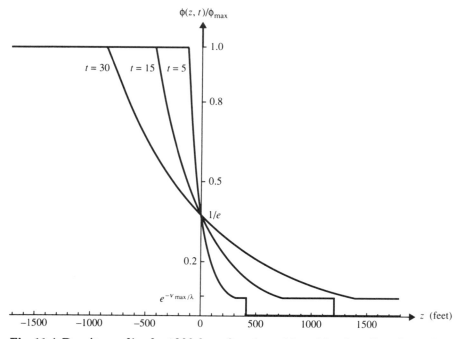

Fig. 11.4 Density profiles for 1800 feet of road on either side of a railroad crossing, within half a minute of the gates opening. Times are quoted in seconds; v_{max} is 55 m.p.h. and λ is 20 m.p.h.

at his station begins to move, and that the observer at $z = z_0$ realizes the gates are open only when the first car passes).

Notice, on using the Chain Rule and (11.27), that

$$\frac{d\phi}{dt} = 0 \tag{11.32}$$

whenever

$$\frac{dz}{dt} = F'(\phi). \tag{11.33}$$

Thus ϕ is constant along curves in the z-t plane defined by (11.33), which must therefore be straight lines. But the z-t plane is a mathematical fiction. What happens in reality is that the point on the road where a given density is found changes with time *as though* it moved (either forward or backward) with velocity $F'(\phi)$. Thus the information that the gates have opened moves back through the queue with speed λ because $F'(\phi_{max}) = -\lambda$, while the same information propagates forward into the open road with velocity v_{max} because $F'(0) = v_{max}$. The curves in the z-t plane along which a given value of ϕ is propagated are known as *characteristics* and are represented schematically in Fig. 11.5. But remember that characteristics are just convenient mathematical fictions. Don't go down to a railroad crossing in the hope of meeting a characteristic. You'll be disappointed.

It's important, in particular, to distinguish between characteristics, along which information propagates, and *trajectories*, along which cars actually move (although the two coincide for the first car in the queue, as you can easily verify). A trajectory is the solution of the ordinary differential

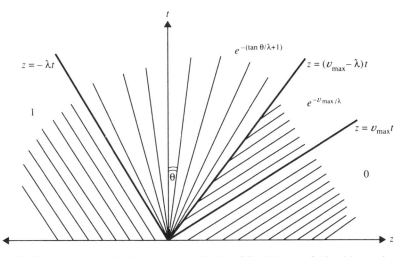

Fig. 11.5 Characteristics in the z-t plane for (11.30). Values of $\phi(z, t)/\phi_{max}$ in each of four regions are also given; $\theta = \arctan(z/t)$.

equation

$$\frac{dz}{dt} = v(z, t), \tag{11.34}$$

subject to the appropriate initial condition. But we have already found that the car at $z = -z_0$ takes time z_0/λ to begin moving. Hence, from Fig. 11.3, we must solve

$$\frac{dz}{dt} = \frac{z}{t} + \lambda, \qquad z\left(\frac{z_0}{\lambda}\right) = -z_0. \tag{11.35}$$

The solution (Exercise 11.5) is

$$z(t) = \begin{cases} \lambda t\left\{\ln\left(\frac{\lambda t}{z_0}\right) - 1\right\}, & \text{if } t \geq \frac{z_0}{\lambda}; \\ -z_0, & \text{if } t \leq \frac{z_0}{\lambda}. \end{cases} \tag{11.36}$$

A typical trajectory in the z-t plane is sketched in Fig. 11.6. Notice how it crosses the characteristics (dashed). Notice also that the car that queued a distance z_0 behind the crossing will ultimately drive over it at time

$$\frac{ez_0}{\lambda} \tag{11.37}$$

after the gates have opened.

How good is this model—does it fit the facts? Our remarks have indicated that there is certainly qualitative agreement with reality. But (11.37) also provides a quantitative test of our model. According to (11.37), after

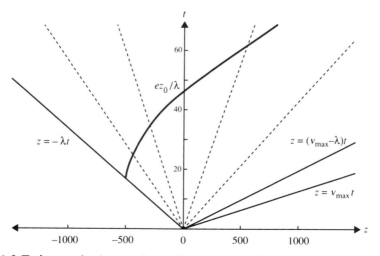

Fig. 11.6 Trajectory in the z-t plane of a car that waited in line at 500 feet from the crossing; $v_{max} = 55$ m.p.h., $\lambda = 20$ m.p.h. Distance is measured horizontally in feet; time is measured vertically in seconds.

the gates have opened, a car will take about 2.7 times as long to reach the crossing as it has to wait to get moving. Is this what actually happens? Perhaps you would like to perform the experiment! Of course, it isn't necessary to perform the experiment at a railroad crossing. You could use a stop light (provided no cars may turn off), a school bus unloading, or indeed any temporary obstruction of sufficient duration in the absence of entrances or exits.

11.4   Avoiding a Crash at the Other End. A Combination

In this section, we will attempt to model the other end of the queue of cars that forms at a railroad crossing. As usual, z will measure distance along the road, but $z = 0$ will now denote the rear of the queue instead of the head.

Suppose that the crossing is on a very lonely road, where the density is ordinarily very small. Nevertheless, the trains in this region are so long and so slow that a long line of cars always accumulates when the crossing gates are closed—so long, in fact, that if we are interested only in the rear of the queue then it is legitimate to pretend that

$$\phi(z,0) = \phi_{max}, \qquad z > 0. \tag{11.38}$$

Now, along you come at time $t = 0$, approaching the queue at speed v_{max} but miles ahead of the following car. It is therefore legitimate to pretend that

$$\phi(z,0) = 0, \qquad z < 0. \tag{11.39}$$

Suddenly, ahead of you, you see the queue for the crossing. What do you do? According to the model of Section 11.3, you either brake to a halt in zero distance or crash into the end of the queue!

The problem with the model of the previous section is this. It says that traffic velocity is determined by the local traffic density, according to $v = v(\phi)$, and takes no account of (spatial) *variations* of density; so that, in $z < 0$, you would drive at $v(0) = v_{max}$ right up until you hit the last car in the queue. In reality, of course, you look ahead, observe the sudden increase in traffic density, and react accordingly. We must find a way to model this.

In Section 11.2 we attempted to model traffic flow by using a flux law of the form

$$F = -\Gamma \frac{\partial \phi}{\partial z}. \tag{11.40}$$

This has the obvious advantage that it takes account of the density gradient, which is exactly what you would do if you were approaching a stationary queue of cars. On the other hand, it says that the effect of congestion ahead is to cause approaching traffic to flow backward, when, in reality, F could never be less than zero. Must we therefore abandon (11.40), or can we modify it to suit our present needs?

The answer lies in the observation that congestion ahead will decrease the flow, and dispersion ahead will increase the flow, not relative to zero but *relative to the flow that exists already*. Let us therefore combine the models of the two previous sections to obtain the hybrid flux law

$$F = \phi v(\phi) - \Gamma \frac{\partial \phi}{\partial z}, \tag{11.41}$$

which couples the effects of convection with those of diffusion. Then, from (11.17), we obtain the partial differential equation

$$\frac{\partial \phi}{\partial t} + \frac{\partial F}{\partial z} = \frac{\partial \phi}{\partial t} + \frac{\partial}{\partial z}(\phi v(\phi)) - \Gamma \frac{\partial^2 \phi}{\partial z^2} = 0. \tag{11.42}$$

We can ensure that F in (11.41) is never negative by making Γ sufficiently small, which is anyhow what we would expect for a diffusivity.

You may be concerned that (11.23) and (11.41) are incompatible if $\Gamma \neq 0$. The correct interpretation of the apparent discrepancy is that (11.23) *defines* velocity; whereas (11.41) is a legitimate possibility for the flux function, in which $v(\phi)$ is an arbitrary function but not the velocity. It will be so close to the velocity if Γ is very small, however, that we prescribe it to be (11.28) in the subsequent development.

Equation (11.42) is difficult to solve in general, but just as the solution of (11.5) subject to (11.10)–(11.12) led to the steady state (11.15), so we might expect that the solution of (11.42) subject to (11.38) and (11.39) would also tend to a steady state. Moreover, because the density begins as steady except at $z = 0$, we might expect that the solution of (11.42) would differ from (11.38) and (11.39) only in a narrow transition layer near $z = 0$. As far as this layer is concerned, the regions $\phi = 0$ and $\phi = \phi_{max}$ will appear to be at infinity. Therefore, on setting $\partial/\partial t = 0$ in (11.42), it is sufficient to seek the solution $\phi = \phi(z)$ of the ordinary differential equation

$$\frac{d}{dz}(\phi v) - \Gamma \frac{d^2 \phi}{dz^2} = 0 \tag{11.43}$$

subject to

$$\phi(\infty) = \phi_{max}, \qquad \phi'(\infty) = 0, \tag{11.44}$$

and

$$\phi(-\infty) = 0, \qquad \phi'(-\infty) = 0. \tag{11.45}$$

If this makes you feel uncomfortable, imagine that z is measured in miles in (11.38) and (11.39), but in inches in (11.43)–(11.45). On integrating (11.43), we obtain

$$F(z) = \phi v - \Gamma \frac{d\phi}{dz} = \text{constant}. \tag{11.46}$$

In view of (11.44) and (11.45), the constant must be zero. Thus the effect of the transition layer is to shift the road from a state in which the flux is zero by virtue of no cars to one in which the flux is zero by virtue of no movement. On using the form of $v(\phi)$ given by (11.28), we readily obtain the solution of the ordinary differential equation (11.46) in the implicit form

$$\frac{\phi(z)}{\phi_{\max}} = \begin{cases} \alpha_1 e^{v_{\max} z/\Gamma}, & \text{if } 0 \le \dfrac{\phi}{\phi_{\max}} \le e^{-v_{\max}/\lambda}, & (11.47a) \\ e^{-\alpha_2 e^{-\lambda z/\Gamma}}, & \text{if } e^{-v_{\max}/\lambda} \le \dfrac{\phi}{\phi_{\max}} \le 1, & (11.47b) \end{cases}$$

where α_1 and α_2 are constants (Exercise 11.7).

The constants α_1 and α_2 are determined from the continuity of ϕ and its derivative at the point where $\phi/\phi_{\max} = \exp(-v_{\max}/\lambda)$, i.e., at the point where $\phi(z)$ switches from (11.47a) to (11.47b). In deriving (11.47), however, we have not yet identified this point; denote it, therefore, by $z = \xi$. Because (11.47a) must satisfy (11.45), it describes the end of an open road. Because (11.47b) must satisfy (11.44), it describes the beginning of a congested road. Hence $\phi_{\max} \exp(-v_{\max}/\lambda)$ is a critical density, at which drivers must begin to adjust their speed in response to congestion ahead; and ξ is the value of z for which this critical density obtains. Requiring both $\phi(z)$ and $\phi'(z)$ to be continuous at $z = \xi$ yields:

$$\alpha_1 e^{v_{\max}\xi/\Gamma} = e^{-\alpha_2 e^{-\lambda\xi/\Gamma}}$$

$$\frac{\alpha_1 v_{\max}}{\Gamma} e^{v_{\max}\xi/\Gamma} = \frac{\alpha_2 \lambda}{\Gamma} e^{-\lambda\xi/\Gamma} e^{-\alpha_2 e^{-\lambda\xi/\Gamma}}. \qquad (11.48)$$

Division yields

$$v_{\max} = \alpha_2 \lambda e^{-\lambda\xi/\Gamma}, \qquad (11.49)$$

whence

$$\phi(\xi) = e^{-v_{\max}/\lambda} \qquad (11.50)$$

for any value of ξ. The value of ξ being thus purely arbitrary, we choose $\xi = 0$ for convenience. Thus, from (11.48) we have

$$\alpha_1 = e^{-v_{\max}/\lambda}, \qquad \alpha_2 = \frac{v_{\max}}{\lambda}; \qquad (11.51)$$

whence, from (11.47) and (11.28), we obtain the explicit expressions

$$\frac{\phi(z)}{\phi_{\max}} = \begin{cases} e^{v_{\max}\left(\frac{z}{\Gamma} - \frac{1}{\lambda}\right)}, & \text{if } z < 0; \\ e^{-\frac{v_{\max}}{\lambda} e^{-\lambda z/\Gamma}}, & \text{if } z \ge 0; \end{cases} \qquad (11.52)$$

$$\frac{v(z)}{v_{\max}} = \begin{cases} 1, & \text{if } z < 0; \\ e^{-\lambda z/\Gamma}, & \text{if } z \ge 0. \end{cases} \qquad (11.53)$$

Notice that, according to (11.52), the beginning of the queue, where ϕ reaches $\phi_{\max}$, is reached only in the limit as $z \to \infty$; but this is quite acceptable because what is short according to the long scale may be extremely long according to the short scale. Naturally, having stipulated that drivers begin to react at $z = 0$, we cannot also have the beginning of the queue at $z = 0$, for drivers require a finite distance in which to brake.

The velocity and density profiles (11.52) and (11.53) are sketched in Fig. 11.7(a). Distance is measured on the horizontal scale in units of Γ/λ. Observe that the effective width of the transition layer is about $5\Gamma/\lambda$, which is small because Γ is small. The suddenness of this transition is sufficient justification, when solving (11.43), for imagining the regions of constant density to be at infinity; the mathematics is simplified but still produces a density profile with effectively constant regions that are but a short distance from $z = 0$. If you view the road from afar (the locality of infinity, as it were) then the profiles predicted by (11.43)–(11.45) will look like those in Fig. 11.7(b); the transition layer will appear as a discontinuity. Because $\phi(z)$ is a steady state, which may have persisted since $t = 0$, we can view the transition layer as being hidden in the discontinuity of (11.38) and (11.39), which are the initial conditions for a distant observer. To this observer, when you approach the queue, you appear to brake from v_{max} to

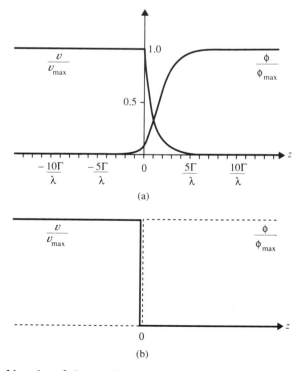

Fig. 11.7 Double take of the steady state density and velocity profiles; $v_{max}/\lambda = 55/20 = 2.75$. (a) Close-up of transition layer; (b) view from afar.

zero in zero distance, but a local observer would know from Fig. 11.7(a) that you brake to a halt over a distance $5\Gamma/\lambda$ and thus avoid crashing without the use of magic. Similarly, in the model of the previous section, the discontinuity of (11.18) hides the transition layer between ϕ_{max} and 0; it becomes the "expansion fan" of Fig. 11.5, in which characteristics diverge from the origin.

The concept of a narrow transition layer is a fruitful one in mathematical modelling (particularly in fluid dynamics, where it is usually dubbed a shock or boundary layer). It may remind you of the metered models of Chapters 1 and 5. There, when we had two different time scales, we were able to use one model for the small time scale and another for the large one. Here, when we have two different length scales, we are able to use one model, Fig. 11.7(a), for the small length scale and another, Fig. 11.7(b), for the large length scale. If you still feel uncomfortable about all this, please rest assured that a formidable arsenal of mathematical artillery exists for formalizing the use of several competing length and time scales. But because our brief is modelling, not methods, we restrict our attention to a heuristic approach. If you wish to pursue this matter more rigorously, then you might begin with Nayfeh (1981, 1973).

We conclude this section by observing that we have used the traffic diffusivity Γ without knowing its value (except that it's small), but the model of Fig. 11.7(a) can be used to obtain a crude estimate. It probably takes a car about 150 feet, or 5/176 mile, to brake to a halt from 55 m.p.h. Taking $\lambda = 20$ m.p.h., and equating this braking distance and $5\Gamma/\lambda$, we obtain the crude estimate

$$\Gamma = \frac{5}{44} \approx 0.1\,\text{m}^2/\text{h}. \tag{11.54}$$

11.5   Spreading Canal Pollution. An Adaptation

Suppose that a finite amount of pollution, P_0, is instantly injected into a canal, or indeed any other body of stagnant water.[3] How do we describe the subsequent diffusion? Now, we have already developed a model for diffusion of contaminant through stagnant water—that of Section 9.5. But we assumed in doing so that, initially, the contaminant was dispersed uniformly throughout the body of water. We must therefore adapt our model for circumstances in which, initially, the pollution is concentrated.

Let y denote distance along our canal and $y = 0$ the place of injection. Let t denote time, $t = 0$ the instant of injection, and $\phi(y, t)$ the concentration of contaminant per unit length of canal, at station y at time t. Because diffusion is so slow, let's assume that the canal extends to infinity in both

---

[3]You may wish to review Section 9.5 before proceeding.

directions, Then, from Section 9.5, the diffusion of contaminant throughout the canal is governed by

$$\frac{\partial \phi}{\partial t} = \gamma \frac{\partial^2 \phi}{\partial y^2}, \qquad -\infty < y < \infty, \tag{11.55}$$

where γ denotes the diffusivity.

From a mathematical viewpoint, what makes this problem different
from that considered in Section 9.5 is simply the initial condition. There,
the pollution was uniformly dispersed initially, yielding an initial condition
of the form $\phi(y, 0) = $ constant for all y. Here, however, the pollution is
initially concentrated at $y = 0$. Thus

$$\phi(y, 0) = 0 \qquad \text{if } y \neq 0, \tag{11.56}$$

and

$$\lim_{y \to 0} \phi(y, 0) = \infty. \tag{11.57}$$

Moreover, because the finite amount of injected pollution stays in the water
forever, we have

$$\int_{-\infty}^{\infty} \phi(y, t) dy = P_0, \tag{11.58}$$

for all y and for all time, including the initial instant.

Is there some way in which we can compress conditions (11.56),
(11.57), and (11.58), which $\phi(y, 0)$ must satisfy, into a single initial condition? Suppose that we could construct a function δ with the properties

$$\delta(y) = 0 \qquad \text{if } y \neq 0,$$

$$\int_{-\infty}^{\infty} \delta(y) dy = 1. \tag{11.59}$$

Then the initial condition could be expressed as

$$\phi(y, 0) = \bar{P}_0 \delta(y). \tag{11.60}$$

Naturally, this is just an extreme mathematical idealization of the fact that
contaminant is injected suddenly, in much the same way as we assumed in
Section 3.4 that a forest could be instantly felled.

But how is such a function to be constructed? Recall from Chapter 6
that the probability density function for a normal distribution with mean
zero and variance σ is given by

$$f(y; \sigma) = \frac{1}{\sigma \sqrt{2\pi}} e^{-y^2 / 2\sigma^2}. \tag{11.61}$$

The smaller the variance, the narrower the hump, and by definition,

$$\int_{-\infty}^{\infty} f(y; \sigma) dy = 1. \tag{11.62}$$

Thus, as $\sigma \to \infty$, we obtain a sequence of probability density functions like those shown in Fig. 11.8. It therefore seems reasonable to define

$$\delta(y) = \lim_{\sigma \to 0} f(y; \sigma). \tag{11.63}$$

The "function" defined by (11.63) is known as the Dirac delta function. Strictly, it is not a function in the ordinary sense but an example of a so-called *generalized function*. For greater rigor, see Lighthill (1962, p. 10).

Now, the initial distribution of contaminant in our canal will resemble the sketch on the far right of Fig. 11.8. At later times, however, we would expect it to resemble one of the sketches toward the left of the diagram, as though the variance of the normal distribution were increasing with time. We therefore seek a solution of (11.55) in the form

$$\phi(y, t) = f(y; \sigma(t)) = \frac{P_0}{\sigma(t)\sqrt{2\pi}} e^{-y^2/2\sigma(t)^2}, \tag{11.64}$$

with

$$\sigma(0) = 0, \qquad \sigma'(t) > 0. \tag{11.65}$$

Observe that (11.64) satisfies both (11.58) and (11.60). It is readily shown (Exercise 11.8) that (11.55) now implies $\sigma(t) = \sqrt{2\gamma t}$, so that diffusion of pollution throughout the canal obeys

$$\phi(y, t) = \frac{P_0}{2\sqrt{\pi \gamma t}} e^{-y^2/4\gamma t}. \tag{11.66}$$

According to (11.66), if diffusion were the *only* cleaning mechanism, then the 5% cleaning time at the dumping site would be a depressing $100/\pi\gamma$!

Be sure you appreciate that our mathematical model is fully deterministic. Like the tree growth matrix S in the model of Section 9.2, it nicely illustrates how mathematical ideas developed in a probabilistic context may later be useful in a deterministic one.

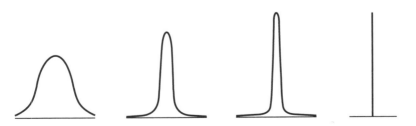

Fig. 11.8 Convergence of probability density function of normal distribution toward Dirac delta function, as the variance tends to zero.

11.6   Flow and Diffusion in a Tube: A Generic Model

We will now present a more general model of flow and diffusion in which several effects can be combined.[4] All of our earlier flow and diffusion models will emerge as special cases.

Suppose that some substance or quantity is suspended in and carried by a medium; we will refer to this substance or quantity as the *suspension*. The suspension might, for example, be a river contaminant or other chemical substance (e.g., oxygen dissolved in water), people (in the river of life), or cars (in a stream of traffic); or it might be a property of the medium itself (e.g., mass, momentum, or energy). We will derive a model for the flow and diffusion of this suspension.

Consider a cylindrical tube of medium, with arbitrary cross-sectional area A, like that depicted in Fig. 11.9. For the sake of simplicity, we will consider only a right cylinder, i.e., a cylinder generated by straight lines perpendicular to the cross section. But our model will also be valid for a tube with slight curvature, slightly varying cross section $A = A(z)$, or both (a river, perhaps).[5] Let z measure distance along the tube, or in the case of a curved tube, arc length along the curve that joins the center of each cross section. Let $\phi(z, t)$ denote the *concentration* of suspension at station z at time t, i.e., the amount of suspension per unit *volume* of medium. If—as in modelling traffic flow, for example—it were more convenient to define concentration as amount per unit *length* of tube, rather than per unit volume, then we would simply set $A = 1$ (making unit length and unit volume equivalent).

We will assume that ϕ does not vary over the tube's cross section (but you are invited to relax this assumption in Exercise 11.15). Thus, at time t, the amount of suspension in an infinitesimal slice of tube of length δz and centered at station z is $\phi(z, t) \cdot A \cdot \delta z + o(\delta z)$; whence the amount of

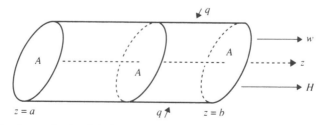

Fig. 11.9 A piece of tube filled by some medium.

---

[4]You may wish to review Sections 1.10, 1.11, and 2.6 before proceeding.
[5]From a mathematical viewpoint, the word "slightly" need not even be inserted. The assumption in the following paragraph, that $\phi(z, t)$ does not vary over the cross section, would hardly be plausible, however, if the tube were highly kinked.

suspension in the fixed region $a \le z \le b$ at time t is

$$x(t) = \int_a^b \phi(z, t)A\,dz; \tag{11.67}$$

the error term vanishes during the limit process of integration. The suspension can enter or leave the cylindrical region $a \le z \le b$ by any of the following three methods:

1. by being created or destroyed in the medium's interior, through chemical reaction;
4. by being transported, either convectively (with the medium) or diffusively (by molecular motions), across the planes $z = a$ or $z = b$; (11.68)
5. by being injected or withdrawn across the cylindrical wall of the tube.

The reason for counting 1, 4, 5, rather than 1, 2, 3, has to do with Exercise 11.15. Note that every method may not be applicable to every suspension. For example, (1) will apply only if the medium is a fluid, the suspension a chemical substance, and ϕ its mass per unit volume. Let L_k denote the rate at which suspension is added to the region $a \le z \le b$ by method (k), i.e., the amount of suspension added per unit time. Then, clearly,

$$\frac{dx}{dt} = L_1 + L_4 + L_5. \tag{11.69}$$

It remains to find expressions for L_1, L_4, and L_5. We'll consider each in turn.

First, suppose that the suspension is a chemical substance that can be created or destroyed in a reaction, and that ϕ is its mass per unit volume; suppose also that the medium is a fluid. Then the amount (mass) destroyed per unit of time, per unit volume, is proportional to $[\phi]$, where $[\phi]$ is the *molar* concentration of suspension (see Section 1.10). But $[\phi] = \phi/m$, where m is the mass of a mole of the suspension. Hence the rate of destruction per unit volume can be written

$$K\phi(z, t), \tag{11.70}$$

where K is independent of ϕ, at least for the most commonly occurring reactions. On the other hand, the rate per unit volume at which the suspension is being created by chemical reaction will depend upon the concentrations of other substances, but not upon ϕ. Let us denote this rate simply by $C(z, t)$. Then the net rate of addition of suspension to $a \le z \le b$ by method (1)—the amount (mass) of suspension added per unit time—is

$$L_1 = \int_a^b (C - K\phi)A\,dz. \tag{11.71}$$

Second, suppose that the suspension has flux $\Pi(z, t)$ at time t, in the positive z direction, through the cross section A with coordinate z. Then the net flux of suspension *into* the region $a \leq z \leq b$ by method (4)—the amount of suspension flowing in per unit time—is

$$L_4 = H(a, t) - H(b, t) = -\int_a^b \frac{\partial H}{\partial z} dz. \qquad (11.72)$$

Third, suppose that the suspension is being injected at a *net* rate $q(z, t)$ per unit length, at station z, at time t; $q < 0$ would correspond to net withdrawal. This injection or withdrawal could arise if the tube were porous to the suspension but not to the medium. Then the net rate of addition of suspension to $a \leq z \leq b$ by method (5)—the amount injected per unit of time—is

$$L_5 = \int_a^b q(z, t) dz. \qquad (11.73)$$

We note in passing that our notation does not require q to be known explicitly as a function of z and t; perhaps $q = Q(\phi)$, with $q(z, t) = Q(\phi(z, t))$.

The region $a \leq z \leq b$ does not vary with time; therefore we have

$$\frac{dx}{dt} = \frac{d}{dt} \left(\int_a^b A\phi \, dz \right) = \int_a^b \frac{\partial(A\phi)}{\partial t} dz, \qquad (11.74)$$

by Leibnitz's rule from the calculus. Hence from (11.69) and (11.71)–(11.73) we have

$$\int_a^b \left\{ \frac{\partial(A\phi)}{\partial t} + A\{K\phi - C\} + \frac{\partial H}{\partial z} - q \right\} dz = 0. \qquad (11.75)$$

But the region $a \leq z \leq b$ is arbitrary and, in particular, can be made arbitrarily small. Hence, for reasons given in Section 9.4, the integrand in (11.75) must be identically zero wherever it is continuous (usually wherever ϕ is continuously differentiable). Thus

$$\frac{\partial(A\phi)}{\partial t} + AK\phi + \frac{\partial H}{\partial z} = AC + q, \qquad (11.76)$$

except perhaps at certain isolated discontinuities.

The form of H in (11.76) will depend upon the suspension denoted by ϕ. Three possibilities will now be considered. Let $w(z, t)$ denote the velocity of the medium in the direction of increasing z. Then a volume $Aw(z, t)$ of medium crosses the section with coordinate z per unit of time. Thus if the suspension is the medium's mass, i.e., if ϕ denotes mass of medium per unit volume, then

$$H = Aw\phi; \qquad (11.77)$$

H is the mass of medium contained in the volume Aw.

If the suspension is a chemical dissolved in a fluid, however, then although it is still true that mass $w\phi$ of chemical is carried by the medium across unit area per unit time, there now exists the additional possibility that the suspension diffuses in the direction of flow.[6] Let γ_3 be the diffusivity. Then from Section 9.5, the diffusive flux per unit area is $-\gamma_3 \partial\phi/\partial z$. When we add the convective and diffusive fluxes, the total flux of suspension (dissolved substance) is given by

$$H = A\left(w\phi - \gamma_3 \frac{\partial\phi}{\partial z}\right). \tag{11.78}$$

The third possibility we consider is that the suspension is not the medium's mass but its *momentum*, so that ϕ denotes momentum of medium per unit volume. In this case also, we shall assume that the medium is a fluid. Now, borrowing from physics, force is just rate of change of momentum, so that any force on the fluid in $a \le z \le b$ will contribute in (11.69) to the rate of change of that region's momentum. The only force that will concern us here arises from the fluid pressure, denoted by $p(z, t)$. Because pressure is a force per unit area, the force due to pressure on the cross section A at $z = a$ is $Ap(a, t)$, and the force due to pressure at $z = b$ is $Ap(b, t)$. Recalling from Section 2.6 that pressure acts equally in all directions, we find that the net pressure force in the z direction on the region $a \le z \le b$ is

$$Ap(a, t) - Ap(b, t) = -\int_a^b \frac{\partial(Ap)}{\partial z} dz. \tag{11.79}$$

Comparing with (11.72), we see that fluid pressure is equivalent to a momentum flux pA in the z direction. The momentum per unit volume is $\phi = \rho w$, where ρ is the fluid density, and so the convective momentum flux is $Aw\phi = \rho Aw^2$. The diffusive momentum flux per unit area is $-\nu \partial\phi/\partial z$, where ν is a constant. But we have already identified $\rho\nu$ as the shear viscosity μ of the fluid (Section 2.6). Hence the diffusive momentum flux per unit area can be written as $-\partial(\mu w)/\partial z$. Combining these three contributions, we see that the flux is given by

$$H = A\left\{\rho w^2 + p - \frac{\partial(\mu w)}{\partial z}\right\}, \tag{11.80}$$

when fluid momentum is the suspension. Note that fluid pressure is merely the macroscopic manifestation of molecular motions of the fluid; therefore both (11.78) and (11.80) can be viewed as the sum of macroscopic and microscopic components.[7]

---

[6]The mass of the fluid itself cannot diffuse, as a direct consequence of the definition of fluid density. For an explanation, see Batchelor (1967, pp. 4–6).

[7]Then the definition of velocity is (macroscopic flux)/(concentration). This does not contradict the remarks following (11.42) because the diffusive term in that traffic flow model is macroscopic.

Once variables have been relabelled, the lake-cleaning models of Section 9.3–9.5 emerge as special cases of model (11.76), with a flux term of the form (11.78). To see this, substitute (11.78) into (11.76) and set A = constant, obtaining the partial differential equation

$$\frac{\partial \phi}{\partial t} + \frac{\partial (w\phi)}{\partial z} - \gamma_3 \frac{\partial^2 \phi}{\partial z^2} = \frac{q}{A} - K\phi + C. \qquad (11.81)$$

Because our lake-cleaning models took no account of chemical reactions, we have $C = 0 = K$ in all three cases. In addition, the pure diffusion model of Section 9.5 has $w = 0 = q$; whereas the pure convection model of Section 9.4 has w = constant and $\gamma_3 = 0 = q$. The model of Section 9.3 has $w = 0 = \gamma_3$. Moreover, perfect mixing is equivalent to uniform injection and withdrawal of pollution, along the entire length $b - a$ of the lake. Pollution flows in at rate $5rx_E(t)/6$ and out at rate $r\phi$, and so we have

$$q = \frac{r}{b-a} \left(\frac{5}{6} x_E(t) - \phi \right). \qquad (11.82)$$

Substituting (11.82) into (11.81) and setting $w = \gamma_3 = K = C = 0$, we obtain

$$V\frac{\partial \phi}{\partial t} + r\phi = \frac{5}{6} r x_E(t), \qquad (11.83)$$

where $V = A(b-a)$ is the volume of the lake. Equation (9.29) is now readily obtained by integrating (11.83) with respect to z, as in (9.39) or (11.67). Now try Exercise 11.9.

11.7   River Cleaning. The Streeter–Phelps Model

In our models of environmental purification, we have assumed hitherto that a body of water is polluted by a single contaminant.[8] But many substances can pollute a lake or river. In general, they will interact both with each other and with other substances, which are not considered contaminants. How can we model this interaction? In this section, we will attempt to supply an answer to that question.

Suppose that a polluted river contains N suspensions, with (mass) concentrations ϕ_i, $i = 1, \ldots, N$. Then a possible approach to modelling river purification would be to apply (11.81) to each of the chemical suspensions:

$$\frac{\partial \phi_i}{\partial t} + \frac{\partial (w\phi_i)}{\partial z} - \gamma_{3i} \frac{\partial^2 \phi_i}{\partial z^2} = \frac{q_i}{A} - K_i\phi_i + C_i, \qquad i = 1, 2, \ldots, N, \quad (11.84)$$

where z measures distance along the river, w is the velocity of the water, A is its cross-sectional area, and quantities pertaining to contaminant i are labelled by subscript i. Because C_i would depend on $\phi_1, \phi_2, \ldots, \phi_{i-1}$, $\phi_{i+1}, \ldots, \phi_N$, however, equations (11.84) would be coupled and therefore

---

[8]You may wish to review your solution to Exercise 8.13 before proceeding.

difficult to solve analytically. On the other hand, we can avoid this difficulty as follows. Let $\phi_1, \ldots, \phi_k$ denote the concentrations of the k most important suspensions; let ϕ_{k+1} denote an appropriate measure of concentration of all other suspensions combined. By this method, the number of dependent variables can be reduced from N to $k + 1$, and we say that the $N - k$ least important suspensions have been *aggregated*. Used judiciously, such aggregation can be very insightful. Indeed almost every model in this book has involved some degree of implicit aggregation.

The most important variable in the purification of a river is the concentration of dissolved oxygen. Let this be denoted by ϕ_1. Fishes and other underwater organisms die without oxygen; hence any substance that consumes oxygen can reasonably be regarded as a pollutant, irrespective of whether it entered the river as manufactured effluent or as a result of natural processes. In 1925, H.W. Streeter and E.B. Phelps devised a way of aggregating all such substances. They defined *biochemical oxygen demand*, BOD, to be the amount of oxygen the pollutants would need for their complete oxidization, per unit volume of river water. In other words, BOD is the maximum amount of oxygen per unit volume that the pollutants could consume, because if they had consumed that much, then there would be no remaining pollutant molecules for the oxygen to combine with. We will denote BOD by ϕ_2. Then a possible purification model is (11.84) with $N = 2$. We have thus taken $k = 1$ and aggregated all but one of the variables.

Let's assume that the river flows with constant velocity W, and that it flows sufficiently rapidly that diffusion along the river can be neglected. We therefore set $\gamma_3 = 0$. (See Section 9.5, in particular the remarks following (9.64).) We will also assume that pollution input has ceased and that BOD can decay only by combining with oxygen or by flowing downstream; pollutants do not, for example, evaporate. Thus $q_2 = 0 = C_2$ in (11.84), and $C_1 = 0$ because oxygen is destroyed by the chemical reactions. With these assumptions, our purification model simplifies to

$$\frac{\partial \phi_1}{\partial t} + W\frac{\partial \phi_1}{\partial z} = \frac{q_1}{A} - K_1\phi_1 \qquad \text{(Oxygen)},$$

$$\frac{\partial \phi_2}{\partial t} + W\frac{\partial \phi_2}{\partial z} = -K_2\phi_2 \qquad \text{(BOD)}. \qquad (11.85)$$

It still remains, however, to find suitable expressions for q_1, K_1, and K_2.

Oxygen diffuses into the river from the air immediately above the water. There is some evidence that the air–water interface behaves like a membrane that is permeable to oxygen, in the sense that vertical diffusion near the surface of the river is much less efficient than at lower levels; see, for example, Rinaldi et al. (1979, p. 75). If so, then the model developed in Exercise 8.13 suggests that the flux of oxygen into the river is

$$\frac{k_0(\omega - \phi_1)}{h} \qquad (11.86)$$

per unit area, where h is the effective depth of the imaginary membrane, k_0 its permeability to oxygen, and ω the concentration of oxygen in the air immediately above the river. To obtain the rate at which oxygen enters the river *per unit length*, we merely multiply (11.86) by the (average) breadth, b, of the river. Hence

$$\frac{q_1}{A} = \frac{bk_0(\omega - \phi_1)}{Ah} = \lambda(\omega - \phi_1), \qquad (11.87)$$

where $\lambda = bk_0/Ah$ is a constant. Expression (11.87) was used by Streeter and Phelps. They called $\omega - \phi_1$ the *oxygen deficit* and λ the *reaeration coefficient*. Note that λ can also be written as k_0/hD, where D is the average river depth. Thus λ has the dimensions of a specific rate (i.e., T^{-1}) because k_0 is a diffusivity.

With q_1 suitably specified, we now turn to K_1 and K_2. The chemical reaction that consumes the river's oxygen may be written symbolically as

$$\text{Oxygen} + \text{BOD} \longrightarrow \text{Product.} \qquad (11.88)$$

From our studies of chemical reactions in Chapters 1 and 10, we suppose that the rate at which the two reactants, namely oxygen and BOD, convert into product is proportional to their concentration. Moreover, by definition of BOD, they must both convert at the same rate. Hence

$$-K_1\phi_1 = -\mu\phi_1\phi_2 = -K_2\phi_2, \qquad (11.89)$$

where μ is a constant. It immediately follows that $K_1 = \mu\phi_2$ and $K_2 = \mu\phi_1$. Combining these results with (11.85) and (11.87), we obtain the pair of coupled nonlinear partial differential equations

$$\begin{aligned} \frac{\partial\phi_1}{\partial t} + W\frac{\partial\phi_1}{\partial z} &= \lambda(\omega - \phi_1) - \mu\phi_1\phi_2, \\ \frac{\partial\phi_2}{\partial t} + W\frac{\partial\phi_2}{\partial z} &= -\mu\phi_1\phi_2. \end{aligned} \qquad (11.90)$$

We will refer to this model as the modified Streeter–Phelps model.

Equations (11.90) are not the Streeter–Phelps model itself because Streeter and Phelps replaced the coupling term $-\mu\phi_1\phi_2$ by a term proportional to ϕ_2 alone. Although this has the convenient effect of linearizing the equations to make them analytically tractable, there appears to be no justification for it—unless ϕ_2 is so small compared to ϕ_1 that ϕ_1 may be regarded as constant. In that case, there would seem to be no need for a ϕ_1 equation. Therefore, in our search for analytical solutions, we prefer to proceed more cautiously.

Let us first denote the oxygen deficit by

$$\psi(z, t) = \omega - \phi_1(z, t). \qquad (11.91)$$

Then (11.90) can be rewritten with oxygen deficit and BOD as dependent variables:

$$\frac{\partial \psi}{\partial t} + W\frac{\partial \psi}{\partial z} + \lambda\psi - \mu\omega\phi_2 = -\mu\phi_2\psi$$

$$\frac{\partial \phi_2}{\partial t} + W\frac{\partial \phi_2}{\partial z} + \mu\omega\phi_2 = \mu\phi_2\psi. \tag{11.92}$$

Now suppose that the BOD and oxygen deficit are both small, so that the river is relatively unpolluted. Then the term $\mu\phi_2\psi$ on the right-hand sides of equations (11.92) is a product of small quantities and is negligibly small compared to terms on the left-hand sides. We will therefore approximate it by zero. For a mildly polluted river, our mathematical model thus becomes

$$\frac{\partial \psi}{\partial t} + W\frac{\partial \psi}{\partial z} = \mu\omega\phi_2 - \lambda\psi,$$

$$\frac{\partial \phi_2}{\partial t} + W\frac{\partial \phi_2}{\partial z} = -\mu\omega\phi_2. \tag{11.93}$$

Let the initial oxygen deficit and BOD be denoted, respectively, by

$$\psi(z, 0) = \psi_0(z),$$

$$\phi_2(z, 0) = \phi_0(z). \tag{11.94}$$

Then you can verify by direct substitution that the solution of equations (11.93) subject to (11.94) is given by[9]

$$\psi(z, t) = e^{-\lambda t}\psi_0(z - Wt) + \frac{\mu\omega(e^{-\mu\omega t} - e^{-\lambda t})}{\lambda - \mu\omega}\phi_0(z - Wt)$$

$$\phi_2(z, t) = e^{-\mu\omega t}\phi_0(z - Wt). \tag{11.95}$$

To interpret (11.95), let's assume that the river is completely pure when, at time $t = 0$, a dollop of pollution is discharged into it. If this injected pollution is instantly mixed with the river water in such a way that there is constant BOD concentration β between $z = z_1$ and $z = z_2$, then $\psi_0(z) \equiv 0$ and

$$\phi_0(z) = \begin{cases} 0, & \text{if } z < z_1; \\ \beta, & \text{if } z_1 < z < z_2; \\ 0, & \text{if } z > z_2. \end{cases} \tag{11.96}$$

We immediately deduce from (11.95) and (11.96) that the BOD at subsequent times is given by

$$\phi_2(z, t) = \begin{cases} 0, & \text{if } z < z_1 + Wt; \\ \beta e^{-\mu\omega t}, & \text{if } z_1 + Wt < z < z_2 + Wt; \\ 0, & \text{if } z > z_2 + Wt; \end{cases} \tag{11.97}$$

---

[9]See Supplementary Note 11.1.

and the oxygen concentration $\phi_1(z, t) \equiv \omega - \psi(z, t)$ is given by

$$
\omega - \begin{cases} 0 & \text{if } z < z_1 + Wt, \\ \dfrac{\mu\omega\beta(e^{-\mu\omega t} - e^{-\lambda t})}{\lambda - \mu\omega} & \text{if } z_1 + Wt < z < z_2 + Wt, \\ 0 & \text{if } z > z_2 + Wt. \end{cases} \tag{11.98}
$$

Thus the pollution is confined to a region of length $z_2 - z_1$, which moves downstream with velocity

$$
\frac{dz}{dt} = W. \tag{11.99}
$$

An observer moving with the velocity of the river, i.e., according to (11.99), would see BOD within the region of confinement decay exponentially at the specific rate $\mu\omega$, as is clear from (11.97). For the same observer, the oxygen concentration would decrease during the interval $0 < t < t_c$, where

$$
t_c = \frac{1}{\lambda - \mu\omega} \ln\left(\frac{\lambda}{\mu\omega}\right), \tag{11.100}
$$

as is clear from (11.98) and Fig. 11.10. Thereafter, ϕ_1 would increase, returning to level ω as $t \to \infty$. The maximum oxygen deficit would be

$$
\psi(t_c) = \beta\left(\frac{\mu\omega}{\lambda}\right)^{\lambda/(\lambda - \mu\omega)}. \tag{11.101}
$$

The curve in Fig. 11.10 is known as an *oxygen sag curve*. A similar curve, though with a deeper sag, obtains for more heavily polluted rivers. To see this, observe that when *following the motion of the river*, i.e., when (11.99) is satisfied, equations (11.90) can be written as the pair of nonlinear

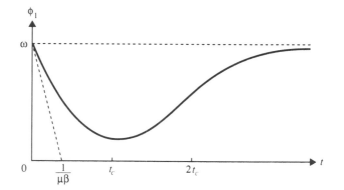

Fig. 11.10 Oxygen sag curve (11.98) for a station within the confinement region on a mildly polluted river. The sloping dashed line depicts the initial slope. The minimum is at $t = t_c$, and there is an inflection point at $t = 2t_c$.

ordinary differential equations

$$\frac{d\phi_1}{dt} = \lambda(\omega - \phi_1) - \mu\phi_1\phi_2$$

$$\frac{d\phi_2}{dt} = -\mu\phi_1\phi_2. \tag{11.102}$$

Thus, following the motion of the fluid, the growth or decay of dissolved oxygen and BOD can be monitored in the ϕ_1-ϕ_2 phase-plane, depicted in Fig. 11.11. The solid curve without arrows represents

$$\phi_2 = \frac{\lambda}{\mu}\left(\frac{\omega}{\phi_1} - 1\right), \tag{11.103}$$

above which $d\phi_1/dt$ is always negative, and below which $d\phi_1/dt$ is always positive; $d\phi_2/dt$ is always negative because $\phi_1\phi_2 > 0$. The trajectory beginning at A_1 in Fig. 11.11 corresponds to our purification theory for a weakly polluted river, with the length B_1D_1 representing (11.101); ϕ_2 remains small, and ϕ_1 close to ω. More generally, the evolution of ϕ_1 and ϕ_2, following the motion of the river, is represented by phase trajectories such as those beginning at A_2 and A_3. A_2 corresponds to a mildly pol-

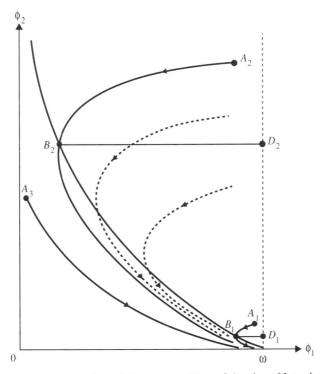

Fig. 11.11   The ϕ_1-ϕ_2 phase-plane following the flow of the river. Note that the tangent to A_2B_2 is vertical at B_2.

luted river that suddenly receives a substantial input of pollution. The point $(\phi_1(t), \phi_2(t))$ follows the phase trajectory through A_2 and B_2, crossing (11.103) vertically at B_2 at time $t = t_c$; B_2D_2 is the maximum oxygen deficit. (The time t_c could be found numerically for any given values of ω, μ, and λ). Notice from the dashed trajectories that the maximum oxygen deficit is an increasing function of $\phi_2(0)$, the initial BOD. A_3 in Fig. 11.11 corresponds to a river that is heavily polluted initially. There is no longer a sag in the oxygen curve, the maximum deficit occurring at $t = 0$.

In Sections 2.1, 9.3–9.5, 11.5, and, finally, in the present section, we have seen numerous developments of the simple lake purification model of Section 1.1. Of all these models, that associated with Streeter and Phelps is the most advanced. I think you'll agree that the modified Streeter–Phelps model, in both linear form (11.93) and nonlinear form (11.102), affords many useful theoretical insights. Nevertheless, its assumptions are too drastic for many lakes and rivers. For example, in a real river, oxygen may be created by photosynthesis, so that $C_1 \neq 0$ in (11.84) and an extra term must be added to (11.85). Again, we have ignored the effect of temperature on the rate of chemical reaction, although, to some extent, it can be acknowledged by choosing appropriate values for the parameters λ and μ. Moreover, if a body of water receives a steady influx of contaminant, then the dissolved oxygen concentration will never return to its ambient value, ω. Perhaps you would like to extend the model (11.85) to incorporate some of these additional effects. If so, then you might begin your study with Chapter 3 of Springer (1986) or Chapter 4 of Rinaldi et al. (1979). If not, then at least attempt Exercise 11.10!

11.8 Why Does a Stopped Organ Pipe Sound an Octave Lower Than an Open One?

A stopped organ pipe sounds an octave lower than an open one because the frequency of oscillation of air in a stopped pipe is only half as high as that in an open one. But why should that be so? In order to answer this question, we will need to know something about sound waves. We will borrow as necessary from the science of physics.

Air is a fluid and an organ pipe a hollow cylindrical tube, so that the model of Section 11.6 applies directly. Let z measure distance along the pipe, and let A be its cross-sectional area, assumed constant. Then (11.76) yields

$$\frac{\partial(A\phi)}{\partial t} + AK\phi + \frac{\partial H}{\partial z} = AC + q, \qquad (11.104)$$

where t denotes time, ϕ the concentration of an appropriate suspension, and H its flux, in the direction of increasing z; C, K, and q have the same meanings as in Section 11.6.

The two properties of air that make sound waves possible are density and momentum. But neither mass nor momentum of a body of air is created or destroyed by chemical reaction. Therefore, regardless of whether the suspension in (11.104) is mass or momentum, we have $C = 0 = K$. Moreover, because an organ pipe is impervious to air, we have $q = 0$, too. Hence

$$\frac{\partial}{\partial t}(A\phi) + \frac{\partial H}{\partial z} = 0. \tag{11.105}$$

Let $\rho(z, t)$ denote mass per unit volume, at time t, of the air at station z in our pipe. Let $w(z, t)$ denote the corresponding air velocity in the direction of increasing z. Then the flux of suspension for mass is

$$H = A\rho w, \tag{11.106}$$

by (11.77); whence, setting $\phi = \rho$ in (11.105) and removing the constant factor A, we obtain

$$\frac{\partial \rho}{\partial t} + \frac{\partial}{\partial z}(\rho w) = 0. \tag{11.107}$$

The momentum per unit volume is simply DENSITY × VELOCITY, i.e., $\rho(z, t)w(z, t)$ at station z at time t. Moreover, from (11.80), the associated flux of suspension is

$$H = A\left\{\rho w^2 + p - \frac{\partial(\mu w)}{\partial z}\right\}, \tag{11.108}$$

where $p(z, t)$ denotes the pressure of the air and μ is its shear viscosity. To keep matters simple, however, we'll simply neglect viscosity; thus $\mu = 0$. Now, setting $\phi = \rho w$ in (11.105), using (11.108), and removing the constant factor A, we have:

$$\frac{\partial}{\partial t}(\rho w) + \frac{\partial}{\partial z}(p + \rho w^2) = 0. \tag{11.109}$$

From our point of view, the ultimate justification for neglecting viscosity is that we will be able to answer the title question without it. There is therefore no reason to suspect that viscosity plays a significant role in determining the pitch of an organ pipe (but see Exercise 11.14).

Equation (11.109) is readily expanded to yield:

$$\frac{\partial \rho}{\partial t}w + \rho\frac{\partial w}{\partial t} + \frac{\partial p}{\partial z} + \frac{\partial(\rho w)}{\partial z}w + \rho w\frac{\partial w}{\partial z} = 0. \tag{11.110}$$

The sum of the first and fourth terms in this equation vanishes, by (11.107). Hence

$$\rho\frac{\partial w}{\partial t} + \rho w\frac{\partial w}{\partial z} + \frac{\partial p}{\partial z} = 0. \tag{11.111}$$

Air is a gas, which is compressible. The greater the density, ρ, the greater the pressure, p. That's why your tooth hurts if you have an abscess:

bacteria, by creating gas, increase the density of gas in a cavity, and the concomitant increase in pressure is felt by your nerves as pain. It therefore seems reasonable to assume that p is an increasing function of ρ, i.e., that $p = p(\rho)$, with $p'(\rho) > 0$. Because $p'(\rho) > 0$, we can define

$$a(\rho) \equiv \sqrt{\frac{dp}{d\rho}}. \qquad (11.112)$$

You should verify that a has the dimensions of velocity. Now (11.107) and (11.111) may be replaced by the following pair of coupled partial differential equations for $\rho(z, t)$ and $w(z, t)$:

$$\frac{\partial \rho}{\partial t} + \rho \frac{\partial w}{\partial z} + w \frac{\partial \rho}{\partial z} = 0, \qquad (11.113a)$$

$$\rho \frac{\partial w}{\partial t} + \rho w \frac{\partial w}{\partial z} + \{a(\rho)\}^2 \frac{\partial \rho}{\partial z} = 0. \qquad (11.113b)$$

Notice that if the air is at rest, i.e., if $w = 0$, then ρ must be independent of both z and t; i.e., $\rho = \bar\rho$, where $\bar\rho$ is a constant. This equilibrium solution of (11.113) would describe the air in an organ that nobody was playing. We will refer to the associated pressure $\bar p \equiv p(\bar\rho)$ as *atmospheric pressure*. Thus atmospheric pressure is the pressure that obtains in undisturbed air.

Now, suppose that the air is disturbed only slightly from rest, so that the velocity magnitude $|w(z, t)|$ is tiny. Let the density corresponding to this slight disturbance be

$$\rho(z, t) = \bar\rho + \eta(z, t), \qquad (11.114)$$

where $|\eta|$ is also tiny. Then, on using Taylor's theorem to expand $a(\rho)$ about $\rho = \bar\rho$, (11.113) yields

$$\frac{\partial \eta}{\partial t} + \bar\rho \frac{\partial w}{\partial z} = -\frac{\partial}{\partial z}(w\eta),$$

$$\bar\rho \frac{\partial w}{\partial t} + \{a(\bar\rho)\}^2 \frac{\partial \eta}{\partial z} = -\eta \frac{\partial w}{\partial t} - (\bar\rho + \eta)w \frac{\partial w}{\partial z} \qquad (11.115)$$

$$-\frac{\partial \eta}{\partial z}\{2\eta a'(\bar\rho) + O(\eta^2)\}.$$

The right-hand sides of (11.115) consist of negligibly small products of small quantities. We therefore approximate (11.115) by

$$\frac{\partial \eta}{\partial t} + \bar\rho \frac{\partial w}{\partial z} = 0, \qquad (11.116a)$$

$$\bar\rho \frac{\partial w}{\partial t} + \bar a^2 \frac{\partial \eta}{\partial z} = 0, \qquad (11.116b)$$

where

$$\bar a \equiv a(\bar\rho) \qquad (11.117)$$

is independent of z and t. Differentiating (11.116a) with respect to t and (11.116b) with respect to z now yields the single partial differential equation:

$$\frac{\partial^2 \eta}{\partial t^2} = \bar{a}^2 \frac{\partial^2 \eta}{\partial z^2}. \tag{11.118}$$

The general solution of (11.118) is (Exercise 11.11)

$$\eta(z,t) = F(z - \bar{a}t) + G(z + \bar{a}t), \tag{11.119}$$

where F and G are arbitrary functions, to be determined by relevant initial and boundary conditions. In (11.119), F represents a wave propagating in the direction of increasing z at speed $\bar{a}$. To see this, first set $G = 0$. Then imagine that our pipe is infinitely long and that the air in it is completely undisturbed until time $t = 0$. At that time, however, the air in section $-\epsilon/2 < z < \epsilon/2$ of the pipe is suddenly compressed, from density ρ to density $\bar{\rho} + \delta$. Thus, because $G = 0$, we have

$$\eta(z,t) = F(z - \bar{a}t) \tag{11.120}$$

with

$$\eta(z,0) = \begin{cases} 0, & \text{if } z < -\dfrac{\epsilon}{2}; \\ \delta, & \text{if } -\dfrac{\epsilon}{2} < z < \dfrac{\epsilon}{2}; \\ 0, & \text{if } z > \dfrac{\epsilon}{2}. \end{cases} \tag{11.121}$$

Setting $t = 0$ in (11.120) yields $F(z) = \eta(z,0)$ for arbitrary argument z. Hence, replacing z by $z - \bar{a}t$, we have

$$\eta(z,t) = \begin{cases} 0, & \text{if } z < \bar{a}t - \dfrac{\epsilon}{2}; \\ \delta, & \text{if } \bar{a}t - \dfrac{\epsilon}{2} < z < \bar{a}t + \dfrac{\epsilon}{2}; \\ 0, & \text{if } z > \bar{a}t + \dfrac{\epsilon}{2}. \end{cases} \tag{11.122}$$

We see that the region of length ϵ, in which density was increased, travels in the positive z direction with constant speed $\bar{a}$. Similarly, you can show that G represents a wave propagating at speed $\bar{a}$ in the direction of decreasing z.

We call these waves sound waves and $\bar{a}$ the sound speed, because the pressure perturbations that accompany the density perturbations can be detected by the human ear. The density perturbations are usually oscillatory, however, and it is their frequency of oscillation that the human ear detects as pitch. To understand this, pretend that our imaginary infinite pipe has a horizontal axis (z increases horizontally) and that its (vertical) cross section A is large enough for a listener to stand at $z = z_0 > 0$; and that beginning at time $t = 0$, a transmitter standing at $z = 0$ causes a tuning fork to oscillate with frequency f. An oscillating prong compresses the air on the side toward which it is moving, and it causes the air on the opposite side to fill a vacuum, by expanding. This is depicted in Fig. 11.12. The net effect is that the air density oscillates; say $\rho(0, t) = \bar{\rho} + \delta \sin(2\pi f t)$, where f is the frequency of oscillation. Thus, from (11.114),

$$\eta(0, t) = \begin{cases} 0, & \text{if } t < 0; \\ \delta \sin(2\pi f t), & \text{if } t > 0. \end{cases} \tag{11.123}$$

The sound wave that propagates in the direction of decreasing z will not affect our listener; therefore we can assume that $G = 0$ in (11.119). Thus, setting $z = 0$ in (11.120) and using (11.123), we deduce that $F(-\bar{a}t) = \delta \sin(2\pi f t)$, for any value of $t > 0$. Replacing t by $t - z_0/\bar{a}$ (when positive) and using (11.120) and (11.123), we thus obtain

$$\eta(z_0, t) = \begin{cases} 0, & \text{if } t < z_0/\bar{a}; \\ \delta \sin\left\{2\pi f\left(t - (z_0/\bar{a})\right)\right\}, & \text{if } t > z_0/\bar{a}. \end{cases} \tag{11.124}$$

This means that the sound of the tuning fork takes time $z_0/\bar{a}$ to reach our listener and is perceived thereafter as a pressure oscillation of frequency f. On using (11.112), (11.114), (11.117), and Taylor's theorem, the excess of pressure $p(z, t)$ over atmospheric pressure $\bar{p} = p(\bar{\rho})$ is given by $p(z, t) - \bar{p} = p(\rho) - \bar{p} = p(\bar{\rho} + \eta) - p(\bar{\rho}) \approx p'(\bar{\rho})\eta = \bar{a}^2\eta$; i.e.,

$$p(z, t) - \bar{p} = \bar{a}^2\eta(z, t), \tag{11.125}$$

to the same degree of approximation as was used to obtain (11.116). Thus, on using (11.124), we see that

$$p(z_0, t) - \bar{p} = \begin{cases} 0, & \text{if } t < z_0/\bar{a}; \\ \bar{a}^2\delta \sin\left\{2\pi f\left(t - (z_0/\bar{a})\right)\right\}, & \text{if } t > z_0/\bar{a}. \end{cases} \tag{11.126}$$

This is the pressure perturbation our listener perceives as sound.

Of course, a real organ pipe has a vertical axis. To determine why a stopped organ pipe sounds an octave lower than an open one, therefore, let z increase vertically downward. Let the upper end of the pipe be at $z = 0$ and the lower end at $z = L$ (Fig. 11.13). Thus the pipe has length L. We will assume that the lower end of the pipe is open to the atmosphere.

Fig. 11.12 Oscillations of a tuning fork's prong; B is the imaginary pivot of the movement.

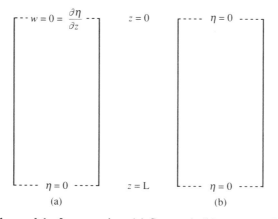

Fig. 11.13 Crude model of organ pipe. (a) Stopped; (b) unstopped.

If the entire atmosphere were undisturbed, then the pressure would be $\bar{p}$ throughout, and this would be true, in particular, at the lower end of the pipe. Now, the sound of the organ must disturb the atmosphere, but if we imagine this disturbance to be slight, then it seems reasonable to suppose that the entire atmosphere is still approximately at atmospheric pressure. Thus the lower end of the pipe is, at least approximately, at atmospheric pressure. Hence $p(L, t) = \bar{p}$; whence, from (11.125),

$$\eta(L, t) = 0. \tag{11.127}$$

Conditions at the upper end of the pipe will depend upon whether it is open to the atmosphere or closed by a pipe stop. If open, then $p(0, t) = \bar{p}$; whence

$$\eta(0, t) = 0, \tag{11.128}$$

by (11.125). If closed, then because air cannot flow through a pipe stop, the air adjacent to $z = 0$ must be at rest at all times:

$$w(0, t) = 0, \qquad 0 \leq t < \infty. \tag{11.129}$$

Hence, because $w(0, t)$ does not change,

$$\frac{\partial w}{\partial t}(0, t) = 0, \qquad 0 \leq t < \infty. \tag{11.130}$$

It now follows from (11.116b) that

$$\frac{\partial \eta}{\partial z}(0, t) = -\frac{\bar{p}}{\bar{a}^2}\frac{\partial w}{\partial t}(0, t) = 0, \qquad 0 \leq t < \infty; \tag{11.131}$$

i.e., $\partial \eta / \partial z = 0$ at $z = 0$, for all t. Thus, combining (11.118) with (11.127) and (11.128), we find that density perturbations in our (model of an) unstopped organ pipe must satisfy the sound wave equation

$$\frac{\partial^2 \eta}{\partial t^2} = \bar{a}^2 \frac{\partial^2 \eta}{\partial z^2}, \qquad 0 \le z \le L, \quad 0 \le t < \infty, \tag{11.132}$$

subject to the boundary conditions

$$\eta(L, t) = 0 = \eta(0, t), \qquad 0 \le t < \infty; \tag{11.133}$$

and, from (11.127) and (11.131), density perturbations in our stopped organ pipe must satisfy (11.132) subject to the boundary conditions

$$\eta(L, t) = 0 = \frac{\partial \eta}{\partial z}(0, t), \qquad 0 \le t < \infty. \tag{11.134}$$

It follows immediately (Exercise 11.12) that a possible density perturbation for our stopped organ pipe is the sinusoidal oscillation

$$\eta(z, t) = A_0 \cos\left(\frac{\pi z}{2L}\right) \cos\left(\frac{\pi \bar{a} t}{2L}\right), \tag{11.135}$$

where A_0 is a constant. For any fixed value of time, say $t = t_0$, η has the form

$$\eta(z) = C \cos\left(\frac{\pi z}{2L}\right), \tag{11.136}$$

where

$$C = A_0 \cos\left(\frac{\pi \bar{a} t_0}{2L}\right) \tag{11.137}$$

is just a constant. Now, the spatial pattern (11.136) goes through one complete cycle for every increase or decrease of $4L$ in z (verify). We say that the pattern has *wavelength* $4L$. But the height of our organ pipe is only L. Hence the oscillation of the air inside it is restricted to a quarter of a wavelength. For any fixed value of z, say $z = z_1$, η has the form

$$\eta(t) = C \cos\left(\frac{\pi \bar{a} t}{2L}\right), \tag{11.138}$$

where

$$C = A_0 \cos\left(\frac{\pi z_1}{2L}\right) \tag{11.139}$$

is just a constant. The temporal pattern (11.138) goes through one complete cycle for every increase of $4L/\bar{a}$ in time (verify). Hence the frequency of oscillation, or the number of cycles per unit of time, is

$$f_0 \equiv \frac{\bar{a}}{4L}. \tag{11.140}$$

Likewise, a possible density perturbation for our unstopped pipe is the sinusoidal oscillation

$$\eta(z, t) = B_1 \sin\left(\frac{\pi z}{L}\right) \cos\left(\frac{\pi \bar{a} t}{L}\right), \tag{11.141}$$

where B_1 is a constant. By arguments similar to those in the previous paragraph, we find that the oscillation of the air is restricted to half a wavelength and has frequency

$$g_1 \equiv \frac{\bar{a}}{2L}. \tag{11.142}$$

Notice that $g_1 = 2f_0$. An answer to the title question has thus emerged: the stopped organ pipe sounds an octave lower than an unstopped one because, by altering the upper boundary condition from (11.128) to (11.131), the stop enables twice as much wavelength to fit into the pipe, with a concomitant halving of frequency.

Although this answer is essentially correct, it is by no means the whole story. To begin with, (11.135) is not unique. For any integer $j = 0, 1, \ldots$, and for any constants A_j and ϵ_j, (11.132) and (11.134) are satisfied (Exercise 11.13) by the pattern

$$\eta(z, t) = A_j \cos\left\{\left(j + \frac{1}{2}\right) \frac{\pi z}{L}\right\} \cos\left\{\left(j + \frac{1}{2}\right) \frac{\pi \bar{a} t}{L} + \epsilon_j\right\}. \tag{11.143}$$

This is known in acoustics as the jth *normal mode*. For any fixed z, it represents a sinusoidal oscillation of constant amplitude and frequency

$$f_j \equiv \left(j + \frac{1}{2}\right) \frac{\bar{a}}{2L}. \tag{11.144}$$

Furthermore, by failing to specify initial conditions for η and w, we have not described how the air in our organ pipe is first set in motion. Now, although we will not do so here, it is readily shown that the general solution of (11.132), subject to (11.134), is the infinite superposition of normal modes

$$\eta(z, t) = \sum_{j=0}^{\infty} A_j \cos\left\{\left(j + \frac{1}{2}\right) \frac{\pi z}{L}\right\} \cos\left\{\left(j + \frac{1}{2}\right) \frac{\pi \bar{a} t}{L} + \epsilon_j\right\}, \tag{11.145}$$

and that the constants A_j and ϵ_j are uniquely determined for all j by appropriate initial conditions, e.g., by specifying $w(z, 0)$ and $\eta(z, 0)$ for all z. Moreover, A_j decreases monotonically with j, no matter how crudely the organ pipe is constructed. But an organ pipe is so designed and the air so set in motion as to make this monotonic decrease extremely rapid. In other words, the dominant term in (11.145), by far, is the normal mode of lowest frequency—namely, (11.135). In acoustics, this is known as the *fundamental* (normal modes of higher frequency are known as overtones). Likewise, although the most general solution of (11.132), subject to (11.133), has the form

$$\eta(z, t) = \sum_{j=0}^{\infty} B_j \sin\left\{\frac{j\pi z}{L}\right\} \cos\left\{\frac{j\pi \bar{a} t}{L} + \theta_j\right\}, \tag{11.146}$$

where B_j and θ_j are determined by initial conditions, and is therefore a superposition of normal modes of frequency

$$g_j \equiv \frac{j\bar{a}}{2L}, \tag{11.147}$$

the dominant term in (11.146) by far is the normal mode of lowest frequency—namely, (11.141).

Moreover, we have already acknowledged that pressure at an open end of a pipe is only approximately atmospheric. It is possible to relax this assumption, though the details would require more space than is appropriate in this text. The end result is that the pitch of an unstopped organ pipe remains g_1, defined by (11.142), and the pitch of a stopped organ pipe remains f_0, defined by (11.140); all this provided that we replace L by an effective pipe-length $L+\epsilon$, where ϵ/L is small. In other words, our answer to the title question, though reached somewhat prematurely, is basically satisfactory. If you wish to study this model improvement for yourself, then see, for example, Lamb (1925, in particular pp. 270–274).

Our simple model, (11.116), has thus satisfactorily explained why stopped organ pipes sound an octave lower than open ones. But it does not explain how the sound of an organ dies away when the organist stops playing. Indeed (11.135) and (11.141) yield the absurd prediction that the sound will last forever! Thus, if we wish to explain why sound dies away, then one of the assumptions that led to (11.116) must be relaxed. Which one? See Exercise 11.14, then try Exercise 11.15.

Exercises

*11.1 It is not uncommon for a resident of Florida to have an "unheated" outdoor swimming pool, one that is heated solely by the sun. In such pools, the temperature of the water near the surface fluctuates with the daily variation in the temperature of the air, while the deeper water maintains a fairly constant temperature. The depth at which this deeper layer begins, i.e., the depth beyond which daily temperature variations fail to penetrate, is called the penetration depth.

We are interested only in the water near the surface; therefore we can pretend that the pool is infinitely deep. Suppose, moreover, that the overnight low has been 21°C and the midday high 35°C for a considerable period of time, so that a steady state has been reached. Then the temperature $T(z,t)$ at depth z at time t is governed by

$$\frac{\partial T}{\partial t} = \kappa \frac{\partial^2 T}{\partial z^2}, \qquad 0 \le z < \infty, \tag{a}$$

$$T(0,t) = 28 + 7\cos\left(\frac{\pi t}{12}\right), \tag{b}$$

where $\kappa \approx 1.4 \times 10^{-3} \text{cm}^2/\text{sec}$ is the thermal diffusivity of water, and where we have approximated the daily surface temperature variation between successive highs at $t = 0$, $t = 24$, $t = 48$, ..., by using a circular function. What, therefore, is the penetration depth?

Hint: Seek a limit-cycle solution of the form

$$T(z,t) = 28 + 7e^{-az} \cos\left(\frac{\pi t}{12} - az\right),$$ (c)

where a is a constant to be determined. In terms of Section 9.5, (c) represents a wave propagating into the water, whose amplitude decreases with depth.

11.2 Show that (11.21) satisfies both (11.18) and (11.20). *Hint*: This exercise is similar to Exercise 9.8. Note that the lower limit of integration in (11.21) tends to $+\infty$ or $-\infty$ according to whether $z > 0$ or $z < 0$.

*11.3 (i) In our first traffic models, drivers reacted only to the car immediately in front of them. What is the advantage of (11.19) compared to such models? (ii) Why could (11.19) not be correct if drivers were meeting increasing congestion?

11.4 Show that the solution of (11.27) subject to (11.18) and (11.29) is (11.30). *Hint*: For $-\infty < z \le -\lambda t$ and $v_{max}t < z < \infty$, the solution follows immediately from (11.32) and (11.33); see Fig. 11.5. For $-\lambda t \le z \le (v_{max} - \lambda)t$, seek a solution of the form $\phi(z,t) = \phi(w)$, where $w = z/t$. Finally, for $(v_{max} - \lambda)t \le z \le v_{max}t$, use (11.32) and (11.33) to continue the solution away from an appropriate characteristic.

11.5 Show that (11.36) is the solution of (11.35). *Hint*: Use a substitution suggested by Exercise 11.4.

*11.6 Compare Fig. 11.5 with the fundamental diagram of Exercise 2.5. In which direction does information propagate (i) in light traffic; (ii) in heavy traffic?

11.7 Obtain the solution of (11.46) subject to (11.44) and (11.45). *Hint*: Evaluate $z = \int \{\phi v(\phi)\}^{-1} d\phi$ for each of the two cases in (11.28). The second quadrature will require the substitution $x = \ln(\phi)$, $dx = d\phi/\phi$.

11.8 Verify that (11.64) and (11.65) imply $\sigma(t) = \sqrt{2\gamma t}$.

*11.9 Some special cases of (11.76) were discussed at the end of Section 11.6. Identify all other models encountered so far that are also special cases of (11.76). Don't forget the exercises.

*11.10 Under what conditions might the Michaelis–Menten law (Section 10.1) adequately describe the rate at which oxygen is taken up by river pollutants? *Hint*: Compare Fig. 11.11 with Fig. 10.1. The rate of oxygen uptake (following the river) is the same as the rate at which BOD decays. You might find it helpful to make equations (11.102) dimensionless.

11.11 Obtain (11.119). *Hint*: By using the Chain Rule for partial differentiation, change the independent variables in (11.118) from z, t to $\xi \equiv z + \bar{a}t$ and $\zeta \equiv z - \bar{a}t$. Instead of equation (11.118) for $\eta(z, t)$, you will obtain the equation

$$\partial^2 \eta / \partial \xi \partial \zeta = 0$$

for $\eta(\xi, \zeta)$, which integrates readily to $F(\zeta) + G(\xi)$.

11.12 (i) Verify that (11.135) satisfies (11.132) and (11.134).
(ii) Verify that (11.141) satisfies (11.132) and (11.133).
(iii) Show that (11.141) has wavelength $2L$ and frequency g_0, defined by (11.142).

11.13 Show that (11.143) satisfies (11.132) and (11.134) for any value of j.

*11.14 The model of Section 11.8 does not explain how the sound of an organ dies away when the organist stops playing, because (11.135) and (11.141) predict that the sound will last forever. By relaxing one of the model's assumptions, extend (11.116) to include a phenomenon that plays an important role in damping sound. Why is this phenomenon not also important in determining the pitch?

*11.15 In Section 11.8 we succeeded in explaining, at least in broad terms, how sound can propagate from transmitter to listener. We did this by considering an imaginary horizontal pipe in which air was constrained to move parallel to the axis; see the analysis leading to (11.126). Later, we succeeded in explaining the pitch of an organ pipe. This time, we considered a vertical pipe, but the air was again constrained to move parallel to its axis. Thus, strictly, the second case did not allow the sound of the organ pipe to be heard by a distant listener, because such propagation of sound requires air to pulse horizontally! By the same token, the first case allowed a listener to hear distant sounds but prevented an organ pipe from making any, because organ music requires air to pulse vertically. In each case, we applied a sound wave model, (11.116), that restricted flow and diffusion to a single spatial dimension. Although each of two natural phenomena was satisfactorily explained in isolation by the model, it could not explain how the phenomena interacted. In a similar vein, by restricting flow and diffusion to a single spatial dimension, convection along a lake and diffusion across a lake were adequately modelled as separate phenomena in Chapter 9, even though the tube model of Section 11.6 is not general enough to include both phenomena simultaneously. For that, we require a generic model for flow and diffusion in more than one spatial dimension. Derive one.

Supplementary Notes

11.1 The solution is obtained by a characteristic method, as in Section 11.3. We first observe that because $d/dt = \partial/\partial t + (dz/dt)\partial/\partial z$, (11.93) reduces to a pair of ordinary differential equations (ODEs) $d\psi/dt = \mu\omega\phi_2 - \lambda\psi$, $d\phi_2/dt = -\mu\omega\phi_2$ on the characteristic lines in the z-t plane, namely, $dz/dt = W$ or $z - Wt =$ constant. On the straight line $z - Wt = z_0$ from $(z_0, 0)$ to (z, t), we can solve the second ODE, with initial condition $\phi_2(z_0, 0)$, to obtain $\phi_2 = e^{-\mu\omega t}\phi_2(z_0)$. Then the first ODE becomes $d\psi/dt + \lambda\psi = \mu\omega e^{-\mu\omega t}\phi_0(z_0)$, which is solved for initial condition $\psi(z_0, 0)$, by using the integrating factor $e^{-\lambda t}$, to yield $\psi = e^{-\lambda t}\psi_0(z_0) + \mu\omega(e^{-\mu\omega t} - e^{-\lambda t})\phi_0(z_0)/(\lambda - \mu\omega)$. By substituting $z_0 = z - Wt$ we obtain (11.95). By construction, the solution above is unique. The case $\lambda = \mu\omega$ is excluded on the usual grounds that only parameter *in*equalities are likely to be valid in the laboratory of nature. The theoretical solution for $\lambda = \mu\omega$ is readily obtained, however; the same methods yield $\phi_2(z, t) = e^{-\lambda t}\phi_0(z - Wt)$ and $\psi(z, t) = e^{-\lambda t}\{\psi_0(z - Wt) + \lambda t\phi_0(z - Wt)\}$.

12 FURTHER OPTIMIZATION

In our final chapter, we further pursue the probabilistic approach to decision making. We also introduce some more advanced ideas about optimization.

Our first four sections concern the human species. Section 12.1, which is a sequel to Section 7.5, serves as an introduction to the method of dynamic programming. This method is then further applied in Sections 12.2–12.4. Our last two sections concern other animal species. Section 12.5 examines the foraging behavior of birds. It also introduces some vector optimization concepts. Finally, in Section 12.6, we combine these concepts with the method of dynamic programming in an attempt to model the behavior of an insect.

12.1 Finding an Optimal Policy by Dynamic Programming

In Section 7.5, we broached the optimization problem that confronts a buyer who places orders for some product at regular intervals.[1] We supposed that the buyer would place N orders at times $t = 0, 1, 2, \ldots, N - 1$. We called N the horizon and supposed that it was arbitrary. But we restricted attention to horizon 2. In this section, we will extend our model to arbitrary horizons. As in the earlier section, we shall continue to assume that demand for each period is uniformly distributed between values a and b, that advertising costs are zero, and that shipping costs have been absorbed into c_0, the cost per item to the buyer.

In Section 7.5, we found the optimal policy $\{u_0^*, u_1^*\}$ for $N = 2$ by starting with the decision at $t = 1$ and then working back to the one at

---

[1] You are advised to review Sections 3.4 and 7.5 before proceeding.

$t = 0$. This suggests that we should find the optimal policy for horizon N by starting with the decision at $t = N - 1$ and working back to the one at $t = 0$. The direction in which decisions are *formulated* in time (backward to 0) is then the opposite of that in which decisions are *implemented* in time (forward to $N - 1$); therefore we shall introduce some new notation. For $k = 0, 1, \ldots, N$, let us define

$$
\begin{aligned}
w_k &= u_{N-k}; & S_k(w_k) &= I_{N-k}(u_{N-k}); \\
Y_k &= X_{N-k}; & \psi_k(w_k) &= \phi_{N-k}(u_{N-k}); \\
y_k &= x_{N-k}.
\end{aligned}
\tag{12.1}
$$

The essential difference between the two notations is that, whereas previously we labelled quantities according to the number of periods that had elapsed since $t = 0$, we now label quantities according to the number of periods remaining until the horizon. Thus the quantity ordered at time $t = N - k$ is now denoted by w_k rather than by u_{N-k}. The surplus stock at time $t = N - k$, *when viewed from* $t < N - k$, is now denoted by the random variable Y_k rather than by X_{N-k}; but, *when viewed from* $t \geq N - k$, by y_k rather than by x_{N-k}. Reward from the interval $N - k < t < N - k + 1$, formerly denoted by $I_{N-k}(u_{N-k})$, is now denoted by $S_k(w_k)$ instead. Optimal reward from the interval $N - k < t < N$, *as viewed from* $t = N - k$, is now denoted by $\psi_k(w_k)$. Previously, we would have denoted this quantity by $\phi_{N-k}(u_{N-k})$ instead (although this notation was actually used in Section 7.5 only for the special case $N = 2$, $k = 1$). Finally, the surplus at time $t = N - k + 1$, *when viewed from the earlier time of* $t = N - k$, is now related to demand Z_{N-k} for the interval $N - k < t < N - k + 1$ by

$$
Y_{k-1} = \begin{cases} w_k + y_k - Z_{N-k}, & \text{if } Z_{N-k} < w_k + y_k; \\ 0, & \text{if } Z_{N-k} > w_k + y_k, \end{cases}
\tag{12.2}
$$

instead of by (7.53).

The advantage of this new notation is that the index k increases in the direction in which the problem is actually solved. We anticipated this point in Chapter 7 when we defined the functions J_1 and J_2. More generally, reward from the interval $N - k < t < N$, *as perceived by a decision maker at time* $t = N - k$, would have been denoted by $J_k(u_{N-k})$ in the old notation. In our new notation, however, we will denote this quantity by $J_k(w_k)$ instead. (To be quite precise, we shall use $J_k(w_k)$ to denote the true worth of reward, but we'll get to that in a moment.)

With these definitions, let us recast the problem of Section 7.5 into fresh perspective. In that section, we found the optimal policy for horizon 2, $\{u_0^*, u_1^*\}$, on the understanding that the final stock $x_2 (= y_0)$ was worthless. For the sake of simplicity, we shall make the same assumption here. Then a few moments of thought reveals that Section 7.5's analysis provides the optimal policy for the *last* two periods when the horizon is N.

In other words, the same analysis provides $\{u^*_{N-2}, u^*_{N-1}\}$; or, in our new notation, $\{w^*_1, w^*_2\}$.[2] From this new perspective, the optimal decision for the final period is still given by (7.64). But we rewrite the result in our new notation as

$$w^*_1(y_1) = (\lambda_1 - y_1)H(\lambda_1 - y_1), \qquad \lambda_1 \equiv \frac{p - c_0}{p + c_1}(b - a), \tag{12.3}$$

where H is the threshold function defined by (7.3), and where p_1 is the selling price and c_1 is the holding cost per item. Furthermore, the maximum reward from the final period is still given by (7.66). But we rewrite the result in our new notation as

$$
\begin{aligned}
\psi_1(y_1) &= J_1(w^*_1(y_1)) \\
&= c_0 y_1 + \frac{1}{2}(p - c_0)(a + \lambda_1) \qquad \text{if } y_1 \leq \lambda_1 \\
&= py_1 - \frac{1}{2}\frac{p + c_1}{b - a}(y_1 - a)^2 \qquad \text{if } y_1 > \lambda_1.
\end{aligned}
\tag{12.4}
$$

We will now derive a recurrence scheme that enables us to complete formulation of the last k steps of an optimal policy if the last $k - 1$ steps have already been formulated. Suppose that $\{w^*_1, w^*_2, \ldots, w^*_{k-1}\}$ has been determined, so that the maximum reward from $N - k + 1 < t < N$, namely, $\psi_{k-1}(y_{k-1})$, is known as a function of y_{k-1}. Consider our buyer at time $t = N - k$, looking forward to $t = N - k + 1$. Although she will always do what is optimal, the surplus y_{k-1} is unknown to her at time $N - k$; only its distribution is known. In other words, when viewed from $t = N - k$, surplus at the later time $t = N - k + 1$ is, not y_{k-1}, but rather the random variable Y_{k-1}. Because $\psi_{k-1}(Y_{k-1})$ is not a reward but a utility, the optimal reward from $N - k + 1 < t < N$, *when viewed from* $t = N - k$, is the expected value $E[\psi_{k-1}(Y_{k-1})]$. As in Section 7.5, we will refer to this as *perceived optimal reward*.

A further complication now arises. The true worth of profits that accrue at different times will be affected by the discount rate (Section 3.4). Let $\bar{\delta}$ be the (real) interest rate per period: assume that each dollar deposited in a bank at time $t = j$ would be worth $1 + \bar{\delta}$ dollars at time $t = j + 1$. Let us assume that prices and costs are assessed at the beginning of each period and are constant in real terms. Then the true worth of the perceived optimal reward at time $N - k$ is, not $E[\psi_{k-1}(Y_{k-1})]$, but rather its present value

---

[2]In Section 7.5 we assumed that there was no surplus stock at time $t = 0$, i.e., $x_0 = 0$; see (7.54). In our new notation, this is equivalent to $y_2 = 0$. Now, for horizons greater than 2, y_2 will not in general be zero, but the effect of nonzero x_0 (or y_2) on the analysis of Section 7.5 is a minor one. In equation (7.67), u_0 is replaced by $u_0 + x_0$, and in equation (7.68), u_0 is replaced by $u_0 + x_0$ and then px_0 is added to the entire expression. You can confirm this by setting $\bar{\delta} = 0$ in (12.10) and (12.11).

$(1 + \bar{\delta})^{-1} E[\psi_{k-1}(Y_{k-1})]$. We will refer to this as the *discounted, perceived optimal reward*.

The reward from $N - k < t < N$ is the sum of the reward from $N - k < t < N - k + 1$ and the reward from $N - k + 1 < t < N$. The former is obtained directly from (7.62) but rewritten in our new notation as

$$S_k(w_k) = p(w_k + y_k) - c_0 w_k - \frac{p + c_1}{2(b - a)}(w_k + y_k - a)^2. \tag{12.5}$$

An expression for the latter has just been derived. Thus the true worth of reward from the interval $N - k < t < N$, as perceived at time $t = N - k$, is

True worth of reward from current decision		True worth of reward from next period		Discounted, perceived optimal reward from next decision until horizon	
	$=$		$+$		(12.6a)
$J_k(w_k)$	$=$	$S_k(w_k)$	$+$	$\dfrac{E[\psi_{k-1}(Y_{k-1})]}{1 + \bar{\delta}}.$	(12.6b)

Our buyer will choose $w_k = w_k^*(y_k)$ so as to maximize this quantity over $0 \leq w_k \leq b - y_k$. The associated maximum reward is $\psi_k(y_k) = J_k(w_k^*(y_k))$. Hence, defining the reward from $t \geq N$ to be zero, we have the complete recurrence scheme

$$\psi_0(y_0) = 0, \tag{12.7}$$

$$\psi_k(y_k) = \max_{0 \leq w_k \leq b - y_k} \left\{ S_k(w_k) + \frac{E[\psi_{k-1}(Y_{k-1})]}{1 + \bar{\delta}} \right\}, \qquad k = 1, \ldots, N; \tag{12.8}$$

and $w_k^*(y_k)$ is the w_k that maximizes (12.6b). Remember that we are attempting to maximize profits from a particular commodity sold by a retailer. Thus even if she stays in business after $t = N$, it is not unreasonable to assume (12.7), and any salvage value associated with surplus stock y_0 at time N could be incorporated into the final period as in Section 9.1. Because we shall eventually take the limit $N \to \infty$ in the following analysis, however, its results would not be altered by a salvage value.

It is important to realize that the relative simplicity of recursion (12.8) ultimately derives from the knowledge that a rational decision maker will proceed optimally at every stage of a decision making process. From a mathematical point of view, this means that the expected value in (12.6) is computed with respect only to the *possibilities arising at the "current" time*, $t = N - k$, i.e., with respect to the sample space of a single random variable. Possibilities arising at earlier times are not yet relevant; possibilities arising at later times have all been incorporated into the perceived optimal reward (which has been possible precisely because we can rely on the decision maker to have proceeded optimally at all later times). Were this not the case, the kth decision from the horizon would have to recognize explicitly the variability associated with k random variables, not simply one. This would mean computing expected values with respect to the joint sample space of all k variables, a formidable task in itself; followed by minimization of k decision variables, a nonlinear programming as opposed to a calculus problem.

Recursion (12.8) is a special case of what has come to be known as a *dynamic programming* recursion. Dynamic programming is a generic description for any procedure, not necessarily of the form (12.6a), by which decision problems of increasing complexity are imbedded in one another in such a way that the optimal policy can be formulated in the direction of increasing complexity but implemented in the opposite direction. Thus with dynamic programming there is a sense in which you are bound to get more than you need. You cannot, for example, formulate the first step in a 26-period policy without knowing the complete 25-period policy, too.

Let us now continue with the problem at hand. The probability distribution of the random variable Y_k is easily deduced from (12.2) and the fact that demand for each period is uniformly distributed between a and b. Indeed one simply uses (12.1) to convert (7.55)–(7.57) to our new notation (Exercise 12.1). Then, by analogy with (7.67), we readily find that

$$E[\psi_{k-1}(Y_{k-1})] = \psi_{k-1}(0)\frac{b - w_k - y_k}{b - a} + \frac{1}{b - a}\int_0^{w_k+y_k-a} \psi_{k-1}(y)dy; \quad (12.9)$$

see Exercise 12.2. On setting $k = 2$ in (12.6) and (12.9) and on using (12.4), we find (Exercise 12.3) that $J_2(w_2)$ is given by

$$J_2(w_2) = (p - c_0)\left\{w_2 + \frac{a + \lambda_1}{2(1 + \delta)}\right\} + py_2$$
$$- \frac{\{(p + c_1)(1 + \delta) - c_0\}(w_2 + y_2 - a)^2}{2(1 + \delta)(b - a)}, \quad (12.10)$$

if $w_2 \le a + \lambda_1 - y_2$, and

$$J_2(w_2) = (p - c_0)\left\{w_2 + \frac{(a + \lambda_1)(b - w_2 - y_2) + a\lambda_1}{2(1 + \delta)(b - a)}\right\} + py_2$$
$$- \frac{(\delta p + (1 + \delta)c_1)(w_2 + y_2 - a)^2}{2(1 + \delta)(b - a)}$$
$$- \frac{(p + c_1)\{(w_2 + y_2 - 2a)^3 - (\lambda_1 - a)^3\}}{6(b - a)^2(1 + \delta)}, \quad (12.11)$$

if $w_2 > a + \lambda_1 - y_2$. Note that this reduces to (7.68) if w_2 is replaced by u_0 and y_2 set equal to zero.

The expression given in (12.10) and (12.11) is a complicated one, unless we can be sure that $w_2 + y_2 < \lambda_1 + a$. This would certainly be true if

$$\lambda_1 + a > b; \quad (12.12)$$

which therefore, for the time being, assume. This condition would hold, for example, if we took $p = 45$, $c_0 = 25$, $c_1 = 5$, $a = 500$, and $b = 1000$; but it is anyhow inessential to the main conclusion of this section, namely, (12.19) below. In these circumstances, it is straightforward to show that the maximum of $J_2(w_2)$ occurs at $w_2 = w_2^*$, where

$$w_2^*(y_2) = (\lambda_2 - y_2)H(\lambda_2 - y_2), \tag{12.13}$$

H is the threshold function defined by (7.3), and λ_2 is defined by

$$\lambda_2 = \lambda_1 + \frac{c_0(p - c_0)(b - a)}{(p + c_1)\{(1 + \delta)(p + c_1) - c_0\}}. \tag{12.14}$$

We see that λ_2 is the level above which the stock at time $N - 2$ should not be raised. It is therefore also the optimal initial order for the two-period problem, when (12.12) is satisfied. Notice that (for reasons discussed in Exercises 7.15 and 7.16) λ_2 is always greater than λ_1 but is a decreasing function of the interest rate. On substituting (12.13) into (12.10), we obtain (verify)

$$\psi_2(y_2) = c_0 y_2 + \frac{(p - c_0)\{a + \lambda_1 + (1 + \delta)(a + \lambda_2)\}}{2(1 + \delta)}, \qquad \text{if } y_2 \leq \lambda_2;$$

$$\psi_2(y_2) = \frac{(p - c_0)(a + \lambda_1)}{2(1 + \delta)} + py_2 - \frac{(p - c_0)(y_2 - a)^2}{2(\lambda_2 - a)}, \qquad \text{if } y_2 > \lambda_2. \tag{12.15}$$

This expression can now be substituted back into (12.8) and the recursion continued. In view of (12.12), you can easily verify (Exercise 12.4) that, for all $k \geq 2$,

$$E[\psi_{k-1}(Y_{k-1})] = \psi_{k-1}(0) + \frac{c_0(w_k + y_k - a)^2}{2(b - a)} \tag{12.16}$$

and

$$\psi_k(y_k) = \max_{0 \leq w_k \leq b - y_k} \left\{ \frac{\psi_{k-1}(0)}{1 + \delta} + py_k + (p - c_0)\left[w_k - \frac{(w_k + y_k - a)^2}{2(\lambda_2 - a)} \right] \right\}. \tag{12.17}$$

Terms independent of w_k do not affect the maximization, and so we deduce that

$$w_k^*(y_k) = (\lambda_2 - y_k)H(\lambda_2 - y_k) \tag{12.18}$$

for all $k \geq 2$. In other words, no matter how long the horizon may be, the optimal policy is to keep raising the inventory level to λ_2 at the start of each period unless it already exceeds λ_2; except at the start of the final period, when the level should be raised to λ_1 instead.

In obtaining (12.18), we have assumed that (12.12) is satisfied. More generally, the optimal policy will be of the form

$$w_k^*(y_k) = \begin{cases} (\lambda_k - y_k)H(\lambda_k - y_k), & \text{if } k < k_\infty; \\ (\lambda_\infty - y_k)H(\lambda_\infty - y_k), & \text{if } k \geq k_\infty, \end{cases} \tag{12.19}$$

where λ_1 is given by (12.3) but the remaining parameters k_∞, λ_2, ..., λ_∞ must be determined. We do not prove this result because such a proof would be against the spirit of the text. Moreover, it falls within the scope of more general results, which are mentioned at the end of the section.

If (12.12) is satisfied then, as we have just established, $k_\infty = 2$ and $\lambda_\infty = \lambda_2$ is defined by (12.14). If (12.12) does not hold, however, then the parameters k_∞, λ_2, ..., λ_∞ are best determined numerically. For example, when

$$a = 0, \quad b = 100, \quad p = 45, \quad c_0 = 30, \quad c_1 = 5, \quad \bar{\delta} = \frac{1}{14},$$
$$(12.20)$$

so that the (real) interest rate per period is a little over 7%, we find

$$k_\infty = 3, \qquad \lambda_1 = 30, \qquad \lambda_2 = 59, \qquad \lambda_\infty = 68, \qquad (12.21)$$

on assuming that the quantity ordered must be an integer.[3]

Thus the optimal policy is to prevent the inventory at the start of each period from falling below λ_∞, except for the last $k_\infty - 1$ periods. Even if the buyer's horizon were 999 periods, she would ensure an inventory of (at least) λ_∞ for the first $1000 - k_\infty$ of them. Accordingly, because k_∞ appears to be small in practice, it seems reasonable to conjecture that the optimal policy for an infinite horizon is the *stationary* policy

$$\text{Order } u^*(x) = \begin{cases} \lambda_\infty - x, & \text{if } x \le \lambda_\infty; \\ 0, & \text{if } x > \lambda_\infty, \end{cases} \qquad (12.22)$$

where x is the surplus stock at the end of any period. The optimality of this policy can be established quite rigorously, though we shall not do so here.

A more general policy than (12.22) would be:

If $x < \lambda$ Then raise stock level to Γ;

If $x \ge \lambda$ Then do not order, $\qquad (12.23)$

where λ and Γ are parameters to be determined. Note that (12.22) is the special case of (12.23) for which $\lambda = \lambda_\infty = \Gamma$. Ordering policies of the form (12.23) are widely used in business and industry, where they are known as *two-bin* policies. The additional parameter Γ is necessary to deal with more general cost functions than we have assumed in this section. Again, provided demand has the same distribution within each period, the optimality of two-bin policies can be established quite rigorously for much more general cost functions than we have employed. See, for example, Whittle (1982, p. 197). Even more general policies are discussed in the growing literature on inventory theory. As with queuing theory (Chapter 6), however, the technical difficulties associated with the derivations of such more general policies far outweigh the additional contribution they make

---

[3]See Supplementary Note 12.1.

to an understanding of the modelling process. Therefore, we do not pursue them; but if you are interested in pursuing this matter for yourself, then see Chapter 13 of Whittle (1982).

We obtained a stationary policy for an infinite horizon because we assumed that prices were constant (in real terms) and that demand Z_j, for the interval $j < t < j+1$, had the same distribution for all values of j. It should be clear, however, that the recursion (12.8) would generate an optimal policy for a finite horizon even if a, b, p, c_0, c_1, and $\bar{\delta}$ were all dependent on j. The arithmetic might become rather tedious, and we would certainly wish to use a computer; but even in such j-dependent cases, the optimal policy could still be determined by dynamic programming. It is principally because of this flexibility and robustness that dynamic programming has enjoyed such widespread acceptance as a decision maker's tool.

We conclude this section by noting that many of the actions taken by individuals, whether in their personal or professional lives, derive from decision rules of the form:

$$\begin{array}{ll} \text{If VARIABLE exceeds THRESHOLD then} & \text{DO} \quad \text{take action;} \\ \text{If VARIABLE is below THRESHOLD then DO NOT take action.} \end{array} \quad (12.24)$$

The cancellation of a sporting event because of the weather, for example, could fall into this category, the variable being precipitation, or degree of frost, or some measure of their consequences. The decision rule (12.19) falls into this category, too, when reformulated as

$$\begin{array}{ll} \text{If } y_k \geq \lambda_k \text{ then} & \text{DO} \quad \text{not order;} \\ \text{if } y_k < \lambda_k \text{ then DO NOT not order.} \end{array} \quad (12.25)$$

I'm sure you can think of many others.

Now, the form of the optimal decision rule (12.19) was something that arose naturally out of the mathematical model and did not need to be assumed. If we had assumed the optimal decision rule to have the form (12.24), however, then formulating the policy would have reduced to determining the threshold sequence $\{\lambda_k\}$. This is reminiscent of Chapter 3, where we solved a number of optimal control problems by assuming the set of potentially optimal controls to be a very limited one. In practice, human action is bound by so many political constraints that options are often very limited, and we are often justified in assuming that (12.24) is the only feasible basis for action. It also makes the mathematics easier. The following application of dynamic programming falls into this category.

12.2 The Interviewer's Dilemma. An Optimal Stopping Problem

A personnel manager is trying to fill a vacancy within her company. The vacancy in question arose very suddenly and must also be quickly filled, quite possibly immediately. To be specific, suppose that a typist became ill overnight and will be absent from work for several days. During that time, much typing needs to be done. It cannot await the return of the typist. Therefore, first thing in the morning, our personnel manager (or P.M.) phones all the local secretarial agencies and asks to interview typists. In the following, we'll talk as though candidates for the vacancy are interviewed in person, though in practice, interviews might be conducted by telephone.

At such an early hour of the morning, there is no shortage of typists seeking work, especially in a large city, where "temps" abound. The P.M. expects to interview several, and would naturally like to appoint the best. She is thus reluctant to appoint too early, lest a later candidate might be better qualified. On the other hand, good typists are so scarce that, if not appointed immediately after interview, they cease to be candidates, being quickly snapped up by other companies. Therefore, immediately after each interview, the P.M. must decide whether or not to appoint the latest candidate; and that decision is irrevocable. How can we help her to solve this dilemma?

Let's suppose that each candidate receives a score between 0 and 1 as a result of her interview, 1 corresponding to the best typist available and 0 to the worst. Then the P.M. would like to appoint the candidate with the highest score; her utility is the score of the appointed candidate. Unfortunately, the scores of the candidates are unknown until (immediately) after interview. We'll assume, however, that the P.M. is sufficiently experienced as to be able to attach a probability distribution to each candidate's score. This makes utility a random variable (for which an explicit expression will shortly be derived). Utility can no longer itself be maximized, and so we must instead maximize a reward, i.e., some statistic associated with utility. As usual, the statistic we choose is expected value, which, for reasons similar to those set forth at the beginning of Section 8.8, is a sensible choice if the interviewer's dilemma is a recurring one. Thus the P.M. wishes to formulate a policy that maximizes the expected value of the appointed candidate's score, i.e., maximizes reward.

With this in mind, let N denote the number of typists available for interview. In this section we'll assume that N is known, but we'll relax that assumption in Section 12.3. The number of candidates is analogous to the horizon in the previous problem. Let the random variable V_k denote utility when k candidates remain, immediately prior to interviewing candidate $N - k + 1$. Because, by assumption, the first $N - k$ typists have now all been rejected, V_k is the score of the candidate appointed from the k who remain. Thus, when k candidates remain, the reward is $E[V_k]$, where the expected value is computed with respect to possibilities arising at step k of the decision making process. As promised, an explicit expression for V_k, and hence $E[V_k]$, will shortly be derived.

Let y_k denote the score immediately *after* interview of the first of the last k candidates, i.e., of candidate $N - k + 1$. Let Y_k denote the score of the same candidate immediately *before* interview. Then Y_k is a random variable; but y_k is fully determined, being the value Y_k actually took. We'll assume that Y_k is continuously distributed between 0 and 1, with probability density function f_k. Thus

$$\text{Prob}\left(y - \frac{1}{2}\delta y < Y_k \le y + \frac{1}{2}\delta y\right) = f_k(y)\delta y + o(\delta y), \tag{12.26}$$

for $0 < y < 1$. These probability distributions are determined subjectively by dint of the P.M.'s vast experience of interviewing typists.

After interviewing candidate $N - k + 1$, $1 \le k \le N$, the P.M. assigns her a score y_k. If y_k is sufficiently close to 1 then the P.M. will appoint her. If y_k is sufficiently close to zero then the P.M. will not appoint, proceeding instead to interview candidate $N - k + 2$. Let us therefore assume that there exists a threshold value λ_k, $0 \le \lambda_k \le 1$, above which the P.M. will accept the candidate and below which she will reject the candidate. The P.M.'s decision rule is therefore of the form (12.24):

If $y_k \ge \lambda_k$ then   DO   appoint candidate $N - k + 1$;

if $y_k < \lambda_k$ then DO NOT appoint candidate $N - k + 1$. $\hspace{1cm}$ (12.27)

Because λ_k can be any number between 0 and 1, there are many sequences $\{\lambda_k\}$ that could be used in conjunction with a decision rule of the form (12.26). Of all possible sequences, however, the P.M. should use the sequence, $\{\lambda_k^*\}$, that will maximize reward. Thus λ_k is a decision variable on which reward depends, and it is appropriate to write the reward from the last k candidates as $J_k(\lambda_k)$. (Thus $J_k(\lambda_k) = E[V_k]$.) Let us define ψ_k to be the optimal reward from the last k candidates, as observed *immediately prior to interview* $N - k + 1$. Then clearly, $\psi_k = J_k(\lambda_k^*)$.

As in Section 12.1, we will derive a recurrence scheme that enables us to complete formulation of the last k steps of the optimal policy when the last $k - 1$ steps have already been formulated. Imagine that the last $k - 1$ steps have been formulated; i.e., suppose that $\lambda_1^*, \lambda_2^*, \lambda_3^*, \ldots, \lambda_{k-1}^*$ are already known. Then $\psi_{k-1} = J_k(\lambda_{k-1}^*)$ is also known. Moreover, because the performance of candidate $N - k + 1$ can in no way influence the optimal reward from the last $k - 1$ candidates—since ψ_{k-1} does not depend upon y_k—ψ_{k-1} does not become a random variable as we move back one step from $k - 1$ to k in the decision-making process. It follows that ψ_{k-1} is also the *perceived optimal reward* from the last $k - 1$ candidates at the time of interview $N - k + 1$. (Or if you prefer, $E[\psi_{k-1}] = \psi_{k-1}$.)

After candidate $N - k + 1$ has been interviewed, her score is known to be y_k. If $y_k \ge \lambda_k$, then the P.M. will appoint her, as agreed in (12.27). If $y_k < \lambda_k$, however, then the P.M. will not appoint her. Thus, immediately *after* candidate $N - k + 1$ has been interviewed, the score of the appointed candidate when k candidates remain is

$$v_k = \begin{cases} y_k, & \text{if } y_k \geq \lambda_k; \\ \psi_{k-1}, & \text{if } y_k < \lambda_k. \end{cases} \qquad (12.28)$$

Before the interview, however, y_k is unknown. Thus the score of the next candidate is the random variable Y_k. Hence the P.M.'s utility is the random variable V_k, where

$$V_k \equiv \begin{cases} Y_k, & \text{if } y_k \geq \lambda_k; \\ \psi_{k-1} & \text{if } y_k < \lambda_k. \end{cases} \qquad (12.29)$$

Recall that the reward is then $E[V_k]$, where the expected value is taken with respect to possibilities arising immediately prior to interview $N - k + 1$.

Now, V_k is a mixed random variable. It takes values y between λ_k and 1 with probability given by (12.26); whereas the discrete value ψ_{k-1} is taken with probability

$$\text{Prob}(V_k = \psi_{k-1}) = \text{Prob}(Y_k < \lambda_k) = \int_0^{\lambda_k} f_k(y)dy. \qquad (12.30)$$

Hence, on using (A.32), the P.M. must maximize the reward

$$J_k(\lambda_k) \equiv E[V_k] = \psi_{k-1} \cdot \text{Prob}(V_k = \psi_{k-1}) + \int_{\lambda_k}^1 yf_k(y)dy$$

$$= \psi_{k-1} \int_0^{\lambda_k} f_k(y)dy + \int_{\lambda_k}^1 yf_k(y)dy. \qquad (12.31)$$

On using Leibnitz's rule from the calculus to differentiate (12.31) with respect to λ_k, we find that

$$J_k'(\lambda_k) = \psi_{k-1}f_k(\lambda_k) - \lambda_k f_k(\lambda_k) \qquad (12.32)$$

and $J_k''(\psi_{k-1}) = -f_k(\psi_{k-1}) < 0$. Hence (12.31) has a maximum at $\lambda_k = \lambda_k^*$, where

$$\lambda_k^* = \psi_{k-1}. \qquad (12.33)$$

Thus, on using (12.31) and (12.33), the optimal reward when k candidates remain is

$$\psi_k \equiv J_k(\lambda_k^*) = \psi_{k-1} \int_0^{\psi_{k-1}} f_k(y)dy + \int_{\psi_{k-1}}^1 yf_k(y)dy. \qquad (12.34)$$

Because f_k is assumed to be known, the right-hand side of (12.34) is a function of ψ_{k-1} alone. Let's assume that the last candidate must be appointed, however appalling her typing skills, i.e., that $\lambda_1^* = 0$. Hence, on using (12.33), we have

$$\psi_0 = 0. \qquad (12.35)$$

The sequence $\{\psi_k\}$ can be generated by recursion from (12.33)–(12.35).

As an illustration, suppose that the quality of a candidate at any stage is equally likely to lie anywhere between the best and worst available. Thus Y_k does not depend upon k and is uniformly distributed between 0 and 1; i.e., $f_k(y) = 1$, $0 < y < 1$, for all values of k. Then you can easily verify that the sequence $\{\psi_k\}$ is generated by the recursion

$$\psi_0 = 0,$$
$$\psi_k = \frac{1}{2}\left(1 + \psi_{k-1}^2\right), \qquad k = 1, 2, \ldots, N. \qquad (12.36)$$

The threshold sequence $\{\lambda_k^*\}$ is deduced from (12.33).

The resultant policy for 20 interviews is shown in Table 12.1. You can easily verify this table by using a calculator. According to this policy, the next candidate when, for example, 10 candidates remain—or, which is exactly the same thing, the first candidate in a field of ten candidates—should not be appointed unless her score is at least $\lambda_{10} = \psi_9 = 0.85$. The table illustrates the sense in which dynamic programming gives more than is absolutely necessary, because the optimal policy for N candidates contains the optimal policy for any number of candidates between 1 and $N - 1$. Notice also that (12.36) implies $\psi_k - \psi_{k-1} = \frac{1}{2}(1 - \psi_{k-1})^2 > 0$. Thus ψ_k increases monotonically toward the limit 1 as $k \to \infty$, as confirmed by Table 12.1 and entirely in agreement with what intuition would suggest: the higher the number of candidates, the higher the quality to be expected from the candidate appointed. Conversely, the smaller the field, the less choosy the P.M. can afford to be, because λ_k decreases as k decreases. But it may

Table 12.1 Optimal thresholds and optimal values of the expected score of the candidate appointed for a field of (up to) 20 candidates, when scores of all candidates are uniformly distributed between 0 and 1.

k	λ_k^*	ψ_k	k	ψ_k
1	0	0.5	11	0.871
2	0.5	0.625	12	0.879
3	0.625	0.695	13	0.886
4	0.695	0.742	14	0.893
5	$\cdots$	0.775	15	0.899
6		0.800	16	0.904
7		0.820	17	0.908
8		0.836	18	0.913
9		0.850	19	0.916
10		0.861	20	0.920

not be very realistic to assume that $f_k(y) = 1$, $0 < y < 1$, for all k. Why? Can you make a more realistic assumption? See Exercises 12.5 and 12.6.

In practice, N may not be known, but its value becomes irrelevant if (i) the field of candidates is very large and (ii) Y_k is independent of k. If (i) is satisfied, then the P.M. can define an *effective* field length N_e, which is simply the maximum number of candidates she might have time to interview before the deadline for filling the position. Because the field is so large, she cannot possibly interview all candidates. On the other hand, even if the P.M. knew probability distributions for each value of k in the actual field of candidates, she could not use this information because she would not know by how much the true number of candidates exceeded the effective number; hence the need for (ii). As suggested by Exercise 12.6, however, the larger the field of candidates, the more realistic it is to assume (ii). In conclusion, therefore, the model of the present section may suffice if either N is known or the field of candidates is so large as to make its value irrelevant. There are circumstances, however, in which neither is true. To these we now turn.

12.3   A Faculty Hiring Model

Let's now examine the other extreme of the interviewer's problem, where the field of candidates is so small that it is not even certain that there will be another candidate. This would be true if the vacant post required highly specialized knowledge; such as might be required by, say, a professor of applied mathematics. The uncertainty in the length of the field adds a new ingredient to the interviewing problem; yet it still has several features in common with the problem we have just discussed. In particular, the interviewer can still assign scores between 0 and 1 to candidates for the post. Furthermore, she must use a decision rule of the form (12.27), because candidates who are not offered the post at the time of interview will be snapped up elsewhere. We therefore expect that we can adapt the model of the previous section to deal with the new circumstances.

For the sake of definiteness, let's suppose that there is a vacancy for the post of Professor of Applied Mathematics, with special responsibility for supercomputing. This post requires highly specialized knowledge of several fields, which few individuals possess; but there have been a few applications, and more may continue to trickle in. On the other hand, the field of candidates might just dry up. After consultation with colleagues, the chairperson of the department has estimated there is probability q that the current candidate will be the last one available. Hence $p = 1 - q$ is the probability that there will be at least one more candidate. The number of candidates is not known for certain but does have a probability distribution, according to which the expected value of the number of candidates is $1/q$ (Exercise 12.8). We must try to adapt the model of Section 12.2 to incorporate these new conditions.

As in the previous section, we will derive a recurrence scheme that enables us to complete formulation of the last k steps of the optimal policy if the last $k - 1$ steps have already been formulated. In the circumstances described above, however, the chairperson cannot possibly know the total number of candidates. Hence she does not know the value of k that corresponds to any particular decision. Nevertheless, taking our cue from Section 12.1, where the dynamic programming recursion converged so quickly to a stationary policy that the number of decisions remaining became irrelevant, we will hope that the recursion we are about to derive converges so rapidly that k becomes irrelevant. There is, however, a price to be paid. Just as the stationary policy in Section 12.1 arose because demand was assumed to have the same distribution in every period, so we must now assume that scores have the same distribution for every candidate. In other words, f_k is independent of k.

Until we reach the point where the value of k becomes irrelevant, we will continue to speak of the remaining candidates as if their number were known for certain. Moreover, we shall refer to the first of the last k candidates as the "current" candidate. Let us therefore imagine that the optimal policy for the last $k - 1$ (potential) candidates has already been formulated; i.e., suppose that the optimal thresholds $\lambda_1^*, \lambda_2^*, \lambda_3^*, \ldots, \lambda_{k-1}^*$ for decision rule (12.27) are already known. Then the optimal reward from the last $k - 1$ candidates, ψ_{k-1}, is already known.

Consider, now, the chairperson's viewpoint *immediately after* she has learned whether or not the "current" candidate was the last. This is clearly a later time than the current candidate's time of interview. What is the reward from the remaining candidates? If there is at least one more candidate, then it is ψ_{k-1}, because the chairperson is a rational decision maker who will always proceed optimally at future times. But there may not be another candidate, in which case the reward from the remaining candidates is zero. Thus, from the chairperson's current viewpoint, the reward from the last $k - 1$ candidates is

$$c_{k-1} = \begin{cases} 0, & \text{if no further candidate;} \\ \psi_{k-1}, & \text{if at least one more candidate.} \end{cases}$$

From the chairperson's current viewpoint, this is not a random variable. As we move backward in time, however, to just *immediately before* the chairperson learns whether or not there is to be another candidate, then reward from the last $k - 1$ candidates becomes the random variable

$$C_{k-1} = \begin{cases} 0, & \text{if no further candidate;} \\ \psi_{k-1}, & \text{if at least one more candidate.} \end{cases} \tag{12.37}$$

This utility cannot itself be maximized, but we can still maximize its expected value. In the terminology we have chosen to adopt, the perceived optimal reward from the last $k - 1$ candidates, at slightly earlier times, is not (12.37) but its expected value $E[C_{k-1}]$. This expectation is taken with respect to all possibilities arising at that time. But the only possibilities in the sample space of C_{k-1} are whether or not there is another candidate. Hence, on using (A.31) we get

$$E[C_{k-1}] = 0 \cdot \text{Prob}(C_{k-1} = 0) + \psi_{k-1} \cdot \text{Prob}(C_{k-1} = \psi_{k-1})$$
$$= 0 \cdot q + \psi_{k-1} \cdot p = p\psi_{k-1}. \tag{12.38}$$

Let us now move even further backward in time to immediately before the current candidate's interview. Because from (12.38) the perceived optimal reward from the last $k - 1$ candidates is now $p\psi_{k-1}$ instead of ψ_{k-1}, the chairperson's utility is not (12.29) but

$$V_k \equiv \begin{cases} Y_k, & \text{if } y_k \geq \lambda_k; \\ p\psi_{k-1}, & \text{if } y_k < \lambda_k. \end{cases} \tag{12.39}$$

It now follows easily that (12.31) is replaced by

$$J_k(\lambda_k) \equiv E[V_k] = p\psi_{k-1} \int_0^{\lambda_k} f_k(y)dy + \int_{\lambda_k}^1 y f_k(y)dy. \tag{12.40}$$

You can readily show that (12.40) has a maximum at $\lambda_k = \lambda_k^*$, where

$$\lambda_k^* = p\psi_{k-1}. \tag{12.41}$$

The sequence $\{\psi_k\}$ is then easily obtained from the recurrence relation

$$\psi_0 = 0,$$
$$\psi_k = J_k(\lambda_k^*) = J_k(p\psi_{k-1}), \qquad k = 1, 2, \ldots. \tag{12.42}$$

The sequence $\{\lambda_k^*\}$ follows immediately from (12.41).

For the sake of illustration, again suppose that the score of each candidate is uniformly distributed between 0 and 1. Then you can show (Exercise 12.9) that (12.42) becomes the recursion

$$\psi_0 = 0,$$
$$\psi_k = \frac{1}{2}\left(1 + p^2\psi_{k-1}^2\right), \qquad k = 1, 2, \ldots. \tag{12.43}$$

The sequence $\{\psi_k\}$ defined by (12.43) converges to ψ_∞ as $k \to \infty$, where ψ_∞ is the only root between 0 and 1 of the equation

$$\psi = \frac{1}{2}\left(1 + p^2\psi^2\right). \tag{12.44}$$

To see this, observe that (12.43) and (12.44) imply

$$|\psi_k - \psi_\infty| = \frac{1}{2}p^2|\psi_{k-1} - \psi_\infty||\psi_{k-1} - \psi_\infty| \leq p^2|\psi_{k-1} - \psi_\infty|. \tag{12.45}$$

Successive application of this inequality yields $|\psi_k - \psi_\infty| \leq p^{2k}|\psi_0 - \psi_\infty|$. Hence $|\psi_k - \psi_\infty|$ tends to zero as k tends to infinity because $p^2 < 1$; and the smaller the value of p, the more rapid the convergence. At the same time, on using (12.41), λ_k^* converges to $\lambda_\infty = p\psi_\infty$. From (12.44) we obtain explicitly

$$\psi_\infty(p) = \frac{1}{1 + \sqrt{1 + p^2}}, \tag{12.46}$$

$$\lambda_\infty(p) = \frac{p}{1 + \sqrt{1 + p^2}}. \tag{12.47}$$

Convergence of the recurrence relation (12.42) for other distributions, such as the family considered Exercise 12.5, is established along similar lines.

In practice, we can usually assume that the remaining field of potential candidates is large enough for ψ_k and λ_k^* to have converged already to $\psi_\infty(p)$ and $\lambda_\infty(p)$, respectively. The chairperson can therefore apply the following stationary policy, where y denotes the score of the candidate most recently interviewed:

> If $y \geq \lambda_\infty(p)$ then   DO   appoint the candidate;
>
> if $y < \lambda_\infty(p)$ then DO NOT appoint the candidate. (12.48)

We drop the subscript k because candidates' scores are assumed to have identical distributions.

The graphs of $\psi_\infty = \psi_\infty(p)$ and $\lambda_\infty = \lambda_\infty(p)$ are plotted in Fig. 12.1 for $0 < p < 1$; ψ_∞ as the dashed curve labelled U and λ_∞ as the solid curve. Notice that the chairperson can expect little more than $E[Y]$ from the optimal policy unless p is very close to 1. To demonstrate that this is not a consequence of assuming a uniform distribution for the candidate's score Y, the analysis of this section is repeated for two other distributions, labelled S and T in Fig. 12.1 and Fig. 12.2 (to understand how these graphs were drawn, see Exercise 12.10). Distribution S is a symmetrical triangular distribution, corresponding to $a = 1/2$ in Exercise 12.5. Distribution T is an asymmetrical triangular distribution, weighted toward high quality, and corresponds to $a = 1$ in Exercise 12.5. You can see from the diagram that distribution S yields the bleakest prospects and distribution T the brightest, these having respectively the lowest and highest probabilities of scores close to 1. But it is distressing to observe that in all three cases the optimal policy implies accepting a below-average candidate unless there is less than 20% chance that the current candidate will be the last one available for interview. Even then the prospects are bleak.

To appreciate this more fully, let's consider a special case. Suppose that distribution U applies and that the chance of the current interview being the last is only 10%. Thus $q = 0.1$, $p = 0.9$, and $\psi_\infty = \psi_\infty(0.9) = 0.696$. Now, for these values of p and q, the expected value of the number of interviews is 10, from Exercise 12.8. But if the chairperson *knew* that there

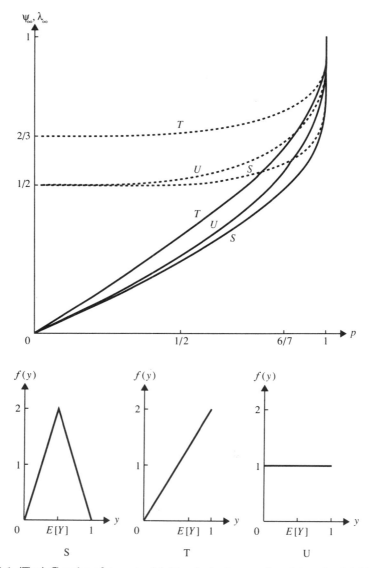

Fig. 12.1 (Top) Graphs of $\psi = \psi_\infty(p)$ (the dashed curves) and $\lambda = \lambda_\infty(p)$ (the solid curves) for the three merit distributions sketched in Fig. 12.2.
Fig. 12.2 (Bottom) Probability density functions of three possible distributions of merit between 0 and 1.

would be 10 candidates, then the maximum expected value of the score of the appointed candidate would, from Table 12.1, be 0.861. There is therefore a high price to be paid for any uncertainty in the length of the field of candidates.

Like the stationary buying policy of Section 12.1, the stationary interviewing policy of this section has the attractive feature that it is particularly easy to implement, the critical score being the same for each interview. It also has the attractive feature of flexibility, as discussed in Section 7.8. Curves like those in Fig. 12.1 could easily be drawn by computer for any distribution of merit, and so the model could be used by each of two college deans who have similar personal preferences (appoint the best candidate) but differ in their opinions of the quality of the field. Moreover, the graphs in Fig. 12.1 afford them the luxury of knowing the optimal policy at a glance, for an arbitrary value of p. On the other hand, it has the unattractive feature that the expected value of the quality of the candidate appointed will be significantly lower than if the number of candidates were known for certain. Thus an interviewer should strive to apply the policy of the previous section whenever possible. Now try Exercise 12.11.

12.4 The Motorist's Dilemma. Choosing the Optimal Parking Space

A motorist is approaching the central crossroads of an old-fashioned town, one that underwent ribbon development in the nineteenth century and hasn't changed much since (Fig. 12.3). She would like to park as close as possible to the central crossroads O, but she knows that the central parking spaces may not be free when she reaches them. She must therefore be prepared to park away from the center. How far from the center should she accept a vacant parking space? In this section, we will attempt to answer that question.

For the sake of definiteness, we will assume that the motorist approaches the crossroads from the west. Because traffic is extremely heavy, she decides that it will be virtually impossible to cross the road for a parking space. She therefore looks for a parking space on only her side of the road. The motorist has undertaken this journey many times before and knows how far each parking space is from the center. Her policy is to take the first available parking space if it is sufficiently close to the center. If she reaches the crossroads without finding a space, however, then she takes the next available space. If we assume that parking spaces are equally likely to be free along any of the four approach roads, then it makes no difference whether the motorist turns right (or left, though the heavy traffic will strongly discourage this) or carries on, if she reaches the crossroads. For the sake of definiteness, however, let's assume that she turns right—south in Fig. 12.3. Thus the qualitative form of the optimal policy is known. Our problem is to determine the optimal interpretation of "sufficiently close."

Let's begin by converting the decision rule to the form (12.24), in which action is taken if a certain variable exceeds a certain threshold. The motorist's potential parking spaces are as labelled in Fig. 12.3. For the

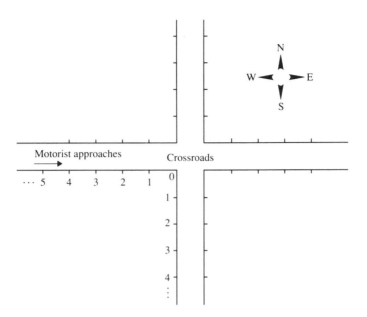

Fig. 12.3 Plan of central crossroads in an old-fashioned town.

sake of simplicity, we will assume that they are equidistant, so that the kth parking space from the center on the western approach road can be assumed to lie a distance k from the center. If they were not equidistant, however, then you could measure distance nonlinearly, as we measured time in Chapters 5 and 10, so that our model could still be applied. Let us agree that parking space k on the western approach road is "sufficiently close" to the center if $k \leq k_c$, where k_c is an arbitrary integer. Then the decision rule can be cast in the form (12.24) by writing

If $k > k_c$ then DO NOT park (whether or not the space is empty);

if $k \leq k_c$ then   DO   park (if the space is empty).

$$(12.49)$$

We will attempt to determine the optimal value of k_c by adapting the model of the previous section.

Let us assume that parking spaces are occupied or unoccupied independently of one another. Let p denote the probability that a parking space is empty; let $q = 1 - p$. We will determine the optimal value of k_c as follows. Let us pretend that the motorist has already set k_c, arbitrarily. Then, for each $k \leq k_c$ along the western approach, she "interviews" parking space k, accepting it if it is empty and rejecting it if it is occupied, as implied by (12.49). Let the random variable V_k denote distance from the center at which the motorist will park, when k spaces remain before O. Thus the sample space of V_k is the set of integers $\{1, 2, \ldots, k\}$, $k \leq k_c$. The motorist wishes to park as close to the center as possible, and so her utility, the quantity she would like to maximize, is $-V_k$. But this is a random variable, which cannot itself be maximized; so we will assume that our motorist agrees instead to maximize its expected value, $E[-V_k]$.[4]This is the same as minimizing $-E[-V_k] \equiv E[V_k]$, so we will refer to V_k as *disutility*. Let $L(k)$ denote $E[V_k]$. As usual, this expected value is computed with respect to possibilities arising at the time of "interview." Because the motorist wishes to minimize it, we will refer to $L(k)$ as her *disreward*, when $k(\leq k_c)$ candidates remain and her strategy is to grab the next available parking space. We will calculate $L(k)$ by recursion. If there is any value of k between 0 and k_c for which $L(k) < L(k_c)$, then k_c has been set too high and should be lowered, because the motorist would then accept parking space k_c even though a nearer one to the center corresponds to a smaller disreward. If there is no value of k between k and k_c for which $L(k) < L(k_c)$, then k_c may have been set too low and may need to be raised, because a further parking space from O might be associated with a lower disreward if the policy of taking the next available space were applied earlier (on the other hand, k_c could have been guessed just exactly right). In practice, the pretense we have just made is not necessary because if we calculate the sequence $\{L(k) : k = 0, 1, \ldots\}$ on the assumption that k_c is at least as great as the current value of k, then $L(k)$ will decrease from $L(0)$ to a minimum $L(k^*)$ at $k = k^*$ and then increase again (see below). Thus k^* is the optimal value of k_c.

To begin our recursion, we must first find $L(0)$. With this in mind, suppose that the motorist reaches O without success, and let the (integer-valued) random variable D denote her parking distance from the center. Then $\text{Prob}(D = j)$ is the probability that the first $j - 1$ spaces are filled and the jth is empty—$q^{j-1}p$—because parking spaces are available or occupied independently of one another. Thus, if she passes O, then her disreward is

$$E[D] = \sum_{j=1}^{\infty} j \cdot \text{Prob}(D = j) = \sum_{j=1}^{\infty} j \cdot q^{j-1} p = \frac{1}{p}, \qquad (12.50)$$

by the same calculation as in Exercise 12.8. We deduce immediately that

$$L(0) = \frac{1}{p}. \qquad (12.51)$$

Now suppose that the motorist is "interviewing" parking space k, where $0 \leq k \leq k_c$. There are two possibilities. Either the space is free, in which case she takes it and her disutility is $V_k = k$, or else the space is occupied in which case her disutility is the perceived disreward from the remaining $k - 1$ parking spaces, i.e., $L(k - 1)$. Hence her disutility is

---

[4]Because she is a frequent visitor to town, this is quite rational; see Section 8.8.

the random variable

$$V_k = \begin{cases} k, & \text{if the space is empty;} \\ L(k-1), & \text{if the space is occupied.} \end{cases} \qquad (12.52)$$

This is a discrete random variable whose expected value over the two possibilities that now arise can be calculated from (A.31). Hence

$$L(k) = E[V_k] = p \cdot k + q \cdot L(k-1). \qquad (12.53)$$

Thus the sequence $\{L(k)\}$ can be calculated from the recursion

$$L(0) = \frac{1}{p}, \qquad L(k) = pk + qL(k-1), \qquad k = 1, 2, \dots. \qquad (12.54)$$

The motorist would like to know the value, k^*, for which the integer-valued function $L(k)$ achieves its minimum.

The values of $L(k)$ for values of k between 0 and 5 are presented in Table 12.2. The table suggests that

$$L(k) = p \sum_{i=1}^{k} i q^{k-i} + \frac{q^k}{p} = pq^k \sum_{i=1}^{k} i \left(\frac{1}{q}\right)^i + \frac{q^k}{p}$$

$$= k - \frac{q}{p} + \frac{q^k(1+q)}{p}, \qquad (12.55)$$

by essentially the same calculation as in Exercise 6.4. The truth of (12.55) can be established by mathematical induction (see Exercise 12.12). Because

$$L(k+1) - L(k) = 1 - (1+q)q^k, \qquad (12.56)$$

we see that L attains its minimum at $k = k^*(p)$, where $k^*(p)$ is the smallest integer exceeding

$$\frac{\ln(2-p)}{-\ln(1-p)} = \frac{\ln(1+q)}{\ln\left(\frac{1}{q}\right)}. \qquad (12.57)$$

The staircase-like function $k = k^*(p)$ is plotted in Fig. 12.4 as the solid curve, with (12.57) the dashed one. Notice that the motorist should hold

Table 12.2   The first few recursions of (12.54).

k	$L(k)$
0	$1/p$
1	$p + q/p$
2	$p(2+q) + q^2/p$
3	$p(3 + 2q + q^2) + q^3/p$
4	$p(4 + 3q + 2q^2 + q^3) + q^4/p$
5	$p(5 + 4q + 3q^2 + 2q^3 + q^4) + q^5/p$

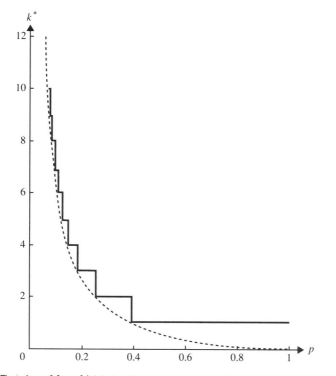

Fig. 12.4 Graphs of $k = k^*(p)$ (solid curve) and (12.57) (dashed curve). A motorist who wishes to park as close as possible to the center of town should not accept an empty parking space if it is more than k^* spaces from the center.

out for the final parking space $k = 1$ unless it is less than about 40% likely to be vacant (the precise percentage is $50\{3 - \sqrt{5}\}$). This is because, in such circumstances, the motorist has a high probability of finding a good consolation parking space on the road going south if the one she really wants is filled. If p is very small, however, then k^* is off the diagram in Fig. 12.4, and the motorist must be prepared to park a long way from town. Naturally, the value of the probability p must be supplied (most likely, subjectively) before the optimal policy can be implemented, but the fact that p can be retained as a parameter is an attractive feature of our model. Variations of it are considered in Exercise 12.13.

12.5   How Should a Bird Select Worms? An Adaptation

Suppose that a bird is out searching for worms. What type of worm should she be looking for—a big, fat, juicy one or a long, thin, skinny one? To be sure, a juicy worm contains more energy (calories) than a skinny one. But what if it takes longer to pull juicy worms out of the ground—perhaps as long as it takes to extract a whole beakful of skinny ones? Even if juicy worms slip out easily, what if they are difficult to find in the first place? Would it be wise, in such circumstances, for the bird to keep turning up her beak at skinny worms? It seems that the matter is getting quite complicated. We shall hardly be able to help the poor bird unless we build a mathematical model.

We will assume throughout that worms are not found in patches. Thus the bird must "interview" worms separately and accept or reject each candidate on its own merits, rather than for its proximity to others in the field. Moreover, for the sake of simplicity, we will assume that the bird can choose between just two types of worm, say $k = 1$ and $k = 2$, and that one of these types is more "profitable" to eat than the other. But what do we mean by profitable? To answer this question, let's suppose that two different meals have presented themselves to the bird, who must choose between them. The first meal consists of a single worm. It contains E calories of energy and takes h minutes to extract and eat. Extraction and eating involve "handling" the worm, and so we will call h the handling time. The second meal consists of two worms, each containing only $E/2$ calories of energy, but each requiring only $h/2$ minutes to handle. Because the choices on this menu offer the same total energy value E for the same total handling time h, how could the bird prefer one to the other? Wouldn't it be better to regard them as equally profitable? It thus appears that a suitable measure of a worm's profitability is its ratio of energy value to handling time, E/h.

We have already assumed that one of the bird's two types of worm is more profitable than the other. Without loss of generality, let $k = 1$ be the more profitable type; hence $k = 2$ is the less profitable one. Moreover, let a worm of type k contain E_k calories and require h_k seconds to handle; $k = 1, 2$. Then

$$\frac{E_1}{h_1} > \frac{E_2}{h_2}. \tag{12.58}$$

Bear in mind that (12.58) does not imply $E_1 > E_2$, because h_2 could be much greater than h_1. To put it another way, if the skinny worms are much easier to handle than the juicy ones, then they may also be more profitable.[5]

Let's suppose that, *if* the bird flew around for a whole minute without eating a thing (no matter how many worms she sighted) then she would spot W_1 worms of type 1 and W_2 worms of type 2. Then W_1 and W_2 are random variables. Nevertheless, over many such minutes, the values realized by W_1 and W_2 would tend to average out at λ_1 and λ_2, respectively, where we define $\lambda_1 = E[W_1]$, $\lambda_2 = E[W_2]$, and where E denotes expected

---

[5]Nevertheless, E_1 did exceed E_2 in the experiment by Krebs et al. (1977) discussed below.

value. Thus λ_1 is the number of type-1 worms that the bird would spot during an average minute if she refused to eat, and similarly for λ_2. We will refer to λ_k as the rate of encounter with worms of type k; $k = 1$, 2.

But this bird is not in the habit of refusing every worm she meets. Rather, she will extract and eat at least some worms. To be precise, let's suppose that she eats u_k of every λ_k worms of type k she encounters, $k = 1$, 2, i.e., that u_k/λ_k is the fraction of type-k worms accepted. Then, certainly,

$$0 \le u_1 \le \lambda_1, \qquad 0 \le u_2 \le \lambda_2. \tag{12.59}$$

We will refer to the pair of values (u_1, u_2) as a decision and to the u_1-u_2 plane as decision space. Then feasible decisions must occupy the rectangle defined by(12.59). It is shaded in Fig. 12.5.

We can also think of u_k as the bird's average acceptance rate per minute of worms of type k. We must be very clear, however, about what we actually mean by this. We do not mean that the bird eats u_k worms of type k during an average minute. That would be true if worms took zero time to extract and eat. But u_1 worms of type 1 and u_2 of type 2 will take $h_1 u_1 + h_2 u_2$ minutes to extract and eat. They will also take a minute to find, on average. Thus, again on average, a total of $1 + h_1 u_1 + h_2 u_2$ minutes must be devoted to their demise. Because they will contribute $E_1 u_1 + E_2 u_2$ calories to the bird's reserves of energy, the average rate at which she takes up energy is Φ, where we define

$$\Phi(u_1, u_2) = \frac{E_1 u_1 + E_2 u_2}{1 + h_1 u_1 + h_2 u_2}. \tag{12.60}$$

For the sake of brevity, we will refer to Φ henceforward as the *energy rate*. It is now clear that u_k is the number of type-k worms accepted during an average minute of time spent actually searching (as opposed to both searching and handling). Moreover, the profitability E_k/h_k can be regarded as the energy rate that would be associated with feeding exclusively on type-k worms, if it were somehow possible to find them in zero time.

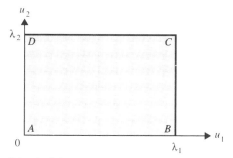

Fig. 12.5 Set of feasible decisions.

Now that we know exactly what u_k means, let's attempt to determine how the bird should select her worms. To achieve this, we will borrow from the science of biology, more specifically, from the science of animal behavior. It has been suggested by some biologists that predators select prey so as to maximize their energy rate. We will call this Hypothesis 1. If this is true, then our bird should pick (u_1, u_2) so as to maximize (12.60), subject to (12.59). This is a constrained nonlinear programming problem of the form (8.70), with $L = -\Phi$, $m = 4$, $g_1(u_1, u_2) \equiv u_1$, $g_2(u_1, u_2) \equiv \lambda_1 - u_1$, $g_3(u_1, u_2) \equiv u_2$, $g_4(u_1, u_2) \equiv \lambda_2 - u_2$, and no equality constraints; and it is readily shown that the optimal decision is given by

$$u_1 = \lambda_1, \qquad u_2 = \begin{cases} \lambda_2, & \text{if } \Delta\lambda_1 < E_2, \\ 0, & \text{if } \Delta\lambda_1 > E_2, \end{cases} \qquad (12.61)$$

where

$$\Delta = h_2 E_1 - h_1 E_2 \qquad (12.62)$$

is positive, by (12.58). If $\Delta\lambda_1 = E_2$, then the optimal u_1 is still λ_1, and the value of u_2 is irrelevant.[6]

There is therefore a critical threshold for λ_1, namely,

$$\lambda_c = \frac{E_2}{\Delta}, \qquad (12.63)$$

below which the bird should gobble up every worm in sight, but above which only the more profitable worms should be eaten. We can interpret this optimal decision by observing that $\Phi < E_2/h_2$ or $\Phi > E_2/h_2$ according to whether $u_1 < \lambda_c$ or $u_1 > \lambda_c$. But $u_1 > \lambda_c$ requires $\lambda_1 > \lambda_c$. Thus, when $\lambda_1 < \lambda_c$, the bird cannot acquire energy at a faster rate than the profitability E_2/h_2 of a type-2 worm. Therefore type-2 worms should not be spurned when $\lambda_1 < \lambda_c$ because the energy rate for the interval during which one is being eaten (i.e., E_2/h_2) is then at least as great as overall. If $\lambda_1 > \lambda_c$, however, then a higher overall rate than E_2/h_2 can be maintained; and the highest such rate is

$$\frac{E_1 \lambda_1}{1 + h_1 \lambda_1} > \frac{E_2}{h_2}, \qquad (12.64)$$

which occurs when all type-2 worms are refused. Note that the value of λ_2 is irrelevant. (Does this strike you as odd? See Exercise 12.18.)

We now know what is optimal, *if,* that is, Hypothesis 1 is correct. But shouldn't what is optimal for birds correspond to what birds actually do, because of "pressure due to natural selection;" i.e., because competition for survival in nature is so fierce that only animal lines whose genes just happen to make them behave optimally would not have become extinct? If so, then (12.61) should agree with observed behavior. Does it?

---

[6]See Supplementary Note 12.2.

An experiment on great tits by John Krebs, Jonathan Erichsen, and Michael Webber of the University of Oxford, and by Eric Charnov of the University of Utah, was designed to test the prediction (12.61). Because they could not accurately have observed birds seeking worms in the wild, these experimenters perched captive birds in their laboratory and brought worms to them on a conveyor belt. In other words, the worms moved and the birds were stationary; whereas, in the wild, the birds move and the worms are stationary. Details of this experiment (for example, how the experimenters manipulated handling times and prevented their birds from knowing the next worm's type in advance) would be out of place in this chapter, and if you are interested, then you can consult the original paper by Krebs et al. (1977). For our purposes, it suffices here to say that the experimenters could fix λ_1, λ_2, h_1, h_2, E_1/E_2, and hence λ_c; and that they were able to record the fractions of type-1 worms and type-2 worms accepted by a bird for various values of λ_1. Let us denote these fractions by p_1 and p_2, respectively. Moreover, let us identify p_1 in the experiment with u_1/λ_1 in our model; and, likewise, let us identify p_2 with u_2/λ_2. Then, from (12.61), our model predicts that

$$p_1 = 1, \qquad p_2 = \begin{cases} 1, & \text{if } \lambda_1 < \lambda_c, \\ 0, & \text{if } \lambda_1 > \lambda_c. \end{cases} \tag{12.65}$$

Is this what Krebs et al. actually observed?

Essentially, they observed that their birds refused nothing at subcritical values of λ_1, i.e., for $\lambda_1 < \lambda_c$, and continued to accept all type-1 worms at supercritical values of λ_1, i.e., for $\lambda_1 > \lambda_c$; but they did not feed *exclusively* on type-1 worms at such supercritical encounter rates. Rather, they accepted a fraction of the type-2 worms. Because Krebs et al. (1977, pp. 35–36) suggest that this fraction was a decreasing function of λ_1, becoming extremely small for very large values of λ_1, it appears that the results of their experiment are more accurately described by

$$p_1 = 1, \qquad p_2 = \begin{cases} 1, & \text{if } \lambda_1 < \lambda_c, \\ \sigma(\lambda_1), & \text{if } \lambda_1 \geq \lambda_c, \end{cases} \tag{12.66}$$

than by (12.65); where, for $\lambda_1 \geq \lambda_c$,

$$\sigma(\lambda_c) = 1, \qquad \frac{\partial \sigma}{\partial \lambda_1} < 0, \qquad \lim_{\lambda_1 \to \infty} \sigma(\lambda_1) = 0. \tag{12.67}$$

Note that the dependence of σ on λ_2, h_1, h_2, E_1, and E_2 has been suppressed by this notation.

Why does this discrepancy exist between (12.65) and (12.66)? Perhaps Hypothesis 1 is false; perhaps birds are not simply trying to maximize their energy rate. But do we really wish to abandon completely the notion that Φ should somehow be maximized? After all, at subcritical values of λ_1, there is excellent agreement between theory and observation. Thus, if we continue to believe that birds behave optimally in some appropriate sense, then mustn't we rather adapt the notion that Φ should be maximized in such a way that the adaptation has no effect at subcritical values of λ_1, but makes a difference at supercritical ones? How can we do this?

Let us reason as follows. Suppose that our bird has just encountered a type-2 worm. Should she refuse it? If she does, then she surely hopes that the next worm encountered will be of type 1. But she does not *know* that the next worm encountered will be of type 1. All she knows is that type-1 worms are being encountered λ_1/λ_2 times as frequently as type-2 worms, on average; or that

$$\text{Prob[Next worm is type 1]} = \frac{\lambda_1}{\lambda_1 + \lambda_2}. \tag{12.68}$$

If λ_1 is very high, far in excess of the critical value λ_c, then (12.68) is so close to 1 that the next worm encountered will almost certainly be of type 1. Then refusing all type-2 worms is an attractive policy. If λ_1 is *just* bigger than λ_c, however, so that (12.68) is not especially close to 1, then there is a significant chance that the next worm encountered will again be of type 2. Indeed there is a significant (though, of course, lesser) chance that the next *two* worms encountered will be of type 2, and so on. Refusing all type-2 worms would not be such an attractive policy now because it could cause the bird to abstain too much and hence waste precious search time.

Thus a possible answer to the question we posed a little while ago is that, in addition to sustaining a high energy rate, it is important for birds to use time efficiently.[7] After all, time is money, even for a bird, and the time not used in searching for worms can be invested in some other beneficial activity. Recalling that each minute of searching generates $h_1 u_1 + h_2 u_2$ minutes of handling, we see that the proportion of time spent actually looking for worms is Ψ, where we define

$$\Psi(u_1, u_2) = \frac{1}{1 + h_1 u_1 + h_2 u_2}. \tag{12.69}$$

For the sake of brevity, it will be convenient to breach linguistic etiquette and refer to Ψ henceforward as the *travel time*, even though Ψ does not have the dimensions of time.

If using time efficiently were the bird's *sole* objective, then we would expect her to select the decision that minimizes (12.69), subject to (12.59). It is obvious that the solution to this constrained nonlinear programming problem is $(u_1, u_2) = (\lambda_1, \lambda_2)$. In other words, if minimizing Ψ were her sole objective, then a bird would never refuse a worm. In the laboratory,

---

[7]This is not, however, the only answer, and Krebs et al. (1977) suggest a different one. See Exercise 12.17.

such behavior would correspond to

$$p_1 = 1, \qquad p_2 = 1. \tag{12.70}$$

Compare (12.66) with (12.65) and (12.70). Doesn't it appear as though what birds actually do involves a compromise between maximizing Φ and minimizing Ψ? Let us therefore conjecture that foraging birds behave in such a way as to achieve an optimal compromise between maximizing energy rate and minimizing travel time. We will dub this conjecture Hypothesis 2. In the jargon of behavioral ecology, Φ and Ψ would be known as currencies. Thus Hypothesis 2 claims that birds are optimizing two currencies; whereas Hypothesis 1 claims that they optimize only a single currency, namely, Φ.

But what do we mean by an optimal compromise? In answering this question, it will be helpful to work in the Φ-Ψ plane, which we shall dub the *currency space*. Now, for given values of E_1, E_2, h_1, and h_2, (12.60) and (12.69) map the set of feasible decisions (12.59) into the quadrilateral

$$\begin{aligned} h_2\Phi + E_2\Psi \geq E_2 \geq h_2\Phi + (E_2 - \lambda_1\Delta)\Psi, \\ h_1\Phi + E_1\Psi \leq E_1 \leq h_1\Phi + (E_1 + \lambda_2\Delta)\Psi \end{aligned} \tag{12.71a}$$

in currency space, and the inverse mapping from (12.71a) to (12.59) is given by

$$\Delta u_1 = E_2 - \frac{E_2 - h_2\Phi}{\Psi}, \qquad -\Delta u_2 = E_1 - \frac{E_1 - h_1\Phi}{\Psi} \tag{12.71b}$$

(Exercise 12.14). The set of points in currency space that correspond, through (12.71), to feasible decisions will be dubbed the *feasible region*. For subcritical values of λ_1, i.e., for $\lambda_1 < \lambda_c$, this feasible region is shaded in Fig. 12.6(a). The vertices A, B, C, and D correspond to the corners of the rectangle in Fig. 12.5. Observe that the line BC, corresponding to decisions for which all worms of type 1 are eaten, has negative slope. The point C is therefore lower and further to the right than any other feasible point in currency space. Hence the bird would surely select the decision corresponding to C, i.e., $(u_1, u_2) = (\lambda_1, \lambda_2)$, if she were maximizing Φ, but she would also choose it if she were minimizing Ψ. In other words, $(u_1, u_2) = (\lambda_1, \lambda_2)$ is clearly her best decision when optimizing either currency alone. It must therefore still be her best decision when she is optimizing both. Thus compromise is unnecessary at subcritical values of λ_1. There is no need to trade one currency for the other because the best feasible value of each is attainable.

At supercritical values of λ_1, however—i.e., when $\lambda_1 > \lambda_c$—the line BC has rotated past vertical in a clockwise direction to acquire a positive slope, as depicted in Fig. 12.6(b), where the feasible region is again shaded. In terms of maximizing Φ, B has now become the optimal feasible point in currency space because it is furthest to the right. Indeed we know this

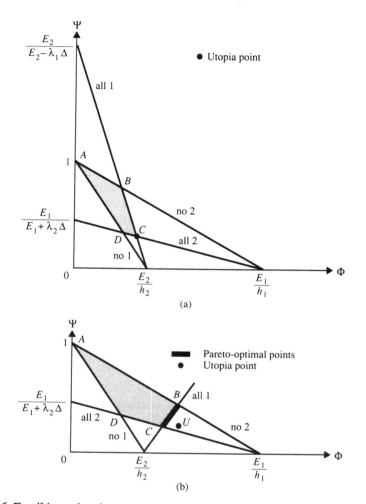

Fig. 12.6 Feasible region in currency space at (a) subcritical and (b) supercritical encounter rates with more profitable type of worm.

already from (12.61). In terms of minimizing Ψ, however, C remains the optimal point in currency space. Because C does not coincide with B, a compromise is needed.

With this in mind, observe that all points on the line segment between B and C in Fig. 12.6(b) have the following property. It is impossible to move from any of these points to another feasible point in currency space without either increasing the value of Ψ or decreasing the value of Φ. In other words, improvement in one currency is impossible without disimprovement in the other. We will describe such points as *unimprovable.*

Unimprovable points are logical candidates for a compromise decision. But there are infinitely many of them. Then which is the optimal compromise? The point U in Fig. 12.6(b), at which the line through C parallel to the Φ-axis meets the line through B parallel to the Ψ-axis, corresponds both to the maximum feasible value of Φ and the minimum feasible value of Ψ. Let us call this point the *utopia point*. If this point were attainable—if it corresponded to a feasible decision—then it would be optimal for both currencies, as in Fig. 12.6(a), where it coincides with C. But it is not attainable because it lies outside the feasible region in currency space.

If the best possible values of two currencies cannot simultaneously be achieved by a feasible decision, i.e., if the utopia point is unattainable, then what is the next best thing? In other words, what is the best nonutopian decision? Wouldn't you agree that the next best thing is the decision corresponding to the closest pair of currency values to the utopia point, because the utopia point would unquestionably be optimal, if attainable?

A difficulty now arises. If energy is measured in calories and time in hours, then Φ has dimensions [ENERGY]/[TIME]; whereas Ψ is dimensionless. If we are going to calculate distances in currency space, however, then our currencies should at least have the same dimensions. We can arrange this by using either E_1/h_1 or E_2/h_2 to scale the energy rate. But which, if either, of these scalings is correct? Because we don't know, let's simply choose one and see what happens. Accordingly, define dimensionless currencies ϕ and ψ by

$$\phi = \frac{h_2 \Phi}{E_2}, \qquad \psi = \Psi. \tag{12.72}$$

It is also convenient to define dimensionless parameters

$$\eta = \frac{h_2}{h_1}, \qquad \beta = \frac{E_1 h_2}{E_2 h_1} = \frac{E_1 \eta}{E_2}, \qquad \epsilon = \frac{\lambda_1 - \lambda_c}{\lambda_c}, \qquad \theta = \frac{\lambda_2}{\lambda_c}, \tag{12.73}$$

where, in view of (12.58), $\beta > 1$. Then (Exercise 12.14) the feasible region in dimensionless currency space, i.e., ϕ-ψ space, is

$$\phi + \psi \geq 1 \geq \phi - \epsilon\psi, \qquad \phi + \beta\psi \leq \beta \leq \phi + (\beta + \eta\theta)\psi, \tag{12.74a}$$

as depicted in Fig. 12.7; and ϕ-ψ space is mapped back to (12.59) by

$$u_1 = \lambda_c \left(1 - \frac{1 - \phi}{\psi} \right), \qquad -\eta u_2 = \lambda_c \left(\beta - \frac{\beta - \phi}{\psi} \right). \tag{12.74b}$$

Clearly, everything we have said about Fig. 12.6 is still true about Fig. 12.7, but with the advantage that ϕ and ψ have common dimensions. For the sake of brevity, we will continue to refer to ϕ as the energy rate.

Now let us agree, at least for the time being, that the closest point to U in currency space should correspond, through (12.74), to the opti-

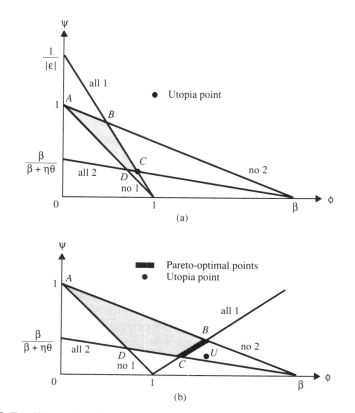

Fig. 12.7 Feasible region in dimensionless currency space when (a) $\epsilon < 0$ and (b) $\epsilon > 0$.

mal compromise. Then, in Fig. 12.7(b), the optimal compromise corresponds to the foot of the perpendicular from U to BC. Let us denote this point by S. As λ_1 increases from its critical value λ_c toward infinity, thereby rotating the line BC clockwise from the vertical position toward the ϕ-axis, S will shift gradually along the line BC from C toward B. In other words, if our bird agrees with us on what is optimal, then she will gradually shift from eating all prey of type 2 toward eating no prey of type 2. This is indicated schematically in Fig. 12.8 (where US is perpendicular to BC). It is a straightforward exercise in trigonometry to calculate the coordinates of S in (dimensionless) currency space, whence (Exercise 12.14) we deduce from (12.74) that the optimal compromise is (u_1^*, u_2^*), where

$$u_1^* = \lambda_1, \qquad \frac{u_2^*}{\lambda_2} = \begin{cases} 1, & \text{if } \lambda \leq \lambda_c, \\ \sigma, & \text{if } \lambda > \lambda_c, \end{cases} \qquad (12.75a)$$

Fig. 12.8 Shift of optimal compromise in the direction of increasing specialization on more profitable type of worm.

and

$$\sigma = \frac{\epsilon + \beta}{(1 + \epsilon^2)(\epsilon + \beta) + \epsilon^2 \eta \theta}. \tag{12.75b}$$

Note that (12.67) is satisfied by (12.75b) because $\lambda_1 = (1 + \epsilon)\lambda_c$. This suggests that Hypothesis 2 is superior to Hypothesis 1 because its predictions are equivalent to those of Hypothesis 1 at subcritical values of λ_1, and because it explains, at least qualitatively, the discrepancy between (12.65) and (12.66) at supercritical values. On the other hand, in adapting our earlier model, we all but guaranteed that the modified version would satisfy (12.67), so that it is debatable whether we have given any more credence to the idea of birds being optimal compromisers than a purely verbal argument. Let us therefore see whether we can subject Hypothesis 2 to a more stringent test, by obtaining data for β, ϵ, η, θ, and σ in (12.75).

The type-1 worms used by Krebs, Erichsen, Webber, and Charnov invariably had twice the calorific value of the type-2 worms. Thus $E_1/E_2 = 2$. In all, Krebs et al. used five captive great tits in their experiment, but only three of these birds experienced supercritical encounter rates with type-1 worms. One of the three was affectionately known as PW.

By watching PW devour about 30 worms, Krebs et al. concluded that her average handling times for worms of type 1 and type 2 were 8.5 and 12.5 seconds, respectively; and it seems reasonable to use these values in our analysis. Incidentally, the average used by Krebs et al. (1977) was not the mean but the median, the value that was exceeded by as many values as it exceeded. Also, in case you are wondering why PW took longer to handle the type-2 worms if they were only half as fat, the reason is that they were cleverly swaddled in plastic tape, which PW had to remove before she could eat them. Anyhow, quoting data correct to only two significant figures, we have $h_1 = 0.14$ (minutes) and $h_2 = 0.21$; whence $\eta = 1.5$. Moreover, because $E_1/E_2 = 2$, we have $\beta = 2.9$ and $\lambda_c = 3.6$ (because $\lambda_c h_1(\beta - 1) = 1$). Now, each bird in the experiment was subjected to five treatments, corresponding to five different pairs of rates of encounter with the two types of worm. Krebs et al. denoted these treatments by A, B, C, D, and E. The corresponding encounter rates (in units of minute^{-1}) are shown in Table 12.3 together with the corresponding values of the dimensionless parameters ϵ and θ, defined by (12.73). Let ζ denote the fraction of type-1 worms in PW's diet. Then Hypothesis 2 predicts that

$$
\zeta = \frac{u_1^*}{u_1^* + u_2^*} =
\begin{cases}
\dfrac{\epsilon + 1}{\epsilon + \theta + 1}, & \text{if } \epsilon \leq 0, \\[4mm]
\dfrac{1}{1 + \{\theta(\epsilon + \beta)/(\epsilon + 1)[(1 + \epsilon^2)(\epsilon + \beta) + \epsilon^2 \eta \theta]\}}, & \text{if } \epsilon > 0,
\end{cases}
$$

$$(12.76)$$

on using (12.75); whereas Hypothesis 1 would predict the same value of ζ for $\epsilon < 0$, but $\zeta = 1$ for $\epsilon > 0$.

Values of ζ predicted by (12.76) are recorded in Table 12.3 together with the observed fractions of type-1 worms in PW's diet. For treatments A and B, the discrepancy between observed and predicted values is exactly the same as would arise from Hypothesis 1. For treatment C, Hypothesis 2 predicts more accurately than Hypothesis 1, although there is still a discrepancy of 12%. For treatments D and E, however, not only does Hypothesis 2 predict much more accurately than Hypothesis 1, but also the discrepancies are less than 4% and 2%, respectively. Thus Hypothesis 2 is a clear winner over Hypothesis 1. Does this mean that Hypothesis 2 should be accepted?

If Gilbert and Sullivan had had time to write a popular song about a scientist, then they would probably have observed that her work is never done, although her lot is surely happier than a policewoman's. Because they died before they got around to it, however, let me remind you that we haven't yet examined all the evidence; for two other birds in the experiment by Krebs et al. experienced supercritical encounter rates with type-1 worms. The birds in question were known as BW and RO.

For BW, the average handling times were $h_1 = 0.18$ and $h_2 = 0.18$, which gives us $\beta = 2$, $\eta = 1$, and $\lambda_c = 5.7$. For RO, the average handling times were $h_1 = 0.083$ and $h_2 = 0.15$, yielding $\beta = 3.6$, $\eta = 1.8$, and $\lambda_c = 4.6$. From these data, we can use (12.76) to predict the eating preferences of

Table 12.3 PW's eating preferences.

Treatment	λ_1	λ_2	ϵ	θ	ζ	Observed fraction of more profitable prey in diet
A	1.5	1.5	-0.59	0.41	0.5	0.52
B	3.0	3.0	-0.18	0.82	0.5	0.44
C	9.0	3.0	1.5	0.82	0.92	0.82
D	9.0	9.0	1.5	2.5	0.83	0.80
E	9.0	18.0	1.5	4.9	0.77	0.78

Table 12.4 BW's eating preferences.

Treatment	λ_1	λ_2	ϵ	θ	ζ	Observed fraction of more profitable prey in diet
A	1.5	1.5	−0.74	0.26	0.5	0.52
B	3.0	3.0	−0.48	0.53	0.5	0.50
C	9.0	3.0	0.58	0.53	0.81	0.86
D	9.0	9.0	0.58	1.6	0.61	0.98
E	9.0	18.0	0.58	3.2	0.46	0.64

BW and RO from Hypothesis 2. They are compared with the observations in Tables 12.4 and 12.5. We see that, although Hypothesis 2 is better than Hypothesis 1 at predicting BW's performance during treatments C and E, the discrepancy between observed and predicted values of ζ is rather large, particularly for E; and Hypothesis 1 is far superior to Hypothesis 2 for treatment D. Though Hypothesis 2 predicts as well as Hypothesis 1 for RO's treatment D, and not much worse for C, the only good thing we can say about its prediction for E is that it isn't as bad as for BW's treatment D! Does this mean that Hypothesis 2 should now be rejected, after all? Perhaps you would like to think about that. We'll return to the matter later.

Irrespective of the extent to which observed and predicted values of ζ in Tables 12.3, 12.4, and 12.5 agree, a possible criticism of our model is that we have used the Euclidean norm to measure distance in ϕ-ψ space, i.e., we have used the formula

$$d_{AB} = \left(|\phi_A - \phi_B|^2 + |\psi_A - \psi_B|^2\right)^{1/2}$$

to define the distance in currency space between the points (ϕ_A, ψ_A) and (ϕ_B, ψ_B). This is merely a special case of the more general distance formula

$$d_{AB} = \left(|\phi_A - \phi_B|^\rho + |\psi_A - \psi_B|^\rho\right)^{1/\rho}, \tag{12.77}$$

where $\rho \geq 1$. Mathematicians usually describe (12.77) as the distance formula for the L_ρ norm. Thus the Euclidean norm, which we have been using,

Table 12.5 RO's eating preferences.

Treatment	λ_1	λ_2	ϵ	θ	ζ	Observed fraction of more profitable prey in diet
A	1.5	1.5	−0.68	0.33	0.5	0.54
B	3.0	3.0	−0.35	0.65	0.5	0.52
C	9.0	3.0	0.95	0.65	0.86	0.94
D	9.0	9.0	0.95	2.0	0.72	0.86
E	9.0	18.0	0.95	3.9	0.62	0.91

is the same thing as the L_2 norm.[8] But what is our justification for using this norm? Why not instead use (12.77) to measure distance in currency space, with $\rho \neq 2$?

The question just raised is an important one, and I haven't yet found a thoroughly convincing answer to it. On the other hand, which value of ρ other than 2 are we going to use? After all, did I hear you complain in Section 8.4, where we used the Euclidean norm to measure distances for curve-fitting? Then why should you object to us using it now? Ultimately, perhaps the best one can hope for is that optimal compromises are not very sensitive to the value of ρ; see Exercise 12.15. If so, then we should certainly choose $\rho = 2$, because it simplifies the mathematics.

In obtaining the optimal compromise (12.75), we have borrowed from the literature on *vector optimization*, optimization of a vector of currencies. The idea of an unimprovable point originates from the work of Italian economist and sociologist Vilfredo Pareto. Indeed unimprovable points are usually described as *Pareto-optimal* in the literature on vector optimization, and we will use this description henceforward. The idea that an optimal compromise should correspond to the nearest feasible point in currency space to the utopia point originates from the work of Soviet academician M. E. Salukvadze. Indeed, in the jargon of vector optimization, (12.75) is a Salukvadze compromise solution.

Salukvadze's method is not the only one for assessing trade-offs between currencies. A popular alternative would be to argue as follows. We have been attempting to maximize ϕ and minimize ψ simultaneously. Because minimizing ψ is the same as maximizing $-\psi$, we have effectively been maximizing the vector currency $(\phi, -\psi)$. Now, let us take the view that our foraging bird should solve this optimization problem by devoting $100\alpha\%$ of her effort to maximizing ϕ and the remaining $100(1-\alpha)\%$ to maximizing $-\psi$, where α, satisfying $0 < \alpha < 1$, is some fixed parameter. Then we can replace the bird's problem by that of maximizing $J(\phi, \psi) = \alpha\phi - (1 - \alpha)\psi$. In doing so, we make the hypothesis that birds are maximizing a single currency, but that it is J rather than ϕ. In such circumstances, it is sometimes said that a vector optimization problem has been "scalarized," or that ϕ and ψ have been reduced to a "common currency," J.

The decision that maximizes $J(u_1, u_2) = \alpha\phi(u_1, u_2) - (1 - \alpha)\psi(u_1, u_2)$, subject to (12.59), is readily found to satisfy

$$u_1 = \lambda_1, \qquad u_2 = \begin{cases} \lambda_2, & \text{if } \lambda_1 < \lambda_c^{\text{new}}, \\ 0, & \text{if } \lambda_1 > \lambda_c^{\text{new}}, \end{cases} \qquad (12.78)$$

---

[8] Apart from L_2, the two most commonly used norms are L_1 and L_∞, called the maximum norm. The corresponding distance formulas are obtained from (12.77) by setting $\rho = 1$ and by taking the limit as $\rho \to \infty$, respectively. Taking $\rho \to \infty$ yields $d_{AB} = \max\{|\phi_A - \phi_B|, |\psi_A - \psi_B|\}$.

where we define

$$\lambda_c^{\text{new}} = \frac{\lambda_c}{\alpha} > \lambda_c. \qquad (12.79)$$

If $\lambda_1 = \lambda_c^{\text{new}}$, then the optimal u_1 is still λ_1, and the value of u_2 is not relevant because $J(\lambda_c^{\text{new}}, u_2) = \alpha$ is independent of u_2. Observe that (12.78) is qualitatively the same as (12.61); the only quantitative difference is that the critical encounter rate with type-1 worms, at which a bird should switch discontinuously from eating every worm in sight to accepting only worms of the more profitable type, is raised from (12.63) to (12.79). But we know that such a discontinuity was not observed in the experiment by Krebs et al.

This does not mean, of course, that birds are not optimizing a single currency. It simply means that the currency is not J. The prevailing view among biologists is that if animals are maximizing anything, then the thing that they maximize is "fitness."[9] What is fitness? Broadly speaking, it is a measure of an animal's long-term future success in surviving and reproducing its genetic material; it is a measure of the extent to which an animal's genes are transmitted to successive generations. But the concept of fitness is extremely difficult to model mathematically. Consider our bird, for example. We might conjecture that the rate at which she takes up energy today affects her future reproductive success, and that the faster she takes up energy today (for fixed values of everything else), the greater her future reproductive success will be. We might also conjecture that the proportion of time she spends travelling to food affects her future reproductive success, and that the smaller this proportion (for fixed values of energy rate, etc.), the greater her future reproductive success will be. We might even suppose that these are the only two factors that affect her future reproductive success or, more accurately, that these two factors will have far greater consequences for the transmission of her genes than any other factors (at least while she is foraging). Under these conditions, fitness can be modelled by a function of the form $f = f(\phi, \psi)$, where

$$\frac{\partial f}{\partial \phi} > 0, \qquad \frac{\partial f}{\partial \psi} < 0. \qquad (12.80)$$

These inequalities are clearly satisfied by $f(\phi, \psi) = \alpha\phi - (1-\alpha)\psi, 0 < \alpha < 1$, but we have already discovered that this is not a suitable fitness function. Can we choose a better one?

We can certainly satisfy (12.80) by choosing

$$f(\phi, \psi) = \alpha\phi^b - (1 - \alpha)\psi^c, \qquad 0 < \alpha < 1, \qquad (12.81)$$

---

[9]The validity of the hypothesis that animals maximize fitness is the subject of a long-standing controversy among biologists. See, for example, the article "Optimalists under fire" in the January, 1988 issue of *Scientific American* (Volume 258, No. 1, pp. 21–22).

where b, c are positive numbers. This agrees with J above when $b = 1 = c$. If this is a suitable fitness function, and if animals behave so as to maximize fitness, then our bird's optimal decision should be the one that minimizes

$$L(u_1, u_2) \equiv (1 - \alpha)[\psi(u_1, u_2)]^c - \alpha[\phi(u_1, u_2)]^b, \qquad 0 < \alpha < 1, \quad (12.82)$$

subject to (12.59) and $b \neq 1$ or $c \neq 1$. Let us define $U^*(\alpha, \epsilon)$ to be the solution of the equation

$$\alpha b \epsilon h_2^{b-1}(1 + h_1(1 + \epsilon)\lambda_c + h_2 u_2^*)^{c-b}(E_1(1 + \epsilon)\lambda_c + E_2 u_2^*)^{b-1} = c(1 - \alpha)E_2^{b-1}$$
$$(12.83a)$$

for u_2^*; and let us define $\lambda_1^-(\alpha)$ and $\lambda_1^+(\alpha)$, respectively, to be the solutions of the following two equations for λ_1:

$$\frac{c\lambda_c E_2^{b-1}}{c\lambda_c E_2^{b-1} + b(\lambda_1 - \lambda_c)h_2^{b-1}(E_1\lambda_1 + E_2\lambda_2)^{b-1}(1 + \lambda_1 h_1 + \lambda_2 h_2)^{c-b}} = \alpha,$$
$$(12.83b)$$

$$\frac{c\lambda_c E_2^{b-1}}{c\lambda_c E_2^{b-1} + b(\lambda_1 - \lambda_c)(h_2 E_1 \lambda_1)^{b-1}(1 + \lambda_1 h_1)^{c-b}} = \alpha. \qquad (12.83c)$$

Our notation suppresses the fact that U^*, λ_1^-, and λ_1^+ depend on parameters other than α or ϵ—for example, b, or c. In particular, when $b = 1 = c$, λ_1^- is the same as λ_1^+ because both are then equal to (12.79). From (12.83), we see that $\lambda_c < \lambda_1^-(\alpha) < \lambda_1^+(\alpha)$, and the optimal decision is readily found to be (u_1^*, u_2^*), where $u_1^* = \lambda_1$ and[10]

$$\frac{u_2^*}{\lambda_2} = \begin{cases} 1, & \text{if } \lambda_1 \leq \lambda_1^-(\alpha); \\ [U^*(\alpha, \epsilon)]/\lambda_2, & \text{if } \lambda_1^-(\alpha) < \lambda_1 < \lambda_1^+(\alpha); \\ 0, & \text{if } \lambda_1 \geq \lambda_1^+(\alpha). \end{cases} \qquad (12.84)$$

By the shift technique of Section 2.7, we can show that $U^*(\alpha, \epsilon)$ is a decreasing function of ϵ. Thus the hypothesis that (12.81) is a suitable fitness function when $b \neq 1$ or $c \neq 1$ predicts the existence of two critical thresholds, namely, λ_1^- and λ_1^+. If $\lambda_1 < \lambda_1^-$, then our bird should never refuse a worm; if $\lambda_1 > \lambda_1^+$, then the bird should specialize on type-1 worms, and as λ_1 increases from λ_1^- to λ_1^+, the bird should decrease her acceptance rate of type-2 worms from the maximum possible to zero.

To illustrate this predicted optimum numerically, let us first define $\epsilon^-(\alpha) = (\lambda_1^-(\alpha) - \lambda_c)/\lambda_c$ and $\epsilon^+(\alpha) = (\lambda_1^+(\alpha) - \lambda_c)/\lambda_c$, so that $\lambda_c < \lambda_1^-(\alpha) < \lambda_1^+(\alpha)$ is replaced by $0 < \epsilon_1^-(\alpha) < \epsilon_1^+(\alpha)$. Thus 100ϵ, $100\epsilon_1^-(\alpha)$ and $100\epsilon_1^+(\alpha)$ are the percentages by which λ_1, $\lambda_1^-(\alpha)$, and $\lambda_1^+(\alpha)$, respectively, exceed the critical value λ_c. Let us also (arbitrarily) choose $b = 2 = c$, and with BW in mind, set $h_1/h_2 = 1 = 2E_2/E_1$, so that $\lambda_c h_1 = 1$. Then you can readily

---

[10]See Supplementary Note 12.3.

show (Exercise 12.14) that

$$\frac{U^*(\alpha, \epsilon)}{\lambda_2} = \frac{2\{1 - \alpha - 2\alpha\epsilon(1 + \epsilon)\}}{\theta\alpha\epsilon}, \qquad \epsilon^-(\alpha) \leq \epsilon \leq \epsilon^+(\alpha),$$

$$\epsilon^-(\alpha) = \frac{2(1 - \alpha)}{(\theta + 2)\alpha + \sqrt{(\theta + 2)^2\alpha^2 + 8\alpha(1 - \alpha)}}, \qquad (12.85)$$

$$\epsilon^+(\alpha) = \frac{1 - \alpha}{\alpha + \sqrt{\alpha(2 - \alpha)}}.$$

During the experiment by Krebs et al., BW experienced the supercritical encounter rate corresponding to $\epsilon = 0.58$ for three values of θ, namely, $\theta = 0.53$, $\theta = 1.6$, and $\theta = 3.2$. (All numbers have been rounded to two significant figures.) In all three cases, the fraction of type-2 worms accepted lay between 0 and 1. Thus (12.85) would be a good approximation only for $\epsilon^-(\alpha) < 0.58 < \epsilon^+(\alpha)$. You can easily check that this is satisfied by all three values of θ only if α lies between about 0.32 and 0.36. But no such value of α yields values of $u_1^*/(u_1^* + u_2^*)$ close to those observed by Krebs et al., for all three values of θ, when we set $u_1^* = \lambda_1$ and $u_2^* = U^*(\alpha, 0.58)\lambda_2$. It therefore appears that (12.81) with $b = c = 2$ is not a suitable fitness function for any value of α *if* birds are compromisers between energy rate and travel time.

By experimenting with other values of b and c, we might well find a more acceptable pair. But by doing so we would merely have forced the hypothesis (that birds maximize fitness) to fit the observations; whereas the point of our model is to test whether the observations fit the hypothesis. It thus appears that even if animals do behave so as to maximize a single currency, a multiple-currency approach to predicting their behavior may still be preferable to a common-currency approach, at least when our only knowledge about f is the qualitative knowledge that the signs of certain derivatives are fixed, as in (12.80). In such circumstances, a multiple-currency model circumvents the need to assign values to several arbitrary parameters, such as α, b, and c in (12.81); yet it can still generate testable predictions, such as (12.75). The norm associated with currency space may well be arbitrary (though see Exercise 12.15), but it is much less arbitrary when we have no quantitative knowledge about fitness than any functions we might use as substitutes. In short, it appears that a vectorial approach to modelling animal behavior may be a good idea.

Or does it? After building a new model, one should always criticize it as thoroughly as possible, but an obvious criticism of our model has yet to be addressed. What is it? See Exercise 12.16.

Finally, one should always bear in mind that an hypothesis is merely one assumption, albeit a most important one; whereas a model consists of many assumptions. If a model yields false predictions, then the fault does not necessarily lie with the particular assumption we have elevated to the rank of hypothesis. It may lie with one of the others. If so, then we must improve our model by relaxing the assumption in question, before our hypothesis can properly be tested.

12.6   Where Should an Insect Lay Eggs? A Combination

In the previous section, we used vector optimization concepts to predict some aspects of the behavior of an animal (specifically, the foraging of a bird). We assumed that factors which affected the animal's fitness (specifically, the bird's energy rate and travel time) were fully determinate. In reality, of course, these factors are not determinate quantities. Rather, they are random variables.

Let us suppose that the fitness of some arbitrary animal depends on m such factors, say $Y_1, Y_2, Y_3, \ldots, Y_m$. Then fitness $f = f(Y_1, Y_2, Y_3, \ldots, Y_m)$ is also a random variable. Ideally, for maximum fitness, f should be made as large as possible, subject to constraints on decisions made by the animal. Now, it may be true that every factor is either wholly favorable or wholly unfavorable to fitness, by which I mean that for every $k = 1, \ldots, m$, it may be true that f either increases or decreases with respect to Y_k for all values of $Y_1, \ldots, Y_{k-1}, Y_{k+1}, \ldots, Y_m$. This would certainly give us some information about f but not an explicit expression for it. In view of the latter, we could propose that f have the form

$$f(Y_1, \ldots, Y_m) = \sum_{k=1}^{m} \alpha_k Y_k^{b_k}, \tag{12.86}$$

where $b_1, \ldots, b_m$ are positive parameters, α_k is positive or negative according to whether Y_k is favorable or unfavorable to fitness, $1 \leq k \leq m$, and $\alpha_1 + \alpha_2 + \cdots + \alpha_m = 1$. But $2m + 1$ of the $2m$ parameters $\alpha_1, \ldots, \alpha_m$, $b_1, \ldots, b_m$ would be arbitrary, and it might be unrealistic to hope that an optimal decision would not be sensitive to their values. Thus, as argued in the previous section, a vectorial approach may be preferable. We will anyhow adopt one.

Therefore, for $1 \leq k \leq m$, Y_k should be made as large as possible or as small as possible according to whether Y_k is favorable or not to the fitness of our arbitrary animal, subject, of course, to constraints on its decisions. But Y_k is a random variable, which cannot itself be optimized. It must therefore be replaced by a reward, as defined in Chapter 7. Which statistic of Y_k's distribution would be most suitable for this reward? An answer derives from arguments similar to those we put forth at the beginning of Section 8.8. There, we argued that expected value of profit is a suitable reward for a buyer in the long term, i.e., over many similar buying periods; so that good (better than expected) and bad (worse than expected) profits can cancel the effects of one another (although it might not be such a suitable reward in the short term). Likewise, if we are interested in the behavior of a typical animal (as opposed to one particular animal), and if random variation of the environment can be averaged over many animals, then $E[Y_k]$ is a suitable reward for factor Y_k, and we shall adopt it henceforward. For consistency with the previous section, however, we will refer to $E[Y_k]$ not as a reward but as a currency, and to $(E[Y_1], E[Y_2], \ldots, E[Y_m])$, not as a vector reward, but as a vector currency. Thus our approach to modelling animal behavior is to optimize the vector currency $(E[Y_1], E[Y_2], \ldots, E[Y_m])$. It should nevertheless be borne in mind that *if* we knew f explicitly, then we would abandon this vectorized approach and instead maximize the scalar currency $E[f(Y_1, Y_2, Y_3, \ldots, Y_m)]$, as described in recent work by Mangel and Clark (1988) and references therein. It is precisely because we do not know f that vector optimization concepts are so useful.

We have talked long enough about arbitrary animals, and arbitrary animals are not very interesting. Let us therefore study a particular one: a tiny wasp, scarcely a couple of millimeters long, which rejoices under the intriguing name of *Anagrus delicatus*, or *A. delicatus*.[11] The behavior of this insect has recently been studied by biologist Don Strong at Florida State University, and the impetus for the following model emerged from conversations between us.

A. delicatus is a parasite, and it kills its hosts by parasitizing them (for which reason, it is called a parasitoid). The unfortunate hosts are the eggs of a different insect, on which the *A. delicatus* adult female lays her own eggs; whereupon her larvae proceed to consume the host eggs. The adult female is really little more than a flying egg-laying machine, because she never lives for more than a day or two and rarely, if ever, pauses to feed. She begins her adult life with 32 fully formed eggs, and she begins immediately to patrol the saltmarshes along Florida's Gulf Coast, searching for egg-laying sites. What do we mean by a site? We mean a collection of suitable (unparasitized) host eggs, clustered together in a particular geographical location. A site may contain many host eggs or as few as one or two; yet no matter how few, the wasp never lays more than a single egg per host. She is a caring parent. There is no danger that her larvae will starve by virtue of having to share food with one another.

Let's assume that the insect will behave so as to maximize her fitness. What will contribute to this fitness—to her future reproductive success? Certainly the number of eggs she lays will be a contributing factor, and the more eggs she lays, the greater her fitness (at least, for fixed amounts of everything else). Let's denote this number of eggs by Y_1. To be more precise, let Y_1 be the number of eggs she lays in her entire adult lifetime. Thus $Y_1 \leq 32$. Then Y_1 is favorable to fitness. We might even suppose that Y_1 is

---

[11] My initial reaction to discovering this name was to suppose that it meant "nice anagram." How about "Use a Dracula-sting?"

the only factor that contributes to fitness, i.e., that fitness can be modelled by a function of the form $f = f(Y_1)$, where f increases with Y_1. Therefore, the insect would like Y_1 to be as large as possible, subject, of course, to any constraints on her actions. What are these constraints? She can't lay more eggs at a given site than there are suitable hosts, and she can't lay more eggs than remain in her body, say x. Suppose, however, that she is zooming along with x eggs aboard, when she encounters an apparently perfect site— x or more delicious host eggs just asking to be parasitized. Wouldn't you think in these circumstances that the insect would be happy to deposit her entire load? But this is almost never what *A. delicatus* does. Rather, she refuses a considerable fraction of the suitable hosts at each encountered site, especially early in life. Is this because she can't find them? No, because direct observations indicate that she often actually touches suitable hosts, but nevertheless refuses them. Why does the lady behave in this way?

A possible answer has been suggested by Professor Strong. Let me paraphrase it for you. Fitness is a matter of survival and reproduction; reproduction is impossible without survival of larvae; and it just isn't good enough for *A. delicatus* to ensure that her larvae have adequate food to survive. She must also attempt to provide them with adequate heat to survive, and therefore to protect them from lethal extremes of cold weather. Now, along Florida's Gulf Coast, temperatures at different sites appear to fluctuate independently of one another. Sometimes the temperature at a particular site will become so low that all the larvae there are killed, while a higher low temperature at a different site will allow all the larvae there to survive. Our insect is quite incapable of forecasting the weather; therefore the number of sites to which she disperses her eggs would seem to be crucial to her larvae's survival and hence to her fitness; and for a given number of parasitized hosts, the more sites she reaches, the greater her fitness. Let's denote this number of sites by Y_2. To be more precise, let Y_2 be the number of sites she reaches in her entire adult lifetime. Then Y_2 is favorable to fitness. We have already agreed that Y_1 is favorable to fitness, and so it seems that fitness involves a trade-off between the number of eggs laid and number of sites reached, or that fitness should be modelled by a function of the form $f = f(Y_1, Y_2)$, where f increases with both Y_1 and Y_2. Thus, in the absence of an explicit expression for f, both Y_1 and Y_2 should be made as large as possible, but because Y_1 and Y_2 are random variables, we will instead maximize their expected values, subject to the constraints on our insect's actions. We will denote $E[Y_1]$ by ϕ and $E[Y_2]$ by ψ. Because a decision that maximizes the egg currency, ϕ, need not simultaneously maximize the site currency, ψ, and vice versa, optimal behavior requires a compromise. We now proceed to model this.

We make several simplifying assumptions. Principally, these are that the adult female takes zero time to lay her eggs or, as we shall say, that handling time is zero; that she does not feed; that unparasitized hosts are equally acceptable at every site; that their number exceeds the number of eggs still carried by the insect; and that her dispersal time to a subsequent site is a random variable with a known probability distribution. We will discover that the optimal compromise depends critically upon the mean of this distribution. We do not have data to test these five assumptions. Nevertheless, Professor Strong has a hunch that the second assumption may not be unreasonable; and the third is quite reasonable because the host eggs are virtually uniform in size and potential to support parasitoid development. Again, although the first and fourth assumptions are not strictly justified, the first should yield to a good approximation if handling time is sufficiently short compared to searching time, and the second should yield a good approximation if our parasitoid is an early invader of virgin territory.

Although the concept of Pareto-optimality was introduced in Section 12.5, we will define it again for the particular purpose we have in mind. Let the interval $[0, \omega]$ correspond to the adult life span of our female wasp. Let a denote her age in suitable time units, with $a = 0$ corresponding to the beginning of her adult life span and $a = \omega$ to her death. Suppose that she is carrying x eggs and has attained age a when she encounters a site. Let $\phi(x, a, u)$ denote the number of eggs she can expect to lay between now and the end of her life, if she lays u eggs at the present site (where $0 \leq u \leq x$). Likewise, let $\psi(x, a, u)$ denote the number of sites at which she can expect to lay, if she lays u eggs at the present site. Then the decision to lay u eggs is Pareto-optimal (or unimprovable) if there is no other decision that will increase either currency without decreasing that of the other, i.e., if there does not exist v, satisfying $0 \leq v \leq x$, such that either $\phi(x, a, v) > \phi(x, a, u)$ and $\psi(x, a, v) \geq \psi(x, a, u)$ or $\phi(x, a, v) \geq \phi(x, a, u)$ and $\psi(x, a, v) > \psi(x, a, u)$. (A numerical illustration of this definition will appear in Table 12.6.)

Let us also redefine utopia point for the same site as in the previous paragraph. You will recall from Section 12.5 that the utopia point lies in currency space, where a typical point has coordinates (ϕ, ψ). Let U_ϕ denote a decision that would maximize the egg currency without any regard for the site currency. Likewise, let U_ψ denote a decision that would maximize the site currency without any regard for the egg currency. Then the point in currency space with coordinates (ϕ_u, ψ_u), where $\phi_u \equiv \phi(x, a, U_\phi)$ and $\psi_u \equiv \psi(x, a, U_\psi)$, is our utopia point. This utopia point is always unique, though U_ϕ or U_ψ need not be. To be quite specific, $U_\phi = x$ is unique; whereas U_ψ is not unique, although we know that $U_\psi \geq 1$ (unless, of course, $x = 0$). Also $\phi_u = x$, but there is no simple formula for the (unique) value of $\psi(x, a, U_\psi)$ because the expected number of future sites depends upon the age-specific mortality of the insect.

If $U_\phi = U_\psi$ then the optimal decision is clearly $u = U_\phi = U_\psi$, for no other decision can lead to a higher value of either currency. In general, however, U_ϕ does not equal U_ψ; i.e., the utopia point is unattainable, as we have seen already in Section 12.5. For such circumstances, we will again define the best nonutopian decision, i.e., the optimal compromise, to be the Salukvadze solution, which you will recall is the Pareto-optimal decision for which the corresponding pair of currency values is closest to the utopia point in currency space. We will denote distance in currency space between the point (ϕ, ψ) and the utopia point (ϕ_u, ψ_u) by $d(\phi, \psi)$. Moreover, we will use the familiar Euclidean norm to measure this distance, as we did in Section 12.5.[12]

Accordingly, suppose that a parasitoid aged a and carrying x eggs has just encountered a site. Let $\phi^*(x, a)$ denote the number of eggs she can expect to lay during the rest of her life if she makes the optimal decision now and, furthermore, makes optimal decisions at every future site. As explained above, by optimal decision we mean the Pareto-optimal decision for which the corresponding pair of currency values is nearest to the utopia point. Likewise, let $\psi^*(x, a)$ denote the number of sites the parasitoid can expect to parasitize if she behaves optimally from now until her death. Because ω is the maximum age to which our insect can live,

$$\phi^*(x, a) = 0 = \psi^*(x, a) \qquad \text{for all } x, \text{if } a \geq \omega,$$
$$\phi^*(0, a) = 0 = \psi^*(0, a) \qquad \text{for all } a. \tag{12.87}$$

Let the (continuous) random variable L denote the life span of the insect, and imagine she knew that upon leaving the present site she would survive to encounter her next site after t units of time had elapsed. Then the number of eggs she could expect to deposit at future sites, if she deposited u eggs at the present site, would be $\phi^*(x - u, a + t)$. But the insect could not know for sure that she would survive to age $a + t$. Hence, by essentially the same argument as yielded (12.38) in Section 12.3, her optimal future addition to egg currency, as presently perceived, would be not $\phi^*(x-u, a+t)$ but rather $\pi(a, t) \cdot \phi^*(x - u, a + t)$, where we have defined the conditional probability

$$\pi(a, t) \equiv \text{Prob}(L > a + t \mid L > a). \tag{12.88}$$

But the insect could not know for sure that she would encounter her next site after a further time t. Rather, the time until the next site is a random

---

[12]Even if you were remiss enough to ignore Exercise 12.16, you will recall from Section 12.5 that if we are going to calculate distances in currency space, then our currencies should have the same dimensions. Now, one could argue that ϕ has the dimensions of egg and ψ the dimensions of site; even though, because we restricted ourselves to six basic dimensions in Section 1.14, they are technically both dimensionless. But if the maximum number of eggs that can be laid during an insect's lifetime is ϕ_{max}, and if the maximum number of sites that can be reached is ψ_{max}, then the proportions ϕ/ϕ_{max} and ψ/ψ_{max} are both dimensionless, no matter how one argues; so that they would have the same dimensions if they were used as currencies. But $\phi_{max} = 32 = \psi_{max}$, and maximizing $(\phi, \psi)/32$ is equivalent to maximizing (ϕ, ψ).

variable, which we shall denote by T. Hence, as presently perceived, the optimal future addition to egg currency is the expected value of $\pi(a, T) \cdot \phi^*(x - u, a + T)$. The present addition to egg currency, of course, is u. Hence the number of eggs the insect can expect to lay between now and the end of her life, if she lays u eggs at the present site, where $0 \leq u \leq x$, is

$$\phi(x, a, u) = u + E[\pi(a, T) \cdot \phi^*(x - u, a + T)], \tag{12.89}$$

where E denotes expected value. Similarly,

$$\psi(x, a, u) = H(u) + E[\pi(a, T) \cdot \psi^*(x - u, a + T)], \tag{12.90}$$

where H denotes the threshold function defined by

$$H(z) \equiv \begin{cases} 0, & \text{if } z = 0; \\ 1, & \text{if } z \leq 1. \end{cases} \tag{12.91}$$

Then $\phi^*(x, a) = \phi(x, a, u^*(x, a))$ and $\psi^*(x, a) = \psi(x, a, u^*(x, a))$, where $u^*(x, a)$ is the optimal compromise, defined to be the Pareto-optimal u for which $d(\phi(x, a, u), \psi(x, a, u))$ is least when the insect is aged a and carrying x eggs.

If the random variable T were allowed to be continuous, then it would be extremely difficult to determine u^* from (12.89)–(12.90) unless we used discretization to approximate the integral that resulted from taking expected values. If discretization is inevitable, then we may as well discretize the random variable itself. Let us therefore assume that T is discrete, taking values $t = 1, 2, 3, \ldots$. Because the unit of time can (at least in principle) be made as small as we please, little generality is lost. Let p_t denote $\text{Prob}(T = t)$. Then (12.89)–(12.90) become

$$\phi(x, a, u) = u + \sum_{t=1}^{\infty} p_t \cdot \pi(a, t) \cdot \phi^*(x - u, a + t),$$
$$\psi(x, a, u) = H(u) + \sum_{t=1}^{\infty} p_t \cdot \pi(a, t) \cdot \psi^*(x - u, a + t). \tag{12.92}$$

To derive an optimal compromise from (12.92), we will now further assume that the insect's life span has distribution given by

$$\text{Prob}(L > s) \equiv \left(1 - \frac{s}{\omega}\right)^{\delta}, \qquad 0 \leq s \leq \omega, \qquad \delta \equiv \frac{\omega - \lambda}{\lambda}. \tag{12.93}$$

We introduced this distribution in Exercise 4.16, whence you can verify that our insect's adult life expectation is $E[L] = \lambda$. Then, from (12.88),

$$\pi(a, t) = \frac{\text{Prob}(L > a + t)}{\text{Prob}(L > a)} = \left(\frac{\omega - a - t}{\omega - a}\right)^{\delta}, \tag{12.94}$$

as you should verify for yourself, whence, on using (12.87), (12.92) reduces to

$$\phi(x,a,u) = u + \sum_{t=1}^{[\omega-a]} p_t \left(\frac{\omega - a - t}{\omega - a}\right)^{\delta} \cdot \phi^*(x - u, a + t), \qquad a < \omega,$$

$$\psi(x,a,u) = H(u) + \sum_{t=1}^{[\omega-a]} p_t \cdot \left(\frac{\omega - a - t}{\omega - a}\right)^{\delta} \cdot \psi^*(x - u, a + t), \qquad a < \omega,$$

$$(12.95)$$

where $[\omega - a]$ is the integer part of $\omega - a$.

It follows immediately from (12.87) and (12.95), on setting $a = \omega - 1$, that $\phi(x, \omega - 1, u) = u$ and $\psi(x, \omega - 1, u) = H(u)$. Thus $U_\phi = x$ and $U_\psi = u$ for any u such that $x \geq u \geq H(x)$. At this age, when the insect has virtually achieved her maximum life span, the utopia point $(x, H(x))$ is attainable, so that $u = x$ is the only Pareto-optimal decision. Hence the optimal compromise is given by $u^*(x, \omega - 1) = x$, with $\phi^*(x, \omega - 1) = x$ and $\psi^*(x, \omega - 1) = H(x)$.

It now follows further, on setting $a = \omega - 2$, that

$$\phi(x, \omega - 2, u) = u + p_1 2^{-\delta}(x - u),$$
$$\psi(x, \omega - 2, u) = H(u) + p_1 2^{-\delta} H(x - u). \qquad (12.96)$$

Thus $u = x$ is Pareto-optimal because any other choice of u would decrease ϕ. If $x \leq 1$, then $u = x$ is the only Pareto-optimal decision and $(\phi_u, \psi_u) = (x, H(x))$, implying that the utopia point is again attainable and that $u^*(\omega - 2, x) = x$. If $x \geq 2$, however, then $U_\phi = x$ and $U_\psi = x - 1$, implying that the utopia point $(\phi_u, \psi_u) = (x, 1 + p_1 2^{-\delta})$ is unattainable. On the other hand, both $u = x$ and $u = x - 1$ are now Pareto-optimal decisions, because increasing u from $x - 1$ to x would decrease ψ and decreasing u from x to $x - 1$ would decrease ϕ. The points in the currency space corresponding to these decisions are, respectively, $(\phi_0, \psi_0) = (x, 1)$ and

$$(\phi_1, \psi_1) = (x - 1 + p_1 2^{-\delta}, 1 + p_1 2^{-\delta}). \qquad (12.97)$$

Observe from Fig. 12.9 that (ϕ_0, ψ_0) is closer to the utopia point than (ϕ_1, ψ_1) if

$$p_1 2^{-\delta} < 1 - p_1 2^{-\delta}; \qquad (12.98)$$

whereas (ϕ_1, ψ_1) is closer to the utopia point than (ϕ_0, ψ_0) if the above inequality is reversed. We ignore as fantasy the intermediate possibility that the parameters of an insect's environment could be so precisely related as to convert (12.98) to an equality and destroy the uniqueness of the optimal compromise. Hence, combining our results, we obtain

$$u^*(x, \omega - 2) = \begin{cases} x, & \text{if } x \leq 1 \text{ or } x \geq 2 \text{ and } p_1 < 2^{\delta-1}; \\ x - 1, & \text{if } x \geq 2 \text{ and } p_1 > 2^{\delta-1}. \end{cases} \qquad (12.99)$$

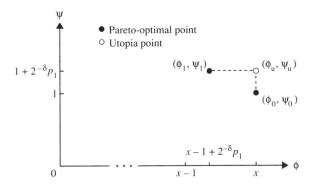

Fig. 12.9 Pareto-optimal points and utopia point for $a = \omega - 2$ and with L distributed according to (12.93). As drawn here, $d(\phi_0, \psi_0) < d(\phi_1, \psi_1)$, so that $u^*(x, \omega - 2) = x$ in (12.99).

Continuing in this manner, we can now compute recursively the optimal compromise for ages $\omega - 3, \omega - 4, \omega - 5, \ldots$, much as in earlier sections of this chapter, for any distribution of dispersal time T. Thus our model is essentially a combination of vector optimization concepts with the idea of dynamic programming. To illustrate the recursion, we will use a Poisson distribution for the dispersal time. Recall from Chapter 5, however, that a random variable with Poisson distribution takes values 0, 1, 2, 3, $\ldots$; whereas our insect's dispersal time takes values 1, 2, 3, $\ldots$. Let us therefore assume that $T - 1$ has a Poisson distribution, with mean $\mu - 1 (> 0)$, so that the mean dispersal time is $E[T] = \mu (> 1)$ and

$$p_t \equiv \mathrm{Prob}(T = t) = \frac{(\mu - 1)^{t-1}}{(t - 1)!} e^{-(\mu - 1)}, \qquad t = 1, 2, \ldots. \tag{12.100}$$

For the computation, we must then assign values to three parameters, namely, the mean life span λ, the maximum life span ω, and the mean dispersal time μ.

I computed the optimal compromise from (12.95) and (12.100) for various values of the three parameters λ, ω, and μ. It always had the form

$$u^*(x, a) = \begin{cases} x - r(a), & \text{if } x \geq r(a) + 1, \\ H(x), & \text{if } x \leq r(a), \end{cases} \tag{12.101}$$

where $r(a)$ is determined by computation and where the dependence of r on the values of λ, ω, and μ is suppressed by the notation. Thus at each age a there is an optimal reserve $r(a)$ of eggs, which the insect should retain in her body. When she encounters a site, if she is carrying more than $r(a)$ eggs then she should lay all except $r(a)$; otherwise she should lay only one egg.

Of the three parameters λ, ω, and μ, the optimal reserve depends most critically on μ, the mean dispersal time. Accordingly, it is convenient to adopt notation that suppresses the dependence of r on λ and ω (i.e., on δ), but makes explicit the dependence of r on μ. Instead of (12.101), let us therefore write

$$u^*(x, a, \mu) = \begin{cases} x - R(a, \mu), & \text{if } x \geq R(a, \mu) + 1; \\ H(x), & \text{if } x \leq R(a, \mu). \end{cases} \qquad (12.102)$$

Thus $u^*(x, a, \mu)$ is the optimal compromise for an insect who encounters a site at age a while carrying x eggs, predicated on the assumption that her mean dispersal time will remain fixed at μ for the remainder of her life span. Note that (12.102) agrees with (12.99) for $a = \omega - 2$, where $R(\omega - 2, \mu)$ is 1 or 0 according to whether μ is greater or less than $1 + (1 - \delta) \cdot \ln 2$.

For $\lambda = 30$ time units and $\omega = 50$ time units (so that $\delta = 2/3$), the optimal reserve is plotted against age in the upper diagram of Fig. 12.10, for $\mu = 2$, $\mu = 3$, $\mu = 4$, and $\mu = 10$. The diagram illustrates, for example, that when $\mu = 2$ the optimal reserve drops from five eggs to four as the age of the insect increases from 26 time units to 27. Points in currency space corresponding to this change of reserve are recorded in Ta-

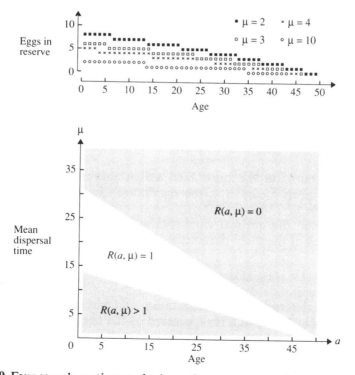

Fig. 12.10 Four sample sections and schematic contour map of the optimal reserve function $R = R(a, \mu)$. In the upper diagram, if the symbol corresponding to a value of μ is not found for some value of a, then its position coincides with that of the adjacent lower value of μ.

ble 12.6 together with their distances from their respective utopia points: namely, $(\phi_u, \psi_u) = (x, 5.111)$ for $a = 26$ and $(\phi_u, \psi_u) = (x, 5.067)$ for $a = 27$. The table corresponds to $x = 10$, but for any other $x \geq 10$, $\phi(x, a, u)$ and $\psi(x, a, u)$ can be obtained simply by adding $x - 10$ to the numbers recorded in columns 2, 3, 5, and 6 of the table. Note that there are six Pareto-optimal decisions at both ages.

For these and other values of μ, points (a, μ) corresponding to particular values of R can be plotted in the a-μ plane. In this way, a contour map of optimal reserves can be constructed. It is depicted schematically in the lower diagram of Fig. 12.10, where the boundaries between regions of constant R are accurate to within a unit of age or time. Because in principle (12.95) can be solved for any value of a, the true boundaries are continuous and the regions of constant R fill the rectangle $0 < a \leq \omega$, $1 < \mu \leq \omega$. (The boundaries are nonlinear, in general, although the two sketched in Fig. 12.10 are almost linear.) Moreover, the restriction $\mu > 1$ is merely a consequence of discretizing the random variable T. Thus if (12.89)–(12.90) could be solved with T continuous, then the regions of constant R would fill the square $0 < a \leq \omega$, $0 < \mu \leq \omega$, with the boundaries converging at $(\omega, 0)$. We see that, for given μ, there exists a set of critical ages $a_0(\mu), a_1(\mu)$, ..., such that the optimal reserve for the next site encountered drops from $j+1$ to j as the insect passes age $a_j(\mu)$. The equations of the boundary curves between regions of constant R can thus be written in the form $a = a_j(\mu)$, the boundaries represented by the lower diagram of Fig. 12.10 being $a = a_0(\mu)$ and $a = a_1(\mu)$. Increasing the value of the mean life span, λ, shifts these curves upward and to the right; decreasing λ has the opposite effect. The equations of the boundary curves between regions of constant R in the a-μ plane can also be written in the form $\mu = \mu_j(a)$. Thus, for each age a at which a site may be encountered, there exists a set of critical mean dispersal times $\mu_0(a), \mu_1(a), \ldots$, such that the optimal reserve will be less than or equal to j if $\mu > \mu_j(a)$. Either way, we see that $R = R(a, \mu)$ satisfies

$$\frac{\partial R}{\partial a} < 0, \qquad \frac{\partial R}{\partial \mu} < 0. \qquad (12.103)$$

Table 12.6   Examples of decision points in currency space when $\mu = 2$.

u	$\phi(10, 27, u)$	$\psi(10, 27, u)$	$d(\phi, \psi)$	$\phi(10, 26, u)$	$\psi(10, 26, u)$	$d(\phi, \psi)$
10	10.000	1.000	4.067	10.000	1.000	4.111
9	9.941	1.941	3.127	9.943	1.943	3.168
8	9.821	2.821	2.253	9.828	2.828	2.289
7	9.637	3.637	1.475	9.653	3.653	1.498
6	9.388	4.388	0.914	9.415	4.415	0.909
5	9.070	5.067	0.930	9.111	5.111	0.889
≤ 4	< 9.070	5.067	> 0.930	< 9.111	5.111	> 0.889

How might an insect determine μ? One can speculate that she would estimate μ by sampling from the lengths of her early dispersal flights. If such an estimate were low, then the optimal reserve would be large, and few eggs would be laid. Suppose, now, that the insect were to pick some inappropriate flight coordinates and undergo an exceptionally long dispersal flight. This might encourage her to revise her estimate of μ. If the revised estimate were sufficiently long, then (12.103) would imply that a much larger fraction, possibly all, of her remaining eggs should be laid immediately. Is this how *Anagrus delicatus* actually behaves? According to Professor Strong, the evidence suggests that the answer is yes. *Anagrus delicatus* rarely deposits her entire load of eggs, but when she does so, it is invariably the case that she has just undertaken a very long dispersal flight.

How good is our model—does it fit the facts? We have hypothesized that, with a view to maximizing her future reproductive success, *Anagrus delicatus* forages to achieve an optimal compromise, in an uncertain environment, between laying as many eggs as possible and reaching as many sites as possible. Our model has predicted that *A. delicatus* will disperse from a site with a reserve $R(a, \mu)$ of eggs, where a is her age, μ the expected value of her dispersal time, and (12.103) is satisfied. This is qualitatively consistent with the available data. Thus our model should not be rejected. On the other hand, there is clearly room for improvement, with a view to applying a more stringent test.

If an hour is taken to be the unit of time, then the values chosen here for ω and λ may not be unrealistic for *A. delicatus*. It is not, of course, realistic to suppose that sites can be located only after precise multiples of one hour have elapsed, but the unit of time is easily reduced, in principle, to a realistic level, although this might lead to higher costs of computation than are justified by the realism of other parts of the model. There is also no reason to suppose that L and T are distributed according to (12.93) and (12.100), but more appropriate distributions are easily substituted, at least in principle. Moreover, again in principle, we could easily extend this model to incorporate the effects of feeding, nonzero handling times, and a lack of suitable hosts.

In practice, it may be considerably easier to substitute an appropriate distribution for L from an insect life table than to determine an appropriate distribution for T. Preliminary experimentation with other theoretical distributions indicates, however, that the form of the optimal compromise is not only qualitatively but also quantitatively insensitive to the form of T's distribution, being sensitive only to its mean. This is not such a surprising result. Because expected value is the statistic being used as a lens to peer into the future and "determinize" the randomness of the environment, we would perhaps be surprised if the optimal compromise weren't far more sensitive to the mean of T than to any other moment of its distribution.

Thus, despite the room for improvement, it seems that our model has at least the makings of a good one. Indeed it seems to me that vector optimization models should be much more widely applicable in predicting behavior from the hypothesis that animals maximize fitness. We have already remarked that *if* a suitable fitness function, f, could be prescribed then we would abandon our use of multiple currencies and instead maximize $E[f]$. But the consequences of an animal's actions, in terms of its future reproductive success, depend not just on what happens to it between now and the end of its life (or until it stops having offspring), but also on what subsequently happens to the offspring and the offspring's offspring. It would therefore seem to be extremely difficult, in general, to obtain quantitative knowledge about f. Yet it is precisely in the absence of such quantitative knowledge that our vectorial approach is so valuable; for, as we have demonstrated in the last two sections, it enables testable predictions to be generated merely from the qualitative knowledge that f varies monotonically with respect to its factors. Moreover, there is no reason in principle why this vectorial approach would not work as well for a higher number of currencies, say m, than the two we have considered in this chapter, although the technical details might be rather more complicated.

In conclusion, it seems to me that a vectorial approach to modelling animal behavior is both necessary and valuable. But you should know by now that you are not obliged to share my opinion! Then what do you have to say about the matter? ... It seems that the matter is still undecided. Which, in mathematical modelling, is how the matter often rests.

Exercises

12.1 Obtain the probability distribution of the surplus Y_k defined by (12.2).

12.2 Obtain (12.9).

12.3 Obtain (12.10) and (12.11).

12.4 Obtain (12.16) and (12.17).

*__12.5__ (i) According to (12.36), a personnel manager should not appoint the first of ten candidates unless her score exceeds 85 out of 100. Does this figure strike you as high? Elucidate.
(ii) Suppose that the scores Y_k in Section 12.2 are independent of k but have the (asymmetric) triangular probability density function

$$f_k(y) = \begin{cases} \dfrac{2y}{a}, & 0 \le y \le a; \\ \dfrac{2(1-y)}{(1-a)}, & a \le y \le 1. \end{cases}$$

Thus scores are more highly grouped about the mean $(1 + a)/3$ than according to the uniform distribution, because the variance $(a^2 - a + 1)/18$ is less than that of the uniform distribution. Show that for this distribution the recursion (12.34) becomes

$$\psi_k = \frac{1}{3}\left(1 + a + \frac{\psi_{k-1}^3}{a}\right), \qquad \text{if } \psi_{k-1} < a;$$

$$\psi_k = \psi_{k-1} + \frac{(1 - \psi_{k-1})^3}{3(1 - a)}, \qquad \text{if } \psi_{k-1} \geq a.$$

Observe the analogue of Table 12.1 for the special (symmetric) case in which $a = 1/2$. Notice how the thresholds are lowered by reduced variance of the population from which the candidates are drawn. In particular, the first of a field of ten candidates should now be accepted if her score is $\lambda_{10} = \psi_9 \approx 0.75$ or greater. The corresponding value for a uniform distribution is 0.85 (see text).

*12.6   (i) Do you think it's reasonable to assume that the distributions of the random variables Y_k in Section 12.2 are independent of k? Why or why not?

(ii) If you thought not, what qualitative features should more reasonable distributions have? What would be the effect of this on the values of the thresholds λ_k?

(iii) Illustrate your answer to (ii) by choosing a particular set of distributions. *Hint:* You may be able to adapt Exercise 12.5 by making a depend suitably on k.

*12.7   (i) Immediately after the interviewer in Section 12.2 has interviewed candidate $N - k + 1$, the score of the appointed candidate is, from (12.28):

$$v_k = \begin{cases} y_k, & \text{if } \lambda_k \leq y_k; \\ \psi_{k-1}, & \text{if } \lambda_k > y_k. \end{cases}$$

Assuming that ψ_{k-1} and y_k are known, v_k is a function of λ_k. Hence sketch the graph of $v_k = v_k(\lambda_k)$. *Hint:* There are three possible cases (or two, if you assume that $y_k \neq \psi_{k-1}$).

(ii) Write down a simple expression for the maximum of v_k and interpret it.

12.8   Show that the expected length of the field of candidates in Section 12.3 is $1/q$.

12.9   Verify (12.43).

12.10   (i) Suppose that the triangular distribution of Exercise 12.5 is used for the faculty hiring model of Section 12.3. Show that (12.42) yields the recursion

$$\psi_k = \frac{1}{3}\left(1 + a + \frac{p^3 \psi_{k-1}^3}{a}\right), \qquad \text{if } p\psi_{k-1} < a;$$

$$\psi_k = p\psi_{k-1} + \frac{(1 - p\psi_{k-1})^3}{3(1-a)}, \qquad \text{if } p\psi_{k-1} \ge a.$$

(ii) For $a = 1/2$, write down an equation that defines $\lambda_\infty(p)$ implicitly. Hence show that $\lambda_\infty(7/6) = 1/2$. How would you evaluate $\lambda_\infty(p)$ for other values of p and sketch the curves marked S in Fig. 12.1?

(iii) Write down an equation that defines $\psi_\infty(p)$, and hence $\lambda_\infty(p)$, implicitly for $a = 1$. This equation is used to sketch the curves marked T in Fig. 12.1.

*12.11 Some personnel officers, particularly those of government agencies, like to set an applications deadline even for very highly skilled positions. Can you interpret their preference in the light of the model of Section 12.3? Can you comment on the distribution of merit?

12.12 (i) Verify (12.55).

(ii) Verify by mathematical induction that (12.54) implies (12.55).

*12.13 (i) In the parking problem of Section 12.4, let the probability that a parking space is free on the southbound road be, not p, but p_s, where $p_s < p$. What is then the optimal policy?

(ii) Is the use of the model restricted to an old-fashioned town? Can you suggest how the model might be used for hotels? What important assumption would have to be made? In which circumstances would searching for a hotel require a more elaborate model?

12.14 (i) Verify (12.71).

(ii) Verify that (12.71) and (12.74) are equivalent, through (12.72).

(iii) Show that the utopia point in Fig. 12.7(b) is (ϕ_u, ψ_u), where

$$\phi_u = \frac{\beta(\epsilon + 1)}{\epsilon + \beta}, \qquad \psi_u = \frac{\beta - 1}{\epsilon + \beta + \eta\theta}.$$

(iv) Show that the point S in Fig. 12.7(b), at which the line through U perpendicular to BC meets BC, is (ϕ_s, ψ_s), where

$$\phi_s = \frac{\epsilon^2 \phi_u + \epsilon\psi_u + 1}{\epsilon^2 + 1}, \qquad \psi_s = \frac{\epsilon\phi_u + \psi_u - \epsilon}{\epsilon^2 + 1}.$$

Hence verify (12.75).

(v) Verify (12.85).

*12.15 In Section 12.5, define the optimal compromise (u_1^*, u_2^*) to be the decision that corresponds to the closest point in currency space to the utopia point, but let distance be measured according to the L_ρ norm, i.e., by using formula (12.77). How sensitive is (u_1^*, u_2^*) to the value of ρ?

*12.16 Criticize the vector optimization model of Section 12.5 from any viewpoint you consider relevant. Which of Hypotheses 1 and 2 should be rejected? One, the other, both, or neither? *Hint*: You may wish to reexamine (12.72).

**12.17 An implicit assumption in the model that predicts (12.61), from the hypothesis that animals are maximizers of energy rate, is that they have perfect (or at least virtually perfect) information about the values of several parameters. In particular, we assumed that birds can instantly distinguish between profitable worms and unprofitable ones. In practice, however, a bird would have to estimate the various parameters by sampling (see Section 8.7). For example, she would have to estimate λ_1 by sampling the distribution of W_1, and similarly for the other parameters. Thus we have implicitly assumed birds to be so quick and accurate at estimating parameters that the effect of sampling on the optimal diet is unimportant.

Toward the end of their paper, however, Krebs et al. (1977, p. 36) suggest that the effect of sampling may be so important as to explain the discrepancy between (12.61) and the results of their experiment. In other words, they suggest that Hypothesis 1 in Section 12.5 is correct, but that the effect of sampling is so important at supercritical values of λ_1 that any model that ignores it completely will make an incorrect prediction. After reading what Krebs et al. have to say about the matter, attempt to build a model that tests their idea.

**12.18 (You will need a computer for this one.) A bird forages for time at most T_{max}, and her capacity for prey is E_{max} units of energy. There are two types of prey, which the bird encounters randomly. For prey of type i, the rate of encounter is λ_i, the handling time is h_i, and the energy content is E_i; $i = 1$, 2. Type-1 prey is more profitable than type-2, as in (12.58). Let foraging last for time $T_{max} - \Psi$, and let Φ be the total energy value of captured prey; both Φ and Ψ are random variables, with $\Phi \leq E_{max}$ and $\Psi \leq T_{max}$.

(i) Suppose that the bird forages either for time T_{max} or until she becomes satiated. To maximize $E[\Phi]$, where E denotes expected value, how should the bird select prey? With reference to Section 12.5, is the optimal proportion of type-1 prey in the diet still independent of λ_2 at supercritical values of λ_1? Comment.

(ii) The rationale behind computing the policy that maximizes $E[\Phi]$ would be that the bird's future reproductive success is an increasing function of Φ. Because Ψ/T_{max} is the proportion of time available for other activities, however, the bird's future reproductive success is also an increasing function of Ψ. Accordingly, suppose that the bird forages to achieve an optimal compromise between maximizing $E[Y_1]$ and maximizing $E[Y_2]$, where

$$Y_1 = \frac{\Phi}{E_{max}}$$

is the satisfied proportion of energy capacity and

$$Y_2 = \frac{\Psi}{T_{max}}$$

is the proportion of time remaining for other activities. How should the bird select prey? With reference to Section 12.5, which do you now consider superior: Hypothesis 1 or Hypothesis 2?

   Hints: Using the data from Krebs et al. (1977) presented in Section 12.5, adapt the model of Section 12.6. Suppose that, on encountering an item of prey, the bird can either reject it or accept it, and then either cease to forage or continue to forage; thus there are four feasible decisions at each encounter. Assume that the bird is certain to survive its foraging mission (unlike the insect in Section 12.6).

Supplementary Notes

12.1 Substituting (12.20) into (12.10) and (12.11), we obtain

$$J_2(w_2) = \begin{cases} 210 + 30y_2 + 15(w_2 + y_2) - \dfrac{11}{100}(w_2 + y_2)^2, & \text{if } w_2 + y_2 \leq 30; \\[2mm] 231 + 30y_2 + 12.9(w_2 + y_2) - \dfrac{1}{25}(w_2 + y_2)^2 \\[2mm] \quad - \dfrac{7}{9000}(w_2 + y_2)^3, & \text{if } w_2 + y_2 > 30. \end{cases}$$

The maximum of this on $0 \leq w_2 \leq 100 - y_2$ occurs at $(59.16 - y_2)H(59.16 - y_2)$, where H is the threshold function (7.3). We assume, however, that the buyer can order only a whole number of items and hence take $w_2^*(y_2) = (59 - y_2)H(59 - y_2)$ or, comparing with (12.19), $\lambda_2 = 59$. Thus we get

$$\psi_2(y_2) = J_2(w_2^*(y_2)) = \begin{cases} \underline{693.1} + 30y_2, & \text{if } y_2 \leq 59; \\[2mm] 231 + 42.9y_2 - \dfrac{1}{25}y_2^2 - \dfrac{7}{9000}y_2^3, & \text{if } y_2 > 59. \end{cases}$$

The underlined number has been rounded to four significant figures (you can determine its exact value from the fact that $\psi_2(y_2)$ is continuous). On using (12.9) with $k = 3$ we get

$$E[\psi_2(Y_2)] = \underline{693.1}\left(1 - \frac{w_3 + y_3}{100}\right) + \frac{1}{100}\int_0^{w_3 + y_3} \psi_2(y)\,dy;$$

whence

$$E[\psi_2(Y_2)] = \underline{693.1} + 0.15(w_3 + y_3)^2, \qquad \text{if } w_3 + y_3 \leq 59;$$

$$E[\psi_2(Y_2)] = \underline{792.2} - 4.621(w_3 + y_3) + 0.2145(w_3 + y_3)^2$$
$$\quad - \frac{1}{7500}(w_3 + y_3)^3 - \frac{7}{36 \times 10^5}(w_3 + y_3)^4, \qquad \text{if } w_3 + y_3 > 59.$$

Continuing in this fashion (and still assuming that only whole-number quantities can be stocked or ordered), we deduce from (12.8) that $\lambda_k = 68$ for

$k \geq 3$, and

$$\psi_3(y_3) = \underline{1158} + 30y_3, \qquad\qquad\qquad\qquad\qquad\text{if } y_3 \leq 68;$$
$$= \underline{739.4} + \underline{40.69}y_3 - 0.0498y_3^2 - \frac{7}{56250}y_3^3 - \frac{49}{27 \times 10^6}y_3^4, \quad \text{if } y_3 > 68;$$

and

$$\psi_k(y_k) - C_k = \begin{cases} 30y_k, & \text{if } y_k \leq 68; \\ 45y_k - 0.11y_k^2 - 511.36, & \text{if } y_k > 68, \end{cases}$$

for $k \geq 4$, where C_k is just a constant ($C_4 = \underline{1592}$, for example). Underlined numbers have been rounded to four significant figures; all ψ_k are continuous.

12.2 The optimal decision (12.61) is most readily obtained by inspection of two identities:

$$\Phi(\lambda_1, 0) - \Phi(u_1, u_2) \equiv \frac{(\Delta\lambda_1 - E_2)u_2 + E_1(\lambda_1 - u_1)}{(1 + h_1\lambda_1)(1 + h_1u_1 + h_2u_2)}, \qquad\text{(a)}$$

$$\Phi(\lambda_1, \lambda_2) - \Phi(u_1, u_2) \equiv \frac{E_1\lambda_1 + E_2\lambda_2 - (E_1 + \lambda_2\Delta)u_1 - (E_2 - \lambda_1\Delta)u_2}{(1 + h_1\lambda_1 + h_2\lambda_2)(1 + h_1u_1 + h_2u_2)}. \qquad\text{(b)}$$

In view of (12.59), (a) ensures that $\Phi(\lambda_1, 0) \geq \Phi(u_1, u_2)$ for all feasible u_1, u_2 when $\Delta\lambda_1 > E_2$; and (b) ensures that $\Phi(\lambda_1, \lambda_2) \geq \Phi(u_1, u_2)$ for all feasible u_1, u_2 when $\Delta\lambda_1 < E_2$, because the latter inequality implies that the numerator of the fraction in (b) cannot be negative. A third identity

$$\Phi \equiv \frac{E_2}{h_2} + \frac{\Delta u_1 - E_2}{(1 + h_1u_1 + h_2u_2)h_2} \qquad\qquad\text{(c)}$$

shows that $\Phi < E_2/h_2$ or $\Phi > E_2/h_2$ according to whether $\Delta u_1 < E_2$ or $\Delta u_1 > E_2$. Moreover, if $\Delta\lambda_1 = E_2$, then the second term on the right-hand side of (c) cannot be positive, by (12.59); while it takes the value zero when $u_1 = \lambda_1$. Hence the maximum of Φ when $\Delta\lambda_1 = E_2$ is E_2/h_2, and this is achieved for any value of u_2; i.e., $\Phi(\lambda_c, u_2) \equiv E_2/h_2$, where λ_c is defined by (12.63). Note that methods virtually identical to the above can be used to establish (12.78).

12.3 Let (u_1^*, u_2^*) minimize (12.82) subject to (12.59). On defining $D = 1 + h_1u_2 + h_1u_2$, we find that

$$\frac{\partial L}{\partial u_1}(u_1, u_2) = -\frac{c(1 - \alpha)h_1}{D^{c+1}} - \left(\frac{h_2}{E_2D}\right)^b \frac{\alpha b(E_1 + u_2\Delta)(E_1u_1 + E_2u_2)^{b-1}}{D}$$

is strictly negative for any u_1, u_2 satisfying (12.59) because $0 < \alpha < 1$. It follows immediately that $u_1^* = \lambda_1$. Now u_2^* can be found by minimizing $L(\lambda_1, u_2)$ subject to $0 \leq u_2 \leq \lambda_2$, an ordinary calculus problem. We have

$$\frac{\partial L}{\partial u_2}(u_1, u_2) = -\frac{c(1 - \alpha)h_2}{D^{c+1}} - \left(\frac{h_2}{E_2D}\right)^b \frac{\alpha b(E_2 - u_1\Delta)(E_1u_1 + E_2u_2)^{b-1}}{D}.$$

Setting $u_1 = \lambda_1$, we find after some manipulation that $\partial L(\lambda_1, u_2)/\partial u_2$ is negative throughout $0 \le u_2 \le \lambda_2$, implying $u_2^* = \lambda_2$, if $\lambda_1 < \lambda_1^-(\alpha)$, and positive throughout $0 \le u_2 \le \lambda_2$, implying $u_2^* = 0$, if $\lambda_1 > \lambda_1^+(\alpha)$; whereas if $\lambda_1^-(\alpha) < \lambda_1 < \lambda_1^+(\alpha)$, then the sign of $\partial L(\lambda_1, u_2)/\partial u_2$ changes from negative at $u_2 = 0$ to positive at $u_2 = \lambda_2$. In the last of the three cases, the minimum of $L(\lambda_1, u_2)$ on $0 \le u_2 \le \lambda_2$ occurs where $\partial L(\lambda_1, u_2)/\partial u_2 = 0$, i.e., at $u_2 = U^*(\alpha, \epsilon)$.

EPILOGUE

The process of mathematical modelling is a ceaseless one of successive refinement. A simple model, if found to be false, is improved by the methods of Chapter 9, then tested again by the methods of Chapter 4. But a book is distinctly finite. It has to end somewhere, and this one ends here.

Although, in the grand scheme of things, this book is merely a port of entry into the vastness of mathematical modelling, I have reason to hope that it will leave you better prepared for the journey. There are hundreds of languages that a monoglot would describe as foreign. Of all these foreign languages, the most difficult to study is the first. Having experienced the process of studying a foreign language once, a second foreign language is studied more readily, a third more readily still. Yet for all that, the languages must still be learned, and one can hardly be said to have mastered them until one has spoken them fluently for several years. Likewise, having once guided you through the process of mathematical modelling, I hope that you will find it easier in future to use and develop models. Yet for all that, the art of applying mathematics must still be learned, and you can hardly be said to have mastered it until you have practiced it confidently for several years.

Writing this textbook has absorbed almost all of my creative energy for at least two years, and a significant part of it for at least three more. Knowing the manuscript so intimately at every stage of its development, I have been all too aware that it would be seen to have shortcomings, and I have spared no effort to remove such as I perceived to be truly removable. When improvement in one respect would have led to disimprovement in another, however—leaving me further, on balance, from achieving my goal—then I resisted the temptation to change. A less structured approach would have left you without a blueprint for applying mathematics, yet a more structured one would not faithfully have captured the modelling experience; and more attention to the technical aspects of approximation and error analysis, as described in texts on numerical analysis and statistics, would have distracted attention from the process of modelling itself. Above all else, more rigor would have meant less substance; and confidence in one's modelling skills comes with practice, not theory. Between such conflicting considerations, I have attempted to achieve an acceptable compromise.

Did I succeed? As a fervent advocate of criticism, it is my duty to remind you that you must decide.

APPENDIX 1:
A Review of Probability
and Statistics

This appendix is intended to jog the memory of someone who has already taken a first course in probability and statistics. With few exceptions, results are stated rather than derived. Moreover, results are restricted to those actually needed in the text. For a more detailed treatment, consult an introductory text on probability, such as Arthurs (1965) or Meyer (1970).

We begin by reviewing some notation from set theory. If some element u is contained in some set U then we write $u \in U$. For example, if U denotes the set of all real numbers satisfying $0 \leq x < 1$, written $[0, 1)$, and if $u = \pi/4$, then $u \in U$. If V is another set then the *union* of U and V, written $U \cup V$, is defined to be the set of all elements u such that $u \in U$ *or* $u \in V$; and the *intersection* of U and V, written $U \cap V$, is defined to be the set of all elements u such that $u \in U$ *and* $u \in V$. For example, if $V = [1/2, 3/2)$ and $U = [0, 1)$ then $U \cap V = \{x : 1/2 \leq x < 1\} = [1/2, 1)$ and $U \cup V = \{x : 0 \leq x < 3/2\} = [0, 3/2)$. Note that 1 does not belong to $U \cap V$ (written $1 \notin U \cap V$). More generally, if M (usually 0 or 1) and $N(> M)$ are integers, and if a sequence of $N - M + 1$ sets is denoted by $\{U_M, U_{M+1}, U_{M+2}, \ldots, U_N\}$, then the union of all those sets, written $U_M \cup U_{M+1} \cup U_{M+2} \cup \cdots \cup U_N$ or

$$\bigcup_{k=M}^{N} U_k, \tag{A.1}$$

is defined to be the set of all elements u such that $u \in U_j$ for *at least one* j, $M \leq j \leq N$. Likewise, the intersection of all $N - M + 1$ sets, written $U_M \cap U_{M+1} \cap U_{M+2} \cap \cdots \cap U_N$ or

$$\bigcap_{k=M}^{N} U_k, \tag{A.2}$$

is defined to be the set of all elements u such that $u \in U_j$ for *every j* satisfying $M \leq j \leq N$. It can be shown that the operation of union is distributive over the operation of intersection, i.e., that if U is an arbitrary set, then

$$\left(\bigcup_{k=M}^{N} U_k \right) \cap U = \bigcup_{k=M}^{N} (U_k \cap U) . \tag{A.3}$$

If set V contains every element in set U, i.e., if $u \in U$ implies $u \in V$, then we say that U is a *subset* of V. If, in addition, there exists $v \in V$ such that $v \notin U$, then U is a proper subset of V. A set that contains no elements is said to be *empty* and is denoted by $\varnothing$; e.g., if $U = (0, 1)$ and $V = (1, 2)$ then $U \cap V = \varnothing$. Every set contains the empty set and satisfies $U \cup \varnothing = U$, $U \cap \varnothing = \varnothing$. Any two sets U, V satisfying $U \cap V = \varnothing$ are said to be *mutually disjoint*. Moreover, if a sequence $\{U_M, U_{M+1}, U_{M+2}, \ldots, U_N\}$ of sets satisfies $U_i \cap U_j = \varnothing$ for all $M \leq i, j \leq N$ such that $i \neq j$, then (A.1) above is said to form a *mutually disjoint union*. The set Ω, which contains all elements under discussion, is the *universal set*. Moreover, for any subset U of Ω, we define the *complement* of U (strictly, with respect to Ω) to be the set of all elements that belong to Ω but not U. We will use an overbar to denote complement; thus $\bar{U} = \{u \in \Omega : u \notin U\}$.

Having dealt with the necessary set notation, let us now suppose that some experiment consists of a sequence of trials—repeated tosses of a coin, for example, or repeated throws of a die. Assume that the outcome of each trial can be associated with a real number. Thus if coin-tossing were the experiment, heads and tails might be associated with zero and one. This association is not in principle unique, but we shall always make it so by specifying it. For example, if die-tossing were the experiment, the number associated with each toss might be the number facing up on the die, the number facing down, or even their difference (why not their sum?). *Having specified the association uniquely*, it is both permissible and convenient to refer to the associated real number as the outcome, and we shall do so henceforward.

Let X denote the outcome of a trial *before* it is performed; let x denote the outcome of the trial *after* it is performed. Then x is known with certainty; whereas all we know about X is that it belongs to the set, S, of all possible outcomes. We say that X is a *random variable, x* is a *realization* of that random variable, and S is the *sample space*. For example, if the number associated with the toss of a die were the one facing up or the one facing down, then we would have $S = \{1, 2, 3, 4, 5, 6\}$; whereas if the associated number were their difference, then we would have $S = \{1, 3, 5\}$. If their sum were the associated number then, of course, we would have $S = \{7\}$ and the experiment would be pointless.

Because $S = \{1, 2, 3, 4, 5, 6\}$ can be associated with either the number facing up on a die or the one facing down, it is immediately clear that the same sample space may be associated with more than one random variable. More generally, however, a given sample space may be associated with an infinite sequence of random variables; then the sequence is said to form a *stochastic process*. For example, suppose that a die is tossed every minute and that the random variable $X(n)$ denotes the number facing up after n minutes. Then $\{X(n): n = 0, 1, 2, \ldots\}$ is a stochastic process.

Once the mechanism for associating real numbers has been specified, S is some subset of the real numbers, i.e., of the interval $\mathcal{R} \equiv (-\infty, \infty)$. Thus $\Omega = (-\infty, \infty)$. S may be a discrete set of points (as above), a continuous interval, or a combination of the two. For illustrations of the last two possibilities, suppose that a trial consists of spinning the arrow in Fig. A.1 (about O) and letting it come to rest at an angle to ON. Let the random variable Θ be that angle measured in radians. Then S is the continuous interval $0 < \theta \leq 2\pi$. Suppose, on the other hand, that a random variable X is defined by

$$X \equiv \begin{cases} \Theta/\pi, & \text{if } 0 < \Theta \leq \pi; \\ 2, & \text{if } \pi < \Theta \leq \dfrac{3}{2}\pi; \\ 3, & \text{if } \dfrac{3}{2}\pi < \Theta \leq 2\pi. \end{cases} \qquad (A.4)$$

Then $S = (0, 1] \cup \{2, 3\}$, so that S is a combination of a discrete set of points with a continuous interval. Note that X is a function of Θ. More generally, any function of a random variable is itself a random variable.

Any subset of S is called an *event*. Thus, in Fig. A.1, the interval $(1/2, 1)$ is the event that the arrow comes to rest making an angle between $\pi/2$ and π with ON. Let us define a *simple outcome* to be a single point from the sample space. Thus a simple outcome is always an event, but an event may contain as much as infinitely many simple outcomes.

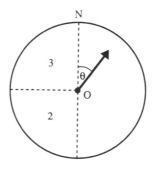

Fig. A.1

Now, let

$$S = \bigcup_{k=M}^{N} U_k \qquad (A.5)$$

be a decomposition of the sample space into a mutually disjoint union of events, i.e., such that $E_i \cap E_j = \varnothing$ whenever $i \neq j$. Then $U_M, U_{M+1}, U_{M+2},$ ..., U_N are said to *exhaust* the sample space. A function $\mathcal{P}$ from S to the real numbers is a *probability function* if, for any such decomposition,

$$0 \leq \mathcal{P}(U_k) \leq 1, \qquad k = M, M+1, \ldots, N, \qquad (A.6a)$$
$$\mathcal{P}(S) = 1, \qquad (A.6b)$$
$$\mathcal{P}\left(\bigcup_{k=i}^{j} V_k\right) = \sum_{k=1}^{j} \mathcal{P}(V_k), \qquad (A.6c)$$

where $\{V_i, \ldots, V_j\}$ is an arbitrary selection of events from the set $\{U_M, U_{M+1}, U_{M+2}, \ldots, U_N\}$, which forms the sample space (for any k satisfying $i \leq k \leq j$, there exists m satisfying $M \leq m \leq N$ such that $V_k = U_m$). Thus a more explicit notation for $\mathcal{P}(U)$, which we adopt exclusively in the main body of text, is $\mathrm{Prob}(U)$.

Although, from a theoretical standpoint, any $\mathcal{P}$ that satisfied (A.6) would be acceptable, meaningful application of the concept of probability depends upon selecting $\mathcal{P}$ such that, for any U and V, if the odds on event U are λ times as good as the odds on event V, then $\mathcal{P}(U) = \lambda \mathcal{P}(V)$. The ideal way to accomplish this is to decompose the sample space into (mutually disjoint) events that are equally likely and then set

$$\mathcal{P}(U_k) = \frac{1}{N - M + 1}. \qquad (A.7)$$

Suppose, for example, that the random variable X denotes the number facing up after the throw of a die. Then the sample space $S = \{1, 2, 3, 4, 5, 6\}$ can be decomposed, according to (A.5), with $M = 1$, $N = 6$, and $U_k = \{k\}$. Then (A.7) yields $\mathcal{P}(U_k) = \mathcal{P}(\{k\}) = 1/6$, $1 \leq k \leq 6$. Note that (A.6a) and (A.6b) are satisfied. Moreover, from (A.6c), we can compute the probability that 2 *or* 5 will face up after a single throw of the die by setting $V_1 = U_2$, $V_2 = U_5$; we obtain

$$\mathrm{Prob}(2 \text{ } or \text{ } 5) = \mathcal{P}(V_1 \cup V_2) = \mathcal{P}(V_1) + \mathcal{P}(V_2) = \frac{1}{6} + \frac{1}{6} = \frac{1}{3}. \qquad (A.8)$$

Use of (A.7) is not always desirable or even possible, however; consider the case where $N \to \infty$! Indeed the assignment of probability to a sample space is a nontrivial problem in mathematical philosophy, which we discuss in Chapter 8. In this appendix, we shall simply assume that a method of assignment has been found.

Now, a sample space may be associated with more than one random variable, and there is no prior reason why the assignment of probability should be the same for each. Rather, to make a probability assignment unique, the sample space S must be associated with a particular random variable, say X. The sample space S and probability function $\mathcal{P}$ are then known collectively as the *distribution* of X. Thus proper assignment of probability requires definition of both sample space and random variable. In particular, probability is assigned to a stochastic process only when infinitely many distributions have been specified.

An event U is *impossible* if $\mathcal{P}(U) = 0$. On setting $j = 2$ and $V_2 = \varnothing$ in (A.6c), we deduce immediately that

$$\mathcal{P}(\varnothing) = 0. \tag{A.9}$$

Thus the event $\varnothing$ is impossible (i.e., nothing can't happen); we therefore call $\varnothing$ the *null event*. But an event is also impossible if it's excluded from the sample space. What this means in practice is that more than one sample space can be used to describe an experiment. Suppose, for example, that the experiment is the toss of a die. Then, instead of using the sample space $S = \{1, 2, 3, 4, 5, 6\}$ with $M = 1$, $N = 6$, and $U_k = \{k\}$ in (A.5), we could define S to be the set of nonnegative integers, i.e., take $M = 0$, $N \to \infty$, $U_k = \{k\}$ in (A.5); and assign zero probability to $U_0 = \{0\}$ and $U_k = \{k\}$, $k \geq 7$. Clearly, this would not in any way alter the probabilities assigned to real events, regardless of how probability were assigned to the events $\{1\}, \ldots, \{6\}$; each sample space would yield equivalent distributions for any random variable. More generally, two sample spaces S_1 and S_2 are *equivalent* if they differ by a set to which probability zero has been assigned, i.e., if

$$\mathcal{P}(S_1 \cap \tilde{S}_2) = 0. \tag{A.10}$$

For the description of any given experiment, S might be chosen to be any of several equivalent sample spaces. Once defined, however, S is unique. Note that equivalence includes equality (set $S_2 = S_1$ and use (A.9)).

Because $\mathcal{P}$ depends upon S, a more complete notation for $\mathcal{P}(U)$ would be Prob$(U \mid S)$, defined as the probability of U given that the sample space is S. If there is no ambiguity concerning the sample space, then this new notation is redundant. But many sample spaces may be associated with a random variable. Suppose, for example, that the toss of a die is counted only if the number facing up is even (if the number facing up should happen to be odd, then the die is simply tossed again). Now, the sample space used for (A.8), namely $S = U_1 \cup U_2 \cup U_3 \cup U_4 \cup U_5 \cup U_6$, can still be used here; we simply alter the probability assignment so that $\mathcal{P}(U_1) = \mathcal{P}(U_3) = \mathcal{P}(U_5) = 0$. We may prefer, however, to make odd numbers impossible by excluding U_1, U_3, and U_5 at the outset, i.e., by restricting the sample space to $S_1 = U_2 \cup U_4 \cup U_6$; see the remarks at the end of the previous paragraph. More generally, any subset V of a sample space might itself be regarded as a sample space. With such ambiguity arising, our new notation ceases to be redundant. We therefore denote the probability of U when the sample space is V by Prob($U \mid V$), and refer to Prob($U \mid V$) as the *conditional probability of U, given V*.

Returning to the die-tossing example of the previous paragraph, if $V = S$ then Prob(U_2) = Prob(U_4) = Prob(U_6) = 1/6. But if $V = S_1$, then we cannot assign the same probabilities because (A.6b) would be violated. We still believe all outcomes equally likely, however, and we therefore assign Prob($U_2 \mid S_1$) = Prob($U_4 \mid S_1$) = Prob($U_6 \mid S_1$) = 1/3, ensuring that (A.6b) is satisfied. What we have done is to recognize that restricting the sample space from S to S_1 makes each outcome $1/\mathcal{P}(S_1)$ times as likely as before; so that $\mathcal{P}(U_k)$ should be replaced by $\mathcal{P}(U_k)/\mathcal{P}(S_1)$, for $k = 2, 4, 6$. Then $\mathcal{P}(S_1) = 1/2$ follows easily, from (A.6c). More generally, the conditional probability of U, given V, is defined by

$$\text{Prob}(U \mid V) \equiv \frac{\mathcal{P}(U \cap V)}{\mathcal{P}(V)} \tag{A.11}$$

whenever $\mathcal{P}(V) > 0$. You should check that this reduces to $\mathcal{P}(U)$ when $V = S$.

Our definition suggests that the right-hand side of (A.11) should be calculated first and the left-hand side deduced from it. But we have just seen that the conditional probability can also be calculated in its own right, so that the right-hand side of (A.11) could also be deduced from the left. This is hardly surprising because a conditional probability differs from any other probability only insofar as the sample space has been restricted. Enlarging upon this idea, we derive one of the most useful formulas in applied probability. Observe that (A.5) is valid for any decomposition of S into disjoint events. Now, if $\{U_M, U_{M+1}, U_{M+2}, \ldots, U_N\}$ is one such decomposition and U is a subset of S, then $\{U \cap U_M, U \cap U_{M+1}, \ldots, U \cap U_N, \bar{U} \cap S\}$ is another decomposition of S into mutually disjoint events. Hence, taking $i = M$, $j = N$, and $V_k = U \cap U_k$, for $M \leq k \leq N$, in (A.6c), we obtain

$$\mathcal{P}(U) = \sum_{k=M}^{N} \mathcal{P}(U \cap U_k), \tag{A.12}$$

because $V_M \cup V_{M+1} \cup V_{M+2} \cup \cdots \cup V_N = (U_M \cup U_{M+1} \cup U_{M+2} \cup \cdots \cup U_N) \cap U = S \cap U = U$ by (A.3) and (A.5). But setting $V = U_k$ in (A.11), we have $\mathcal{P}(U \cap U_k) = \text{Prob}(U \mid U_k) \cdot \mathcal{P}(U_k) = \text{Prob}(U \mid U_k) \cdot \text{Prob}(U_k)$. Hence

$$\text{Prob}(U) = \sum_{k=M}^{N} \text{Prob}(U \mid U_k) \cdot \text{Prob}(U_k) \tag{A.13}$$

for any event U, and for any events $U_M, U_{M+1}, U_{M+2}, \ldots, U_N$ that together exhaust the sample space.

There is an analogous formula to (A.13) in which the sample space S is restricted to the subset V. Because (A.12) is true for any subset U of the sample space, and because $U \cap V$ is such a subset, we have

$$P(U \cap V) = \sum_{k=M}^{N} P(U \cap V \cap U_k) = \sum_{k=M}^{N} P(U \cap U_k \cap V). \qquad (A.14)$$

Hence

$$
\begin{aligned}
\text{Prob}(U \mid V) &= \frac{P(U \cap V)}{P(V)} = \sum_{k=M}^{N} \frac{P(U \cap U_k \cap V)}{P(V)} \\
&= \sum_{k=M}^{N} \frac{P(U \cap U_k \cap V)}{P(U_k \cap V)} \cdot \frac{P(U_k \cap V)}{P(V)} \\
&= \sum_{k=M}^{N} \text{Prob}(U \mid U_k \cap V) \cdot \text{Prob}(U_k \mid V) \qquad (A.15)
\end{aligned}
$$

on using (A.11) repeatedly; and provided, of course, that $P(V)$ and $P(U_k \cap V)$ are both positive. Note that (A.15) is identical to (A.13) if $V = S$. Note also that (A.13) and (A.15) are statements about the sample space, not about random variables. Thus the events U and U_k need not be associated with the same random variable (although, when these formulas are applied in the text, U and U_k are often associated with random variables from the same stochastic process). See, moreover, the important remarks between (A.56) and (A.57).

For a sample space S associated with the random variable X, it will be sufficiently general for the purposes of this textbook to assume that $S = \Delta \cup I$, where Δ is a subset of the real numbers, I is a subset of the nonnegative integers, and $\Delta \cap I = \varnothing$. Either Δ or I, but not both, may be empty. If $\Delta = \varnothing$ then we define $I = \{x_0, x_1, \ldots, x_m\}$ and say that the random variable X has a *discrete distribution* over the sample space I; or, for short, that the random variable X is discrete. We assume that $i > j \Rightarrow x_i > x_j$, and we allow for the possibility that m may be infinite. If $I = \varnothing$ then we define $\Delta = (a, b)$ and say that X has a *continuous distribution* over the sample space Δ; or, for short, that the random variable X is continuous. Note that (a, b) may be the whole real line $\mathcal{R} = (-\infty, \infty)$. In this book, however, the most frequent choice of sample space for a continuously distributed random variable is $(0, \infty)$. If neither I nor Δ is empty then we define I to be $\{x_0, x_1, \ldots, x_m\}$, as before, but define

$$\Delta = (a, b) \cap \bar{I}, \qquad (A.16)$$

so that $\Delta \cap I = \varnothing$ remains valid. We say that X has a *mixed distribution* over S, or that the random variable X is mixed. Note that any integer $x_k \in I$ such that $a < x_k < b$ is excluded from Δ by definition (A.16). Thus a mixed distribution is partly discrete and partly continuous, the random variable X having a purely discrete distribution when S is restricted to I and a purely continuous one when S is restricted to Δ.

For a discrete random variable, (A.6) is satisfied by specifying nonnegative numbers $\pi_0, \pi_1, \pi_2, \ldots$, which sum to 1, and asserting that $P(x_j) = \pi_j$. To remind ourselves of the random variable's identity, however, we often prefer to write

$$\text{Prob}(X = x_j) = \pi_j. \qquad (A.17)$$

This defines the probability of an outcome x_j. The probability of any event can then be deduced from (A.6c), as indicated below. Now, for the main body of text, it is sufficiently general to assume that I contains only the nonnegative integers from 0 to m inclusive (with m allowed to approach infinity); which therefore, henceforward, assume. Thus $x_k = k$ and, for X discrete, we can take $U_k = \{k\}$ in (A.5). Moreover, the event that interests us is usually the event that X lies between i and j, for suitable $0 \leq i \leq j \leq m$. From (A.6c) with $V_k = \{k\}$, we obtain

$$\text{Prob}(i \leq X \leq j) = \sum_{k=i}^{j} P(V_k) = \sum_{k=i}^{j} \text{Prob}(X = k) = \sum_{k=i}^{j} \pi_k. \qquad (A.18)$$

Of special interest is the case in which $i = 0$. We define the *cumulative distribution function* F of the (discrete) random variable X by writing

$$F(j) \equiv \text{Prob}(0 \leq X \leq j) = \sum_{k=0}^{j} \pi_k, \qquad 0 \leq j \leq m. \qquad (A.19)$$

Because X can never be less than 0, negative integers being excluded from the sample space, one often writes simply $F(j) = \text{Prob}(X \leq j)$. Note that

$$F(j) - F(j-1) = \pi_j. \qquad (A.20)$$

For a continuous random variable X, (A.6) is satisfied by specifying a function f, which is differentiable except at isolated discontinuities (i.e., is "piecewise smooth"), such that

$$f(x) \geq 0, \qquad a \leq x \leq b, \qquad (A.21)$$

$$\int_a^b f(x)dx = 1; \qquad (A.22)$$

and then asserting that

$$\text{Prob}(\alpha < X \leq \beta) = P((\alpha, \beta]) = \int_\alpha^\beta f(x)dx \qquad (A.23)$$

for any α, β such that $a \leq \alpha \leq \beta \leq b$. We call f the *probability density function* of the (continuous) random variable X. Note that for any α,

$$\text{Prob}(X = \alpha) = \text{Prob}(\alpha < X \leq \alpha) = P(\varnothing) = 0. \qquad (A.24)$$

The probability of any *simple* outcome is therefore zero! Thus isolated points can be removed from, or added to, the sample space of a continuous random variable at will because the resulting sample space will always be an equivalent one. For example, the sample space $0 < \theta \leq 2\pi$ for the random variable Θ in Fig. A.1 could be replaced by the equivalent sample space $0 < \theta < 2\pi$ without in any way altering our description of the experiment of spinning the arrow. This is why we can always achieve $\Delta \cap I = \varnothing$ for a mixed distribution, if necessary by removing isolated points from the interval (a, b); see (A.16). Thus it is of no real consequence whether the sample space is defined to be the open interval (a, b), the closed interval $[a, b]$ or one of the corresponding half-open intervals. It also explains why f is allowed to have isolated discontinuities. The fact that any *simple* outcome associated with a continuous random variable is impossible should be interpreted as merely a mathematical detail. A continuous random variable X is often used to approximate one that, in reality, can take only integer values. Then

$$\text{Prob}(k - 0.5 < X \leq k + 0.5) = \int_{k-0.5}^{k+0.5} f(x)dx \qquad (A.25)$$

would usually be interpreted as the probability of $\{k\}$, and this may well be positive.

From (A.23), we readily deduce that

$$\text{Prob}(x - \frac{1}{2}\delta x < X \leq x + \frac{1}{2}\delta x) = f(x)\delta x + o(\delta x), \qquad (A.26)$$

on using the little oh notation introduced in Chapter 1. Thus $f(x)\delta x$ is the probability that X will fall within an infinitesimal interval of length δx containing x. By analogy with (A.19), we define the cumulative distribution function F of the (continuous) random variable X by writing

$$F(x) \equiv \text{Prob}(a < X \leq x) = \int_a^x f(\xi)d\xi, \qquad a \leq x \leq b. \qquad (A.27)$$

Again, because X can never be less than a, one often writes simply $F(x) = \text{Prob}(X \leq x)$. F is a continuous, nondecreasing function; $F(a) = 0$ and $F(b) = 1$. Moreover, from (A.27), we have

$$f(x) = F'(x) = \frac{d}{dx}\text{Prob}(X \leq x) \qquad (A.28)$$

wherever F is differentiable. This is the continuous analogue of (A.20).

For a mixed distribution, (A.6) is satisfied as follows. We first assign probabilities $\mathcal{P}(\Delta)$, $\mathcal{P}(I)$ to Δ and I such that $\mathcal{P}(\Delta) + \mathcal{P}(I) = 1$. We then restrict the sample space to I and assign probabilities $p_0, p_1, p_2, \ldots, p_m$ to it by the above method for purely discrete distributions. We thus determine the conditional probabilities $\text{Prob}(X = k \mid I)$, $0 \leq k \leq m$. Next, we restrict the sample space to Δ and assign a probability density function ϕ to it by the above method for purely continuous distributions. We thus determine the conditional probabilities $\text{Prob}(\alpha < X \leq \beta \mid \Delta)$, $a \leq \alpha \leq \beta \leq b$. Then we define

$$\pi_k \equiv \mathcal{P}(I)\rho_k, \qquad k = 0, 1, 2, \ldots, m; \qquad\qquad \text{(A.29a)}$$
$$f(x) = \mathcal{P}(\Delta)\phi(x), \qquad\qquad\qquad\qquad\qquad \text{(A.29b)}$$

note that any event in S must have the form $U = \{i, i+1, \ldots, j\} \cup (\alpha, \beta]$ and assert that

$$\text{Prob}(X \in U) = \mathcal{P}(U) = \sum_{k=i}^{j} \pi_k + \int_{\alpha}^{\beta} f(x)dx. \qquad \text{(A.30)}$$

It is easy to see how this definition could be extended to include compound events consisting of nonconsecutive sets of integers or nonadjacent intervals, but (A.30) is adequate for the purposes of this textbook.

The *expected value* or *mean* of a random variable X, $E[X]$, is usually denoted by μ. For a discrete random variable, $E[X]$ is defined by

$$\mu = E[X] \equiv \sum_{k=0}^{m} k\pi_k. \qquad\qquad\qquad \text{(A.31)}$$

For a continuous random variable, $E[X]$ is defined by

$$\mu = E[X] \equiv \int_{a}^{b} xf(x)dx. \qquad\qquad\qquad \text{(A.32)}$$

Now, as we remarked much earlier, if g is a (real-valued) function and X a random variable, then $Y = g(X)$ is also a random variable. If X, and hence Y, is continuous, then Y has its own probability density function $f(y)$, with which (A.32) can be used to calculate the mean $E[Y]$. But it can also be shown that $E[Y]$ can be calculated directly from the distribution of X, as

$$E[g(X)] = \int_{a}^{b} g(x)f(x)dx; \qquad\qquad\qquad \text{(A.33)}$$

and we shall invariably use this formula.

The *variance* of a random variable X, $\text{Var}(X)$, is usually denoted by σ^2; it is defined by

$$\sigma^2 = \text{Var}[X] \equiv E[(X - \mu)^2] = E[X^2] - \mu^2, \qquad \text{(A.34)}$$

where μ is defined by (A.31) or (A.32) according to whether X is discrete or continuous. The equality of the last two terms in (A.34) is readily established in either case. Thus if X is discrete, we have

$$\sigma^2 = \text{Var}[X] \equiv \sum_{k=0}^{m}(k - \mu)^2\pi_k = \sum_{k=0}^{m} k^2\pi_k - \mu^2, \qquad \text{(A.35)}$$

and if X is continuous, we have

$$\sigma^2 = \text{Var}[X] \equiv \int_{a}^{b}(x - \mu)^2f(x)dx = \int_{a}^{b} x^2f(x)dx - \mu^2. \qquad \text{(A.36)}$$

The latter is the special case of (A.33) for which $g(X) \equiv (X - \mu)^2$.

The *standard deviation* of a random variable, σ, is simply the square root of the variance. The standard deviation is a measure of a distribution's dispersion from the mean. For any distribution,

$$\text{Prob}(\mu - K\sigma < X < \mu + K\sigma) \le \frac{1}{K^2}, \tag{A.37}$$

a result known as Chebyshev's inequality. Thus, in particular, the probability that X will lie further than three standard deviations from the mean is less than 1/9 for any distribution. For the normal distribution, however, it's less than 1/100; see Chapter 6.

The expected value of a mixed random variable is defined by

$$E[X] = \sum_{k=0}^{m} k\pi_k + \int_{a}^{b} xf(x)dx, \tag{A.38}$$

where π_k and f are defined by (A.29). You should check that (A.31) agrees with the value obtained by setting $S = I$, $\Delta = \varnothing$ in (A.38), and that (A.32) agrees with the value obtained by setting $I = \varnothing$ and $S = \Delta$. In each case, you will need to use (A.6) and (A.9). As an example of (A.38), consider the random variable X defined by (A.4). We have $I = \{2, 3\}$ and $\Delta = (0, 1)$. Let's assume that Θ is equally likely to be any angle between 0 and 2π, or is *uniformly distributed* between 0 and 2π. We thus assign $\mathcal{P}(I) = 1/2$, $\mathcal{P}(\Delta) = 1/2$. Now restrict the sample space to I. Because X is then equally likely to be either 2 or 3, we have $\rho_2 = \text{Prob}(X = 2) = 1/2$ and $\rho_3 = \text{Prob}(X = 3) = 1/2$. Now restrict the sample space to Δ. Because X is uniformly distributed over $(0,1)$, we take ϕ to be constant; and equating the integral of ϕ over $(0,1)$ to 1 yields $\phi = 1$. On using (A.29), we thus obtain $\pi_2 = \text{Prob}(X = 2) = 1/4$, $\pi_3 = \text{Prob}(X = 3) = 1/4$ and $f(x) = 1/2$. Then (A.38) yields the expected value

$$E[X] = 2\pi_2 + 3\pi_3 + \int_{0}^{1} \frac{x}{2}dx = \frac{3}{2}. \tag{A.39}$$

This result can also be obtained from (A.32). Because the random variable Θ is purely continuous, with $f(\theta) = 1/2\pi$, $0 < \theta < 2\pi$, we have

$$E[X] = E[g(\Theta)] = \int_{0}^{2\pi} g(\theta)f(\theta)d\theta$$

$$= \int_{0}^{\pi} \frac{\theta}{\pi} \frac{1}{2\pi} d\theta + \int_{\pi}^{3\pi/2} 2 \cdot \frac{1}{2\pi} d\theta + \int_{3\pi/2}^{2\pi} 3 \cdot \frac{1}{2\pi} d\theta = \frac{3}{2}, \tag{A.40}$$

which agrees with (A.39).

Any pure number (as opposed to set of numbers) associated with a random variable is called a *statistic*. Thus $\text{Prob}(X \le \mu)$, $\text{Prob}(X > \mu + 3\sigma)$, μ, and σ are all statistics of the random variable X. Indeed μ and σ are statistics of a particular kind, known as *moments*. The kth moment of X

about the origin is the expected value of the random variable X^k, and the kth moment of X about its mean is the expected value of the random variable $(X - \mu)^k$. Hence μ is the first moment about the origin, and σ^2 is the second moment about the mean (the first moment about the mean is identically zero).

We have assumed until now that a sample space S is associated with a single random variable, namely, X. Let us make the identity of the random variable explicit, denoting the sample space by S_X. But a sample space can also be associated with two random variables, say X and Y. Let S_Y denote the sample space associated with Y alone. Then the sample space associated with X and Y together is simply $S = S_X \times S_Y$, where we define

$$A \times B \equiv \{(\xi, \psi) : \xi \in A, \psi \in B\} \tag{A.41}$$

for arbitrary subsets A and B of $\mathcal{R}$. An event now corresponds to a region in two-dimensional space. It will be sufficiently general for the purposes of the textbook to assume that X and Y are either both discrete or both continuous.

In the first instance, when both X and Y are discrete, we take $S_X = I$, $S_Y = I$ and hence $S = I \times I$, where I contains the nonnegative integers from 0 to m inclusive (but we may allow m to become infinite). We replace (A.5) by

$$S = \bigcup_{i=0}^{m} \bigcup_{j=0}^{m} U_{ij}, \tag{A.42}$$

where $U_{ij} = \{(i, j)\}$, (A.42) being a mutually disjoint union of points in the positive quadrant of the two-dimensional plane. We replace k by ij in (A.6a) and the single summation in (A.6c) by a double summation. We then satisfy (A.6) by writing $\omega_{ij} = \mathcal{P}(U_{ij})$ and choosing numbers ω_{ij} such that $\omega_{ij} \geq 0$, $0 \leq i, j \leq m$, and

$$\sum_{i=1}^{m} \sum_{j=1}^{m} \omega_{ij} = 1. \tag{A.43}$$

Thus

$$\omega_{ij} = \text{Prob}(X = i, Y = j), \tag{A.44}$$

where "$X = i$, $Y = j$" should be interpreted as "$X = i$ and $Y = j$." We call $\mathcal{P}$ the *joint probability function* for X and Y; S and $\mathcal{P}$ together form the *joint probability distribution* of the random variables X and Y. The probability of the event $\{i_1 \leq X \leq i_2, j_1 \leq Y \leq j_2\}$ is then given by

$$\text{Prob}(i_1 \leq X \leq i_2, j_1 \leq Y \leq j_2) = \sum_{r=i_1}^{i_2} \sum_{s=j_1}^{j_2} \omega_{rs}. \tag{A.45}$$

Let $\pi_i = \text{Prob}(X = i)$ and $\sigma_j = \text{Prob}(Y = j)$. Then, applying (A.18) to X and Y in turn, we obtain

$$\text{Prob}(i_1 \leq X \leq i_2) = \sum_{r=i_1}^{i_2} \pi_r, \qquad \text{Prob}(j_1 \leq Y \leq j_2) = \sum_{s=j_1}^{j_2} \sigma_s. \tag{A.46}$$

If

$$\omega_{ij} = \pi_i \cdot \sigma_j, \quad 0 \leq i, j \leq m, \tag{A.47}$$

then it follows immediately from (A.43) and (A.44) that

$$\text{Prob}(i_1 \leq X \leq i_2, j_1 \leq Y \leq j_2) = \text{Prob}(i_1 \leq X \leq i_2) \cdot \text{Prob}(j_1 \leq Y \leq j_2) \tag{A.48}$$

for all i_1, i_2, j_1, and j_2. In these circumstances, we say that the random variables X and Y are *independent*. In other words, X and Y are independent if, for *any* event U in S associated with X and for *any* event V in Y associated with Y, the joint probability of the event $U \cap V$ in S is given by

$$\text{Prob}(U \cap V) = \text{Prob}(U) \cdot \text{Prob}(V). \tag{A.49}$$

It is possible that (A.49) is true for particular events in the joint sample space S of X and Y, even though the random variables are dependent. Then those particular events are said to be independent.

In the second instance, when both X and Y are continuous, we take $S_X = \Delta$, $S_Y = \Delta$ and hence $S = \Delta \times \Delta$, where $\Delta = (a, b)$. We retain (A.5) and (A.6) but interpret U_k, V_k as regions in the plane (and subregions of $\Delta \times \Delta$). We then satisfy (A.6) by specifying a function f such that

$$f(x, y) \geq 0, \qquad a \leq x, \, y \leq b,$$

$$\int_S f(x, y)dxdy = \int_a^b \int_a^b f(x, y)dxdy = 1 \tag{A.50}$$

and then asserting that

$$\text{Prob}(W) = \int_W f(x, y)dxdy \tag{A.51}$$

for any event W in the sample space S. We call f the joint probability density function of the (continuous) random variables X and Y. In particular, the probability of the event that $\alpha < X \leq \beta$ and $\gamma \leq Y \leq \delta$ is given by

$$\text{Prob}(\alpha < X \leq \beta, \gamma < Y \leq \delta) = \int_\gamma^\delta \int_\alpha^\beta f(x, y)dxdy. \tag{A.52}$$

Let f_X and f_Y, respectively, denote the probability density functions of the random variables X and Y. Then, from (A.23),

$$\text{Prob}(\alpha < X \leq \beta) = \int_\alpha^\beta f_X(x)dx, \qquad \text{Prob}(\gamma < Y \leq \delta) = \int_\gamma^\delta f_Y(y)dy. \tag{A.53}$$

Let U be the event that $\alpha < X \leq \beta$ in S, the joint sample space; and let V be the event that $\gamma < Y \leq \delta$ (also in S). If

$$f(x, y) = f_X(x) \cdot f_Y(y), \qquad a \leq x, y \leq b, \tag{A.54}$$

then it follows immediately from (A.52) and (A.53) that

$$\text{Prob}(U \cap V) = \text{Prob}(U) \cdot \text{Prob}(V) \tag{A.55}$$

for *all* events U associated with X and V associated with Y, because any region in the plane is the limit of a disjoint union of rectangles. In these circumstances, we say that the random variables X and Y are *independent*. Moreover, it can be shown that if X and Y are independent, then (A.54) must be satisfied; i.e., the joint probability density function of independent random variables is the product of their individual probability density functions. As in the discrete case, regardless of their association with any random variables, events U and V in S (or indeed any sample space) are independent if (A.55) is satisfied.

More generally still, a sample space can be associated with L random variables $X_1, X_2, \ldots, X_L$, where L may be as large as we please. The joint sample space is then a subset of L-dimensional space $\mathcal{R} \times \mathcal{R} \times \cdots \times \mathcal{R}$, where we define

$$A_1 \times A_2 \times \cdots \times A_L \equiv \{(x_1, x_2, \ldots, x_L): x_i \in A_i, \ 1 \leq i \leq L\}. \tag{A.56}$$

In particular, the joint sample space of more than two discrete random variables is $I \times I \times \cdots \times I$. Joint sample spaces of more than two continuous variables do not arise in the text.

Expressions (A.42)–(A.55) are readily generalized to L-dimensional space. We omit the details, however, because we do not require them. We require only two results. First, we observe that (A.13) and (A.15) are quite general statements about *any* sample space, even the joint sample space of several random variables. This observation is of greatest utility in Chapter 5 and in Sections 10.4 and 10.5.

Second, if the random variables $X_1, X_2, \ldots, X_L$ are all independent, meaning that any pair of them is independent, then

$$\text{Prob}\left(\bigcap_{k=1}^{L} U_k\right) = \text{Prob}(U_1) \times \cdots \times \text{Prob}(U_L) \equiv \prod_{k=1}^{L} \text{Prob}(U_k), \tag{A.57}$$

where U_k is any event in the joint sample space of $X_1, X_2, \ldots, X_L$ that is associated solely with X_k. This result is the obvious generalization of (A.49) and (A.55). The expected value of the sum of the independent random variables $X_1, X_2, \ldots, X_L$ is the sum of their expected values, and similarly for variance, i.e.,

$$E\left[\sum_{i=1}^{L} X_i\right] = \sum_{i-1}^{L} E[X_i], \qquad \text{Var}\left[\sum_{i=1}^{L} X_i\right] = \sum_{i=1}^{L} \text{Var}[X_i]. \tag{A.58}$$

Furthermore, the first (but not the second) equality in (A.58) would hold even if the random variables were not independent.

APPENDIX 2:

Models, Sources, and Further Reading Arranged by Discipline

The purpose of this appendix is threefold: it identifies sources, it groups models within scientific disciplines, and it offers suggestions for further reading. I have placed each model in one of nine categories: biology (B), chemistry (C), decision theory (Y), demography (D), economics (E), operations research (O), physics (P), politics and sociology (S), and traffic studies (T). Several models straddle boundaries between disciplines, making the classification arbitrary. Therefore, after naming each model or class of model, I have indicated alternative classification in parentheses.

Biology

1 One-species population model

A one-species population model is introduced in Section 1.8, further developed in Section 2.4, and criticized in Section 4.13. The model is due to Ricker (1954), although I was introduced to it by Clark (1976) and May (1976b).

2 Two-species population models

A simple model of a predator-prey system is introduced in Section 1.4 and criticized in Section 4.12. A simple model of a competitive ecosystem is introduced in Section 1.5 and criticized in Section 4.11. The first model, due to Alfred J. Lotka and Vito Volterra, is discussed by Lotka (1956, pp. 88–92) in his classical treatise on mathematical biology. The second model, though implicit in Lotka's (1956, p. 78) fundamental equations of kinetics, is not actually discussed by him and is usually attributed to G. F. Gause.

With lynx-hare cycles in mind, the predator-prey model is further developed in Section 10.2, which is based on May (1976c). Other attempts to explain lynx-hare cycles are discussed by Renshaw (1991, pp. 253–257). Renshaw's book is an excellent introduction to more advanced biological population modelling. Renshaw adopts both deterministic and probabilistic approaches; whereas Edelstein-Keshet (1988), though also excellent, treats only deterministic models.

3 Animal behavior

A simple model for prey selection by birds is described in Section 12.5, then adapted in an original manner. This adaptation is based on the model in Chapter 3 of Krebs and Davies (1987), which appears to have had its roots in the work of MacArthur and Pianka (1966) and Emlen (1966), and to have been developed most significantly by Charnov (1976) and Pulliam (1974). A model for searching and oviposition by an insect, *Anagrus delicatus*, appears in Section 12.6. This section, which describes an original adaptation of Model 1 from Mangel (1987), is based on Mesterton-Gibbons (1988c). Godfray (1994, p. 75) critiques the model. Godfray (1994, Section 3.3) also critiques pertinent stochastic models of superparasitism and its avoidance.

For more about *Anagrus delicatus* in particular, see Cronin and Strong (1993). For more about insects in general, see Evans (1978). For more about animal behavior in general, see Ehrlich (1986) and Krebs and Davies (1987).

4 Island biogeography

A model for island species density is suggested by Exercise 4.12 and discussed in Section 8.3. These developments are based on Diamond and May (1976) and Gilpin and Diamond (1976).

5 Pest management (E)

A simple pest management model, suggested by Plant (1985), is developed in Section 3.6 and criticized in Section 4.9.

6 Plant growth

A simple model of plant growth, due to Thornley (1976), is introduced in Section 1.3 and criticized in Section 4.3.

Chemistry

1 Biochemistry (B)

A model for glucose flow across a placenta is developed in Section 10.1. The model is based on Bray and White (1966).

2 Physical Chemistry

The simple classical models for first order and second order chemical kinetics are developed in Sections 1.10 and 1.11. These models are discussed in all substantial texts on physical chemistry, for example, Ander and Sonnessa (1965). They are incorporated into the Streeter–Phelps water purification model in Section 11.7.

Decision theory (O)

I have all but ignored the question of how decision makers construct a utility function. This is discussed, for example, by Holloway (1979) and Lindley (1982, 1985). Nevertheless, in Section 8.8, by providing a concrete example of a decision model incorporating attitudes to risk, I have illustrated many of the fundamental concepts.

Demography

1 Models with age distribution

Models for the structure of a population with births are developed in Sections 9.6 and 9.7. The first model is due to Alfred J. Lotka and discussed by him in his classical treatise (Lotka, 1956, pp. 100–120). The second model is usually attributed to P. H. Leslie (1945, 1948), although as Leslie acknowledges in a footnote (1948, p. 213), it was formulated independently by E. G. Lewis. Leslie's work has since evolved into sophisticated matrix population models; see, e.g, Cushing (1988), Caswell (1989) and Logofet (1993).

I have ignored the theoretical foundations (in eigenanalysis) of these distributed models, because my students have invariably found that each additional ounce of modelling insight so offered is accompanied by several pounds of mathematical bewilderment. This is not, of course, to say that they are unimportant. They are fully discussed by Impagliazzo (1985) and Keyfitz (1968); a simpler treatment can be found in Cullen (1985).

2 Model with age distribution but no births

The "life-table" model is introduced in Exercises 4.15–4.17 and further discussed in Section 8.6. These sections are based on Keyfitz (1968), Keyfitz and Flieger (1971), Smith and Keyfitz (1977), and Fries and Crapo (1981).

3 Undistributed (logistic growth) model (B)

A simple model of U.S. population growth is developed in Section 1.9 and criticized in Section 4.1. The model is due to P.-F. Verhulst, Raymond

Pearl, and L.J. Reed. It is discussed by several authors, among them Lotka (1956) and Olinick (1978), whose initial approach is conceptual (whereas mine is empirical).

Economics

1   Economic growth

A simple model of economic growth is developed in Section 1.6, which is based on Douglas (1934). The associated Cobb-Douglas production function is further applied in Section 3.3. The relationship between interest rates and money supply is explored in Section 10.3, which is based on Brems (1980). The ideas behind all of these models are discussed in texts on economics, for example, Gordon (1978), Nicholson (1979), and Samuelson (1976).

2   Fisheries management (B)

An optimal measure of harvesting effort for a fishing fleet is obtained in Section 3.7 and criticized in Section 4.14. This model is based on Clark (1976). My approach, using ordinary calculus, differs from Clark's (which is much more general, but requires the calculus of variations).

3   Forest management

A model for clearcutting is introduced in Section 3.4 and improved in Section 4.8. The model is due to M. Faustmann, but I was introduced to it by Clark (1976, Chapter 8).

A model for partial harvesting is developed in Sections 5.8 and 9.2. Each of these sections is based on Usher (1966). For a more sophisticated model, which extends the work of Usher, see Buongiorno and Michie (1980). For a survey of forest management models see Reed (1986).

Operations Research

1   Inventory models

A single-period inventory model is introduced in Section 7.1, applied in Section 7.2, further discussed in Section 7.3, extended to two periods in Section 7.5, adapted in Section 9.1, and extended to N periods in Section 12.1. This inventory model is part of the core curriculum of operations research. I was introduced to it by Lange (1971) and subsequently found Whittle (1982) a valuable reference. The model is discussed by several authors, among whom are Taha (1976), Moskowitz and Wright (1979),

Hillier and Lieberman (1980), and Beaumont (1983). For an application of the single-period model to the airline industry, see Kondo (1958).

2   Markov processes, stationary queues, and semi-Markov processes

Birth and death processes are introduced in Sections 5.1–5.3 and further developed in Sections 5.5–5.7. Stationary queuing models are discussed in Sections 6.2–6.5. Like inventory models, queues and Markov processes are part of the core curriculum of operations research and are discussed in detail by Hoel et al. (1972), by Ross (1980), by the authors cited immediately above, and in most other texts on applied probability or operations research.

A semi-Markov process is introduced in Section 10.4 and further developed in Section 10.5. These two sections are based on Howard (1971) and Ross (1983). For a list of applications of semi-Markov models, see the bibliography by Teugels (1986).

3   Small business (Y)

A bookseller's advertising model appears in Section 7.4; a barber shop model appears in Section 7.7. Each of these models is an original variation on the core curriculum of operations research. A fast-food advertising model appears in Section 7.6, which is based on Mesterton-Gibbons (1988b). An interviewing model appears in Section 12.2. This is based on the classical "secretary problem," to which I was introduced by Howard (1971). The model is also discussed by Whittle (1982). For an extension of the model, see Mitsushi (1984).

4   University administration (Y)

A simple model to help determine pay increases is developed in Section 3.2 and criticized in Section 4.15. The model is based on Bruno (1971) and Fabozzi and Bachner (1979). A simple faculty hiring model appears in Section 12.3. This model is based on Howard (1971) and Whittle (1982).

Physics

1   Continuum mechanics

The Hagen-Poiseuille model for flow in a pipe is introduced in Section 2.6 and criticized in Section 4.10. A generic pipe flow model is developed in Section 11.6, then applied to organ pipes in Section 11.8. Each model is based on classical fluid dynamics, as discussed, for example, by Lamb (1925, 1932). For modern fluid dynamics I recommend Acheson (1990).

2   Heat conduction

A model for heat flow is developed in Sections 2.5 and 11.1. The later section is a piece of classical applied mathematics, as discussed, for example, by Morse and Feshbach (1953) and Jeffreys and Jeffreys (1956). The earlier section is a variation on the same theme.

3   Motion of a body submerged in a fluid (B)

A simple rowing model is introduced in Section 1.12, further discussed in Section 2.2, and criticized in Section 4.4. The model is a dynamic version of one due to McMahon (1971), to which I was introduced by Bender (1978).

A simple model of avian flight is introduced in Section 3.1, then extended in Section 4.5. Both this and the rowing model are based on classical aerodynamics, as discussed in the present context by Alexander (1983).

4   Radioactive decay (C)

The simple classical model of radioactive decay is briefly discussed in Section 1.2.

5   Water purification (C)

A simple purification model for a perfectly mixed lake is developed in Sections 1.1 and 2.1, criticized in Sections 1.7 and 4.2, and modified in Section 9.3. The model is due to Rainey (1967); I was introduced to it by Bender (1978).

A simple convective model for water purification is explored in Section 9.4; the role of diffusion is discussed in Sections 9.5 and 11.5. Each of these models is a straightforward application of classical applied mathematics.

The Streeter-Phelps purification model is developed in Section 11.7. I was introduced to it by Rinaldi et al. (1979); the model is also discussed by Jørgensen (1986), by Springer (1986), and by the International Commission on Irrigation and Drainage (1980).

Politics and Sociology

A model of social mobility is introduced in Section 6.6 and further discussed in Section 8.6. The model, to which I was introduced by Champernowne (1980) and Bartholomew (1982), is due to David V. Glass and John Hall and is fully discussed by Glass (1954).

A model of power sharing in the United Kingdom is developed in Sections 8.7 and 10.6.

Table A.1

Section	Code	Section	Code	Section	Code	Section	Code
1.1	P5	3.5	T2	6.1	O2	9.3	P5
1.2	P4	3.6	B5	6.2	O2	9.4	P5
1.3	B6	3.7	E2	6.3	O2	9.5	P5
1.4	B2	4.1	D3	6.4	O2	9.6	D1
1.5	B2	4.2	P5	6.5	O2	9.7	D1
1.6	E1	4.3	B6	6.6	S		
1.7	P5	4.4	P3	6.7	T3	10.1	C1
1.8	B1	4.5	P3			10.2	B2
1.9	D3	4.6	T2	7.1	O1	10.3	E1
1.10	C2	4.7	T2	7.2	O1	10.4	O2
1.11	C2	4.8	E3	7.3	O1	10.5	O2
1.12	P3	4.9	B5	7.4	O3	10.6	S
1.13	T2	4.10	P1	7.5	O1		
1.14	M	4.11	B2	7.6	O3	11.1	P2
		4.12	B2	7.7	O3	11.2	T1
2.1	P5	4.13	B1	7.8	M	11.3	T1
2.2	P3	4.14	E2			11.4	T1
2.3	T2	4.15	O4	8.1	M	11.5	P5
2.4	B1			8.2	M	11.6	P1
2.5	P2	5.1	O2	8.3	B4	11.7	P5
2.6	P1	5.2	O2	8.4	M	11.8	P1
2.7	M	5.3	O2	8.5	M	11.9	P1
2.8	T2	5.4	T3	8.6	D2,S		
		5.5	O2	8.7	S,T3	12.1	O1
3.1	P3	5.6	O2	8.8	Y	12.2	O3
3.2	O4	5.7	O2			12.3	O4
3.3	E1	5.8	E3	9.1	O1	12.4	T4
3.4	E3			9.2	E3	12.5	B3
						12.6	B3

Traffic Studies

1 Continuum models

A continuum approach to traffic flow is adopted in Sections 11.2–11.4. The approach is due to Lighthill and Whitham (1955), although I was introduced to it by Whitham (1974), on which these three sections are variations. Related models are discussed by Haberman (1977).

2 Leader–follower models

A "leader–follower" model for a platoon of cars is introduced in Section 1.13 and developed further in Sections 2.3, 2.8, 3.5, and 4.6. The model is due to Denos C. Gazis, Robert Herman, and Renfrey B. Potts (1959); I was introduced to it by Gazis (1972), Prigogine and Herman (1971), and Whitham (1974). Section 4.7 is an original variation.

3   Models for crossing a stream of traffic

Probabilistic models of traffic flow appear in Sections 5.4 and 6.7. The model for a pedestrian crossing in Section 5.4 is a variation on Noble (1967) and Gerlough and Barnes (1971). The T-junction model in Section 6.7 is based on Mesterton-Gibbons (1987b).

4   Parking model (Y,O)

A model for choosing an optimal parking space appears in Section 12.4. The model is suggested by an exercise in Whittle (1982).

The arrangement of models is summarized in Table A.1. The symbol M is used to denote miscellaneous sections which do not belong to any of the categories defined above. Dimensionality, scaling and units are discussed in Section 1.14. Equilibrium shifts are discussed in Section 2.7. The subjectiveness of decision making is the subject of Section 7.8. In the first three sections of Chapter 8, some purely pedagogical inventions are used to illustrate model choice. Linear and nonlinear regression are discussed in Section 8.4. Section 8.5 is a brief overview of probability assignment. The ubiquitous problem of choosing and fitting distributions is broached, but no more than broached, in Section 8.7. Indeed I have attempted only to provide concrete examples of how it relates to the modelling process at large. Statistical modelling is specifically addressed in the monumental work by Kendall and Stuart (1967), and more recently, by Gilchrist (1984).

SOLUTIONS TO SELECTED EXERCISES

Chapter 1

1.1 Let r in (1.9) depend on t:

$$\frac{1}{x}\frac{dx}{dt} = \frac{r(t)}{V}. \tag{a}$$

Then, integrating between 0 and t:

$$\ln\left(\frac{x(t)}{x(0)}\right) = [\ln(x(s))]_0^t = \int_0^t \frac{1}{x}\frac{dx}{ds}ds = -\int_0^t \frac{r(s)}{V}ds = -\frac{1}{V}\int_0^t r(s)ds. \tag{b}$$

If $r(s) = \rho(1 + \epsilon\sin(2\pi s))$, then (b) yields

$$x(t) = x(0)\exp\left\{-\frac{1}{V}\int_0^t r(s)ds\right\}$$

$$= x(0)\exp\left\{-\frac{\rho}{V}\left[t + \frac{\epsilon}{2\pi}(1 - \cos(2\pi t))\right]\right\}. \tag{c}$$

Pollution is reduced to 5% of its initial level when $x(t) = 0.05x(0)$. Thus (c) implies that

$$\frac{\epsilon}{2\pi}\cos(2\pi t_{0.05}) = t_{0.05} + \frac{\epsilon}{2\pi} + \frac{V}{\rho}\ln(0.05). \tag{d}$$

Setting $\epsilon = 0$ in (c) and (d) yields (1.10) and (1.11), on identifying r with ρ. More generally, $t_{0.05}$ satisfies

$$-\ln(0.05)\frac{V}{\rho} - \frac{|\epsilon|}{\pi} < t_{0.05} < -\ln(0.05)\frac{V}{\rho} + \frac{|\epsilon|}{\pi} \tag{e}$$

and hence differs by less than $1/\pi$ years (less than 4 months) from the value predicted by using mean flow in (1.11). Result (e) follows either from Fig. S1.1 or from using the trigonometric identity $\cos(2\theta) \equiv 1 - 2\sin^2(\theta)$ to simplify (d).

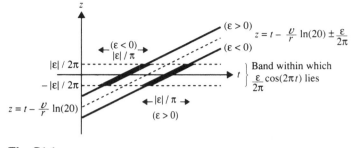

Fig. S1.1

1.2 Use separation of variables; i.e., observe that

$$dt = \frac{dx}{k(x_f - x)x} = \frac{1}{kx_f}\left(\frac{1}{x_f - x} + \frac{1}{x}\right)dx.$$

Then integrate this equation from 0 and t and rearrange.

1.3 Use the fact that

$$\frac{dy}{dx} = \frac{dy/dt}{dx/dt}$$

to obtain the ordinary differential equation

$$\frac{dy}{dx} = \frac{y(-a_2 + b_2 x)}{x(a_1 - b_1 y)}$$

or

$$\left(b_2 - \frac{a_2}{x}\right)dx = \left(-b_1 + \frac{a_1}{y}\right)dy,$$

then integrate.

1.5 (i) From (1.16) we obtain $x(t) = (x(t), y(t))^T$, where

$$x(t) = \frac{x(0)x_f}{x(0) + (x_f - x(0))\exp(-kx_f t)},$$

$$y(t) = \frac{y(0)y_f}{y(0) + (y_f - y(0))\exp(-ky_f t)}.$$

(ii) I used the 37 initial values $x(0) = (7, 10)^T + 6(\sin\theta, \cos\theta)^T$, where $\theta = 2\pi k/37$, $k = 1, 2, \ldots, 37$, and obtained the set of curves depicted in Fig. S1.2. Arrows indicate the direction in which time increases. These trajectories converge to a "stable node" at $(7, 10)^T$, as established in Exercise 2.3. Only the rectangle $0 < x < 7, 0 < y < 10$ is physically meaningful.

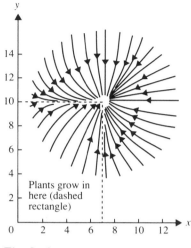

Fig. S1.2

1.7 Equation (1.37) can be written in the form

$$\frac{d}{dt}\{K^\gamma\} = \frac{\sigma a}{\rho}\{L(0)\}^\gamma \frac{d}{dt}\{e^{\rho\gamma t}\},$$

from which (1.38) follows readily. From (1.36) and (1.37):

$$\frac{1}{K}\frac{dK}{dt} - \frac{1}{L}\frac{dL}{dt} = \sigma a\{L(0)\}^\gamma e^{\rho\gamma t}\{K(t)\}^{-\gamma} - \rho$$
$$= \{K(t)\}^{-\gamma}\{\sigma a(L(0))^\gamma e^{\rho\gamma t} - \rho(K(t))^\gamma\}$$
$$= \{K(t)\}^{-\gamma}\{\sigma a(L(0))^\gamma - \rho(K(0)^\gamma\},$$

on using (1.38); whence (1.41) follows because $I(0) = \sigma Q(0) = \sigma a L(0)^\gamma \times K(0)^{1-\gamma}$. It is clear from (1.38) that $\{K(t)\}^\gamma \to \infty$ as $t \to \infty$. Its reciprocal must therefore tend to zero as $t \to \infty$; hence so must (1.41).

1.10 Equation (1.15) reduces to (1.62) if we replace x_f by K and kx_f by R. Hence, from (1.16), the solution to (1.62) is

$$x(t) = \frac{Kx(0)}{x(0) + (K - x(0))e^{-Rt}}.$$

*1.11 The 1980 datum enables us to calculate:

$$D(18) = \frac{1}{2}(x(19) - x(17)) = \frac{1}{2}(226.5 - 179.3) \approx 23.6;$$

whence $D(18)/x(18) \approx 0.12$. We can therefore add the data point with coordinates $(203, 0.12)$ to Fig. 1.7. Now fit a line by eye between the data points for 1950, 1960, and 1970. Its intercept on the vertical axis is about

0.28, and its slope is approximately -0.00075. Comparing to (1.62), we have $R = 0.28$ and $R/K = 0.00075$; whence $K = 373$ (millions). From Exercise 1.10, therefore, a possible model for 1940 onward is

$$x(t) = \frac{373}{1 + 1.48e^{-0.28(t-16)}}.$$

To two significant figures, this model makes correct predictions, namely, 130, 150, 180, 200, and 230 (millions), for the years 1940–1980. The apparent change in the capacity of the land is interpreted in Chapter 4.

1.12 Differentiating

$$\frac{dx}{dt} = Rx \left(1 - \frac{x}{k}\right)$$

twice with respect to t (and substituting for dx/dt), we obtain

$$\frac{d^3x}{dt^3} = R^3 x \left(1 - \frac{x}{K}\right) \left(1 - \frac{6x}{K} + \frac{6x^2}{K^2}\right) = R^3 \phi \left(\frac{x}{K} - \frac{1}{2}\right), \qquad \text{(a)}$$

where

$$\phi(z) \equiv \left(6z^2 - \frac{1}{2}\right) \left(\frac{1}{4} - z^2\right).$$

This symmetric function has local minimum $-1/8$ at $z = 0$ and local maxima $1/24$ at $z = +\sqrt{1/6}$ and $z = -\sqrt{1/6}$. Thus the maximum of $|\phi(z)|$ on $-1/2 \le z \le 1/2$ is $1/8$. Hence the maximum of (a) on $0 \le x \le K$ is $R^3/8$. By Taylor's theorem from the calculus, there exist ξ, satisfying $t < \xi < t+1$, and η, satisfying $t - 1 < \eta < t$, such that

$$x(t+1) = x(t) + 1 \cdot x'(t)/1! + 1^2 \cdot x''(t)/2! + 1^3 \cdot x'''(\xi)/3!,$$
$$x(t-1) = x(t) + (-1)x'(t)/1! + (-1)^2 \cdot x''(t)/2! + (-1)^3 \cdot x'''(\eta)/3!,$$

where a prime denotes differentiation. Hence in approximating dx/dt by $D(t)$ in (1.59), the error $x'(t) - D(t) = -\left\{x'''(\xi) + x'''(\eta)\right\}/12$ can never exceed $R^3/48$, which is less than a thousandth for the value of R in Section 1.9.

1.18 To obtain (1.72) from Exercise 1.8(i), simply replace x by ξ, P_{in}/V by $k_1 a_0 - k_2 b_0$ and r/V by $k_1 + k_2$. Hence directly from (2.5), we get

$$\xi(t) = \frac{k_1 a_0 - k_2 b_0}{k_1 + k_2} + \left\{\xi(0) - \frac{k_1 a_0 - k_2 b_0}{k_1 + k_2}\right\} e^{-(k_1 + k_2)t}.$$

1.19 (i) The equation can be separated and, on using partial fraction decomposition, rewritten in the form

$$\frac{1}{3V} \left\{\frac{1}{V - u} + \frac{u - V}{u^2 + uV + V^2}\right\} du = \frac{bS}{M} dt.$$

The result now follows readily, by integrating from 0 to t and using $u(0) = 0$.

(ii) The indicated substitution reduces (1.77) to the equation

$$\frac{1}{3}M\frac{d(u^3)}{dx} = 8P - bSu^3,$$

which is linear in u^3 and can be solved by using an integrating factor. On using $u(0) = 0$, we obtain

$$u(x) = V\left(1 - e^{-3bSx/M}\right)^{1/3}.$$

Hence $\lim_{x\to\infty} u(x) = V$.

(iii) d/V.

(iv) By combining (i) and (ii) and setting $x = d$, we obtain

$$G(q) = q\left\{\ln\left(\frac{U^2 + U + 1}{(U-1)^2}\right) + \frac{\pi}{\sqrt{3}} - 2\sqrt{3}\tan^{-1}\left(\frac{1+2U}{\sqrt{3}}\right)\right\},$$

where

$$U(q) = \left(1 - e^{-1/2q}\right)^{1/3}.$$

(v) This is just a messy calculus problem, the limit requiring l'Hopital's rule. You can get the general idea from the following table:

q	$G(q)$	q	$G(q)$	q	$G(q)$
1.00	1.95	0.15	1.22	0.04	1.06
0.5	1.60	0.1	1.15	0.03	1.04
0.25	1.35	0.05	1.07	0.02	1.03
0.2	1.29				

(vi) 29% and 3%, because $G(0.2) = 1.29$ and $G(0.02) = 1.03$ from the table above.

1.20 (i) Mass/length^3, measured as, say, kilograms per cubic meter.

(ii) Defining $\hat{u} = u/V$ and $\hat{t} = (VbSt)/M$, we obtain $d\hat{u}/d\hat{t} = (1/\hat{u}) - \hat{u}^2$, $\hat{u}(0) = 1$.

(iii) $$\ln\left\{\frac{\hat{u}^2 + \hat{u} + 1}{(\hat{u}-1)^2}\right\} + \frac{\pi}{\sqrt{3}} - 2\sqrt{3}\tan^{-1}\left\{\frac{1+2\hat{u}}{\sqrt{3}}\right\} = 6\hat{t},$$

agreeing with the expression obtained in Exercise 1.19.

Chapter 2

2.3 (i) From (1.20) we have $u(x,y) = x(a_1 - b_1 y)$, $v(x,y) = y(-a_2 + b_2 x)$. Restricting attention to $x > 0$, $y > 0$ implies $x^* = a_2/b_2$, $y^* = a_1/b_1$. Thus

(2.14) implies $g_{11} = 0$, $g_{12} = -b_1 a_2/b_2$, $g_{21} = b_2 a_1/b_1$, $g_{22} = 0$. Hence (2.19) becomes $\omega^2 + a_1 a_2 = 0$. Thus ω_1, ω_2 are purely imaginary, whence, from (2.21), the equilibrium is a center.

(ii) The matrix $\mathbf{G}$ is the same as in (i) except that g_{21} changes sign. Thus (2.19) becomes $\omega^2 - a_1 a_2 = 0$, whence the equilibrium is a saddle point, by (2.21).

(iii) Now $u(x,y) = kx(x_f - x)$, $v(x,y) = ky(y_f - y)$. Restricting attention to $x > 0$, $y > 0$ implies $x^* = x_f$, $y^* = y_f$. We deduce from (2.14) that $g_{11} = -kx_f$, $g_{22} = -ky_f$, whence (2.19) becomes $(\omega + kx_f)(\omega + ky_f) = 0$. It follows from (2.21) that the equilibrium is a stable node.

*2.4 Instead of (1.80), (2.26), and (2.28) we would have:

$$\text{Braking force} = A(z_j'(t) - z_{j-1}'(t)),$$

$$\frac{dz_j(t + \tau)}{dt} = \lambda |z_j(t) - z_{j-1}(t)| + \alpha_j,$$

$$v = \lambda(d + L) + \alpha$$

and

$$v(x) = \begin{cases} v_{\max}, & \text{if } 0 \leq x \leq x_c; \\ v_{\max}\dfrac{x_c(x_{\max} - x)}{x(x_{\max} - x_c)}, & \text{if } x_c < x \leq x_{\max}. \end{cases}$$

2.6 From (2.37) we have

$$z = ye^{r(1-y)} = ze^{r(1-z)} \cdot e^{r(1-y)} = ze^{r(2-z-y)},$$

whence

$$z(e^{r(2-z-y)} - 1) = 0.$$

Thus either $z = 0$ or $2 - z - y = 0$; i.e., on using (2.37) and (2.38), $2 - z = g(z)$. Similarly, $2 - y = g(y)$.

2.7 The necessary expression for f' is given by

$$f'(r) = 2\left\{ ry(r) \cdot \{y(r) - 2\} + r^2 y'(r) \cdot \{y'(r) - 1\} + 1 \right\},$$

where $y'(r)$ is given in the statement of the problem. With $R_0 = 2.5$ (from the graph), the Newton–Raphson iteration

$$R_{k+1} = R_k - \frac{f(R_k)}{f'(R_k)}, \qquad k = 0, 1, \ldots,$$

converges to $r_2 = 2.5625$ in two steps (to seven significant figures, $R_1 = 2.526401$ and $R_2 = 2.526468 = R_3$). The corresponding value of y is $y(r_2) = 0.2777$.

2.13 We deduce the dimensions of ν from the dimensional consistency of (2.58) and (2.59). We have [MOMENTUM] $= ML/T^2$. Hence the left-hand side of (2.59) has dimensions $[N] =$ [MOMENTUM]$/L^2 = M/LT^2$, implying that $[\mu] \cdot [w] = [\mu w] = L[N] = M/T^2$. But $[w] = L/T$. Hence $[\mu] = M/LT$, whence (2.58) implies that $[\rho] \cdot [\nu] = M/LT$. But $[\rho] = M/L^3$. Therefore $[\nu] = L^2/T$.

2.14 To establish (2.71), we must show that

$$I = \iint\limits_{\frac{x^2}{a^2} + \frac{y^2}{b^2} \leq 1} \left(1 - \frac{x^2}{a^2} - \frac{y^2}{b^2}\right) dxdy = \frac{1}{2}\pi ab.$$

Change variables as suggested. The area magnification factor is

$$\det \begin{bmatrix} \frac{\partial x}{\partial r} & \frac{\partial x}{\partial \theta} \\ \frac{\partial y}{\partial r} & \frac{\partial y}{\partial \theta} \end{bmatrix} = rab.$$

Hence

$$I = \int_0^{2\pi} \int_0^1 (1 - r^2) rab\, dr d\theta = \frac{1}{2}\pi ab,$$

as required.

*2.16 (i) tonne$^{-1} \cdot$ year^{-1}.
(ii) The harvest is governed by the dynamical equation

$$\frac{dx}{dt} = x\left(ax\left(1 - \frac{x}{K}\right) - qu\right). \tag{a}$$

For $x \neq 0$, dx/dt vanishes where

$$x\left(1 - \frac{x}{K}\right) = \frac{qu}{a}. \tag{b}$$

Because $x(1 - x/K)$ has a maximum $K/4$ where $x = K/2$, Fig. S2.1 shows that there are two equilibria if $qu/a < K/4$.

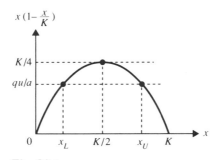

Fig. S2.1

(iii) If $u > aK/4q$ then $\mu(x) < 0$ for all values of x. Thus $x \to 0$ as $t \to \infty$, corresponding to the extinction of the species. There are too many fishermen: the species is *overfished*. In practice, of course, the fishermen would go away if they were catching nothing, giving the endangered species a chance to recover, but this might not actually happen because the fishermen could be harvesting a second species while driving the first one to extinction. See Exercise 9.18 for a possible model of this phenomenon.

(iv) See Fig. S2.1. Differentiating (b) with respect to u, for example, gives

$$\left(1 - \frac{2x_U}{K}\right) \frac{\partial x_U}{\partial u} = \frac{q}{a} > 0. \tag{c}$$

Because $x_U > K/2$, (c) implies that $\partial x_U / \partial u < 0$. Similarly for q and a.

(v) You can see from Fig. S2.1 that dx/dt is negative if $x < x_L$ or $x > x_U$ and positive if $x_L < x < x_U$. Thus x_L is unstable, because a slight push to the left will send x toward zero and a slight push to the right will send x toward x_U. On the other hand, x_U is stable because x approaches it from anywhere in the interval $x_L < x < K$. Because x_L is unstable, the shift technique must not be applied to it. As an example of an absurd prediction, it would tell us that increasing u would increase x ($\partial x_L / \partial u > 0$); whereas that would give x just precisely the push to the left it needed to begin an unfortunate descent to zero.

*2.19 According to (2.111), the eighteenth car will collide with the seventeenth car after 420 units of time (52.5 seconds).

Chapter 3

3.4 (i)

$$\epsilon_U(\lambda) - \epsilon_L(\lambda) = \frac{(3\lambda + 13)(\lambda - 1)}{(113\lambda + 68)(58\lambda + 123)} > 0,$$

because $\lambda > 1$.

(ii) By differentiation with respect to λ, you can show that $\epsilon_U(\lambda)$ is an increasing function, bounded above by $\epsilon_U(\infty) = 2/113$. Hence $\epsilon < \epsilon_U(\lambda)$ implies, in particular, that $\epsilon < 2/113$, and hence that $\epsilon < 1/55$. Thus u_1^* and u_3^* are positive. By the same method as in (i), you can show that $\epsilon_L(\lambda) > (\lambda + 1)/(58\lambda + 68)$; thus $\epsilon > \epsilon_L(\lambda)$ ensures $u_2^* > 0$.

3.7 (i) Differentiating (3.37) with respect to t, $f'(t) = e^{-\delta t}(V'(t) - \delta V(t))$ and $f''(t) = e^{-\delta t}(V''(t) - 2\delta V'(t) + \delta^2 V(t))$. Thus (3.40) and (3.41) are equivalent to $f'(s) = 0$, $f''(s) < 0$. With V given by (3.42),

$$V'(t) = \frac{kbpN\xi_f^2 \exp(-k\xi_f t)}{[1 + b\exp(-k\xi_f t)]^2} = \frac{k(V(t) + c)(V_\infty - V(t))}{pN},$$

on using (3.42) to eliminate $b\exp(-k\xi_f t)$. Now (3.43) follows immediately from (3.40), and (3.44) is simply a rearrangement of (3.42). Higher differentiation yields

$$V''(t) = \frac{kV'(t)}{pN}\left(V_\infty - c - 2V(t)\right);$$

whence, on using (3.40),

$$\begin{aligned}
\frac{V''(s)}{V(s)} - \delta^2 &= \frac{\delta k}{pN}\{V_\infty - c - 2V(s)\} - \delta^2 \\
&= \frac{\delta k}{pN}\left\{V_\infty - c - 2V(s) - \frac{\delta pN}{k}\right\} \\
&= -\frac{\delta k}{pN}\left\{V(s) + \frac{cV_\infty}{V(s)}\right\} < 0,
\end{aligned}$$

as required, on using (3.43).

(ii) Either by substituting $\delta = \nu - r$, $V(t) = e^{-rt}\bar{V}(t)$ into (3.40) or by differentiating the expression for f in (3.33) directly, we obtain

$$\frac{\bar{V}'(s)}{\bar{V}(s)} = \nu.$$

3.8 We must show that $\partial s/\partial \delta < 0$, where $s = s(\delta)$ is defined implicitly by (3.40). Differentiating $V'(s(\delta)) = \delta V(s(\delta))$ with respect to δ, we obtain

$$V''(s)\frac{\partial s}{\partial \delta} = V(s) + \delta V'(s)\frac{\partial s}{\partial \delta} = V(s) + \delta^2 V(s)\frac{\partial s}{\partial \delta},$$

whence

$$\frac{\partial s}{\partial \delta} = \frac{V(s)}{V''(s) - \delta^2 V(s)} < 0,$$

by (3.41). Alternatively, proceeding graphically, an increase in the interest rate means an anticlockwise rotation of the line through 0 in Fig. 3.5, which reduces the value of $V(s)$, and hence the value of s.

*3.9 (i) See Section 4.3.

(ii) Use

$$V(t) \approx \frac{1}{2}\{V(t+5) + V(t-5)\}, \qquad V'(t) \approx \frac{1}{10}\{V(t+5) - V(t-5)\}.$$

See Table 4.1 in Section 4.8.

(iii) See Fig. 4.7 in Section 4.8. The graph suggests that $\theta = 20$ when $t \approx 65$, suggesting that the optimal rotation period is about 65 years.

3.10 From (3.54) and (3.58),

$$\begin{aligned}
\int_0^T x(t)dt &= \int_0^s x(t)dt + \int_s^T x(t)dt \\
&= \left(x(0) + \frac{q}{R}\right)\frac{e^{Rs} - 1}{R} - \frac{qs}{R}
\end{aligned}$$

$$+ \left(x(s) + \frac{q}{R} \right) \frac{e^{R(T-s)} - 1}{R} - \frac{q(T-s)}{R}.$$

On substituting for $x(s)$ from (3.57) and using (3.55) and (3.56), we obtain (3.63).

3.11 Differentiating (3.63) with respect to $U(> U_0)$ and holding s constant, we obtain

$$\frac{\partial f}{\partial U} = c_1 - \alpha p b e^{-\alpha(U - U_0)} \frac{e^{R(T-s)} - 1}{R} \left\{ \left(x(0) + \frac{q}{R} \right) e^{Rs} - \frac{q}{R} \right\}, \quad \text{(a)}$$

$$\frac{\partial^2 f}{\partial U^2} = \alpha^2 p b e^{-\alpha(U - U_0)} \frac{e^{R(T-s)} - 1}{R} \left\{ \left(x(0) + \frac{q}{R} \right) e^{Rs} - \frac{q}{R} \right\} > 0. \quad \text{(b)}$$

Thus $\partial f / \partial U$ vanishes where

$$U = U_0 + \frac{1}{\alpha} \ln \left\{ \frac{\alpha p b}{c_1} \frac{e^{R(T-s)} - 1}{R} \left\{ \left(x(0) + \frac{q}{R} \right) e^{Rs} - \frac{q}{R} \right\} \right\}, \quad \text{(c)}$$

which is greater than U_0 provided

$$\frac{\alpha p b}{c_1} \frac{e^{R(T-s)} - 1}{R} \left\{ \left(x(0) + \frac{q}{R} \right) e^{Rs} - \frac{q}{R} \right\} > 1. \quad \text{(d)}$$

In view of (b), f has a minimum at (c) for any given s provided (d) is satisfied; otherwise the minimum is at U_0. Differentiating (3.63) with respect to $s(\geq 0)$ and holding U constant, we obtain

$$\frac{\partial f}{\partial s} = p b \left\{ \left(x(0) + \frac{q}{R} \right) e^{Rs} - \frac{q}{R} e^{R(T-s)} \right\} (1 - e^{-\alpha(U - U_0)}) \quad \text{(e)}$$

$$\frac{\partial^2 f}{\partial s^2} = p b \left\{ (R x(0) + q) e^{Rs} + q e^{R(T-s)} \right\} (1 - e^{-\alpha(U - U_0)}) > 0 \quad \text{(f)}$$

for $U > U_0$. Thus $\partial f / \partial s$ vanishes where

$$s = \frac{1}{2} T - \frac{1}{2R} \ln \left(1 + \frac{R x(0)}{q} \right), \quad \text{(g)}$$

which is greater than zero if

$$q > \frac{R x(0)}{e^{RT} - 1}. \quad \text{(h)}$$

In view of (f), f has a minimum at (g) for any given $U(> U_0)$ provided (h) is satisfied; otherwise the minimum is at zero. Combining these results, if (h) is satisfied then the optimal spraying time is (g) and the optimal concentration of pesticide is

$$U^* = U_0 + \frac{1}{\alpha} \ln \left(\frac{p b \alpha}{c_1} \left(\frac{\sqrt{q + R x(0)} e^{RT/2} - \sqrt{q}}{R} \right)^2 \right), \quad \text{(i)}$$

provided that the logarithm has argument greater than 1 and provided $f(s^*, U^*) < f(s, 0)$; otherwise it's better not to spray at all. If (h) is not satisfied then the optimal spraying time is the beginning of the growing season and the optimal concentration of pesticide is

$$U^* = U_0 + \frac{1}{\alpha} \ln \left\{ \frac{pb\alpha}{c_1} \frac{e^{RT} - 1}{R} x(0) \right\}, \tag{j}$$

provided that the logarithm has argument greater than 1 and provided $f(0, U^*) < f(s, 0)$; otherwise it's better not to spray. Straightforward algebraic manipulation reveals that the conditions $f(s^*, U^*) < f(s, 0)$ and $f(0, U^*) < f(s, 0)$ mentioned above can be combined as

$$c_0 + c_1 U_0 + \frac{c_1}{\alpha} \left\{ 1 + \ln \left(\frac{\alpha \Delta}{c_1} \right) \right\} < \Delta, \tag{k}$$

where Δ is defined by (3.68). So far, we have found the optimal solution for $U > U_0$, but the law requires that $U \leq U_{max}$. Thus if all the above conditions for the optimal U to be positive are satisfied but $U^* > U_{max}$, i.e., if

$$U_0 + \frac{1}{\alpha} \ln \left(\frac{\alpha \Delta}{c_1} \right) > U_{max}, \tag{l}$$

then the optimal strategy is to apply pesticide with concentration U_{max}. Referring to Section 3.6, (k) corresponds to either (3.65) or (3.70) and (i), (j), and (l) are collected together in (3.71).

Chapter 4

4.1 From (4.22) we have

$$J'(\tilde{u}) = \left(\frac{\eta E}{bS u_0^{*2}} \right) \frac{2(1 - \tilde{u}^3) - 3\tilde{u}^2 \epsilon}{(\tilde{u}^3 + 2)^2}.$$

Hence on using (4.23) we have

$$J'(\tilde{u}^*(\epsilon)) = 0, \qquad J''(\tilde{u}^*(\epsilon)) = -6 \left(\frac{\eta E}{bS u_0^{*2}} \right) \frac{\tilde{u}^*(\epsilon)(\tilde{u}^*(\epsilon) + \epsilon)}{[\tilde{u}^*(\epsilon)^3 + 2]^2} < 0.$$

4.2 Equation (4.23) can be written $2\tilde{u} + 3\epsilon = 2/\tilde{u}^2$. The graphs of the functions of $\tilde{u}$ defined by the two sides of this equation are readily sketched. For positive $\tilde{u}$, the graphs intersect precisely once; and increasing ϵ from 0, which raises the straight line, moves the point of intersection to the left. Hence (4.23) has a unique solution and $\tilde{u}^*(\epsilon) < \tilde{u}^*(0)$.

4.3 (ii) Differentiating (4.23) with respect to ϵ yields

$$\tilde{u}^{*\prime}(\epsilon) = \frac{-\tilde{u}^*(\epsilon)}{2(\tilde{u}^*(\epsilon) + \epsilon)}.$$

Further differentiation with respect to ϵ yields

$$\tilde{u}^{*\prime\prime}(\epsilon) = \frac{2\epsilon + (\epsilon + 2)\tilde{u}^*(\epsilon)}{4(\tilde{u}^*(\epsilon) + \epsilon)^3} > 0,$$

as required.

4.4 It is clear from Fig. S4.1 that $\ln(1 + w) > bw$ whenever $\alpha < w < 0$, where α is the negative root of the equation $\ln(1 + w) - bw = 0$. By the Newton–Raphson method, α is the limit of the sequence defined by $w_0 = -0.6$ (guessed from diagram) and

$$w_{k+1} = \frac{w_k/(1 + w_k) - \ln(1 + w_k)}{1/(1 + w_k) - b}$$

if $k \geq 0$. Thus $\alpha = -0.6321206$.

4.5 For $\epsilon > 0$, s must be such that $V'(s) - \delta V(s) > 0$: too short according to (4.37). Hence $s^* < s$: the optimal rotation time predicted by Section 4.8 is shorter than that predicted by Section 3.4.

4.6 This follows from

$$f'(t) = \frac{V'(t) - \delta V(t) - \delta f(t)}{e^{\delta t} - 1}, \qquad f''(s^*) = \frac{V''(s^*) - \delta V'(s^*)}{e^{\delta s^*} - 1}.$$

4.7 Comparing (4.66) to (2.12), we find that $u(x, y) = x(a_1 - c_1 x - b_1 y)$ and $v(x, y) = y(a_2 - b_2 x - c_2 y)$. Thus the matrix $\mathbf{G}$ defined by (2.14) becomes

$$\mathbf{G} = \begin{bmatrix} g_{11} & g_{12} \\ g_{21} & g_{22} \end{bmatrix} = \begin{bmatrix} a_1 - 2c_1 x^* - b_1 y^* & -b_1 x^* \\ -b_2 y^* & a_2 - b_2 x^* - 2c_2 y^* \end{bmatrix}.$$

Thus, for equilibrium (4.67), it follows from definition (4.71) that (2.19) becomes $(w + a_2)(w + b_1 I_y) = 0$. Similarly, for (4.68), (2.19) becomes $(w + a_1)(w + b_2 I_x) = 0$; and for (4.69), (2.19) becomes

$$w^2 + (c_1 x^* + c_2 y^*)w + \frac{b_1 b_2 c_1 c_2 I_x I_y}{c_1 c_2 - b_1 b_2} = 0,$$

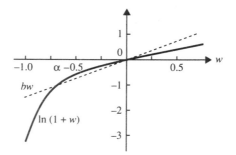

Fig. S4.1

where x^* and y^* are defined by (4.69). Observing that (c) implies $c_1 c_2 > b_1 b_2$ in Table 4.2, and (d) implies $c_1 c_2 < b_1 b_2$, we can now confirm that both w_1 and w_2 are negative for the attracting equilibria.

Chapter 5

5.1 For (5.40), use the integrating factor $e^{d_j t}$.

5.2 For (5.46), use the integrating factor $e^{b_j t}$. To show that the Poisson distribution (5.49) has mean Bt, simply replace 1 by t, and hence B by Bt, in the calculation for (5.51).

5.3 The result

$$r_{ij}(t) = \frac{(j-1)!}{(i-1)!(j-i)!} e^{-Bit}(1 - e^{-Bt})^{j-i}, \qquad j \geq i \geq 1, \qquad \text{(a)}$$

is clearly true for $j = i$, because $r_{ii}(t) = e^{-Bit}$ from (5.46). Assume the truth of (a) for $j = i + k$, where $k \geq 1$. Then

$$r_{i,i+k}(t) = \frac{(i+k-1)!}{(i-1)!k!} e^{-Bit}(1 - e^{-Bt})^k.$$

Substituting into (5.46) with $j = i + k + 1$ then yields

$$r_{i,i+k+1}(t) = b_{i+k} \int_0^t e^{-b_{i+k+1}(t-s)} r_{i,i+k}(s) ds$$

$$= (i+k)B \int_0^t e^{-(i+k+1)B(t-s)} \frac{(i+k-1)!}{(i-1)!k!} e^{-Bis}(1 - e^{-Bs})^k ds$$

$$= \frac{(i+k)!}{(i-1)!k!} e^{-(i+k+1)Bt} B \int_0^t e^{(k+1)Bs}(1 - e^{-Bs})^k ds$$

$$= \frac{(k+1)(i+k)!}{(i-1)!(k+1)!} e^{-(i+k+1)Bt} \int_0^t (e^{Bs} - 1)^k (Be^{Bs} ds)$$

$$= \frac{(i+k)!}{(i-1)!(k+1)!} (k+1) e^{-(i+k+1)Bt} \left[\frac{(e^{Bs} - 1)^{k+1}}{k+1} \right]_0^t$$

$$= \frac{(i+k)!}{(i-1)!(k+1)!} e^{-(i+k+1)Bt}(e^{Bt} - 1)^{k+1}$$

$$= \frac{(i+k)!}{(i-1)!(k+1)!} e^{-Bit}(1 - e^{-Bt})^{k+1}$$

$$= \frac{(j-1)!}{(i-1)!(j-i)!} e^{-Bit}(1 - e^{-Bt})^{j-i}.$$

Hence (a) is true for $j = i + k + 1$ if it is true for $j = i + k$. Its truth for all j now follows by the principle of mathematical induction.

5.4 Set $x = 1 - e^{-\lambda T}$ (which satisfies $0 < x < 1$ for $T > 0$) in

$$\frac{1}{(1-x)^2} = \sum_{j=1}^{\infty} jx^{j-1}.$$

5.5 Because

$$r_{i0}(t) + r_{i1}(t) + r_{i2}(t) = 1 \tag{a}$$

for all $t \geq 0$, one of the equations (5.64) is redundant. Ignore the third, using (a) to eliminate r_{i2} from the second. We then have

$$\frac{d\mathbf{x}}{dt} = \mathbf{A}\mathbf{x} + \begin{bmatrix} 0 \\ d_2 \end{bmatrix}, \tag{b}$$

where

$$\mathbf{x}(t) = \begin{bmatrix} r_{i0}(t) \\ r_{i1}(t) \end{bmatrix}, \qquad \mathbf{A} \equiv \begin{bmatrix} -b_0 & d_1 \\ b_0 - d_2 & -b_1 - d_1 - d_2 \end{bmatrix}. \tag{c}$$

The solution to (b) is $\mathbf{x} = -\mathbf{A}^{-1}(0, d_2)^T + \mathbf{y}$, where $\mathbf{y}$ is the solution to

$$\frac{d\mathbf{y}}{dt} = \mathbf{A}\mathbf{y}. \tag{d}$$

The solution to (d) is

$$\mathbf{y}(t) = c_1 \mathbf{e}_1 e^{\omega_1 t} + c_2 \mathbf{e}_2 e^{\omega_2 t}, \tag{e}$$

where $\mathbf{e}_1, \mathbf{e}_2$ are eigenvectors of $\mathbf{A}$, the constants c_1, c_2 are determined from

$$\mathbf{y}(0) = c_1 \mathbf{e}_1 + c_2 \mathbf{e}_2 \tag{f}$$

and ω_1, ω_2 are the eigenvalues of $\mathbf{A}$ associated, respectively, with $\mathbf{e}_1$ and $\mathbf{e}_2$; i.e., $\mathbf{A}\mathbf{e}_1 = \omega_1 \mathbf{e}_1$, $\mathbf{A}\mathbf{e}_2 = \omega_2 \mathbf{e}_2$, where ω_1, ω_2 are the roots of the equation

$$(\omega + b_0)(\omega + b_1 + d_1 + d_2) = d_1(b_0 - d_2). \tag{g}$$

For $b_0 = d_2 = \alpha$ we obtain $\omega_1 = -\alpha$, $\mathbf{e}_1 = (1, 0)^T$, $\omega_2 = -\alpha - b_1 - b_2$, $\mathbf{e}_2 = (-d_1, b_1 + d_1)^T$ and

$$\mathbf{A}^{-1} = \begin{bmatrix} -\frac{1}{\alpha} & -\frac{d_1}{\alpha(\alpha+b_1+d_1)} \\ 0 & -\frac{1}{\alpha+b_1+d_1} \end{bmatrix}. \tag{h}$$

Thus, on substituting $\mathbf{y}(0) = \mathbf{x}(0) + \mathbf{A}^{-1}(0, \alpha)^T$ into (e) we obtain

$$c_1 - c_2 d_1 = \delta_{i0} - \frac{d_1}{\alpha + b_1 + d_1},$$

$$c_2(b_1 + d_1) = \delta_{i1} - \frac{\alpha}{\alpha + b_1 + d_1}, \tag{i}$$

where δ_{ij} is defined by (5.36). On solving these equations for c_1 and c_2 we finally obtain

$$\begin{bmatrix} r_{i0}(t) \\ r_{i1}(t) \end{bmatrix} = -\mathbf{A}^{-1} \begin{bmatrix} 0 \\ \alpha \end{bmatrix} + \left\{ \delta_{i0} - \frac{d_1}{b_1 + d_1}(1 - \delta_{i1}) \right\} \begin{bmatrix} 1 \\ 0 \end{bmatrix} e^{-\alpha t}$$

$$+ \frac{1}{b_1 + d_1} \left\{ \delta_{i1} - \frac{\alpha}{\alpha + b_1 + d_1} \right\} \begin{bmatrix} -d_1 \\ b_1 + d_1 \end{bmatrix} e^{-(\alpha + b_1 + d_1)t}; \quad (j)$$

and $r_{i2}(t)$ follows from (a). You can easily check that (j) agrees with (5.65).

5.9 The first few powers of $\mathbf{P}$ are:

$$\mathbf{P}^2 = \begin{bmatrix} \frac{1}{4} & 0 & \frac{3}{4} & 0 & 0 \\ 0 & \frac{5}{8} & 0 & \frac{3}{8} & 0 \\ \frac{1}{8} & 0 & \frac{3}{4} & 0 & \frac{1}{8} \\ 0 & \frac{3}{8} & 0 & \frac{5}{8} & 0 \\ 0 & 0 & \frac{3}{4} & 0 & \frac{1}{4} \end{bmatrix}, \quad \mathbf{P}^3 = \begin{bmatrix} 0 & \frac{5}{8} & 0 & \frac{3}{8} & 0 \\ \frac{5}{32} & 0 & \frac{3}{4} & 0 & \frac{3}{32} \\ 0 & \frac{1}{2} & 0 & \frac{1}{2} & 0 \\ \frac{3}{32} & 0 & \frac{3}{4} & 0 & \frac{5}{32} \\ 0 & \frac{3}{8} & 0 & \frac{5}{8} & 0 \end{bmatrix},$$

$$\mathbf{P}^4 = \begin{bmatrix} \frac{5}{32} & 0 & \frac{3}{4} & 0 & \frac{3}{32} \\ 0 & \frac{17}{32} & 0 & \frac{15}{32} & 0 \\ \frac{1}{8} & 0 & \frac{3}{4} & 0 & \frac{1}{8} \\ 0 & \frac{15}{32} & 0 & \frac{17}{32} & 0 \\ \frac{3}{32} & 0 & \frac{3}{4} & 0 & \frac{5}{32} \end{bmatrix}, \quad \mathbf{P}^5 = \begin{bmatrix} 0 & \frac{17}{32} & 0 & \frac{15}{32} & 0 \\ \frac{17}{128} & 0 & \frac{3}{4} & 0 & \frac{15}{128} \\ 0 & \frac{1}{2} & 0 & \frac{1}{2} & 0 \\ \frac{15}{128} & 0 & \frac{3}{4} & 0 & \frac{17}{128} \\ 0 & \frac{15}{32} & 0 & \frac{17}{32} & 0 \end{bmatrix},$$

$$\mathbf{P}^6 = \begin{bmatrix} \frac{17}{128} & 0 & \frac{3}{4} & 0 & \frac{15}{128} \\ 0 & \frac{65}{128} & 0 & \frac{63}{128} & 0 \\ \frac{1}{8} & 0 & \frac{3}{4} & 0 & \frac{1}{8} \\ 0 & \frac{63}{128} & 0 & \frac{65}{128} & 0 \\ \frac{15}{128} & 0 & \frac{3}{4} & 0 & \frac{17}{128} \end{bmatrix}, \quad \mathbf{P}^7 = \begin{bmatrix} 0 & \frac{65}{128} & 0 & \frac{63}{128} & 0 \\ \frac{65}{512} & 0 & \frac{3}{4} & 0 & \frac{63}{512} \\ 0 & \frac{1}{2} & 0 & \frac{1}{2} & 0 \\ \frac{63}{512} & 0 & \frac{3}{4} & 0 & \frac{65}{512} \\ 0 & \frac{63}{128} & 0 & \frac{65}{128} & 0 \end{bmatrix}.$$

The pattern that suggests itself is that, for integer k:

$$\lim_{k \to \infty} \mathbf{P}^{2k+1} = \begin{bmatrix} 0 & \frac{1}{2} & 0 & \frac{1}{2} & 0 \\ \frac{1}{8} & 0 & \frac{3}{4} & 0 & \frac{1}{8} \\ 0 & \frac{1}{2} & 0 & \frac{1}{2} & 0 \\ \frac{1}{8} & 0 & \frac{3}{4} & 0 & \frac{1}{8} \\ 0 & \frac{1}{2} & 0 & \frac{1}{2} & 0 \end{bmatrix}, \quad \lim_{k \to \infty} \mathbf{P}^{2k} = \begin{bmatrix} \frac{1}{4} & 0 & \frac{3}{4} & 0 & \frac{1}{8} \\ 0 & \frac{1}{2} & 0 & \frac{1}{2} & 0 \\ \frac{1}{8} & 0 & \frac{3}{4} & 0 & \frac{1}{8} \\ 0 & \frac{1}{2} & 0 & \frac{1}{2} & 0 \\ \frac{1}{8} & 0 & \frac{3}{4} & 0 & \frac{1}{8} \end{bmatrix}.$$

Notice that $\mathbf{P}^k$ does not itself approach a limit. This Markov chain is said to be periodic with period 2. See Section 6.1 for further discussion.

Chapter 6

6.1 The stationary distribution (6.20), i.e., $\pi_0^* = \pi_1^* = \pi_2^* = 1/3$ is obtained as the unique solution of the four equations

$$\pi_0^* = 0.4467499\pi_0^* + 0.3296303\pi_1^* + 0.2236198\pi_2^*,$$
$$\pi_1^* = 0.3296303\pi_0^* + 0.3407393\pi_1^* + 0.3296303\pi_2^*,$$
$$\pi_2^* = 0.2236198\pi_0^* + 0.3296303\pi_1^* + 0.4467499\pi_2^*,$$
$$\pi_0^* + \pi_1^* + \pi_2^* = 1.$$

Of course, the first three are linearly dependent and (any) one of them should be eliminated.

6.2 Simply verify that (5.89) and (5.92) satisfy (6.10).

6.4 From (6.29) and (6.30):

$$\sum_{j=1}^{N} j\pi_j = \rho\pi_0 \sum_{j=1}^{N} j\rho^{j-1} = \rho\pi_0 \frac{d}{d\rho}\left\{\sum_{j=0}^{N} \rho^j\right\}$$

$$= \rho\pi_0 \frac{N\rho^{N+1} - (N+1)\rho^N + 1}{(1-\rho)^2}. \tag{a}$$

The result now follows immediately from (6.29).

6.5 From (6.25) and (6.34),

$$C_j = \left(\frac{1}{2}\right)^{j-1} \rho^j, \quad j \geq 1. \tag{b}$$

Hence, by (6.26b),

$$\sum_{j=1}^{N} j\pi_j = \rho\pi_0 \sum_{j=1}^{N} j\left(\frac{\rho}{2}\right)^{j-1}.$$

Replacing ρ by $\rho/2$ in (a) now yields (6.35). From (b) and (6.26a),

$$\frac{1}{\pi_0} = 1 + \rho\sum_{j=1}^{N}\left(\frac{\rho}{2}\right)^{j-1} = 1 + \rho\sum_{k=0}^{N-1}\left(\frac{\rho}{2}\right)^{k} = 1 + \rho\frac{1-(\rho/2)^N}{1-(\rho/2)},$$

by the formula for the sum of a finite geometric series (given in Exercise 6.4), whence (6.36) is immediate.

6.10 (i) First integrate by parts (or use a table of integrals) to obtain the formula

$$\int_0^y x^n e^{-ax}dx = \left[-\frac{e^{-ax}}{a^{n+1}}\sum_{k=0}^{n}(ax)^k\frac{n!}{k!}\right]_0^y = \frac{n!}{a^{n+1}} - \frac{e^{-ay}}{a^{n+1}}\sum_{k=0}^{n}(ay)^k\frac{n!}{k!},$$

where n is an integer and $a > 0$. Then applying this formula twice, first with $a = \lambda$, $n = j-1$, $y = t$ and later with $a = \lambda + \mu$, $n = j - k$, $y \to \infty$, we find that

$$\int_0^\infty \int_0^t \mu e^{-\mu t}\frac{\lambda e^{-\lambda v}(\lambda v)^{j-1}}{(j-1)!}dvdt = \int_0^\infty \frac{\mu e^{-\mu t}\lambda^j}{(j-1)!}\left\{\int_0^t e^{-\lambda v}v^{j-1}dv\right\}dt$$

$$= \int_0^\infty \frac{\mu e^{-\mu t}\lambda^j}{(j-1)!}\left\{\frac{(j-1)!}{\lambda^j} - e^{-\lambda t}\sum_{k=1}^{j}\frac{t^{j-k}(j-1)!}{\lambda^k(j-k)!}\right\}dt$$

$$= \int_0^\infty \mu e^{-\mu t}\left\{1 - e^{-\lambda t}\sum_{k=1}^{j}\frac{t^{j-k}\lambda^{j-k}}{(j-k)!}\right\}dt$$

$$= 1 - \mu\int_0^\infty e^{-(\lambda+\mu)t}\sum_{k=1}^{j}\frac{t^{j-k}\lambda^{j-k}}{(j-k)!}dt$$

$$= 1 - \sum_{k=1}^{j} \frac{\mu \lambda^{j-k}}{(j-k)!} \int_0^\infty e^{-(\lambda+\mu)t} t^{j-k} dt$$

$$= 1 - \sum_{k=1}^{j} \frac{\mu \lambda^{j-k}}{(j-k)!} \cdot \frac{(j-k)!}{(\mu+\lambda)^{j-k+1}}$$

$$= 1 - \frac{\mu}{\mu+\lambda} \sum_{k=1}^{j} \left(\frac{\lambda}{\mu+\lambda} \right)^{j-k}$$

$$= 1 - \frac{\mu}{\mu+\lambda} \sum_{k=0}^{j-1} \left(\frac{\lambda}{\mu+\lambda} \right)^{k}.$$

The result in (6.70) now follows from the formula for the sum of a finite geometric series, which is given in Exercise 6.4.

(ii) Just sum the infinite geometric series in (6.73); see Exercise 5.4.

(iii) This is just straightforward algebraic manipulation.

Chapter 7

7.1 From definition (7.3), (7.7) is equal to

$$p \int_0^\infty zf(z)dz + p \int_u^\infty (u-z)f(z)dz - c_1 \int_0^u (u-z)f(z)dz - c_0 u - k$$

$$= p \left\{ \int_0^\infty zf(z)dz - \int_u^\infty zf(z)dz \right\}$$

$$+ pu \int_u^\infty f(z)dz + c_1 \int_0^u zf(z)dz - c_1 u \int_0^u f(z)dz - c_0 u - k$$

$$= p \int_0^u zf(z)dz + pu \left\{ 1 - \int_0^u f(z)dz \right\}$$

$$+ c_1 \int_0^u zf(z)dz - c_1 u \int_0^u f(z)dz - c_0 u - k,$$

which easily reduces to (7.8).

7.2 (i) Define

$$G(u^*) \equiv \frac{c_0 + c_1}{p + c_1} - \int_{u^*}^\infty f(z)dz. \tag{a}$$

Then, because $f(z) \geq 0$ for $0 \leq z < \infty$ but $f(z) = 0$ for $-\infty < z < 0$,

$$G(0) = \frac{c_0 - p}{c_1 + p} < 0, \qquad G(\infty) = \frac{c_0 + c_1}{p + c_1} > 0, \tag{b}$$

and $G'(u^*) = f(u^*) > 0$. Thus a unique solution to (7.11) exists. If Z is normally distributed then, strictly, it is no longer true that $f(z) = 0$ for $z < 0$. As we remarked in Section 6.5, however, $\text{Prob}(Z \leq 0)$ is negligibly small for a normally distributed random variable satisfying

$$\text{Mean} \geq 3 \times \text{Standard deviation} > 0. \qquad (c)$$

It can thus be used as a "statistical model" (see Section 8.7) for random variables that are intrinsically positive (but satisfy (c)). If Z is normally distributed then, because

$$\int_{u^*}^{\infty} f(z)dz = 1 - \Phi\left(\frac{u^* - \mu}{\sigma}\right)$$

by (7.15), where μ is the mean and σ the standard deviation, it follows from (a) that

$$G(0) = \Phi\left(\frac{-\mu}{\sigma}\right) - \frac{(p - c_0)}{p + c_1}.$$

The uniqueness of the solution to (7.11) is thus guaranteed if

$$\Phi\left(\frac{-\mu}{\sigma}\right) < \frac{(p - c_0)}{p + c_1},$$

a condition that is invariably satisfied when the normal distribution is a suitable statistical model for demand.

(ii) Define $\xi = (z - \mu)/\sigma$, so that $d\xi = dz/\sigma$; then change the variable of integration in (7.13) from z to ξ.

7.4 We have $c_0 = 10, c_1 = 1, p = 15$, and Z is uniformly distributed over (0, 480); i.e., $f(z) = H(480 - z)/480$. From (7.10) we obtain $u^*/480 = 5/16$, whence $u^* = 150$. Then, from (7.18),

$$k_c = 5 \left\{ \int_0^{150} \frac{xdx}{480} \middle/ \int_0^{150} \frac{dx}{480} \right\} = 375 \text{ (dollars)}.$$

Hence, in case (i), the bookseller should order 150 copies and, from (7.12), expect a profit of $k_c - k = \$175$; whereas, in case (ii), it would be better not to promote the book at all, because the expected profit is merely zero— which is a guaranteed profit if the venture is not undertaken.

7.5 With the same values for c_0, c_1, p, and the same distribution for Z, we obtain, from (7.30):

$$\int_{u^*}^{480} \left(1 - \frac{u^*}{16x}\right) \cdot \frac{dx}{480} = \frac{11}{16},$$

whence

$$u^* + \frac{u^*}{16} \ln\left(\frac{480}{u^*}\right) = 150.$$

The solution of this equation is found, by the Newton–Raphson method, to be the limit of the sequence defined recursively by

$$u_{k+1} = \frac{2400 - u_k}{15 + \ln\left(480/u_k\right)},$$

i.e., $u^* = 139.2$ (dollars). From (7.32),

$$k_c = \frac{u^{*2}}{960}\left\{\ln\left(\frac{480}{u^*}\right) + \frac{31}{2}\right\} = 338 \text{ (dollars)}.$$

Hence, in the case (i), the bookseller should order 139 copies of the book and expect a profit of $k_c - k \approx \$138$; whereas, in case (ii), the venture should not be undertaken.

7.7 Differentiating (7.11) with respect to p we obtain, from Leibnitz's rule of the calculus,

$$-f(u^*)\frac{\partial u^*}{\partial p} = -\frac{c_0 + c_1}{(p + c_1)^2},$$

whence $\partial u^*/\partial p > 0$. To see that this agrees with Fig. 7.3, observe that the right-hand side of (7.16) increases with p. Thus increasing p raises the horizontal dashed line in Fig. 7.3, moving the vertical one to the right. This increases γ, which increases u^*, by (7.17). Similarly for c_0 and c_1 (don't forget that $c_0 < p$).

7.8 The derivative of (7.41) with respect to k is

$$\begin{cases} -1, & \text{if } 0 \leq k \leq k_0; \\ \dfrac{\alpha_0}{\alpha} - 1, & \text{if } k_0 < k \leq k_1; \\ \dfrac{\alpha_0}{\alpha}\left\{\dfrac{\alpha(B - b_0)}{k - k_1 + \alpha(B - b_0)}\right\}^2 - 1, & \text{if } k \geq k_1. \end{cases}$$

This is clearly never positive if $\alpha \geq \alpha_0$, in which case, the maximum will occur at zero. If $\alpha < \alpha_0$, on the other hand, then the expression for the derivative in $k_1 \leq k < \infty$ will vanish at $k = k^*$, a local maximum because the second derivative of $J(u^*(k))$ is negative. This will also be the global maximum if $\alpha < \alpha_c$, where α_c is defined by (7.42); whereas, if $\alpha_c < \alpha < \alpha_0$, then the possibility arises that $J(u^*(k^*)) < 0$, in which case the global maximum is still at $k = 0$. You can easily check that this situation would arise in Exercise 7.9 if a brochure cost \$1 instead of 50¢ to print and mail.

*7.13 From (7.4)–(7.6), with $k = 0$, and (7.20):

$$P_u = \begin{cases} pZ - c_0 u, & \text{if } Z \leq u; \\ (p - c_0)u, & \text{if } Z > u. \end{cases}$$

Suppose that the rose vendor attempts to maximize $\text{Prob}(P_u \geq r)$; i.e., suppose that $J(u) = \text{Prob}(P_u \geq r)$. Then, because r must not exceed the maximum profit attainable, $r \leq (p - c_0)u$, so that either $u = r/(p - c_0)$ or

$u > r/(p - c_0)$. In the former case,

$$J(u) = \text{Prob}(P_u \geq r) = \text{Prob}(P_u = r)$$
$$= \text{Prob}(Z > u) = \int_u^\infty f(z)dz = \int_{r/(p-c_0)}^\infty f(z)dz.$$

In the second case,

$$J(u) = \text{Prob}(P_u \geq r) = \text{Prob}(pZ - c_0 u \geq r)$$
$$= \text{Prob}\left(Z \geq \frac{c_0 u + r}{p} \right) = \int_{(c_0 u + r)/p}^\infty f(z)dz.$$

This is clearly a decreasing function of u and will therefore take its maximum where u is as low as possible, i.e., where $u = r/(p-c_0)$. Combining the two cases, we see that the optimal order quantity is $w^* = r/(p-c_0)$, that is, DESIRED PROFIT ÷ PROFIT PER ITEM, as common sense would suggest. One way of comparing this policy to that described in Section 7.2 is via an analysis of the worst possible outcome. Suppose that a demand of zero were realized. Then the first rose vendor would suffer a loss of $-c_0 u^*$, where $u^* = B + \sqrt{B}\gamma(c_0/p, 0)$ is defined by (7.24); whereas the second one would suffer a loss of $-c_0 w^*$. Hence the second vendor would have a better worst outcome than the first one if $w^* < u^*$ or

$$r < \left\{ B + \sqrt{B}\gamma(\frac{c_0}{p}, 0) \right\} (p - c_0). \tag{a}$$

If this condition were satisfied, then we might describe expected value as a riskier definition of reward than $\text{Prob}(P_u \geq r)$.

Cautious attitudes to risk will be studied more systematically in Section 8.8. We shall argue there that a one-time buyer might choose a reward other than expected value as a means of *reducing* risk. In such circumstances, the one-time rose vendor maximizing $\text{Prob}(P_u \geq r)$ would certainly wish to ensure (a).

7.14 (ii) $J_2(u_0)$ is continuous and has a continuous derivative, its value at $u_0 = a + \lambda_1$ being $J_2'(a + \lambda_1) = (p + c_1)\delta$, where δ is defined by (7.69). For $u_0 > a + \lambda_1$,

$$J_2'(u_0) = \Delta - \left(\frac{u_0 - 2a}{b - a} + \frac{c_1}{p + c_1} \right)^2,$$

which cannot vanish unless $\Delta > 0$ and $J_2'(a + \lambda_1) > 0$; but if these conditions are satisfied, then the zero of $J_2'(u_0)$ corresponds to a maximum of $J_2(u_0)$ on $[a, b]$, as is easily verified by examining the second derivative. This accounts for u_0^* when $\delta > 0$. Note that $\delta > 0$ *and* $\Delta \leq 0$ is impossible, because

$$2\delta - \Delta = -\left(\frac{p - c_0 + c_1}{p + c_1} \right)^2.$$

If $\delta < 0$, then $J'(u_0) < 0$ for $u_0 > a + \lambda_1$ and the maximum occurs instead to the left of $a + \lambda_1$, at the point where the parabola has zero slope; if $\delta = 0$, i.e., if

$$\frac{a}{b-a} = \frac{(p - c_0)c_0}{(p + c_1)(p - c_0 + c_1)},$$

then

$$\Delta = \left(\frac{p - c_0 + c_1}{p + c_1}\right)^2 > 0$$

and the maximum occurs at $u_0 = a + \lambda_1$, both parts of (7.71) giving the correct formula in this instance. Note that because $p > c_0$, both δ and Δ are positive if $a = 0$.

7.17 From (7.80) we have

$$J'(u) = \frac{u_0(r - \sigma)}{\gamma u^2}\left[\alpha\left(\frac{u}{s} + 1\right)e^{-u/s} - 1\right],$$

which vanishes where $u = U^*$. On using (7.80) we find, after some manipulation, that $J''(U^*) = -\{sU^*(U^* + s)\}^{-1} < 0$. The maximum must satisfy $U^* > su_0/\gamma$ because it follows from (7.81), and the fact that $r > \sigma$, that $J'(su_0/\gamma) > 0$.

Chapter 8

8.8 Kendall and Stuart's data furnish the sample frequency distribution. Our first task is to plot the cumulative frequency distribution of this sample, so that its shape can suggest the identity of a known distribution, and hence a statistical model. Let $F_s(x)$ denote the frequency with which proportions were less than or equal to x. Technically, Kendall and Stuart's data enable us to compute the frequencies with which proportions were *greater* than or equal to x; but this is without consequence because it is safe to assume that all proportions fell in the interiors of their respective ranges. We thus obtain Table S8.1.

This cumulative frequency distribution is plotted in Fig. S8.1. By comparison with Fig. 6.3(b), its shape suggests that we use a normal distribution with mean μ and variance σ^2 as our statistical model. We do not

Table S8.1

x	$F_s(x)$	x	$F_s(x)$	x	$F_s(x)$	x	$F_s(x)$
0.05	1	0.30	12	0.55	254	0.80	549
0.10	1	0.35	41	0.60	343	0.85	569
0.15	1	0.40	69	0.65	433	0.90	574
0.20	1	0.45	128	0.70	493	0.95	576
0.25	4	0.50	190	0.75	530	1.00	576

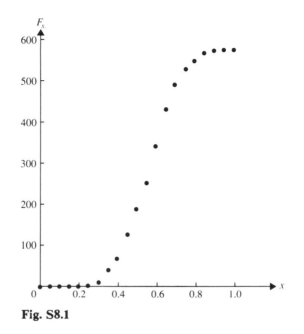

Fig. S8.1

know the individual proportions realized in the various constituencies, $\{x_1, x_2, \ldots, x_{576}\}$; we know only the frequencies with which they fell in certain ranges. Let us therefore approximate each proportion by the mid-point of the range in which it is known to lie. Thus, for example, the three proportions that fell in the range $0.2 \leq x < 0.25$ are all approximated by 0.225. In this way, we can approximate $\{x_1, x_2, \ldots, x_{576}\}$ by Table S8.2.

Then, on setting $n = 576$ in (8.75)–(8.76) and after straightforward calculations, we obtain

$$\bar{x} = \frac{1}{576} \sum_{i=1}^{576} x_i = \frac{323.15}{576} = 0.561$$

Table S8.2

i	x_i	i	x_i	i	x_i
1	0.025	$129 \leq i \leq 190$	0.475	$494 \leq i \leq 530$	0.725
$2 \leq i \leq 4$	0.225	$191 \leq i \leq 254$	0.525	$531 \leq i \leq 549$	0.775
$5 \leq i \leq 12$	0.275	$255 \leq i \leq 343$	0.575	$550 \leq i \leq 569$	0.825
$13 \leq i \leq 41$	0.325	$344 \leq i \leq 433$	0.625	$570 \leq i \leq 574$	0.875
$42 \leq i \leq 69$	0.375	$434 \leq i \leq 493$	0.675	$575 \leq i \leq 576$	0.925
$70 \leq i \leq 128$	0.425				

and

$$s^2 = \frac{1}{575} \sum_{i=1}^{576} (x_i - \bar{x})^2 = \frac{10.68}{575} = 0.01857,$$

whence $s = 0.136$. Note that, to three significant figures, we would have obtained the same value for s even if we had used 576, rather than 575, as divisor in the variance calculation. According to the method of moments (and, for this particular distribution, also the method of maximum likelihood), we approximate μ by $\bar{x}$ and σ by s. We therefore propose

$$\text{Prob}(X \le x) = 2.93 \int_{-\infty}^{x} e^{-27(\xi - 0.561)^2} d\xi$$

as our statistical model of the distribution of X. Note, as usual when applying the normal distribution to an intrinsically positive random variable, that $\text{Prob}(X \le 0)$ is not strictly zero but negligibly small.

Rather than use x_i to approximate values of x known only to lie in the interval $(x_i - h/2, x_i + h/2)$, we could assume that such values are equally likely to fall anywhere within the interval. It is then possible to show that a more appropriate estimate of the true variance of X than s^2 is $s^2 - (1/12)h^2$; see, for example, Eisenhart et al. (1947, p. 195). This correction, known by statisticians as a Sheppard correction, is also valid under other assumptions (Eisenhart et al., p. 193), but it is generally not known whether the assumptions are satisfied. The correction is therefore controversial. Moreover, in an age of high-speed computers, there is scarcely a need for it because h can be made as small as data permit. In our particular case, where $h = 0.05$, you can easily verify that a Sheppard correction does not alter the model correct to three significant figures.

The goodness of fit of this statistical model can be tested by applying the Chi-squared test, with $m = 2$ in (8.85).

**8.12 Here is a possible approach. Let X be a random variable taking values 0 and 1, with 0 denoting that the U.S. president is a Democrat and 1 that she is a Republican. Then we can regard $X(n)$ as the state of a 2-state Markov chain, in which transitions correspond to presidential elections; provided we assume that the U.S. electorate casts its votes largely by ignoring history and deciding whether it wants the incumbent party to continue in office or be replaced by the opposition. Because a presidential election occurs exactly once every four years, linear and nonlinear time coincide.

The data required to estimate the transition matrix **P** can be found in many places; for example, the back pages of Webster's *New Twentieth Century Dictionary*. Because the two-party system did not stabilize until the middle of the nineteenth century, I am going to use data only from 1860 onward. The president was a Democrat (James Buchanan) prior to 1860 and a Republican (Abraham Lincoln) thereafter, and so a transition from state 0 to state 1 took place in 1860. Transitions that took place between 1860 and 1984 are given in Table S8.3.

Table S8.3

Year	Transition	Year	Transition	Year	Transition	Year	Transition
1860	$0 \to 1$	1892	$1 \to 0$	1924	$1 \to 1$	1956	$1 \to 1$
1864	$1 \to 1$	1896	$1 \to 1$	1928	$1 \to 1$	1960	$1 \to 0$
1868	$1 \to 1$	1900	$1 \to 1$	1932	$1 \to 0$	1964	$0 \to 0$
1872	$1 \to 1$	1904	$1 \to 1$	1936	$0 \to 0$	1968	$0 \to 1$
1876	$1 \to 1$	1908	$1 \to 1$	1940	$0 \to 0$	1972	$1 \to 1$
1880	$1 \to 1$	1912	$1 \to 0$	1944	$0 \to 0$	1976	$1 \to 0$
1884	$1 \to 0$	1916	$0 \to 0$	1948	$0 \to 0$	1980	$0 \to 1$
1888	$0 \to 1$	1920	$0 \to 1$	1952	$0 \to 1$	1984	$1 \to 1$

Thus between 1860 and 1984 there were 6 transitions from state 0 to itself and 6 transitions from state 0 to state 1. According to the estimation procedure described in Section 8.6, the best estimate of the first row of $\mathbf{P}$ is therefore given by $p_{00} = 6/12 = 0.5$ and $p_{01} = 0.5$. Likewise, because there have been 14 transitions from state 1 to itself but only 6 from state 1 to state 0, we take $p_{10} = 6/20 = 0.3$ and $p_{11} = 14/20 = 0.7$. Thus

$$\mathbf{P} = \begin{bmatrix} 0.5 & 0.5 \\ 0.3 & 0.7 \end{bmatrix}. \tag{a}$$

The stationary distribution of the Markov chain with transition matrix (a) is readily shown to be

$$\boldsymbol{\pi}^* = \begin{bmatrix} 0.375 \\ 0.625 \end{bmatrix}.$$

Thus if our assumptions are correct, then in the long run, the Oval Office will be occupied by a Republican about 62.5% of the time. Including the two most recent transitions ($1 \to 1$ in 1988, $1 \to 0$ in 1992) would reduce this figure only slightly (to 61.1%).

Chapter 9

9.3 Use the integrating factor $\exp(t/T)$.

9.4 $7.8 \ln \left(\dfrac{25(5\lambda + 12)}{3\alpha} \right).$

9.5 From (9.50), the solution is constant along the lines

$$t - \frac{A(z - a)}{R} = \text{constant},$$

the family of parallel lines with slope

$$\frac{dt}{dz} = \frac{A}{R}.$$

By following the line with slope A/R through the point (z, t) toward the boundary marked by a thick solid line in Fig. S9.1, we deduce that $\phi(z, t)$

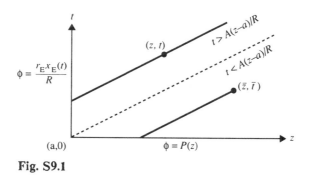

Fig. S9.1

is the (given) value of ϕ at the intersection of this line with the boundary. For (z, t) as drawn in Fig. S9.1, the line $t - A(z - a)/R = c = $ constant intercepts the boundary $z = a$ at $t = c = t - A(z - a)/R$; whence

$$\phi(z, t) = \phi(a, c) = \frac{r_E}{R} x_E \left(t - \frac{A(z - a)}{R} \right).$$

Similarly, the line with slope A/R through $(\bar{z}, \bar{t})$ intercepts the boundary $t = 0$ at $\bar{z} - R\bar{t}/A$; whence

$$\phi(\bar{z}, \bar{t}) = P \left(\frac{\bar{z} - R\bar{t}}{A} \right).$$

The dashed line through $(a, 0)$ in Fig. S9.1 distinguishes between (9.50a) and (9.50b).

9.8 (i) Because $t > 0$, $\phi(0, t) = \phi_0 \, \mathrm{Erf}(0) = 0$. Because $y > 0$, $\phi(y, 0) = \phi_0 \, \mathrm{Erf}(\infty)$. But $\mathrm{Erf}(\infty) = 1$ (from the hint). Hence $\phi(y, 0) = \phi_0$.
(ii) Let $\theta(y, t) = y/\sqrt{4\gamma t}$. Because

$$\frac{d}{d\theta} \{\mathrm{Erf}(\theta)\} = e^{-\theta^2},$$

we have

$$\frac{\partial \phi}{\partial y} = \phi_0 e^{-\theta^2} \frac{\partial \theta}{\partial y} = \phi_0 \frac{e^{-\theta^2}}{2\sqrt{\gamma t}},$$

$$\frac{\partial^2 \phi}{\partial y^2} = -\frac{\phi_0 \theta e^{-\theta^2}}{\sqrt{\gamma t}} \frac{\partial \theta}{\partial y} = -\frac{y \phi_0 e^{-\theta^2}}{4(\gamma t)^{3/2}},$$

and

$$\frac{\partial \phi}{\partial t} = \phi_0 e^{-\theta^2} \frac{\partial \theta}{\partial t} = -\frac{\phi_0 e^{-\theta^2} y}{4\sqrt{\gamma} t^{3/2}};$$

from which (9.65) follows readily.

9.9 If we approximate $\exp\{-\xi^2\}$ by $1 - \xi^2$, then (9.68) becomes

$$\beta - \frac{1}{3}\beta^3 = 0.025\sqrt{\pi}.$$

On using the "repeated substitution" algorithm

$$\beta_0 = 0.025\sqrt{\pi}, \qquad \beta_{k+1} = \frac{1}{3}\beta_k^3 + 0.025\sqrt{\pi}, \qquad k = 0, 1, 2, \ldots,$$

we quickly obtain the approximation $\beta \approx 0.04434$. As it happens, the first three significant figures are given correctly by β_0, and approximating $\exp\left\{-\xi^2\right\}$ by $1 - \xi^2 + \xi^4/2$ leads to no useful improvement (altering only digits of low significance).

9.10 (ii) Use the fact that, along lines $z - t =$ constant,

$$\frac{d\phi}{dz} = \frac{\partial\phi}{\partial z} + \frac{\partial\phi}{\partial t} \cdot \frac{dt}{dz} = \frac{\partial\phi}{\partial z} + \frac{\partial\phi}{\partial t}.$$

(iii) Age and time increase at the same rate.

*9.11 Let $y(z, t)$ denote the immigration rate at age z. Then add

$$\int_{z_1}^{z_2} y(z, t)dz$$

to the right-hand side of (9.80). Thus $y(z, t)$ replaces zero on the right-hand side of (9.81), implying that (9.92a) must be replaced by

$$\frac{\partial\phi}{\partial t} + \frac{\partial\phi}{\partial z} + \mu(z)\phi = y(z, t), \qquad 0 \le z \le \omega, 0 \le t < \infty.$$

We do not attempt to solve this equation because an alternative method for modelling immigration is introduced in Section 9.7.

*9.14 (i) One assumption implicit in the model of Section 9.7 is that immigrants are subject to the same fertility and mortality as the previously resident population.
(ii) Define $W(k) = M(k + 1) - M(k)$. Subtracting the equations

$$M(k + 2) = AM(k + 1) + Qz$$
$$M(k + 1) = AM(k) + Qz, \qquad k = 0, 1, 2, \ldots$$

from one another yields the recurrence relation $W(k + 1) = AW(k)$, $k = 0$, $1, \ldots$. The solution of this is known from (9.154) to be

$$W(k) = A^k W(0), \qquad k = 1, 2, \ldots.$$

Hence

$$M(k + 1) - M(k) = A^k\{M(1) - M(0)\} = A^k\{AM(0) + Qz - M(0)\}$$
$$= A^k(A - I)M(0) + A^k Qz.$$

Replacement of $M(k + 1)$ by $AM(k) + Qz$ and multiplication by -1 yield

$$(I - A)M(k) = A^k(I - A)M(0) + (I - A^k)Qz,$$

as required.

(iii) From a practical viewpoint, (a) is more useful because it is easier to program, but (b) is useful theoretically.

(iv) On setting $\mathbf{z} = (1,1,1,1,1,1)^T$ and using (9.144), (9.151), and (9.152), we obtain Table S9.1 in place of Table 9.10. The asymptotic age structure is the same (although the approach is slower and the population larger at every k). The reason for this is that the magnitude of $M(k)$, and hence of $AM(k)$, increases without bound, and that of $\mathbf{z}$, and hence of $Q\mathbf{z}$, remains fixed. After many recursions, the magnitude of $Q\mathbf{z}$ is so small compared with that of $AM(k)$ that recursion (a) is effectively $M(k+1) = AM(k)$, as when the population is closed to immigration.

Currently, official immigration into *both* male and female U.S. populations is fixed at about 0.6 million per annum. If the sex ratio is roughly 1, then only about three official units of female population immigrate per unit of time in our model. Nobody seems to know the true figures. (For the sake of the exercise, I cavalierly doubled the rate and distributed the influx uniformly between classes.)

(v) We can illustrate the primary and secondary effects of a baby boom as follows. Imagine that the U.S. female population has remained closed to immigration and undergone fixed mortality since 1966. Imagine also that fertility suddenly doubled in 1976, remained at twice its 1966 level throughout the subsequent decade and then suddenly halved, just as mysteriously, to return to its 1966 level in 1986 (of course this didn't happen, but it might have). The immediate effect of this fertility hike would have been a dramatic 1986 increase in class 0. This "primary bulge" would have propagated into the higher age classes at later times, its effect being somewhat attenuated by mortality. After two or three decades, however, if (as we have imagined) fertility were still at its 1966 level, then the boom babies would have had significantly more babies of their own than if there hadn't been a boom, thus creating a "secondary bulge." We can describe

Table S9.1

k	$u_0(k)$	$u_1(k)$	$u_2(k)$	$u_3(k)$	$u_4(k)$	$u_5(k)$	$M(k)$	$\frac{1}{10}\ln\left(\frac{M(k)}{M(k-1)}\right)$
0	0.1995	0.1836	0.1283	0.1152	0.1236	0.2498	99.94	
1	0.1812	0.1775	0.1635	0.1164	0.1042	0.2572	117.6	0.1628×10^{-1}
2	0.1912	0.1603	0.1566	0.1443	0.1035	0.2441	138.8	0.1657×10^{-1}
3	0.1856	0.1693	0.1424	0.1388	0.1268	0.2370	162.1	0.1555×10^{-1}
4	0.1819	0.1651	0.1506	0.1270	0.1224	0.2530	187.8	0.1467×10^{-1}
5	0.1824	0.1625	0.1475	0.1344	0.1127	0.2606	215.6	0.1383×10^{-1}
6	0.1811	0.1635	0.1457	0.1320	0.1192	0.2585	245.8	0.1310×10^{-1}
7	0.1801	0.1625	0.1466	0.1305	0.1172	0.2631	279.2	0.1272×10^{-1}
8	0.1799	0.1619	0.1459	0.1314	0.1159	0.2649	315.7	0.1230×10^{-1}
9	0.1794	0.1619	0.1456	0.1309	0.1167	0.2655	355.8	0.1196×10^{-1}
10	0.1791	0.1616	0.1456	0.1306	0.1163	0.2669	400.0	0.1170×10^{-1}
	...	...	...	...	...	...	...	...
∞	0.1768	0.1600	0.1444	0.1297	0.1151	0.2740	∞	0.9631×10^{-2}

all this by setting y(k) = 0 in (9.158) and defining A(k) so that it equals (9.151) unless $k = 2$, in which case the top row is doubled because θ_i is doubled, $1 \leq i \leq 4$, as is clear from inspection of (9.150). In place of Table 9.10 we then obtain Table S9.2. The solid underlines indicate the primary bulge. The dashed underlines indicate the secondary bulge, which is more diffuse.

The model (9.158) is extremely flexible. It can be adapted easily to describe the effects on age structure of an endless variety of hypothesized changes in fertility, mortality, and/or immigration.

*9.18 One way to proceed would be as follows. Adopting the notation of Section 3.7, let $x(t)$ denote the stock at time t of the first species, R its maximum specific growth rate, and K its capacity. Let $y(t)$, S, and L denote corresponding quantities for the second population. Let $u(t)$ be the number of boats, c the owner's cost per boat; and let p_i be the unit selling price and $h_i(t)$ the harvest rate for species i, $i = 1, 2$. Then, in place of (3.77), the owner's true return from now until eternity is

$$\int_0^\infty e^{-\delta t} \left[p_1 h_1(t) + p_2 h_2(t) - cu(t) \right] dt, \tag{a}$$

where δ is the discount rate; in place of (3.78), it seems fair to assume that

$$h_1(t) = q_1 u(t)x(t), \qquad h_2(t) = q_2 u(t)y(t), \tag{b}$$

where the constant of proportionality q_i is the catchability for species i, $i = 1, 2$.

What we assume in place of (3.79) will depend upon the biological relationship between the two species. If the species are ecologically independent, and if each would grow logistically in the absence of harvesting,

Table S9.2

k	$u_0(k)$	$u_1(k)$	$u_2(k)$	$u_3(k)$	$u_4(k)$	$u_5(k)$	$M(k)$	$\frac{1}{10} \ln\left(\frac{M(k)}{M(k-1)}\right)$
0	0.1995	0.1836	0.1283	0.1152	0.1236	0.2498	99.94	
1	0.1772	0.1791	0.1643	0.1143	0.1014	0.2637	111.0	0.1046×10^{-1}
2	<u>0.3105</u>	0.1338	0.1348	0.1231	0.0846	0.2132	146.5	0.2777×10^{-1}
3	0.1673	<u>0.2783</u>	0.1195	0.1199	0.1082	0.2068	162.9	0.1063×10^{-1}
4	0.2027	0.1435	<u>0.2379</u>	0.1018	0.1008	0.2134	189.3	0.1502×10^{-1}
5	0.2030	0.1741	0.1228	<u>0.2028</u>	0.0857	0.2116	219.7	0.1488×10^{-1}
6	0.1802	0.1794	0.1533	0.1077	0.1756	0.2038	247.8	0.1206×10^{-1}
7	0.1826	0.1588	0.1576	0.1341	0.0931	0.2739	280.3	0.1231×10^{-1}
8	0.1823	0.1640	0.1422	0.1405	0.1181	0.2530	311.0	0.1040×10^{-1}
9	0.1781	0.1637	0.1468	0.1267	0.1237	0.2611	345.2	0.1043×10^{-1}
10	0.1783	0.1603	0.1468	0.1311	0.1118	0.2716	382.2	0.1020×10^{-1}
	...	...	...	...	...	...	...	...
∞	0.1768	0.1600	0.1444	0.1297	0.1151	0.2740	∞	0.9631×10^{-2}

then:

$$\frac{dx}{dt} = Rx(t)\left(1 - \frac{x(t)}{K}\right) - h_1(t), \qquad \frac{dy}{dt} = Sy(t)\left(1 - \frac{y(t)}{L}\right) - h_2(t). \quad \text{(c)}$$

If the species interact, either as competitors or as predator and prey, then we might assume that

$$\frac{dx}{dt} = Rx\left(1 - \frac{x}{K}\right) - \alpha xy - h_1, \qquad \frac{dy}{dt} = Sy\left(1 - \frac{y}{L}\right) + \beta xy - h_2, \quad \text{(d)}$$

where $\alpha > 0$ and β are interaction parameters. If the species compete, then $\beta < 0$. If $\beta > 0$, then y is a predator and x its prey. If $\alpha = \beta = 0$, then (d) reduces to (c).

The mathematics of maximizing (a) subject to (b) and either (c) or (d) is not necessarily trivial. The problem of maximizing (a) subject to (b) and (c) is discussed by Clark (1976, pp. 303–311) and by Mesterton-Gibbons (1987a). The problem of maximizing (a) subject to (b) and (d) is discussed by Chaudhuri (1986) for $\beta < 0$, by Mesterton-Gibbons (1988a) for $\beta > 0$, and by Ragozin and Brown (1985) for $\beta > 0$ and $p_1 = 0 = q_1$. If you are interested in pursuing these problems, then consult these references.

Chapter 10

10.2 Use the Taylor expansion

$$\frac{g}{g + \alpha} = \left(1 + \frac{\alpha}{g}\right)^{-1} = 1 - \frac{\alpha}{g} + o\left(\frac{\alpha}{g}\right).$$

*10.4 Eliminating $I(t)$, $\nu(t)$, $\delta(t)$, and $r(t)$ from (10.38), (10.39), and (10.44)–(10.46), we obtain

$$\{\sigma a_3 + a_1 a_2 - a_1 a_3 a_4\} Q(t) = a_1 S(t) + a_3 I_0 - a_1 D_0 + a_1 a_3 (r_0 - a_4 Q_m),$$

whence

$$\frac{\partial Q(t)}{\partial S(t)} = \frac{a_1}{\sigma a_3 + a_1 a_2 - a_1 a_3 a_4}.$$

Similarly,

$$\frac{\partial r(t)}{\partial S(t)} = \frac{a_4 a_1}{\sigma a_3 + a_1 a_2 - a_1 a_3 a_4},$$

$$\frac{\partial \delta(t)}{\partial S(t)} = -\frac{\sigma}{\sigma a_3 + a_1 a_2 - a_1 a_3 a_4},$$

and

$$\frac{\partial I(t)}{\partial S(t)} = \frac{\sigma a_1}{\sigma a_3 + a_1 a_2 - a_1 a_3 a_4}.$$

It therefore appears that if $\sigma a_3 + a_1 a_2 > a_1 a_3 a_4$, then increasing the money supply will increase inflation, output, and investment but decrease the real interest rate; whereas if $\sigma a_3 + a_1 a_2 < a_1 a_3 a_4$, then increasing the money supply will decrease inflation, output, and investment but increase the real interest rate. The condition $\sigma a_3 + a_1 a_2 > a_1 a_3 a_4$ is most likely to be satisfied in a depressed economy, $\sigma a_3 + a_1 a_2 < a_1 a_3 a_4$ in a booming one. Because falling output and investment are characteristic of a depressed economy, the model seems to say that the money supply should not be increased in a booming economy, because that would send the economy into depression.

As remarked in the text, however, this model is far too simple to describe a real economy. Among other things, we have implicitly assumed that after any change in monetary demand or supply, the economy evolves so quickly toward equilibrium between the two that their relationship is adequately modelled by the quasi-equilibrium $D(t) = S(t)$. The short-term changes in interest rate that might be necessary to reconcile the two are completely ignored. Thus any changes in ν predicted by the model are assumed to be greater and to take place over a longer term. By making S an exogenous variable and all other variables endogenous, we have also assumed implicitly that the real money supply is controllable. In practice, however, governments can control only the nominal money supply, $S_\nu(t) \equiv P(t)S(t)$. Perhaps S should be replaced by S_ν/P. But we have still ignored, among other things, the relationship between production, labor, and capital (which might be described by a Cobb–Douglas production function), and the effect of excess capacity on unemployment. Our model would have to be extended to include all these effects before it could describe a real economy; but Section 10.3 is at least a beginning, and perhaps you would like to extend the model as a case study (Exercise 10.13).

*10.8 Let's first define the probability density function of the waiting time in state i by

$$g_i(t) = \frac{d}{dt}[\text{Prob}(W_i \le t)] = \frac{1}{\mu_i} e^{-t/\mu_i}. \tag{a}$$

For the sake of generality, let's assume that there are $N + 1$ states, namely, $i = 0, 1, \ldots, N$ (previously we had $N = 4$). We will use t to denote linear time instead of l. Hence the probability of being *in* state j at time t, having been in (*not* necessarily entered) state i at time 0, is denoted by $r_{ij}(t)$; and the state distribution at time t, namely, $\pi(t)$, is given by the matrix equation

$$\pi(t)^T = \pi(0)^T \mathbf{R}(t). \tag{b}$$

Suppose that $i \ne j$. Then, to be in state j at time t, the system must wait time $\xi (0 < \xi < t)$ in state i, make a transition to some other state (say m), then move from there to state j in time $t - \xi$ with as many intervening transitions as it pleases. Summing over m, by analogy with (10.67), the probability that the system makes some transition at time ξ and subsequently finds itself in state j at time t is

$$\sum_{m=0}^{N} p_{im} r_{mj}(t - \xi). \tag{c}$$

Now, from (a), the probability that the system makes its first transition during the interval $\xi - \delta\xi/2 < t < \xi + \delta\xi/2$ is $g_i(\xi)\delta\xi + o(\delta\xi)$, on using little oh notation; hence the probability that the system makes its first transition during this interval *and* subsequently finds itself in state j at time t is

$$g_i(\xi) \sum_{m=0}^{N} p_{im} r_{mj}(t - \xi)\delta\xi + o(\delta\xi). \tag{d}$$

Hence the probability that the system goes from i to j in time t, when $i \neq j$, is

$$\int_0^t g_i(\xi) \left\{ \sum_{m=0}^{N} p_{im} r_{mj}(t - \xi) \right\} d\xi. \tag{e}$$

If $i = j$, however, then there is the additional possibility that the system fails to leave state i by time t; the probability of this is

$$\text{Prob}(W_i > t) = e^{-t/\mu_i}. \tag{f}$$

Hence, combining (e) with (f), we have

$$r_{ij}(t) = \delta_{ij} e^{-t/\mu_i} + \int_0^t g_i(\xi) \left\{ \sum_{m=0}^{N} p_{im} r_{mj}(t - \xi) \right\} d\xi, \tag{g}$$

whether $i = j$ or not. This is analogous to (10.68). Changing the integration variable from ξ to $\eta = t - \xi$ and rearranging yields

$$r_{ij}(t) = \delta_{ij} e^{-t/\mu_i} + \sum_{m=0}^{N} p_{im} \int_0^t g_i(t - \eta) r_{mj}(\eta) d\eta$$

$$= e^{-t/\mu_i} \left\{ \delta_{ij} + \frac{1}{\mu_i} \sum_{m=0}^{N} p_{im} \int_0^t e^{\eta/\mu_i} r_{mj}(\eta) d\eta \right\}, \tag{h}$$

on using (a). Differentiating with respect to time, using Leibnitz's rule, and rearranging we obtain

$$\frac{d}{dt}[r_{ij}(t)] = \frac{1}{\mu_i} \left\{ \sum_{m=0}^{N} p_{im} r_{mj}(t) - r_{ij}(t) \right\}, \qquad i, j = 0, \ldots, N. \tag{i}$$

Together with the conditions $r_{ij}(0) = \delta_{ij}$, $i, j = 0, \ldots, N$, (i) forms a set of $(N + 1)^2$ coupled linear differential equations and initial conditions, which may be written more neatly in terms of the matrix $\mathbf{R}(t)$ as

$$\frac{d\mathbf{R}}{dt} = \mathbf{D}(\mathbf{P} - \mathbf{I})\mathbf{R}(t), \qquad \mathbf{R}(0) = \mathbf{I}, \tag{j}$$

where $\mathbf{I}$ is the $(N + 1) \times (N + 1)$ identity matrix and

$$\mathbf{D} = \text{diag}\left(\frac{1}{\mu_0}, \frac{1}{\mu_1}, \ldots, \frac{1}{\mu_N} \right) \tag{k}$$

is the $(N + 1) \times (N + 1)$ diagonal matrix of which the ith entry along the diagonal is $\{E[W_i]\}^{-1}$. Equations (j) are known in the literature on probabilistic processes as the Kolmogorov equations; see, for example, Ross (1980, p. 205 or 1983, p. 148 and p. 151).

You can verify by substitution that the formal solution of (j) is

$$R(t) = I + \sum_{m=1}^{\infty} \frac{t^m}{m!} \{D(P - I)\}^m. \tag{1}$$

For finite N, this series always converges. Moreover, it is clear from (1) that $D(P - I)R = RD(P - I)$; hence (j) yields an alternative form of the Kolmogorov equations, namely,

$$\frac{dR}{dt} = R(t)D(P - I), \qquad R(0) = I. \tag{m}$$

It is usual to call (j) the Kolmogorov *backward* equations and (m) the Kolmogorov *forward* equations. For an explanation of this choice of terminology, see Ross (1983, p. 149).

Equations (j), (1), and (m) are three entirely equivalent descriptions of a continuous-time Markov jump process with a finite number of states. See Exercise 10.9 for an illustration.

10.9 Let's define

$$c = \frac{b_1}{b_1 + d_1}, \qquad a = \alpha + b_1 + d_1.$$

Then, because $b_0 = d_2 = \alpha$ for the elevator model, (k) above yields $D = \text{diag}(\alpha, a - \alpha, \alpha)$, and (5.86) yields

$$P = \begin{bmatrix} 0 & 1 & 0 \\ 1-c & 0 & c \\ 0 & 1 & 0 \end{bmatrix}.$$

Hence

$$D(P - I) = \begin{bmatrix} -\alpha & \alpha & 0 \\ (1-c)(a-\alpha) & \alpha - a & (a-\alpha)c \\ 0 & \alpha & -\alpha \end{bmatrix}.$$

You can prove by mathematical induction that $[D(P - I)]^m =$

$$\begin{bmatrix} -\alpha\{(-\alpha)^{m-1}c + (-a)^{m-1}(1-c)\}, \, \alpha(-a)^{m-1}, \, -\alpha\{(-a)^{m-1} - (-\alpha)^{m-1}\}c \\ (-a)^{m-1}(1-c)(a-\alpha), \, -(-a)^{m-1}(a-\alpha), \, (-a)^{m-1}c(a-\alpha) \\ -\alpha(1-c)\{(-a)^{m-1} - (-\alpha)^{m-1}\}, \, \alpha(-a)^{m-1}, \, -\alpha\{(-a)^{m-1}(1-c) + (-\alpha)^{m-1}c\} \end{bmatrix}$$

Hence, utilizing the Taylor series expansion for the exponential function, (5.65) yields

$$R(t) = \frac{1}{a} \begin{bmatrix} d_1 & \alpha & b_1 \\ d_1 & \alpha & b_1 \\ d_1 & \alpha & b_1 \end{bmatrix}$$

$$+ \sum_{m=0}^{\infty} \left\{ \frac{(-\alpha t)^m}{m!} \begin{bmatrix} c & 0 & -c \\ 0 & 0 & 0 \\ c-1 & 0 & 1-c \end{bmatrix} \right.$$

$$\left. + \frac{(-at)^m}{m!} \cdot \frac{1}{a} \begin{bmatrix} \alpha(1-c) & -\alpha & \alpha c \\ -d_1 & b_1+d_1 & -b_1 \\ \alpha(1-c) & -\alpha & \alpha c \end{bmatrix} \right\}$$

$$= \mathbf{I} + \sum_{m=1}^{\infty} \frac{t^m}{m!} \left\{ (-\alpha)^m \begin{bmatrix} c & 0 & -c \\ 0 & 0 & 0 \\ c-1 & 0 & 1-c \end{bmatrix} \right.$$

$$\left. - (-a)^{m-1} \begin{bmatrix} \alpha(1-c) & -\alpha & \alpha c \\ (c-1)(a-\alpha) & a-\alpha & -c(a-\alpha) \\ \alpha(1-c) & -\alpha & \alpha c \end{bmatrix} \right\}$$

$$= \mathbf{I} + \sum_{m=1}^{\infty} \frac{t^m}{m!} [\mathbf{D}(\mathbf{P} - \mathbf{I})]^m,$$

as required.

*10.10 We can extend the model of Section 5.5 by allowing the elevator to jump from any floor to any other floor. Let us modify (5.58) by defining U_k to be the event that $X(t) = k$. Let λ_k be the transition rate from floor k. Then if $X(t) = k$, the probability that the elevator will jump floors during the interval $[t, t + \delta t)$ is $\lambda_k \delta t + o(\delta t)$. (In Section 5.5, we had $\lambda_k = b_k + d_k$.) If $X(t) = k$, where $k \neq j$, then $X(t + \delta t) = j$ only if the elevator makes a transition during $[t, t + \delta t)$ *and* that transition is to floor j. Thus:

$$\text{Prob}(U \mid U_k) = (\lambda_k \delta t + o(\delta t)) \cdot p_{kj} = \lambda_k p_{kj} \delta t + o(\delta t), \qquad k \neq j.$$

If $X(t) = j$, on the other hand, then $X(t + \delta t) = j$ only if the elevator makes no transition during $[t, t + \delta t)$. Thus $\text{Prob}(U \mid U_j) = 1 - \lambda_j \delta t + o(\delta t)$. Hence, instead of (5.59), we obtain

$$r_{ij}(t + \delta t) = \text{Prob}(U) = \sum_{k=0}^{N} \text{Prob}(U \mid U_k) \cdot \text{Prob}(U_k)$$

$$= \sum_{\substack{k=0 \\ k \neq j}}^{N} \left\{ \lambda_k p_{kj} \delta t + o(\delta t) \right\} r_{ik}(t) + \left\{ 1 - \lambda_j \delta t + o(\delta t) \right\} r_{ij}(t).$$

Rearranging in the obvious manner and taking the limit as $\delta t \to 0$, we obtain:

$$\frac{d}{dt} r_{ij} = \sum_{\substack{k=0 \\ k \neq j}}^{N} \lambda_k p_{kj} r_{ik} - \lambda_j r_{ij}, \qquad 0 \leq j \leq N,$$

in place of (5.61)–(5.63). Because $p_{jj} = 0$, $0 \leq j \leq N$, the matrix form of these equations is

$$\frac{d\mathbf{R}}{dt} = \mathbf{R}(t)\mathbf{D}(\mathbf{P} - \mathbf{I}),$$

where $\mathbf{D} = \mathrm{diag}(\lambda_0, \lambda_1, \ldots, \lambda_N)$ and $\mathbf{I}$ is the $(N + 1) \times (N + 1)$ identity matrix. Setting $\lambda_k = 1/\mu_k$, we see that this agrees with (m), in the solution to Exercise 10.8.

Chapter 11

11.4 For the second of the four regions in (11.30), the recommended substitution yields $F'(\phi(w)) = w$, whence $\phi(z, t)$ follows from (11.29b). For the third of the four regions in (11.30), the solution must be

$$\frac{\phi}{\phi_{\max}} = e^{-v_{\max}/\lambda},$$

because all characteristics in that region have slope determined by (11.29a) and must intersect the line $z = (v_{\max} - \lambda)t$, on which the solution has just been determined.

11.5 The substitution $w = z/t$ reduces the differential equation to the easily integrable $t\,dw/dt = \lambda$.

11.8 Straightforward partial differentiation of (11.64) shows that

$$\frac{\partial^2 \phi}{\partial y^2} = \frac{P_0}{\sigma^5\sqrt{2\pi}}(y^2 - \sigma^2)e^{-y^2/2\sigma^2}, \qquad \frac{\partial \phi}{\partial t} = \frac{P_0}{\sigma^4\sqrt{2\pi}}(y^2 - \sigma^2)e^{-y^2/2\sigma^2}\sigma'(t).$$

Hence (11.55) is satisfied if $\sigma'(t) = \gamma/\sigma(t)$ or $\frac{d}{dt}(\sigma^2) = 2\gamma$. The result $\sigma(t) = \sqrt{2\gamma t}$ now follows immediately from (11.65).

11.13 You can verify directly that (11.143) satisfies (11.134); (11.143) satisfies (11.132) because it is identically equal to

$$\frac{1}{2}A_j \cos\left\{(j + \frac{1}{2})\frac{\pi}{L}(z + \bar{a}t) + \epsilon_j\right\} + \frac{1}{2}A_j \cos\left\{(j + \frac{1}{2})\frac{\pi}{L}(z - \bar{a}t) - \epsilon_j\right\},$$

which is a special case of (11.119).

*11.15 Let (x, y, z) denote a typical point in a three-dimensional Cartesian space $Oxyz$. Let some region R of this space be filled with fluid in which a certain substance or quantity is suspended, and let $\phi(x, y, z, t)$ denote the amount per unit volume of this suspension, at (x, y, z) at time t. Then the amount of suspension in the infinitesimal region with volume $\delta V = \delta x \delta y \delta z$, depicted in Fig. S11.1(a), is $\phi(x, y, z, t)\delta V + o(\delta V)$; whence the amount of suspension in the finite fixed rectangular region

$$V \equiv \{(x, y, z): a_1 \leq x \leq a_2, b_1 \leq x \leq b_2, c_1 \leq x \leq c_2\},$$

depicted in Fig. S11.1(b), is

$$M(t) = \int_V \phi(x,y,z,t)dV = \int_{c_1}^{c_2} \int_{b_1}^{b_2} \int_{a_1}^{a_2} \phi(x,y,z,t)dxdydz, \qquad \text{(a)}$$

the error term vanishing in the limit of integration. You should note that V is an imaginary region completely surrounded by fluid. There is absolutely no suggestion that fluid is contained by a rectangular box. In other words, V is strictly a proper subset of R.

There are five methods by which suspension can enter or leave the imaginary region V, not all of which will apply to every suspension:

1. by being created or destroyed in a chemical reaction;

2. by being transported across the faces perpendicular to Ox, namely, $ABCD$ and $PQRS$;

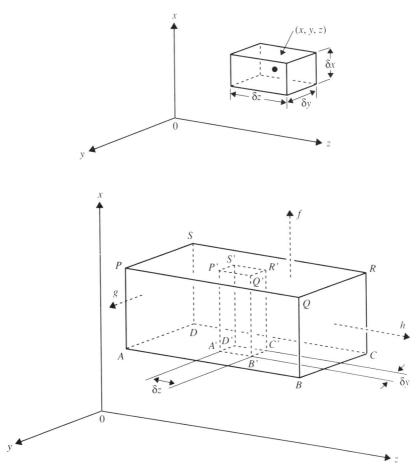

Fig. S11.1

3. by being transported across the faces perpendicular to Oy, namely, $ABQP$ and $DCRS$;
4. by being transported across the faces perpendicular to Oz, namely, $ADSP$ and $BCRQ$;   (b)
5. by being injected into or withdrawn from the interior of V, as, for example, when the suspension is excreted by miniature organisms swimming around in V.

Let L_k denote the *net* rate of addition of suspension to V by method (k). Then, clearly,

$$\frac{dM}{dt} = L_1 + L_2 + L_3 + L_4 + L_5. \tag{c}$$

We will now obtain expressions for $L_1, \ldots, L_5$ in turn.

The first contribution is similar to its counterpart in the tube model of Section 11.6. Let $C(x,y,z,t) - K(x,y,z,t)\phi(x,y,z,t)$ denote the *net* amount of suspension (if any) created by chemical reaction at station (x,y,z) at time t, per unit volume, per unit time. Then

$$L_1 = \int_{c_1}^{c_2} \int_{b_1}^{b_2} \int_{a_1}^{a_2} (C - K\phi) dx dy dz. \tag{d}$$

You should verify that this expression is identical to (11.71) for flow in a tube with rectangular cross section $ADSP$ in Fig. S11.1(b), when ϕ is independent of x and y and c_1, c_2 are substituted for a, b.

Let $f(x,y,z,t)$ denote the flux of suspension per unit area, at time t, in the direction of increasing x, across the plane with coordinate x. Then the flux *into* the rectangular prism $A'B'C'D'S'R'Q'P'$ in Fig. S11.1(b), across the face $A'B'C'D'$, is $f(a_1,y,z)\delta y\delta z + o(\delta y\delta z)$; and the net flux *out* across the face $P'Q'R'S'$ is $f(a_2,y,z)\delta y\delta z + o(\delta y\delta z)$. Hence the net influx is

$$\{f(a_1,y,z) - f(a_2,y,z)\}\,\delta y\delta z + o(\delta y\delta z).$$

Thus the *net* flux of suspension into V in the direction Ox is

$$L_2 = \int_{c_1}^{c_2} \int_{b_1}^{b_2} \{f(a_1,y,z) - f(a_2,y,z)\}\, dy dz = -\int_{c_1}^{c_2} \int_{b_1}^{b_2} \int_{a_1}^{a_2} \frac{\partial f}{\partial x} dx dy dz, \tag{e}$$

the error term vanishing in the limit of integration.

Let $g(x,y,z,t)$ be the flux of suspension per unit area across the plane with coordinate y in the direction of Oy, and let $h(x,y,z,t)$ be the flux of suspension per unit area across the plane with coordinate z in the direction Oz. Then, by arguments similar to those that yielded (e), the net rates of addition of suspension to V by methods (3) and (4) are, respectively,

$$L_3 = \int_{c_1}^{c_2} \int_{a_1}^{a_2} \{g(x,b_1,z) - g(x,b_2,z)\}\, dx dz$$

$$= -\int_{c_1}^{c_2} \int_{b_1}^{b_2} \int_{a_1}^{a_2} \frac{\partial g}{\partial y} dx dy dz \tag{f}$$

and

$$L_4 = \int_{b_1}^{b_2} \int_{a_1}^{a_2} \{h(x,y,c_1) - h(x,y,c_2)\} \, dxdy$$

$$= -\int_{c_1}^{c_2} \int_{b_1}^{b_2} \int_{a_1}^{a_2} \frac{\partial h}{\partial z} dxdydz. \tag{g}$$

Again, (g) agrees with (11.72) when h is independent of x and y and the rectangular cross section $ADSP$ is Fig. S11.1(b) has area A; H must be substituted for Ah, and a,b for c_1, c_2.

Lastly, if the net rate of addition of suspension to the fluid by method (5), per unit volume, per unit time, is $I(x,y,z,t)$ at station (x,y,z) at time t, then the net rate of addition of suspension to V by method (5) is simply

$$L_5 = \int_{c_1}^{c_2} \int_{b_1}^{b_2} \int_{a_1}^{a_2} I dxdydz. \tag{h}$$

Once again, in the appropriate limit, (h) agrees with (11.73), where the net rate of injection per unit *length* is $q = AI$.

Because V is independent of time, (a) implies that

$$\frac{dM}{dt} = \int_{c_1}^{c_2} \int_{b_1}^{b_2} \int_{a_1}^{a_2} \frac{\partial \phi}{\partial t} dxdydz. \tag{i}$$

Hence, on substituting (d)–(h) into (c), we obtain

$$\int_{c_1}^{c_2} \int_{b_1}^{b_2} \int_{a_1}^{a_2} \left\{ \frac{\partial \phi}{\partial t} + K\phi + \frac{\partial f}{\partial x} + \frac{\partial g}{\partial y} + \frac{\partial h}{\partial z} - C - I \right\} dxdydz = 0. \tag{j}$$

But the volume V is arbitrary and, in particular, can be made as small as we please. Hence, for reasons similar to those given in Sections 9.4 and 11.6, the integrand in (j) must vanish identically:

$$\frac{\partial \phi}{\partial t} + K\phi + \frac{\partial f}{\partial x} + \frac{\partial g}{\partial y} + \frac{\partial h}{\partial z} = C + I, \tag{k}$$

wherever the integrand is continuous. An important special case occurs when the suspension is neither created nor destroyed in a chemical reaction, nor injected into or withdrawn from the interior of the fluid. Then $C = K = I = 0$, and

$$\frac{\partial \phi}{\partial t} + \frac{\partial f}{\partial x} + \frac{\partial g}{\partial y} + \frac{\partial h}{\partial z} = 0. \tag{l}$$

A wide variety of phenomena in fluid dynamics can be modelled by (l), by defining suspension to be either mass or momentum of the fluid itself. At station (x,y,z) at time t, let $\rho(x,y,z,t)$ denote fluid mass per unit volume, and let the vector $(u(x,y,z,t), v(x,y,z,t), w(x,y,z,t))$ denote fluid velocity. Then the fluid momentum per unit volume is the vector $(\rho u, \rho v, \rho w)$; i.e., ρu is the momentum per unit volume in direction Ox, ρv the momentum per unit volume in direction Oy, and ρw momentum per unit volume in direction Oz.

To begin with, let mass be the suspension, so that ϕ is density; i.e., set $\phi = \rho$ in (l). Then the flux of suspension per unit area in direction Oz is simply momentum per unit volume in that direction; i.e., $h = \rho w$, as is also clear from dividing (11.77) by A. Similarly, the flux of suspension per unit area in direction Ox is $f = \rho u$, the momentum per unit volume in that direction; and the flux of suspension per unit area in direction Oy is $g = \rho v$. Thus (l) yields the partial differential equation

$$\frac{\partial \rho}{\partial t} + \frac{\partial}{\partial x}(\rho u) + \frac{\partial}{\partial y}(\rho v) + \frac{\partial}{\partial z}(\rho w) = 0 \tag{m}$$

for mass. This *mass conservation* equation is an obvious generalization of (11.107) in Section 11.8.

Now let suspension be fluid momentum per unit volume in direction Oz; i.e., set $\phi = \rho w$. We will refer to this quantity as the z-momentum. Then, from (11.80), the flux of z-momentum per unit area in direction Oz is

$$h = \rho w^2 + p - \frac{\partial(\mu w)}{\partial z}, \tag{n}$$

where μ is the fluid's shear viscosity and $p(x, y, z, t)$ denotes fluid pressure at station (x, y, z) at time t. To obtain the correct expression for f, we recall from Section 2.6 that z-momentum ρw is associated with flux $-\partial/\partial x(\mu w)$ per unit area in direction Ox. If the fluid is constrained to move only in direction Oz, as in Section 2.6, then this is the only flux of z-momentum. But our new fluid moves in direction Ox with velocity u, convecting the z-momentum with it. Hence there is an additional flux $u \cdot \rho w$ of z-momentum in direction Ox. Combining the two contributions, we find that the net flux per unit area in direction Ox of z-momentum is

$$f = u \cdot \rho w - \frac{\partial}{\partial x}(\mu w). \tag{o}$$

Similarly, the net flux per unit area in direction Oy of z-momentum is

$$g = v \cdot \rho w - \frac{\partial}{\partial y}(\mu w). \tag{p}$$

Substituting $\phi = \rho w$ and (n)–(p) into (l) now yields

$$\frac{\partial}{\partial t}(\rho w) + \frac{\partial}{\partial x}\left(\rho u w - \frac{\partial}{\partial x}(\mu w)\right) + \frac{\partial}{\partial y}\left(\rho v w - \frac{\partial}{\partial y}(\mu w)\right)$$
$$+ \frac{\partial}{\partial z}\left(\rho w^2 + p - \frac{\partial}{\partial z}(\mu w)\right) = 0, \tag{q}$$

that is,

$$\rho\left\{\frac{\partial w}{\partial t} + u\frac{\partial w}{\partial x} + v\frac{\partial w}{\partial y} + w\frac{\partial w}{\partial z}\right\} + w\left\{\frac{\partial \rho}{\partial t} + \frac{\partial}{\partial x}(\rho u) + \frac{\partial}{\partial y}(\rho v) + \frac{\partial}{\partial z}(\rho w)\right\}$$
$$= -\frac{\partial p}{\partial z} + \frac{\partial^2}{\partial x^2}(\mu w) + \frac{\partial^2}{\partial y^2}(\mu w) + \frac{\partial^2}{\partial z^2}(\mu w). \tag{r}$$

But the second term in brackets on the left-hand side vanishes, by the mass conservation equation (m). Hence

$$\rho\left\{\frac{\partial w}{\partial t} + u\frac{\partial w}{\partial x} + v\frac{\partial w}{\partial y} + w\frac{\partial w}{\partial z}\right\} = -\frac{\partial p}{\partial z} + \frac{\partial^2}{\partial x^2}(\mu w) + \frac{\partial^2}{\partial y^2}(\mu w) + \frac{\partial^2}{\partial z^2}(\mu w).$$

(s)

This is our equation for conservation of z-momentum.

Now let suspension be fluid momentum per unit volume in direction Ox; i.e., set $\phi = \rho u$. Naturally, we will refer to this quantity as the x-momentum. Either by symmetry or by the same arguments as yielded (n)–(p), the three fluxes per unit area of x-momentum are

$$f = \rho u^2 + p - \frac{\partial(\mu u)}{\partial x}$$

$$g = v \cdot \rho u - \frac{\partial}{\partial y}(\mu u)$$

(t)

$$h = w \cdot \rho u - \frac{\partial}{\partial z}(\mu u).$$

Substitution of (t) into (l) now yields, on using (m):

$$\rho\left\{\frac{\partial u}{\partial t} + u\frac{\partial u}{\partial x} + v\frac{\partial u}{\partial y} + w\frac{\partial u}{\partial z}\right\} = -\frac{\partial p}{\partial x} + \frac{\partial^2}{\partial x^2}(\mu u) + \frac{\partial^2}{\partial y^2}(\mu u) + \frac{\partial^2}{\partial z^2}(\mu u).$$

(u)

This is our equation for conservation of x-momentum. Moreover, by setting $\phi = \rho v$ and continuing to argue in the fashion above, the conservation equation for y-momentum is readily found to be

$$\rho\left\{\frac{\partial v}{\partial t} + u\frac{\partial v}{\partial x} + v\frac{\partial v}{\partial y} + w\frac{\partial v}{\partial z}\right\} = -\frac{\partial p}{\partial y} + \frac{\partial^2}{\partial x^2}(\mu v) + \frac{\partial^2}{\partial y^2}(\mu v) + \frac{\partial^2}{\partial z^2}(\mu v).$$

(v)

Although sound waves cannot exist in the absence of density fluctuations, there are many other phenomena in fluid dynamics that are adequately described by assuming that the fluid's density, ρ, is constant. The fluid is then said to be incompressible.[1] In these circumstances, (m), (v), (u), and (s) reduce to:

$$\frac{\partial u}{\partial x} + \frac{\partial v}{\partial y} + \frac{\partial w}{\partial z} = 0,$$

(w)

$$\frac{\partial u}{\partial t} + u\frac{\partial u}{\partial x} + v\frac{\partial u}{\partial y} + w\frac{\partial u}{\partial z} = -\frac{\partial}{\partial x}\left(\frac{p}{\rho}\right) + \nu\left(\frac{\partial^2 u}{\partial x^2} + \frac{\partial^2 u}{\partial y^2} + \frac{\partial^2 u}{\partial z^2}\right),$$

(x)

$$\frac{\partial v}{\partial t} + u\frac{\partial v}{\partial x} + v\frac{\partial v}{\partial y} + w\frac{\partial v}{\partial z} = -\frac{\partial}{\partial y}\left(\frac{p}{\rho}\right) + \nu\left(\frac{\partial^2 v}{\partial x^2} + \frac{\partial^2 v}{\partial y^2} + \frac{\partial^2 v}{\partial z^2}\right),$$

(y)

---

[1] If ρ is constant, then the fluid is incompressible. If the fluid is incompressible, meaning that changing the pressure won't change the density, then it does not follow that ρ is constant, because the fluid may have had an initial density variation which it preserves (in theory) for the rest of time. This initial density variation, or "stratification," is the basis for many atmospheric flows (which are, nevertheless, incompressible). See, for example, Turner (1973).

$$\frac{\partial w}{\partial t} + u\frac{\partial w}{\partial x} + v\frac{\partial w}{\partial y} + w\frac{\partial w}{\partial z} = -\frac{\partial}{\partial z}\left(\frac{p}{\rho}\right) + \nu\left(\frac{\partial^2 w}{\partial x^2} + \frac{\partial^2 w}{\partial y^2} + \frac{\partial^2 w}{\partial z^2}\right). \quad \text{(z)}$$

These four equations, known in fluid dynamics as the Navier–Stokes equations, form the dynamical system of which Hagen–Poiseuille pipe flow is an equilibrium solution. To see this, set $u = 0 = v$ for pipe flow, then *deduce* from (w) that w is independent of z. For equilibrium, w is also independent of t. Hence w is a function of x and y. Because $u = 0$, (x) tells us that p is independent of x and, because $v = 0$, (y) tells us that p is independent of y. Because p must be independent of t in equilibrium, it follows that p is a function of z alone. This is consistent with (z), in which the last term is a function of x and y only if $\partial p/\partial z$ is a constant; whereupon (z) reduces to (2.69).

Chapter 12

12.1

$$\text{Prob}(Y_k = 0) = \text{Prob}(Z_{N-k-1} > w_{k+1} + y_{k+1})$$

$$= \int_{w_{k+1}+y_{k+1}}^{\infty} f_{N-k-1}(z)dz = \frac{b - w_{k+1} - y_{k+1}}{b - a};$$

and for y lying between 0 and $w_{k+1} + y_{k+1} - a$,

$$\text{Prob}\left(y - \frac{1}{2}\delta y < Y_k < y + \frac{1}{2}\delta y\right) = \frac{\delta y}{b - a} + o(\delta y).$$

For completeness, we append that

$$\text{Prob}(Y_k < 0) = 0 = \text{Prob}(Y_k > w_{k+1} + y_{k+1} - a).$$

*12.7 (i) See Fig. S12.1.

(ii) $\max(y_k, \psi_{k-1})$. You want the current candidate if better than the rest; otherwise the best of the rest.

12.10 (ii) Set $a = 1$ and take the limit of (i) as $k \to \infty$. Because $\lambda_k^* = p\psi_{k-1}$, we deduce that $\lambda_\infty(p)$ is the root of the equation

$$\frac{\lambda_\infty}{p} = \begin{cases} \lambda_\infty + \frac{2}{3}(1 - \lambda_\infty)^3, & \text{if } \lambda_\infty \geq \frac{1}{2}, \\ \frac{1}{2} + \frac{2}{3}\lambda_\infty^3, & \text{if } \lambda_\infty < \frac{1}{2}, \end{cases}$$

such that $\psi_\infty(p) = \lambda_\infty(p)/p$ lies between 0 and 1. It is a straightforward exercise to show that λ_∞ is greater or less than 1/2 according to whether p is greater or less than 6/7 (and, in particular, that $\lambda_\infty(6/7) = 1/2$).

(iii) $$\psi_\infty = \frac{2}{3} + \frac{1}{3}p^3\psi_\infty^3.$$

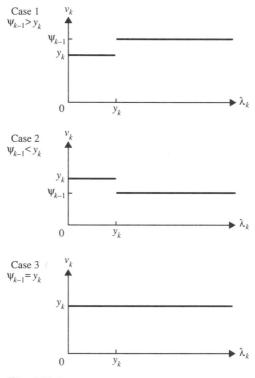

Case 1
$\psi_{k-1} > y_k$

Case 2
$\psi_{k-1} < y_k$

Case 3
$\psi_{k-1} = y_k$

Fig. S12.1

*12.13 (i) Now $L(0)$ is replaced by $1/p_s$; but the rest of the analysis is virtually unaltered, k^* being the smallest integer exceeding

$$\frac{\ln\{p_s/[p_s + (1 - p_s)p]\}}{\ln(1 - p)}.$$

(ii) No. In principle, you could put the hotels in sequence and decide at which point you would be prepared to settle for the next available one. But you would have to assume the cost of travelling time between hotels (the "interviewing" cost, as it were) to be negligible compared to the cost of distance from the town center. Moreover, the model assumes that each space is equally desirable in every respect, except its distance from a certain location. Thus you would have to assume, among other things, that all hotels in your sequence charged the same rate for a room. If you wished to balance distance against room rate or "interviewing" cost, then you would have to develop a more elaborate model. Perhaps you would like to pursue this as a case study.

*12.16 As remarked in Section 12.5, there is more than one way in which Φ and Ψ can be scaled to yield dimensionless currencies. If (12.72) is replaced by

$$\phi = \frac{h_1 \Phi}{E_1}, \qquad \psi = \Psi, \tag{a}$$

for example, then (12.75a) is still valid, but with σ defined by

$$\sigma = \frac{\beta^2(\epsilon + \beta)}{(\beta^2 + \epsilon^2)(\epsilon + \beta) + \epsilon^2 \eta \theta},$$

rather than by (12.75b). Moreover, because

$$\zeta = \frac{1 + \epsilon}{1 + \epsilon + \theta \sigma}$$

from (12.76), the fractions appearing in the last three rows and last two columns of Tables 12.3 and 12.5 are altered as follows (correct to two significant figures):

Treatment	ζ	PW Observed	ζ	BW Observed	ζ	RO Observed
C	0.80	0.82	0.77	0.86	0.77	0.94
D	0.59	0.80	0.53	0.98	0.53	0.86
E	0.45	0.78	0.37	0.64	0.37	0.91

Although, ironically, the prediction for PW's treatment C is better than it was in Table 12.3, we see that all evidence in favor of Hypothesis 2 has disappeared; and for $\lambda_1 > \lambda_c$, we never did have evidence in favor of Hypothesis 1. Does this mean that both hypotheses should be rejected?

We must certainly either reject or improve the *model* developed in Section 12.5 because it does not explain the data. As remarked toward the end of that section, however, it is important to distinguish between rejecting a model and rejecting a hypothesis. Any discrepancy between predictions and observations could always be due to an assumption other than the hypothesis, perhaps even an "implicit" assumption, one we hadn't really thought about. Such an implicit assumption is the topic of Exercise 12.17; other ways to improve the model in Section 12.5 are about to be discussed. Thus it might be premature to reject either hypothesis. Then all we can safely say is that neither hypothesis should be accepted.

We have seen that the optimal compromise (u_1^*, u_2^*) is affected by our choice of scaling. If (12.72) and (a) yield different results, then which is correct? Or is neither correct? There is no way of knowing for sure. By tinkering with scaling factors, we could probably find one that gave us excellent agreement with the data supplied by Krebs et al. (1977), and so our vectorial model ultimately suffers the same deficiency as (12.81).

This is not such an unfortunate discovery, because our model is already deficient in one other respect. Because it predicts optimal behavior in terms of maintaining some fraction of type-1 worms in the diet, our model is incapable of telling the bird whether to accept or reject the particular worm it has just located (unless, of course, the fraction in question is precisely 0 or 1). Thus, even if we disregard the scaling problem, the model still requires improvement.

By good fortune, each of these deficiencies can be eliminated by combining our vector optimization ideas with those of dynamic programming. The difficulty with Φ in Section 12.5 is that it is the rate of change of a quantity, rather than the quantity itself, so that it fails to have a natural capacity, that is, an upper bound that does not depend in any way on assumptions built into our model. If Φ and Ψ were quantities instead of rates or proportions, however, then they could be made dimensionless in a natural way by using their capacities as scaling factors.

Suppose, for example, that Φ calories were the energy value of all prey captured by the bird during a bout of foraging that lasted Ψ minutes. Suppose, moreover, that the bird never foraged for more than T_{max} minutes in a single bout, and that she could not ingest more than E_{max} calories without becoming satiated. Then Φ and $T_{max} - \Psi$ would be quantities of energy acquired and time remaining, respectively; the bird would like them to be as large as possible. By scaling each with respect to its natural capacity, we could define dimensionless variables Y_1 and Y_2 by

$$Y_1 = \frac{\Phi}{E_{max}}, \qquad Y_2 = \frac{T_{max} - \Psi}{T_{max}},$$

so that $0 \leq Y_1, Y_2 \leq 1$. Then we could model fitness by $f = f(Y_1, Y_2)$, where $\partial f/\partial Y_1 > 0$, $\partial f/\partial Y_2 > 0$, and use our vectorial approach to make Y_1 and Y_2 as large as possible, subject to any constraints. We would have no need to agonize over whether Φ should be scaled with respect to E_1 or E_2, or Ψ with respect to h_1 or h_2, because the two quantities would have been normalized as merit functions.

More generally, an animal's fitness might depend on several quantities; say m, where m is not necessarily 2. Then we would define dimensionless factors $Y_1, Y_2, \ldots, Y_{m-1}$, and Y_m by

$$Y_j = \frac{j\text{th QUANTITY}}{j\text{th CAPACITY}}. \tag{b}$$

We would still have to make assumptions about which quantities should be included in the fitness function or, which is the same thing, make an hypothesis about which m goals should be most important to our animal's future reproductive success, at least for the part of its lifetime under scrutiny. Once this hypothesis had been made, however, the factors would be uniquely defined merit functions. There would be no further need to scale them arbitrarily to make them dimensionless.

If $Y_1, Y_2, \ldots, Y_m$ were determinate, then we would predict animal behavior by optimizing the vector $(Y_1, Y_2, \ldots, Y_m)^T$. If $Y_1, Y_2, \ldots, Y_m$ were random variables, on the other hand, then we would predict animal behavior by optimizing the vector $(E[Y_1], E[Y_2], \ldots, E[Y_m])^T$ instead. This idea is illustrated in Section 12.6, where $m = 2$, Y_1 is the number of eggs an insect can lay, and Y_2 is the number of sites at which she can lay them. Because the natural capacities of these two quantities are numerically equal, however, the scaling (b) has no effect.

Section 12.6 also demonstrates how to remedy the second of the two deficiencies described above. After you have read it, you will be well prepared to put things right in Section 12.5. Thus you may wish to tackle Exercise 12.18 as a case study.

**12.18 See Mesterton-Gibbons (1989).

REFERENCES

Acheson, D. J. (1990) *Elementary Fluid Dynamics*. Oxford University Press, Oxford.

Alexander, R. McN. (1983) *Animal Mechanics*. 2nd edition, Blackwell Scientific Publications, Oxford.

Alexander, R. McN. (1990) *Animals*. Cambridge University Press, Cambridge.

Ames, William F. (1977) *Numerical Methods for Partial Differential Equations*. Academic Press, New York.

Ander, Paul and Anthony J. Sonnessa (1965) *Principles of Chemistry: An Introduction to Theoretical Concepts*. Macmillan, New York.

Arthurs, Arnold M. (1965) *Probability Theory*. Routledge and Kegan Paul, London.

Bartholomew, D. J. (1982) *Stochastic Models for Social Processes*. 3rd edition, Wiley, New York.

Batchelor, G. K. (1967) *An Introduction to Fluid Dynamics*. Cambridge University Press, London.

Beaumont, G. P. (1983) *Introductory Applied Probability*. Ellis Horwood, Chichester, U.K.

Beddington, J. R. and D. B. Taylor (1973) "Optimum age specific harvesting of a population," *Biometrics* **29**, 801–809.

Beeton, Alfred M. (1980) "Great lakes," in *Encyclopaedia Britannica,* 15th edition, Vol. **8**, 301–304.

Bender, E. A. (1978) *An Introduction to Mathematical Modeling*. Wiley, New York.

Boyce, William E. and Richard C. DiPrima (1965) *Elementary Differential Equations and Boundary Value Problems.* Wiley, New York.

Bray, H. Geoffrey and Kenneth White (1966) *Kinetics and Thermodynamics in Biochemistry.* Academic Press, New York.

Brems, Hans (1980) *Inflation, Interest and Growth.* Lexington Books, Lexington, Mass.

Bruno, James E. (1971) "Compensation of school district personnel," *Management Science* **17**, B569–B587.

Buongiorno, Joseph and Bruce R. Michie (1980) "A matrix model of uneven-aged forest management," *Forest Science* **26**, 609–625.

Carslaw, H. S. (1930) *An Introduction to the Theory of Fourier's Series and Integrals.* 3rd edition, Dover Publications, New York.

Caswell, H. J. (1989) *Matrix Population Models.* Sinauer Associates, Sunderland, Mass.

Champernowne, David Gawen (1980) "Distribution of wealth and income," in *Encyclopaedia Britannica*, 15th edition, Vol. **19**, 673–680.

Charnov, Eric L. (1976) "Optimal foraging: attack strategy of a mantid," *American Naturalist* **110**, 141–151.

Chaudhuri, Kripasindhu (1986) "A bioeconomic model of harvesting a multispecies fishery," *Ecological Modelling* **32**, 267–279.

Childs, David (1979) *Britain Since 1945.* Ernest Benn, London.

Clark, Colin W. (1976) *Mathematical Bioeconomics: The Optimal Management of Renewable Resources.* Wiley, New York.

Clark, Colin W. (1990) *Mathematical Bioeconomics: The Optimal Management of Renewable Resources.* 2nd edition, Wiley, New York.

Cronin, James T. and Donald R. Strong (1993) "Substantially submaximal oviposition rates by a Mymarid egg parasitoid in the laboratory and field," *Ecology* **74**, 1813–1825.

Cullen, Michael R. (1985) *Linear Models in Biology.* Ellis Horwood, Chichester, U. K.

Cushing, J. M. (1988) "Nonlinear matrix models and population dynamics," *Natural Resource Modeling* **2**, 539–580.

Denison, Edward F. (1974) *Accounting for United States Economic Growth, 1929–1969.* The Brookings Institution, Washington, D. C.

Diamond, Jared M. and Robert M. May (1976) "Island biogeography and the design of natural reserves," in May (1976b), Chapter 9, 163–186.

Douglas, P. H. (1934) *The Theory of Wages.* Macmillan, New York.

Dresher, Melvin (1981) *The Mathematics of Games of Strategy.* Dover Publications, New York.

Dugdale, J. S. (1966) *Entropy and Low Temperature Physics*. Hutchinson, London.

Edelstein-Keshet, Leah (1988) *Mathematical Models in Biology*. Random House, New York.

Ehrlich, Paul R. (1986) *The Machinery of Nature*. Simon and Schuster, New York.

Eisenhart, Churchill, Millard W. Hastay, and W. Allen Wallis (1947) *Selected Techniques of Statistical Analysis*. McGraw-Hill, New York.

Emlen, J. Merritt (1966) "The role of time and energy in food preference," *American Naturalist* **100,** 611–617.

Evans, Howard Ensign (1978) *Life on a Little-Known Planet*. E. P. Dutton, New York.

Fabozzi, Frank J. and Alfred W. Bachner (1979) "Mathematical programming models to determine civil service salaries," *European Journal of Operational Research* **3**, 190–198.

Finney, D. J. (1952) *Probit Analysis*. Cambridge University Press, London.

Fries, James F. and Laurence M. Crapo (1981) *Vitality and Aging*. W. H. Freeman, San Francisco.

Fröberg, Carl-Erik (1969) *Introduction to Numerical Analysis*. 2nd edition, Addison-Wesley, Reading, Mass.

Fry, Thornton C. (1965) *Probability and Its Engineering Uses*. Van Nostrand, New York.

Gazis, Denos C. (1972) "Traffic flow and control: theory and applications," *American Scientist* **60**, 414–424.

Gazis, Denos C., Robert Herman, and Renfrey B. Potts (1959) "Car-following theory of steady-state traffic flow," *Operations Research* **7,** 499–505.

Gerlough, Daniel L. and Frank C. Barnes (1971) *Poisson and Other Distributions in Traffic*. Eno Foundation for Transportation, Saugatuck, Conn.

Gilchrist, Warren (1984) *Statistical Modelling*. Wiley, New York.

Gilpin, Michael E. and Jared M. Diamond (1976) "Calculation of immigration and extinction curves from the species-area-distance relation," *Proceedings of the National Academy of Sciences U.S.A.* **73**, 4130–4134.

Glass, D. V. (1954) *Social Mobility in Britain*. Routledge and Kegan Paul, London.

Godfray, H. C. J. (1994) *Parasitoids*. Princeton University Press, Princeton, New Jersey.

Gompertz, Benjamin (1825) "On the nature of the function expressive of the law of mortality." Relevant sections are reprinted in Smith and Keyfitz, Section 30, 279–282.

Gordon, Robert J. (1978) *Macroeconomics*. Little, Brown and Company, Boston.

Haberman, Richard (1977) *Mathematical Models*. Prentice-Hall, Englewood Cliffs, N.J.

Hansen, Arthur G. (1967) *Fluid Mechanics*. Wiley, New York.

Harte, John (1985) *Consider a Spherical Cow*. William Kaufmann, Los Altos, Calif.

Hillier, F. S. and G. J. Lieberman (1980) *Introduction to Operations Research*. 3rd edition, Holden-Day, San Francisco.

Hoel, Paul G., Sidney C. Port, and Charles J. Stone (1972) *Introduction to Stochastic Processes*. Houghton Mifflin, Boston.

Holloway, Charles A. (1979) *Decision Making Under Uncertainty*. Prentice-Hall, Englewood Cliffs, N.J.

Howard, Ronald A. (1971) *Dynamic Probabilistic Systems*. Vol. 2, Wiley, New York.

Impagliazzo, John (1985) "Deterministic aspects of mathematical demography," *Biomathematics* **13**, Springer-Verlag, Berlin.

International Commission on Irrigation and Drainage (1980) *The Application of Systems Analysis to Problems of Irrigation, Drainage and Flood Control*. Pergamon Press, Oxford.

Janssen, Jacques (ed) (1986) *Semi-Markov Models: Theory and Applications*. Plenum Press, New York.

Jeffreys, Sir Harold and Lady Jeffreys (1956) *Methods of Mathematical Physics*. 3rd edition, Cambridge University Press, London.

Jeter, Melvyn (1986) *Mathematical Programming*. Marcel Dekker, Inc., New York.

Jørgensen, S. E. (1979) *Fundamentals of Ecological Modelling*. Elsevier, New York.

Kao, Edward P. C. (1974) "Modeling the movement of coronary patients within a hospital by semi-Markov processes," *Operations Research* **22**, 683–699.

Kendall, Maurice G. and Alan Stuart (1950) "The law of the cubic proportion in election results," *The British Journal of Sociology* **1**, 183–196.

Kendall, Maurice G. and Alan Stuart (1967) *The Advanced Theory of Statistics*, Vol. II, *Inference and Relationship*. 2nd edition, Hafner, New York.

Keyfitz, Nathan (1968) *Introduction to the Mathematics of Population*. Addison-Wesley, Reading, Mass.

Keyfitz, Nathan and Wilhelm Flieger (1971) *Population: Facts and Methods of Demography*. W. H. Freeman, San Francisco.

Kondo, Jiro (1958) "Determination of the optimum number of seats of a passenger transport plane," *Journal of the Operations Research Society of Japan* **1**, 127–138.

Krebs, J. R., J. T. Erichsen, M. I. Webber, and E. L. Charnov (1977) "Optimal prey selection in the great tit (*Parus major*)," *Animal Behaviour* **25**, 30–38.

Krebs, J. R. and N. B. Davies (1987) *An Introduction to Behavioural Ecology.* Blackwell Scientific Publications, Oxford.

Lack, David L. (1980) "Population, Biological," in *Encyclopaedia Britannica*, 15th edition, Vol. **14**, 824–838.

Lamb, Sir Horace (1925) *The Dynamical Theory of Sound.* Edward Arnold, London.

Lamb, Sir Horace (1932) *Hydrodynamics.* 6th edition, Dover Publications, New York.

Landau, L. D., A. I. Akhiezer, and E M. Lifshitz (1967) *General Physics: Mechanics and Molecular Physics.* Translated from Russian by J. B. Sykes, A. D. Petford, and C. L. Petford, Pergamon Press, Oxford.

Lange, Oskar (1971) *Optimal Decisions.* Pergamon Press, Oxford.

Leslie, P. H. (1945) "On the uses of matrices in certain population mathematics," *Biometrika* **33**, 183–212.

Leslie, P. H. (1948) "Some further notes on the use of matrices in population mathematics," *Biometrika* **35**, 213–245.

Lighthill, M. J. (1958) *Introduction to Fourier Analysis and Generalized Functions.* Cambridge University Press, London.

Lighthill, M. J. and G. B. Whitham (1955) "On kinematic waves: II. A theory of traffic on long crowded roads," *Proceedings of the Royal Society* **A229**, 317–345.

Lindley, Dennis V. (1982) "The subjectivist view of decision-making," *European Journal of Operational Research* **9**, 213–222.

Lindley, Dennis V. (1985) *Making Decisions*, Wiley, New York.

Lindley, D. V. and J. C. P. Miller (1953) *Cambridge Elementary Statistical Tables.* Cambridge University Press, London.

Liu, Mu-Shieung and Jing-Shing Yao (1977) "The capacity of a traffic flow on a T form traffic intersection," *Tam Kang Journal of Mathematics* **8**, 135–143.

Logofet, Dmitrii O. (1993) *Matrices and Graphs: Stability Problems in Mathematical Ecology.* CRC Press, Boca Raton, Florida.

Lotka, Alfred J. (1956) *Elements of Mathematical Biology.* Dover Publications, New York.

Luenberger, David G. (1984) *Linear and Nonlinear Programming.* 2nd edition, Addison-Wesley, Reading, Mass.

MacArthur, Robert H. and Eric R. Pianka (1966) "On the optimal use of a patchy environment," *American Naturalist* **100**, 603–609.

Makeham, William M. (1867) *On the Law of Mortality*. Relevant sections are reprinted in Smith and Keyfitz, Section 31, 283–288.

Mangel, Marc (1987) "Oviposition site selection and clutch size in insects," *Journal of Mathematical Biology* **25**, 1–22.

Mangel, Marc and Colin W. Clark (1988) *Dynamic Modeling in Behavioral Ecology*. Princeton University Press, Princeton, New Jersey.

Maron, Melvin J. (1982) *Numerical Analysis: a Practical Approach*. Collier-Macmillan, New York.

May, Robert M. (ed) (1976a) *Theoretical Ecology*. W. B. Saunders, Philadelphia.

May, Robert M. (1976b) "Models for single populations," in May (1976a, ed), Chapter 2, 4–25.

May, Robert M. (1976c) "Models for two interacting populations," in May (1976a, ed), Chapter 4, 49–70.

Maynard Smith, J. (1974) *Models in Ecology*. Cambridge University Press, London.

McKelvey, Robert W. (ed) (1985) "Environmental and natural resource mathematics," *Proceedings of Symposia in Applied Mathematics* **32**, American Mathematical Society, Providence, R.I.

McMahon, T. A. (1971) "Rowing: a similarity analysis," *Science* **173**, 349-351.

Messenger, P. S. (1980) "Biotic interactions," in *Encyclopaedia Britannica*, 15th edition, Vol. **2**, 1044–1052.

Mesterton-Gibbons, Michael (1987a) "On the optimal policy for combined harvesting of independent species," *Natural Resource Modeling* **2**, 109-134.

Mesterton-Gibbons, Michael (1987b) "When does a T-junction require a left-turn lane?" *Mathematical and Computer Modelling* **9**, 625–629.

Mesterton-Gibbons, Michael (1988a) "On the optimal policy for combined harvesting of predator and prey," *Natural Resource Modeling* **3**, 63–90.

Mesterton-Gibbons, Michael (1988b) "How fast can fast food be served?," *Mathematical and Computer Modelling* **10**, 405–407.

Mesterton-Gibbons, Michael (1988c) "On the optimal compromise for a dispersing parasitoid," *Journal of Mathematical Biology* **26**, 375–385.

Mesterton-Gibbons, Michael (1989) "On compromise in foraging and an experiment by Krebs et al.," *Journal of Mathematical Biology* **27**, 273–296.

Mesterton-Gibbons, Michael (1992) *An Introduction to Game-Theoretic Modelling*. Addison-Wesley, Redwood City, California.

Meyer, Paul L. (1970) *Introductory Probability and Statistical Applications.* Addison-Wesley, Reading, Mass.

Morse, Philip M. and Herman Feshbach (1953) *Methods of Theoretical Physics.* McGraw-Hill, New York.

Moskowitz, Herbert and Gordon P. Wright (1979) *Operations Research Techniques for Management.* Prentice-Hall, Englewood Cliffs, N.J.

Murthy, D. N. P. and E. Y. Rodin (1987) "A comparative evaluation of books on mathematical modelling," *Mathematical and Computer Modelling* **9**, 17–28.

Nayfeh, Ali Hasan (1973) *Perturbation Methods.* Wiley, New York.

Nayfeh, Ali Hasan (1981) *Introduction to Perturbation Techniques.* Wiley, New York.

Nicholson, Walter (1979) *Intermediate Microeconomics and Its Applications.* 2nd edition, the Dryden Press, Division of Holt, Rinehart and Winston, Hinsdale, Illinois.

Noble, Ben (1967) *Applications of Undergraduate Mathematics in Engineering.* MAA/Macmillan, New York.

Olinick, Michael (1978) *An Introduction to Mathematical Models in the Social and Life Sciences.* Addison-Wesley, Reading, Mass.

Pielou, E. C. (1977) *Mathematical Ecology.* Wiley, New York.

Plant, Richard E. (1985) "Applications of mathematics in insect pest management," in McKelvey (1985, ed), 1–17.

Prigogine, Ilya and Robert Herman (1971) *Kinetic Theory of Vehicular Traffic.* Elsevier, New York.

Pulliam, H. Ronald (1974) "On the theory of optimal diet," *American Naturalist* **108**, 59–74.

Ragozin, David L. and Gardner Brown, Jr. (1985) "Harvest policies and nonmarket valuation in a predator-prey system," *Journal of Environmental Economics and Management* **12**, 155–168.

Rainey, R. H. (1967) "Natural displacement of pollution from the Great Lakes," *Science* **155**, 1242–1243.

Reed, William J. (1986) "Optimal harvesting models in forest management—a survey," *Natural Resource Modeling* **1**, 55–79.

Renshaw, Eric (1991) *Modelling Biological Populations in Space and Time.* Cambridge University Press, London.

Ricker, W. E. (1954) "Stock and recruitment," *Canadian Journal of Fisheries and Aquatic Sciences* **11**, 559–623.

Rinaldi, S., R. Soncini-Sessa, H. Stehfest, and H. Tamura (1979) *Modeling and Control of River Quality.* McGraw-Hill, New York.

Ritter, Lawrence S. and William L. Silber (1977) *Money*. 3rd edition, Basic Books, New York.

Ross, Sheldon M. (1980) *Introduction to Probability Models*. 2nd edition, Academic Press, New York.

Ross, Sheldon M. (1983) *Stochastic Processes*. Wiley, New York.

Salukvadze, M. E. (1979) *Vector-Valued Optimization Problems in Control Theory*. Academic Press, New York.

Samuelson, Paul A. (1976) *Economics*. 10th edition, McGraw-Hill, New York.

Scime, Earl (1986) "A Markovian approach to epidemic modeling," unpublished mathematical modelling project, Florida State University.

Shubik, Martin (1982) *Game Theory in the Social Sciences*. MIT Press, Cambridge, Mass.

Shubik, Martin (1984) *A Game-Theoretic Approach to Political Economy* (Volume 2 of *Game Theory in the Social Sciences*, op cit.). MIT Press, Cambridge, Mass.

Smith, David and Nathan Keyfitz (1977) "Mathematical demography–selected papers," *Biomathematics* **6**, Springer-Verlag, Berlin.

Springer, Allan M. (1986) *Industrial Environmental Control (Pulp and Paper Industry)*, McGraw-Hill, New York.

Taha, Hamdy A. (1976) *Operations Research: An Introduction*. 2nd edition, Macmillan, New York.

Tamaki, Mitsushi (1984) "The secretary problem with optimal assignment," *Operations Research* **32**, 847–858.

Teugels, Josef L. (1986) "A second bibliography on semi-Markov processes," in Janssen (1986, ed), 505–584.

Thornley, J. H. M. (1976) *Mathematical Models in Plant Physiology*. Academic Press, New York.

Tucker, Vance A. (1971) "Flight energetics in birds," *American Zoologist* **11**, 115–124.

Turner, John Stewart (1973) *Buoyancy Effects in Fluids*. Cambridge University Press, London.

Usher, M. B. (1966) "A matrix approach to the management of renewable resources, with special reference to selection forests," *Journal of Applied Ecology* **3**, 355–367.

Vincent, Thomas L. and Walter J. Grantham (1981) *Optimality in Parametric Systems*. Wiley, New York.

Whitham, G. B. (1974) *Linear and Nonlinear Waves*. Wiley, New York.

Whittle, Peter (1982) *Optimization Over Time*. 2 volumes, Wiley, New York.

INDEX

A

a posteriori justification, 42
absorbing chain, 201
absorbing state, 201
acceleration, 32, 37
active constraint, 93
adaptation, 327
advertising, 247
aerodynamics, 133
age structure of population, 350, 360
aggregation, 441
airport, 271
Anagrus delicatus, 497
animal behavior, 482
aperiodic, 212
assigning probabilities, 291
asymptotic stability, 49
atmospheric pressure, 448
attitude to risk, 317
augmented Lagrangian, 302
automobile industry, 414
autonomous dynamical system, 73
Avogadro's number, 45

B

backward rate, 31
barber shop, 264

base year, 20
behavioral response function, 263, 266, 268, 281
bifurcation, 156
big oh, 4
biochemical oxygen demand, 441
biochemistry, 383
biogeography, 282
bird, 93, 132, 479
birth and death chain, 199
birth and death process, 189
birth cohort, 166
birth process, 185
bookseller, 247
boundary condition, 66
boundary layer, 433
boundary layer thickening, 150
Britain, 311, 409
business intensity, 216
buyer, 237, 245, 253

C

canal, 433
capacity, 28, 44, 84, 283, 390
capital, 14
caribou, 156
catalyst, 384
catchability, 116

center, 49
CGS unit, 38
characteristic, 374, 375, 427
Chebyshev's inequality, 526
checkout, supermarket, 215, 217, 219, 224
chemical conductivity, 384
chemical dynamics, 29, 30, 171, 324, 325, 437, 440, 442
Chi-squared distribution, 308
Chi-squared test, 309
choice, 275
choice of utility function, 145, 162, 316
clearcutting, 104
Cobb-Douglas production function, 17, 103, 170
combination, 328
complement (of set), 517
compromise, 485
concentration, 5, 37
concentration gradient, 419
conditional probability, 521
conduction, 66, 419
constrained nonlinear programming, 302, 482
constraint, 91
continuous distribution, 522
continuous-time Markov jump process, 413
convection, 66, 346, 350, 419
convergence of recurrence relation, 473
coronary care, 300
crop spraying, 111, 146
crop, 43, 111, 146
crossing control, 188
"crude" birth rate, 351
"crude" death rate, 352
cumulative distribution function, 523
cumulative frequency distribution of sample, 305
currency, 485
curve-fitting, 285

D
DDT, 344
death process, 183
deer, 378
deflation, 21
degree of freedom, 308
demography, 166
Denmark, 359
depreciation, 41
descriptive model, 163, 268
diet selection, 480
differential equation, 335, 386
differential-delay equation, 79, 139
diffusion, 345, 350, 416, 419, 433, 436
diffusivity, 348, 433
dimension, 36
Dirac delta function, 123, 435
direct method (for identifying equilibria), 52
discount factor, 107
discount rate, 160, 258, 271
discounted, perceived optimal reward, 461
discrete distribution, 522
dispersal, 498
disreward, 477
distribution, 520
disutility, 477
domain of attraction, 50
double-glazing, 63
drag, 32, 67, 133
drainage time, 38
dynamic equilibrium, 387, 424
dynamic programming, 458
dynamical system, 13, 152, 155, 386

E
economics, 13, 103, 170, 394, 414
ecosystem, 8, 11, 12, 151, 155
effective pipe-length, 454
effective power, 33
eigenvalue, 211

eigenvector, 211
elasticity of production, 17
election, 312, 323, 410
elevator, 189, 193, 213, 414
empirical probability assignment, 293
endogenous, 18
energy, 32, 37
entry region, 149
enzyme, 384
epidemic, 206
equilibrium, 46, 51, 209
equilibrium shift technique, 73
equivalent sample space, 520
ergodic, 212
error function, 375
estimating parameters, 285, 304
estimation, 275
Euclidean norm, 491
evaporation, 343, 375
event, 518
exhaustive, 519
exogenous, 18
expectation of life, 167
expected value, 525
exponential decay, 7
exponential distribution, 180, 182, 306
extension, 327
extinction, 88, 164, 282

F

factor of production, 17
factor in evaluation, 96
faculty hiring, 470
fast-food restaurant, 258, 322
feasible control, 99
fertility, 353
fishing, 115, 161, 379
fitness, 493
flexibility, 162, 268, 475
flow, 416, 436
fluid dynamics, 66, 148, 436, 456

focus, 48
foraging, 496, 510
force of mortality, 166
force, 37
forest, 106, 142, 200, 210, 322, 329
forest rotation, 106, 142
forward rate, 31
Fourier series, 420
friction, 67
functional model, 281
fundamental, 453
Fundamental diagram (of traffic flow), 84, 426, 455

G

gamma distribution, 224
gamma function, 308
Gause's equations, 11
Gause's principle of competitive exclusion, 151
generalized function, 123, 435
glucose, 383
Gompertz law of mortality, 168
goodness of fit, 307
gradient of concentration, 346
gravity, 32
Great Lakes, 6, 129, 335

H

half-life, 7
handling time, 480
harvesting model, 43, 104, 115, 331
health care, 323, 353
heat flow, 63, 85, 417
heat flux, 63
heavy traffic, 84
holding cost, 238
horizon, 253
hospital, 300
hotel, 509
hypothesis, 495

I

immigration, 112, 164, 283, 358, 377
inactive constraint, 93
independent, 528
index (of labor etc.), 14
index, price, 21
indirect method (for identifying equilibria), 52
induced power, 94, 134
induced velocity, 95
inertia, 32
inflation rate, 21, 396
insect, 42, 43, 111, 496
instability, 48, 51
instantaneous annual birth rate, 351
instantaneous annual death rate, 352
instantaneous annual output, 13
insurance, 271, 281
interarrival time, 187
interest rate, 104, 393
interference, 154
intersection, 516
interview, 466
inventory, 463
investment rate, 18, 395
irreducible, 212

J

Jacobian, 88
joint probability density function, 229, 528
joint probability distribution, 527
joint probability function, 527

K

Keynesian, 394
kinematic viscosity, 68
Kronecker delta, 183

L

Lake Ontario, 129
lake purification, 5, 52, 128, 334, 337
land value, 144
Leslie model, 372
life table, 168, 203, 293
life-line, 376
light traffic, 84
likelihood, 314
limit cycle, 50, 58, 157, 389, 420, 455
limiting distribution, 405
linear programming, 100, 162, 333, 379
linear time, 197
little oh, 5
logistic growth, 8, 175, 275, 289
Lotka–Volterra equations, 9
lynx, 155

M

Makeham's law, 289
marginal interpretation, 118, 121, 122, 143, 252
Markov chain, 198, 225, 230, 297, 363
Markov model, 176
Markov property, 176, 190
mass, 32
maximum life potential, 169
maximum likelihood estimate, 314
maximum likelihood, 313
mean, 525
measles epidemic, 206
medium (of motion), 416
memorylessness, 176
merit function, 317
merit matrix, 97
metastability, 49, 51
metered model, 23, 194, 331
method of extremes, 281
method of moments, 310

Michaelis–Menten law, 389, 390, 456
migration, 94, 132
mixed distribution, 522
mixed random variable, 240, 254, 522
mixing, 341, 350
mole, 29, 45
moment, 526
momentum, 67, 439
Monetarist, 394
money supply, 393
mortality, 166, 293, 352
mule deer, 156
multiplier method, 302
mutually disjoint, 517

N

natural selection, 482
net productivity, 43
net specific growth rate, 3
neutral stability, 49
Newton–Raphson method, 336
node, 48
nominal interest rate, 104
nominal output, 21
nominal price, 106
nominal value, 107
nondifferential dynamical system, 18, 75
nondifferential model, 394
nonlinear programming, 288
nonlinear time, 197
norm, 491
normal distribution, 222, 241, 434
normal mode, 453
null event, 520
numerical instability, 89

O

objectivist, 292
oil extraction, 279
opportunity cost, 238

optimal control problem, 92
optimal harvesting model, 279
optimal policy, 258
optimal stopping problem, 465
ordinary differential equation, 445
organ pipe, 446
output, instantaneous annual, 13
overtone, 453
oxygen deficit, 442
oxygen sag curve, 444

P

paradigm, 158, 359
parameter estimation, 131, 137
parasitoid, 497
Pareto-optimal, 492, 499
parking, 475
partial differential equation, 70, 339, 376, 417, 442, 449
penetration depth, 454
perceived optimal reward, 257
percentage error, 28
perturbation, 47, 51, 78
pesticide, 42, 43, 111, 146
phase trajectory, 46
phase-plane, 50, 445
philosophy of science, 127
planet, 86
plant growth, 7, 107, 129
Poiseuille's formula, 71, 86, 149
Poisson distribution, 186, 503
Poisson distribution, normal approximation to the, 244
Poisson process, 186, 187, 228, 243, 259, 373
pollution, 5, 53, 128, 334, 337, 433
population growth, 26, 350, 360
positive recurrent, 212
power, 32, 37
predator–prey oscillation, 155, 389
prescriptive model, 163, 268
present value, 105
pressure, 37, 69, 439

price index, 21
probabilistic model, 175
probability density function, 523
probability function, 519
product (of chemical reaction), 29
production function, 17
propagating wave, 342, 449, 455

Q

quasi-equilibrium, 75, 387
quasi-stable, 355
queue, 185
queuing model, 259, 265
queuing theory, 222

R

radioactive decay, 7
radix, 168
railroad crossing, 421, 423, 429
rainfall, 343, 375
random variable, 176, 517
reactant, 29
reaeration coefficient, 442
real interest rate, 106
real output, 21
real price, 106
real value, 107
realization (of random variable),
    517
recurrence relation, 22, 79, 194,
    196, 472
recursion, 461, 469, 472, 478, 503
reducible, 212
reward, 237
Reynolds number, 150
river purification, 440
Rocky Mountain lion, 156
rotation, forest, 106, 142, 329
rowing, 32, 53, 130

S

S.I. unit, 38
saddle-point, 49

salary increase, 95
salmon, 23, 42, 58, 158
Salukvadze compromise, 492
salvage value, 328, 461
sample, 304
sample mean, 304
sample space, 517
sample variance, 304
sampling, 506, 510
scaling, 38
semi-Markov process, 398, 406
semi-Markov property, 400
separatrix, 12, 152
service time, 223
sex ratio, 353
shear viscosity, 68, 439, 447
shock, 433
simple outcome, 518
site (for oviposition), 497
snowshoe hare, 155
social mobility, 225
solute, 31
solvent, 31
sound speed, 449
sound wave, 446, 449
specific growth rate, 3
specific heat, 417
stability, 46, 48, 51, 209
stability, analytical method for, 55,
    59
stable age structure, 355
standard deviation, 526
state of system, 115
static equilibrium, 47
stationary birth and death process,
    214
stationary distribution, 209, 405
stationary policy, 464, 473
statistic, 526
statistical inference, 310
statistical model, 305
steady state, 50
stimulus, 263, 266
stochastic matrix, 179, 184,
    186

stochastic process, 176, 193, 518
storm window, 85
street vendor, 243, 328
Streeter-Phelps model, 440
structural stability, 157
subjectivist, 292
subset, 517
supermarket checkout, 215, 217, 219, 224
surface area, 33
suspension (in medium), 436
sustained harvest, 122, 332
sustained profit, 122
sustained yield, 122

T

T-junction, 204, 227
tailback, 228
thermal conductivity, 64
threshold function, 239
tractive force, 32
traffic density, 57
traffic flow, 34, 57, 76, 109, 136, 138, 187, 204, 227, 421
traffic flux, 83
trajectory, 46, 427
transient, 420
transition, 178
transition layer, 432
transition matrix, 196, 226
triangular distribution, 473
true worth, 107
tunnel, 57, 109, 136

turbulence, 350
two-bin policy, 464

U

U.S. population growth, 128
uniformly distributed, 526
unimprovable points, 487
union (of sets), 516
unit (of measurement), 38
universal set, 517
unsteady heat flow, 417
utility, 91, 316
utopia point, 487, 499

V

validation, 127
variance, 525
vector optimization, 492
vendor, 243, 328
Venezuela, 359
viscosity, 68, 439, 447

W

waiting time, 219, 221, 259
waiting time distribution, 400
water purification, 5, 52, 129
wave, propagating, 342, 449, 455
wavelength, 452
window, heat flow through, 63, 85, 418
wolf, 156
work force, 225, 297